Climate Change and Managed Ecosystems

Climate Change
and
Managed Ecosystems

Edited by
J.S. Bhatti
R. Lal
M.J. Apps
M.A. Price

CRC Press
Taylor & Francis Group
Boca Raton London New York

CRC Press is an imprint of the
Taylor & Francis Group, an **informa** business

CRC Press
Taylor & Francis Group
6000 Broken Sound Parkway NW, Suite 300
Boca Raton, FL 33487-2742

First issued in paperback 2019

© 2006 by Taylor & Francis Group, LLC
CRC Press is an imprint of Taylor & Francis Group, an Informa business

No claim to original U.S. Government works

ISBN-13: 978-0-8493-3097-1 (hbk)
ISBN-13: 978-0-367-39148-5 (pbk)

Library of Congress Card Number 2005028910

Library of Congress Cataloging-in-Publication Data

Climate change and managed ecosystems / edited by J.S. Bhatti ... [et al.].
        p. cm.
    Includes bibliographical references and index.
    ISBN-10: 0-8493-3097-1 (hardcover)
    1. Climatic changes. 2. Ecosystem management. I. Bhatti, J. S. (Jagtar S.)

QC981.8.C5C5113823 2006
333.95--dc22   2005028910

**Visit the Taylor & Francis Web site at
http://www.taylorandfrancis.com**

**and the CRC Press Web site at
http://www.crcpress.com**

Preface

The idea for this book arose during the planning phases of an International Conference in Edmonton, Canada in July 2004 entitled "The Science of Changing Climates — Impacts on Agriculture, Forestry and Wetlands." The conference was organized jointly by the Canadian Societies of Animal Science, Plant Science and Soil Science with support from Natural Resources Canada/Canadian Forest Service because they saw climate change as one of the most serious environmental problems facing the world. The United Nations Convention on Climate Change (UN 1992, article 2) called for a "... stabilization of greenhouse gas concentrations in the atmosphere at a level that would prevent dangerous anthropogenic interference with the climate system ..." For agriculture, forestry and wetlands, these potentially dangerous interferences include changes in ecosystems boundaries, loss of biodiversity, increased frequency of ecosystem disturbance by fire and insects, and loss and degradation of wetlands. Regional temperature increases, precipitation increases and decreases, change in soil moisture availability, climatic variability and the occurrence of extreme events are all likely to influence the nature of these impacts. The book is organized into five main parts.

Part 1: Climate Change and Ecosystems (Chapters 1–3). We discuss the fragility of ecosystems in the face of changing climates, particularly through human-caused increases in atmospheric GHGs. Chapter 2 details how and why the climate has changed in the past; and what can be expected to occur in the foreseeable future. The implications of climate change to agriculture, forestry and wetland ecosystems in Canada are discussed in Chapter 3, and potential adaptation responses to reduce the impacts of a changing climate are identified.

Part 2: Managed Ecosystems — State of Knowledge (Chapters 4–15). We explore what is known about the impacts of climate on our agricultural, forested and wetland ecosystems. This section illustrates the importance of terrestrial ecosystems in the global carbon cycle and focuses on discussions of the potential interaction between terrestrial and atmospheric carbon pools under changing climatic conditions. Our current understanding of the impact of climate change on food and fiber production as well as the potential role of the different ecosystems in carbon source/sink relationships has been discussed in detail here.

Part 3: Knowledge Gaps and Challenges (Chapters 16–18). We attempt to identify what needs to be known and done to ensure continued stability in these ecosystems. This part includes a description of some of the activities that have been undertaken in the past to identify gaps in our understanding of GHGs emissions from agriculture, forest and wetland and their mitigation, as well as current research initiatives to address these gaps.

Part 4: Economics and Policy Issues (Chapter 19). This provides an overview of economic reasoning applied to climate change and illustrates how terrestrial carbon-uptake credits (offset credits) operate within the Kyoto Protocol framework. Attention is focused on the potential of terrestrial carbon sinks to slow the rate of CO_2 buildup in the atmosphere.

Part 5: Summary and Recommendations (Chapters 20–21). We give an overall view of the knowledge gained from the conference and identify research needs to achieve reduced atmospheric carbon levels. The first chapter (Chapter 20) synthesizes the major findings of all the previous chapters and examines the implications for different ecosystems. The second chapter (Chapter 21) identifies key knowledge gaps relating to climate and climate-change effects on agriculture, forestry, and wetlands. It further points toward the needs to make management of these ecosystems part of a global solution, by identifying gaps in the current understanding of adaptation or mitigation strategies for terrestrial ecosystems.

While we are confident that the material contained in this book will be helpful to anyone seeking up-to-date information, we are also aware that in such a rapidly evolving field it is inevitable that material will quickly become dated. With that in mind we encourage you, the reader, to contact the chapter authors for their current views and information on the topics covered.

J.S. Bhatti

R. Lal

M.J. Apps

M.A. Price

Acknowledgments

This book would not have been possible without the assistance of a great many people and organizations. We would like to acknowledge in particular our Platinum Sponsors: Alberta Agriculture, Food and Rural Development; Canadian Adaptation and Rural Development Fund; Canadian Climate Impacts and Adaptation Research Network; National Agroclimate Information Service; Natural Resources Canada, Canadian Forest Service; Poplar Council of Canada; Prairie Adaptation Research Collaborative; and University of Alberta; our Gold Sponsor: Ducks Unlimited; and our Silver Sponsors: Agrium and MERIAL/igenity. We also want to gratefully acknowledge the thorough work of our anonymous group of reviewers, who helped to ensure that the manuscripts met the highest scientific standards. Thanks, too, to Cindy Rowles for her invaluable clerical assistance and advice. And finally, thanks are due to the managers and staff of Taylor & Francis Group for their careful attention to detail in publishing this book.

About the Editors

J.S. Bhatti, Ph.D., is a research scientist and project leader with Natural Resources Canada, Canadian Forest Service, Northern Forestry Centre, in Edmonton, Alberta. He received his Ph.D. in soil science from University of Florida and started working for Natural Resources Canada, where he concentrated on nutrient dynamics in boreal forests under various harvesting practices and moisture regimes.

Dr. Bhatti's interest in climate change moved him to Northern Forestry Centre in 1997, where his focus has been on carbon dynamics under changing climate and disturbance regimes both in upland and low land boreal forest ecosystems. His scientific publications deal with improving the precision of carbon stock and carbon stock estimates, changes in forest carbon dynamics in relation to disturbances, moisture, nutrient and climate regimes, and understanding the influence of bio-physical processes on forest dynamics. He is coordinating a national effort to monitor forest carbon dynamics to understand and quantify the prospective impacts of climate change on Canadian forests.

R. Lal, Ph.D., is a professor of soil physics in the School of Natural Resources and Director of the Carbon Management and Sequestration Center, FAES/OARDC at The Ohio State University. He was a soil physicist for 18 years at the International Institute of Tropical Agriculture, Ibadan, Nigeria. In Africa, Professor Lal conducted long-term experiments on land use, watershed management, methods of deforestation, and agroforestry. Since joining The Ohio State University in 1987, he has worked on soils and climate change. Professor Lal is a fellow of the Soil Science Society of America, American Society of Agronomy, Third World Academy of Sciences, American Association for the Advancement of Sciences, Soil and Water Conservation Society and Indian Academy of Agricultural Sciences.

Dr. Lal is the recipient of the International Soil Science Award, the Soil Science Applied Research Award and Soil Science Research Award of the Soil Science Society of America, the International Agronomy Award and Environment Quality Research Award of the American Society of Agronomy, the Hugh Hammond Bennett Award of the Soil and Water Conservation Society, and the Borlaug Award. He is the recipient of an honorary degree of Doctor of Science from Punjab Agricultural University, India, and of the Norwegian University of Life Sciences, Aas, Norway. He is past president of the World Association of the Soil and Water Conservation and the International Soil Tillage Research Organization. He was a member of the U.S. National Committee on Soil Science of the National Academy of Sciences (1998–2002). He has served on the Panel on Sustainable Agriculture and the Environment in the Humid Tropics of the National Academy of Sciences. He has authored and co-authored more than 1100 research publications. He has written 9 books and edited or co-edited 43 books.

M.J. Apps, Ph.D., retired as senior scientist, carbon and climate change, from Natural Resources Canada, Canadian Forest Service in 2005, but continues to work part time on various international projects. He obtained his Ph.D. in physics from the University of Bristol, and continued in solid state physics as a research associate at Simon Fraser University before moving to the University of Alberta to take a research position in the Faculty of Pharmacy and Pharmaceutical Sciences, where he set up the Neutron Activation Analysis system for trace element analysis at the Slowpoke Nuclear Reactor. His interest in environmental issues led him to join the Canadian Forest Service in 1980, where he initiated research on trace pollutants and radionuclides in the terrestrial environment. He moved into climate change and carbon cycling as a focus for his work in forest ecosystem modeling in 1990, and spearheaded the development of the Carbon Budget Model of the Canadian Forest Sector, now used for Canada's reporting under the Kyoto Protocol.

Dr. Apps is the author or co-author of more than 200 published manuscripts, has served as lead or convening lead author on many reports of the Intergovernmental Panel on Climate Change, and sits on several international and national scientific steering committees on global change issues. He has received significant national and international recognition, including the International Forestry Achievement Award presented at the World Forestry Congress, an honorary diploma issued by the International Boreal Forest Research Association in St. Petersburg, designated Leader of Sustainable Development by the five natural resource departments of the government of Canada, and the 2005 Award of Excellence by the Public Service of Canada.

M.A. Price, Ph.D., P.Ag., FAIC, is professor emeritus of livestock growth and meat production at the University of Alberta and was, until his retirement in 2004, research director at the university's Beef Cattle Research Ranch at Kinsella, Alberta. He was born and raised on the family farm in the U.K., and farmed there after high school. He received his post-secondary education at the University of Zimbabwe (B.Sc., agriculture), University of New England, Australia (M.Rur.Sc. and Ph.D. in livestock production) and University of Alberta, Canada (NRC post-doctoral fellowship in animal science).

Dr. Price served as chairman of the Department of Animal Science at the University of Alberta from 1987 to 1995. His areas of research concentrate mainly on sustainable methods of increasing efficiency and decreasing costs of production in meat production systems. He has published more than 115 scientific papers in peer-reviewed journals, and more than 130 extension articles in trade and industry magazines. He is the editor of the *Canadian Journal of Animal Science*.

CONTRIBUTORS

M.J. Apps
Canadian Forest Service
Natural Resources Canada
Pacific Forestry Centre
Victoria, BC, Canada

T. Asada
Wetlands Research Centre
University of Waterloo
Waterloo, ON, Canada

J.K.A. Atakora
Department of Agricultural, Food
 & Nutritional Science
Agriculture/Forestry Centre
University of Alberta
Edmonton, AB, Canada

R.O. Ball
Department of Agricultural, Food
 & Nutritional Science
Agriculture/Forestry Centre
University of Alberta
Edmonton, AB, Canada

V.S. Baron
Crops & Soils Research Station
Agriculture and Agri-Food Canada
Lacombe, AB, Canada

I.E. Bauer
Canadian Forest Service
Northern Forestry Centre
Edmonton, AB, Canada

P.Y. Bernier
Canadian Forest Service
Natural Resources Canada
Saint-Foy, PQ, Canada

J.S. Bhatti
Natural Resources Canada
Canadian Forest Service
North Forestry Centre
Edmonton, AB, Canada

D. Burton
Nova Scotia Agricultural College
Truro, NS, Canada

J. Casson
Alberta Agriculture, Food & Rural
 Development
Agriculture Centre
Lethbridge, AB, Canada

O.G. Clark
Department of Agricultural, Food
 & Nutritional Science
Agriculture/Forestry Centre
University of Alberta
Edmonton, AB, Canada

W.A. Dugas
Texas Agricultural Experiment Station
Texas A&M University
College Station, TX, USA

I. Edeogu
Technical Services Division
Alberta Agriculture, Food & Rural
 Development
Edmonton, AB, Canada

J.J. Feddes
Department of Agricultural, Food
 & Nutritional Science
Agriculture/Forestry Centre
University of Alberta
Edmonton, AB, Canada

H. Hengeveld
Environment Canada
Scientific Assessment and Integration
Downsview, ON, Canada

G. Hoogenboom
Department of Biological &
  Agricultural Engineering
The University of Georgia
Griffin, GA, USA

B.C. Joern
Department of Agronomy
West Lafayette, IN, USA

C. La Bine
Campbell Scientific (Canada) Corp.
Edmonton, AB, Canada

R. Lal
School of Natural Resources
College of Food, Agricultural
  & Environmental Sciences
The Ohio State University
Columbus, OH, USA

D.B. Layzell
BIOCAP Canada Foundation
Queen's University
Kingston, ON, Canada

J.J. Leonard
Department of Agricultural, Food
  & Nutritional Science
University of Alberta
Edmonton, AB, Canada

P.C. Mielnick
Blackland Research Center
Texas A&M University
Temple, TX, USA

S. Moehn
Department of Agricultural, Food
  & Nutritional Science
University of Alberta
Edmonton, AB, Canada

B. Morin
Technical Services Division
Alberta Agriculture, Food & Rural
  Development
Edmonton, AB, Canada

L.D. Mortsch
Adaptation and Impacts Research
  Group
Meteorological Service of Canada
Environment Canada
University of Waterloo
Waterloo, ON, Canada

K.H. Ominski
Department of Animal Science
University of Manitoba
Winnipeg, MB, Canada

J.D. Price
Technical Services Division
Alberta Agriculture, Food & Rural
  Development
Edmonton, AB, Canada

M.A. Price
Department of Agricultural, Food
  & Nutritional Science
University of Alberta
Edmonton, AB, Canada

B.T. Richert
Department of Animal Sciences
Purdue University
West Lafayette, IN, USA

M.A. Sanderson
USDA-ARS
Pasture Systems & Watershed
  Management Research Unit
University Park, PA, USA

W.C. Sauer
Department of Agricultural,
  Food & Nutritional Science
University of Alberta
Edmonton, AB, Canada

J. Sauvé
Alberta Agriculture
Food & Rural Development
Edmonton, AB, Canada

R.H. Skinner
USDA-ARS
Pasture Systems and Watershed
  Management Research Unit
University Park, PA, USA

J. Stephen
BIOCAP Canada Foundation
Queen's University
Kingston, ON, Canada

J.M.R. Stone
Environment Canada
Meteorological Service of Canada
Hull, PQ, Canada

A.L. Sutton
Department of Animal Sciences
Purdue University
West Lafayette, IN, USA

G.C. van Kooten
Department of Economics
University of Victoria
Victoria, BC, Canada

D.H. Vitt
Department of Plant Biology
Southern Illinois University
Carbondale, IL, USA

G.C. Waghorn
Dexcel Ltd.
Hamilton, New Zealand

B.G. Warner
Department of Geography
University of Waterloo
Waterloo, ON, Canada

K.M. Wittenberg
Department of Animal Science
University of Manitoba
Winnipeg, MB, Canada

S.L. Woodward
Dexcel Ltd.
Hamilton, New Zealand

D.G. Young
Crops & Soils Research Section
Agriculture & Agri-Food Canada
Lacombe, AB, Canada

Y. Zhang
Department of Agricultural, Food &
  Nutritional Science
University of Alberta
Edmonton, AB, Canada

R.T. Zijlstra
Department of Agricultural, Food
  & Nutritional Science
University of Alberta
Edmonton, AB, Canada

Contents

PART III Knowledge Gaps and Challenges

PART IV Economics and Policy Issues

PART V Summary and Recommendations

Part I

Climate Change and Ecosystems

1 Interaction between Climate Change and Greenhouse Gas Emissions from Managed Ecosystems in Canada

J.S. Bhatti, M.J. Apps, and R. Lal

CONTENTS

1.1   INTRODUCTION

The world's terrestrial ecosystems are being subjected to climate change on an unprecedented scale, in terms of both rate of change and magnitude. Understanding the ability of terrestrial ecosystems to adapt to change requires fundamental knowledge of the response functions. The changes under consideration in this book include not only the climatic change from increased concentration of greenhouse gases (GHGs) and consequent warming trends especially in the north, but also land use, land-use changes, and alterations in disturbance patterns, both natural and human induced. The interactive nature of climate change is complex and nonlinear because the variables of change are strongly interactive (Figure 1.1) and not independent. To remain viable, agricultural

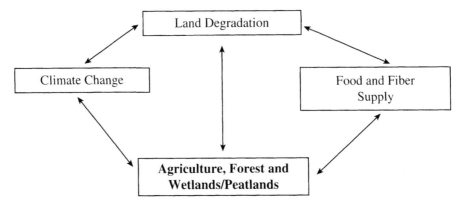

FIGURE 1.1 Linkage between various climate change issues and different ecosystems.

and forest production systems will need to change rapidly to meet the challenge of the inevitable changes in the mosaic of ecosystems across the landscape.

Agricultural ecosystems (including crop and animal production, pastures and rangelands), forest ecosystems, and wetlands (including peatlands) can be regarded as one dimension of the problem, and climate change as another. This book synthesizes our current understanding of the processes of climate change and its impacts on different managed ecosystems. From a human perspective the impacts of climate change lie in the interactions among these different ecosystems: vulnerability must be assessed in terms of the collective impact on these terrestrial ecosystems that supply essential goods and services to society. Humans depend on these ecosystems for food, fiber, and clean air and water, and the adverse impacts of climate change are likely to have far-reaching effects on human lives and livelihoods.

An important indicator of the human interaction with the global climate system is the human perturbations to global carbon cycle. Carbon is exchanged between terrestrial ecosystems and the atmosphere through photosynthesis, respiration, decomposition, and combustion. Terrestrial ecosystems have the capacity for either accelerating or slowing climate change depending on whether these systems act as a net source or a net sink of carbon. This source or sink status is, however, not a static characteristic of the ecosystem, but will change over time as a result of changes in the physical, chemical, and biological processes of these systems,[1] all of which are influenced by human activity. Data on global carbon stocks in major biomes are presented in Table 1.1. Response to climate change will alter these carbon stocks, changing the fluxes among terrestrial ecosystems and the atmosphere differently in different geographical regions. There is a strong need to quantify these fluxes both in relation to different management options and to different environmental pressures. The basic challenge is the detection of very small changes relative to the size of the pools. Thus, it is important to understand the dynamics of these different pools, and identify factors that make these pools either sinks or sources of GHGs.

TABLE 1.1
Global Carbon Stocks and Net Primary Productivity of the Major Terrestrial Biomes

Biome	Area (109 ha)[18]	Carbon Stock (Pg C)[18]	Total C NPP (Pg C yr^{-1})[19]
Tropical forests	1.76	428	21.9
Temperate forests	1.04	159	8.1
Boreal forests	1.37	290	2.6
Northern peatlands	0.26	419	—
Arctic tundra	0.95	127	0.50
Crops	1.60	131	4.1
Tropical grasslands	2.25	330	14.9
Temperate grasslands	1.25	304	4.4
Deserts	4.55	199	3.5

1.2 PAST AND FUTURE CLIMATE CHANGE

Changes in climate are not new: Earth has long been subjected to sequential glacials, interglacials, and warm periods, and all parts of Canada have been warmer, cooler, wetter, and drier at various times in the past. A number of natural factors control climatic variability, including Earth's orbit, changes in solar output, sunspot cycles, and volcanic eruptions (Chapter 2). However, the present climatic change is unprecedented in character: it cannot be explained by these factors alone. The recently observed increase in global temperature is strongly related to increases in the concentration of GHGs in the recent past,[2] increases that are directly attributable to human activities. Over the course of the 20th century global mean temperature has risen by about 0.6°C, and is projected to continue to rise at an average rate of 0.1 to 0.2°C per decade for the next few decades then increase to a rate of warming of between 1.4 and 5.8°C per decade by 2100.[2] Average temperatures across Canada are expected to rise at twice the global rate. In general, Canadian temperatures have been increasing steadily over the last 58 years, with winter temperatures above normal between 1985 and 2005 (Figure 1.2). At the same time, in general, over the last 58 years, winter precipitation has been decreasing (Figure 1.3) across Canada. In southern Canada, surface temperatures have increased by 0.5 to 1.5°C during the 20th century. The greatest warming has occurred in western Canada, with up to 6°C increase in the minimum temperature. In addition, the frequency of days with extreme temperature, both high and low, is expected to increase, snow and ice cover to decrease, and heavy precipitation events to increase.[3] During the second half of the 21st century, heat sums, measured in growing degree days, across southern Canada are expected to increase by between 40 and 100%.

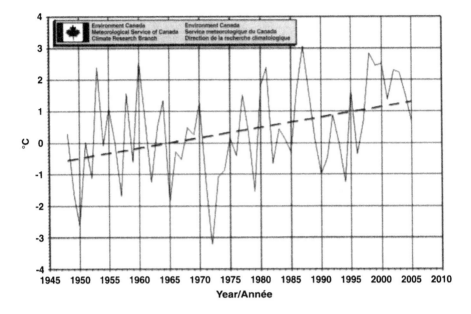

FIGURE 1.2 Canadian winter temperature deviation with weight running mean between 1948 and 2005. (Courtesy of Environment Canada.[7])

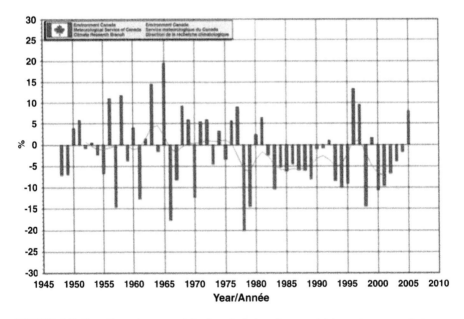

FIGURE 1.3 Canadian winter precipitation deviation from weight running mean between 1948 and 2005. (Courtesy of Environment Canada.[7])

1.3 GREENHOUSE GAS EMISSIONS FROM AGRICULTURE, FORESTRY, AND WETLAND ECOSYSTEMS

Natural processes such as decomposition and respiration, volcanic eruptions, and ocean outgassing are continuously releasing greenhouse gases such as water vapor, carbon dioxide (CO_2), methane (CH_4), and nitrous oxide (N_2O) into the atmosphere. Molecule for molecule, CO_2 is a weak GHG in terms of global warming potential (GWP); most other GHGs have a stronger GWP. Compared to CO_2 on a 100-year timescale, the GWP is 21 times greater for CH_4, 310 times greater for N_2O, and 900 or more times greater for chlorofluorocarbons and hydrochlorofluorocarbons.[4] However, the net contribution of each gas to the greenhouse effect depends on four factors: the amount of the gas released into the atmosphere per year, the length of time that it stays in the atmosphere before being destroyed or removed, any indirect effect it has on atmospheric chemistry, and the concentration of other GHGs. In taking into account all these factors, the net contribution of CO_2 to the greenhouse effect is two to three times higher than that of CH_4 and about 15 times higher than that of N_2O.[4] Over the 20th century there has been a significant increase in GHGs in the atmosphere due to human activities such as fossil fuel burning and land use change. For example, there has been about a 30% increase in the concentration of CO_2 since the pre-industrial era: from 280 ppm in the late 18th century to 382 ppm in 2004.[5]

A global CO_2 emission rate of approximately 23.9 gigatonnes (Gt) has recently been estimated by the Carbon Dioxide Information and Analysis Centre.[6] Deforestation, land use, and ensuing soil oxidation have been estimated to account for about 23% of human-made CO_2 emissions. CH_4 emissions generated from human activities, amounting to ~360 Mt per year, are primarily the result of activities such as livestock and rice cultivation, biomass burning, natural gas delivery systems, landfills, and coal mining. Total annual emissions of N_2O from all sources are estimated to be within the range of 10 to 17.5 Mt N_2O, expressed as nitrogen (N).[4] While Canada contributes only about 2% of total global GHG emissions, it is one of the highest per capita emitters, largely the result of its resource-based economy, climate (i.e., energy demands), and size.[7] The change in Canada's GHG emissions between 1990 and 2002 by a number of different sectors (specifically, energy, transportation, industrial processes, agriculture, land-use change and forestry, and waste and landfills) is presented in Figure 1.4.[7] Total Canadian emissions of all GHGs in 2002 were 20.1% more than the 1990 level of 609 Tg of C. This growth in emissions appears to be mainly the result of increased energy production and fossil fuel consumption for heating in the residential and commercial sectors, as well as increases in the transportation, mining, and manufacturing sectors. The average annual growth of emissions over the 1990–2002 period was 1.7%.

Historically, agricultural activities have been a source of atmospheric enrichment of GHGs. In 2002, agriculture-related GHG emissions totaled 59 Tg of C, representing 8% of total Canadian emissions.[7] This sector accounted for 66% of Canada's total emissions of N_2O and 26% of CH_4 emissions. On a category basis, agricultural soils contributed 50% of the sector's emissions (29.6 Tg of C) in 2002 with the other half coming from domestic animals (32% or 18.8 Tg) and manure management

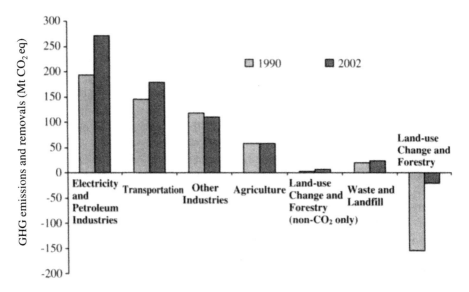

FIGURE 1.4 Change in GHG emissions and sinks for Canada between 1990 and 2002 for different sectors. (Courtesy of Environment Canada.[7])

(17% or 10.2 Tg of C). While total sector emissions rose 2% between 1990 and 2001, emissions from manure management rose 22% and enteric fermentation emissions increased by 18%. Net CO_2 emissions from agricultural soils partially offset these increases, changing from a net source of 7.6 Tg of C in 1990 to a net sink of 0.5 Tg of C in 2002. The N_2O emissions from soils, however, rose 15% over the same period.[7]

The profile of GHG emissions from the agricultural sector is very different from other sectors. For this sector, N_2O emissions associated primarily with N sources (fertilizer and animal manure) represent 61% of GHG emissions, CH_4 from ruminants and other sources represent another 38%, while net CO_2 emissions account for less than 1% of agricultural GHG emissions. N_2O is released during the biological process of denitrification, and CH_4 is released from enteric fermentation by ruminants, most specifically cattle, grazing the forage produced on these grasslands (Chapter 12). Digestive processes involving the breakdown of plant materials under conditions that are oxygen free, or oxygen limited, result in CH_4 production and account for 28% of agricultural emissions. Indirect emissions from livestock operations such as handling, storage, and land application of farm manure account for 14% of agricultural emissions. Microbial decomposition of manure can result in CO_2, CH_4, and N_2O emissions, with their relative contributions dependent on factors such as manure dry matter, C and N contents, as well as temperature and oxygen availability during storage.

The forest sector, limited to productive managed forest lands in Canada, was a net sink in 2002, as it removed 15 Tg of C from the atmosphere.[7] This estimate represents the sum of the net CO_2 flux and non-CO_2 (CH_4 and N_2O) emissions. The net CO_2 flux alone amounted to a sink of 21 Tg, which reduced total Canadian emissions in 2002 by 3%. Non-CO_2 emissions were about 6.0 Tg in 2002. However, the source/sink relationship for the forest sector is strongly influenced by disturbances, especially fire and insect outbreaks, which makes the GHG uptake or emissions of Canadian forests in a given year hard to predict[8] (Chapter 9).

In terms of greenhouse gases, wetlands can either be sources or sinks. Due to the complex biogeochemistry of peatlands/wetlands, they may function as sinks for one gas while acting as sources for others (Chapters 4 and 10). Peatland/wetlands may also change from sinks to sources due to anthropogenic impacts such as increased nutrient loading, drainage, flooding, burning, and vegetation change.

1.4 CLIMATE CHANGE IN RELATION TO AGRICULTURE, FORESTRY, AND WETLANDS

1.4.1 AGRICULTURAL ECOSYSTEMS

Arable agriculture occurs on only 7% of Canada's landmass due to climatic and soil limitations, and about 70% of Canada's arable acreage is located in Alberta and Saskatchewan.[9] Even under current conditions, climate has a major influence on the year-to-year variation in agricultural productivity in this region. Climate change can be expected to lead to more extreme weather conditions (i.e., conditions outside the range of previous norms), increases in weed and pest problems, and severe water shortage. On the other hand, these impacts will vary on a regional basis,[10] and some Canadian agricultural regions will benefit from a warmer climate and longer growing season, while others will be adversely affected.

With agricultural intensification to meet increases in food demand, soil degradation emerges as a major threat under climate change[11] (Chapters 4 and 6). Degradation of soil quality under climate change could result from decreases in soil organic matter (SOM), nutrient leaching, and soil erosion. Soil erosion is a major threat to agricultural productivity and sustainability as well as having adverse effects on air and water quality (Chapters 4 and 6). Wind and water erosion may increase significantly in agricultural soils due to increases in extreme weather condition such as heavy precipitation events or prolonged droughts.[12] Warmer winters may result in lower snow cover, and the reduction in soil moisture content could further increase the risk of wind erosion. Land-use change from natural vegetation to croplands potentially exacerbates these impacts due to increased vulnerability of the landscape to erosion.

1.4.2 FOREST ECOSYSTEMS

Forests cover more than one third of the land surface of the Earth. Almost half (410 Mha) the total landmass of Canada is forestland.[9] Boreal forests are the dominant forest type, spanning the complete width of the country (Figure 1.5). About 51% of Canada's forests are deemed suitable for timber production. The productivity of forest ecosystems largely depends upon the climate, nutrient, and moisture regimes.[13] Climate affects the distribution, health, and productivity of the forest and has a strong influence on the disturbance regime. The realization

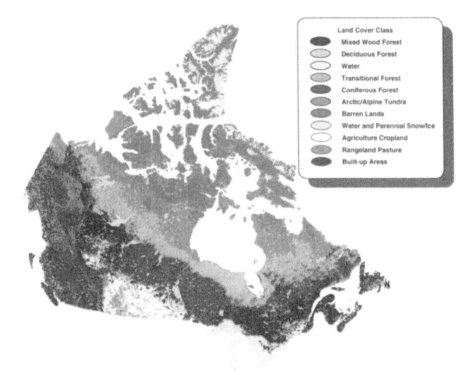

Land Cover Class
- Mixed Wood Forest
- Deciduous Forest
- Water
- Transitional Forest
- Coniferous Forest
- Arctic/Alpine Tundra
- Barren Lands
- Water and Perennial Snow/Ice
- Agriculture Cropland
- Rangeland Pasture
- Built-up Areas

FIGURE 1.5 Land cover map of Canada. (Courtesy of Natural Resources Canada.[9])

of the potential increase in plant productivity due to climate change depends on a variety of factors such as changes in species and competitive interactions, water availability, and the effect of temperature increase on photosynthesis and respiration (Chapter 16). In addition to the direct influence of climate change, other variables such as land-use change and existing land cover have profound influences on the forest distribution and productivity.

The future C balance of the forest will largely depend on the type and frequency of disturbances, changes in species composition, and alterations to the nutrient and moisture regimes under changing climate conditions (Chapter 9). It will also depend on forest management practices that affect both the disturbance regime and nutrient status. Projected climate change scenarios for the boreal forest generally predict warmer and somewhat drier conditions, posing questions about regeneration as well as productivity. In addition, the disturbance patterns are also expected to change. With more frequent disturbances (Figure 1.6), more of the stands will move into younger age classes where the uptake by regrowth is initially more than offset by CO_2 efflux from the decomposition of soil pools and elevated detritus left by the disturbance. This situation is expected to worsen as climatic change proceeds, especially if the conditions for successful regeneration are adversely altered. Altered boreal forest disturbance regimes — especially increases in frequency, size, and severity — may release CO_2 from vegetation, forest floors, and soils at higher rates than the rate of C accumulation in the regrowing vegetation.[8]

FIGURE 1.6 Boreal forest under fire. (Courtesy of Canadian Forest Service.)

The precise balance of C uptake and release depends on the detailed processes, and especially the outcome of interactions among climate, site variables, and vegetation over the changing life cycles of forest stands. Quantifying life-cycle dynamics at the stand level is essential for projecting future changes in forest level C stocks (Chapter 9). Forest management options to enhance or protect C stocks include reducing the regeneration delay through seeding and planting, enhancing forest productivity, changing the harvest rotation length, the judicious use of forest products, and forest protection through control and suppression of disturbance by fire, pests, and disease.

1.4.3 WETLAND/PEATLAND ECOSYSTEMS

Canada contains the world's second largest area of peatlands (after Russia). In Canada these peatlands cover approximately 13% of the land area and 16% of the soil area (Figure 1.7).[14] The largest area of peatlands (96%) occurs in the Boreal and Subarctic peatland regions. The dominant peatland types are bogs (67%) and fens (32%), with swamps and marshes together accounting for less than 1% of the Canadian peatlands. Overall, the most important controls of the carbon cycle in peatlands are plant community, temperature, hydrology, and chemistry of plant tissues and peat.[15] Limited data and understanding of the influence of changing environmental conditions and disturbance (including fires and permafrost melting) on the carbon cycle of peatlands over short and medium timescales (10 to 100 years)

FIGURE 1.7 Land-use change from natural peatlands to agricultural activities. (Courtesy of Steve Zoltai, Canadian Forest Service.)

hinder predictions of the changes in the carbon sink/source relationships under a changing climate. The projected warming and associated changes in precipitation will influence both net primary production and decomposition in peatlands, but how global warming will directly influence peatland carbon dynamics remains uncertain (Chapters 10 and 17). Melting of permafrost tends to increase peatland carbon stocks through increased bryophyte productivity but also appears to increase heterotrophic respiration.[16] Peatland fires result in decreased net primary production and elevated post-fire decomposition rates, but little is known about the recovery of the carbon balance after peatland fires (Chapters 10 and 17).

1.5 PURPOSE OF THIS BOOK

The major objective of this book is collation and synthesis of the current state-of-knowledge of the impacts of climate change on agriculture, livestock, forestry, and wetlands. Although many of the specific examples draw on Canadian studies, these examples have lessons that are useful in other parts of the world, especially in the Northern Hemisphere. The sustainable management of northern regions is a critical objective in terms of human needs for food and fiber. Climate change, along with land-use change and increased disturbance regimes, will be a significant threat to meeting this objective, and will require significant improvements in understanding and modification to present management practices.

Another objective of global importance is to help policy makers and land managers to reach informed choices regarding the relationships between carbon sources and sinks and the potential to increase sink capacity and reduce emissions for the regional landscapes under their jurisdiction. Terrestrial ecosystems are very diverse, ranging from highly managed agricultural systems to natural northern peatlands. How will these ecosystems respond to climate change and how can these ecosystems be managed under changing conditions? Are there management options that could be implemented to increase the sink capacity or to minimize the GHGs emissions? Are there strategies that minimize future vulnerabilities, while maintaining present supply needs?

Different components of the terrestrial ecosystems as well as issues of climate change are discussed in detail to address the following points:

- The vulnerability of the systems to climate change
- Specific impacts of changing climate on agriculture, forestry, or wetland systems
- Forms or methods of mitigation
- Adaptation measures or options to reduce the impacts on different eco-systems, and the goods and services we require from them

This synthesis presents the current scientific understanding of carbon dynamics in different ecosystems for Canada and identifies major knowledge gaps that hinder our ability to forecast responses to future climate change. Understanding the sink/source relationships and quantifying the contributions of different ecosystems are essential steps toward bridging the gap between policy need (increased sinks of atmospheric carbon) and science.

1.6 SUMMARY AND CONCLUSIONS

There is growing evidence that climate change is already occurring. At the global scale, average surface temperatures rose about 0.6°C over the 20th century and is expected to increase by another 1.4 to 5.8°C by the year 2100 — a rapid and profound change, with only dimly perceived consequences. The major factor responsible for this unprecedented increase in temperature and change in climate is an increase in the concentration of GHGs in the atmosphere, caused by human activities. Climate change is expected to bring both advantages and disadvantages for the agricultural and forest sectors in Canada. For example, although warmer temperatures would increase the length of the growing season, they may also increase crop damage due to heat stress and water and pest problems. Changes in the frequency and intensity of extreme events (e.g., droughts, floods, and storms) have been identified as the greatest challenge that would face the agricultural industry as a result of climate change. Drought and extreme heat have also been shown to affect livestock produc-tion operations.

Climate change has the potential to greatly influence Canadian forests and wetlands/peatlands, since even small changes in temperature and precipitation can significantly affect tree regeneration, survival, and growth. For example, the 1°C increase in temperature over the last century in Canada has been associated with longer growing seasons, increased plant growth, shifts in distribution, permafrost melting, and changes in plant hardiness zones. Future climate change is expected to further affect species distribution, forest productivity, and disturbance regimes. Regenerating trees today will grow and mature in a future climate that will be very different, and one to which they may be poorly adapted. Therefore, understanding the vulnerability of terrestrial ecosystems to these changes is essential for management of resources and need for food and fiber.

REFERENCES

1. Kauppi, P.E., Sedjo, R.A., Apps, M.J., Cerri, C.C., Fujimori, T., Janzen, H., Krankina, O.N., Makundi, W., Marland, G., Masera, O., Nabuurs, G.J., Razali, W., and Ravindranath, N.H., Technical and economic potential of options to enhance, maintain and manage biological carbon reservoirs and geo-engineering. In *IPCC Working Group III Contribution to the Third Assessment Report on Mitigation of Climate Change,* van der Linden, P.J., Dai, X., Maskell, K., and Johnson, C.A., Eds., Cambridge University Press, New York, 2001, 301–323.

2. IPCC (Intergovernmental Panel on Climate Change), *Climate Change 2001: The Scientific Basis, Contribution of Working Group I to the Third Assessment Report of the Intergovernmental Panel on Climate Change,* Houghton, J.T., Ding, Y., Griggs, D.J., Noguer, M., van der Linden, P.J., Dai, X., Maskell, K., and Johnson, C.A., Eds., Cambridge University Press, New York, 2001.

3. Folland, C.K., Karl, T.R., Christy, R., Clarke, R.A., Gruza, G.V., Jouzel, J., Mann, M.E., Oerlemans, J., Salinger, M.J., and Wang, S.W., Observed climate variability and change. In *Climate Change 2001: The Scientific Basis, Contribution of Working Group I to the Third Assessment Report of the Intergovernmental Panel on Climate Change,* Houghton, J.T., Ding, Y., Griggs, D.J., Noguer, M., van der Linden, P.J., Dai, X., Maskell, K., and Johnson, C.A., Eds., Cambridge University Press, New York, 2001, 99–182.

4. IPCC (Intergovernmental Panel on Climate Change), *Greenhouse Gas Inventory Reporting Instructions,* Vol. 1; and Greenhouse Gas Inventory Reference Manual, Vol. 3, Revised 1996 IPCC Guidelines for National Greenhouse Gas Inventories, 1997.

5. Keeling, C.D. and Whorf, T.P., Atmospheric CO_2 records from sites in the SIO air sampling network. In *Trends: A Compendium of Data on Global Change,* Carbon Dioxide Information Analysis Center, Oak Ridge National Laboratory, U.S. Department of Energy, Oak Ridge, TN, 2005.

6. Marland, G., Boden, T.A., and Andres, R.J., Global, regional, and national CO_2 emissions. In Trends: A Compendium of Data on Global Change, Carbon Dioxide Information Analysis Center, Oak Ridge National Laboratory, U.S. Department of Energy, Oak Ridge, TN, 2005.

7. Environment Canada, *Canada's Greenhouse Gas Inventory 1990–2002,* submission to the UNFCCC Secretariat, Ottawa.

8. Kurz, W.A. and Apps, M.J., A 70-year retrospective analysis of carbon fluxes in the Canadian forest sector, *Ecol. Appl.,* 9, 526, 1999.

9. Natural Resources Canada, *The State of Canada's Forests 2003–2004,* Natural Resources Canada, Ottawa, 2004.

10. Cohen, S., et al., North America. In *Climate Change 2001: Impacts, Adaptation and Vulnerability, Contribution of Working Group II to the Third Assessment Report of the IPCC,* McCarthy, J.J., Canziani, O.F., Leary, N.A., Dokken, D.J., and White, K.S., Eds., Cambridge Press, New York, 2001, chap. 15.
11. Gitay, H., Brown, S., Easterling, W.E., Jallow, B., Antle, J., Apps, M. J., Beamish, R., Chapin, T., Cramer, W., Franji, J., Laine, J., Erda, L., Magnuson, J.J., Noble, I., Price, C., Prowse, T.D., Sirotenko, O., Root, T., Schulze, E.-D., Sohngen, B., and Soussana, J.-F., Ecosystems and their services. In *IPCC,* Noguer, M., van der Linden, P.J., Dai, X., Maskell, K., and Johnson, C.A., Eds., Cambridge University Press, New York, 2001, 235–342.
12. Lal, R., Soil erosion and global carbon budget, *Environ. Int.,* 29, 437, 2003.
13. Kimmins, J.P., Importance of soil and role of ecosystem disturbance for sustained productivity of cool temperate and boreal forest, *Soil Sci. Soc. Am. J.* 60, 1643–1654, 1996.
14. Environment Canada, Wetlands in Canada: A Valuable Resource. Fact Sheet 86-4. Lands Directorate, Ottawa, Ontario, 1986.
15. Moore, T.R., Roulet, N.T., and Waddington, J.M., Uncertainty in predicting the effect of climatic change on the carbon cycle of Canadian peatlands, *Climatic Change* 40, 229–245, 1998.
16. Turetsky, M.R., Weider, R.K., Halsey, L.A., and Vitt, D.H., Current disturbance and the diminishing peatland C sink, *Geophys. Res. Lett.* 29(11), 1526–1526, 2002.
17. Turetsky, M.R., Wieder, R.K., Williams, C.J., and Vitt, D.H., Organic matter accumulation, peat chemistry, and permafrost melting in peatlands of boreal Alberta, *Ecoscience* 7, 379–392, 2000.
18. Intergovernmental Panel on Climate Change (IPCC), 2000. *Land Use, Land-Use Change, and Forestry,* Watson, R.T., Novel, I.R., Bolin, N.H., Ravindranath, N.H., Verardo, D.J., and Dokken, D.J., Eds., Cambridge University Press, New York, 2000.
19. Saugier, B., Roy, J., and Mooney, H., Estimations of global terrestrial productivity: converging towards a single number? In *Terrestrial Global Productivity,* Roy, J., Saugier, B., and Mooney, H.A., Eds., Academic Press, New York, 2001.

2 The Science of Changing Climates

H. Hengeveld

CONTENTS

2.1 INTRODUCTION

Climate is commonly defined as average weather. That is, the climate of a particular locale or region is the average of the day-to-day variations in temperature, precipitation, cloud cover, wind, and other atmospheric conditions that normally occur within that region over an extended period of time (usually three decades or more). For the Edmonton (Alberta, Canada) international airport, for example, the statistical climate "normals" calculated on the basis of past weather during the 1971–2000 time period indicate mean annual temperatures of 2.4°C, with an average daily temperature range of 12.3°C. Average annual precipitation was 483 mm, 25% of which fell as snow.

But climate is more than just the aggregate of these average values. It is also defined by the variability of individual climate elements and by the frequency with which various kinds of weather conditions occur. Indeed, any factor that is characteristic of a particular location's typical weather behavior is part of its climate.

The notion of climate as described above assumes a long-term consistency and stability in regional weather behavior. Nevertheless, climate is also a changeable phenomenon. It always has been. That is because Earth's climate system is dynamic, continuously responding to forces, both internal and external, that alter the delicate balances that exist within and between each of its components. Often, these changes are relatively small in magnitude and short in duration — like a period of cool climate conditions following a large volcanic eruption, or a few decades of dry conditions caused by a temporary shift in global atmospheric circulation patterns. However, evidence from the Earth's soils, its ocean and lake bottom sediments, its coral reefs, its ice caps, and even its vegetation indicate that such forces can cause major, long-term shifts in climate. Over long timescales of hundreds of thousands of years or more, for example, these changes include very large shifts from glacial to interglacial conditions and back again — changes that caused massive redistributions of flora and fauna around the planet. However, during the pre-industrial period of the past 10,000 years, such changes have been of relatively small magnitude. While from time to time regionally disruptive, they have allowed global vegetation to flourish over most landmasses. There is general agreement that interglacial conditions will persist for many more millennia — perhaps another 50,000 more years. Hence these natural changes are expected to remain modest within the foreseeable future.[1]

As early as 8000 years ago, humans began to interfere with these natural processes of change. Until about 100 years ago, this interference was primarily caused by gradual changes in land use, and the effects on climate were generally local. However, during the past century, a rapidly growing and increasingly industrialized society has significantly enhanced this influence. Much more worrisome changes are expected in the decades and centuries to come. Early warnings about the related risks were already issued in 1985, when international experts meeting in Villach, Austria cautioned policy makers that "many important economic and social decisions are being made today on long-term projects ... all based on the assumption that past climate data, without modification, are a reliable guide to the future. This is no longer a good assumption."[2]

The focus of this chapter is consideration in greater detail of how and why the climate has changed in the past and what can be expected to occur over the next few decades and centuries.

2.2 CHANGING CLIMATES — THE PAST

2.2.1 RECONSTRUCTING AND OBSERVING PAST CLIMATES

2.2.1.1 Paleo Records

Within the rich diversity of living species around the world there are some that thrive in hot climates and others that prefer cooler and even cold climates. Some like it wet, and some like it dry. The result is a tremendous range in the composition of regional ecosystems, with the characteristics of each largely determined by its prevailing climate. Therefore, if the climate of a particular region changes over time, so will its ecological composition. As species grow, reproduce, and eventually die within these ecosystems, they also leave vestiges of their presence in the surrounding ice, soils, rocks, corals, and/or lake and ocean sediments. These traces allow paleo-climatologists, who analyze these repositories, to determine which species were present at any given location and time, and thus to reconstruct the historical evolution of the environment at that location.

There are also other nonbiotic proxies for past climate, such as the vertical heat profile in the Earth's crust and the isotopic composition of ice buried in polar or alpine ice sheets. For more recent times, there are also human anecdotal records — like information on dates of harvest, the types of crops grown, major weather catastrophes — that can help the climate detective reconstruct patterns of the past.

Each type of paleo and proxy data provides only part of the climate story, and has its own values and limitations. Some are reliable indicators of detailed fluctuations in climate variables, and others only provide filtered information. Some provide information about growing seasons only, while others are most valuable for estimating winter conditions. Hence, where possible, paleoclimatologists use multiple types of proxies for each location that complement one another in providing a more complete picture of past local climate. When aggregated over space, such site-specific reconstructions can also be used to determine how regional, hemispheric, and even global climates have changed. Caution must be used in interpreting these reconstructed records of past climates, since they are based on many different indicators of varying reliability. However, many decades of work by paleoclimate experts have helped to extract from these varied data sources valuable information on both how the Earth's climate has changed and why.[3–5]

2.2.1.2 Recent Climate Observations Using Instrumentation

Although historical human anecdotal information and the Earth's natural environment have been valuable sources of proxy climate data, they have major limitations in terms of spatial and temporal details and provide little information on aspects of climate other than temperature and precipitation.

With the advent of instrumental climate record keeping in Europe several centuries ago, systematic observations of temperature, precipitation, and many other climate variables began to remove some of these limitations. Initially the spatial coverage of climate monitoring systems was sparse, particularly in polar regions and parts of North America, Africa, China, and Russia. However, global coverage was much improved by the mid-20th century. The advent of satellite observing systems some 25 years ago has further added to this coverage.

However, there are also some significant challenges in analyzing these instrumental data records for trends and variations in regional and global climate conditions. For example, changes in observing coverage and density over time have in some cases introduced systematic biases in measurements that need to be corrected when analyzing the data for trends. Furthermore, land-use change such as deforestation or increased urbanization has caused a significant bias in many local temperature records. Various research groups have worked meticulously to identify and remove possible biases in these records. Although there continue to be uncertainties in the success of these corrective measures, the high level of consistency among the various independent analyses undertaken to date and between corrected sea and land data where they abut along coastlines lends considerable confidence in the significance of the trends observed, particularly at the global scale.[6]

Analyzing global trends in precipitation and other hydrological variables (including cloud characteristics) is even more problematic, since hydrological variables can be significantly influenced by local factors. Furthermore, there is relatively little information for monitoring trends in precipitation over oceans. Hence, while good estimates for precipitation trends are available for some land regions with long records and a reasonably dense network of monitoring stations, there are no reliable estimates of global trends.[7]

In addition to the networks for monitoring near surface temperature and precipitation, over the past 50 years there has been an increasing array of complementary measurements of meteorological conditions within the atmosphere provided by balloon-borne and satellite-based instrument packages. These data have helped us to better understand global trends in atmospheric conditions, including cloud cover, humidity, and atmospheric temperatures.

Finally, there are many indirect indicators of recent and current trends in climate provided by monitoring of the global cryosphere (snow cover, sea ice, and glaciers) and of behavior of flora and fauna.

2.2.2 MAJOR CLIMATE REGIMES OF THE PAST 420,000 YEARS

Analyses of oxygen and hydrogen isotopes within the ice sheets of Antarctica are particularly valuable in reconstructing regional temperature fluctuations over the past 420,000 years. Temperatures during much of this period seem to have followed a cycle of long-term, quasi-periodic variations. Periods of cold temperatures, corresponding to major global glaciations, appear to have occurred at roughly 100,000-year intervals. Each of these extended glacial periods has been followed by a dramatic 8 to 10°C warming to an interglacial state. Within this 100,000-year cycle, smaller anomalies have occurred with regularity. Similar patterns are found in data extracted from Greenland ice cores and from ocean sediments. However, the latter suggest that, when averaged around the planet, the change in temperature during a glacial-interglacial cycle may be a more moderate 4 to 6°C.[8,9]

More detailed polar temperature data for the Holocene (approximately the past 10,000 years) indicate that mid- to high-latitude temperatures peaked slightly during the middle of the Holocene, some 5000 to 6000 years before present. This warm peak of the interglacial is commonly referred to as the Holocene maximum. During this period, Canada's climate was generally warmer, drier, and windier than that of today. In contrast, European climates during that period were initially warmer and wetter, then became drier. Climates in arid regions of Africa and Asia were also significantly wetter than today. However, both paleo data and model studies suggest that mid-Holocene temperatures may have been slightly cooler than today in low-latitude regions. Hence, when averaged on a hemispheric scale, mean global surface temperatures appear to have been remarkably stable during the entire Holocene. Several "little ice ages," or short periods of cooling, appear superimposed upon the Holocene record at approximately 2500-year intervals, the latest having occurred between about A.D. 1400 and 1900.[10–13]

2.2.3 CLIMATES OF THE 20TH CENTURY

2.2.3.1 Temperature Trends

Globally, average surface temperatures (Figure 2.1) have increased by about 0.7°C (±0.2°C) over the past century. However, the observed global trends in temperature have not been uniform in time. While average temperatures changed very little between 1860 and 1920, they increased relatively rapidly over the next two decades. The climate cooled moderately from mid-century until the early 1970s, then warmed rapidly at about 0.15°C/decade during the past 30 years. During the more recent warming period, nighttime minimum temperatures have been increasing at a rate about twice that of daytime maximum temperatures, thus decreasing the diurnal temperature range. Land surface temperatures have also been rising at about twice the rate of sea surface temperatures. Together, these factors have contributed to a lengthening of the frost-free period over lands in mid to high latitudes.[14,15]

When compared with the proxy data for climate variations of the past two millennia, it seems likely that the 20th century is now the warmest over that time period, and that the 1990s was the warmest decade. Furthermore, the *rate* of warming in recent decades appears to be unprecedented over that time period.[7,16]

Although the monitoring of temperatures within the Earth's atmosphere has a much shorter history than that for surface temperatures, climatologists now have some 45 years of data directly recorded by radiosondes borne aloft by balloons and almost 25 years of information obtained indirectly by instruments onboard satellites, particularly the microwave sounding unit (MSU). Comparison of the longer radio-sonde records with surface observations show that the long-term trend of globally averaged temperatures in the lower atmosphere since 1957 is very similar to that at the surface. However, there are significant differences in trends on decadal time-scales. For example, the lower atmosphere warmed more rapidly than the surface

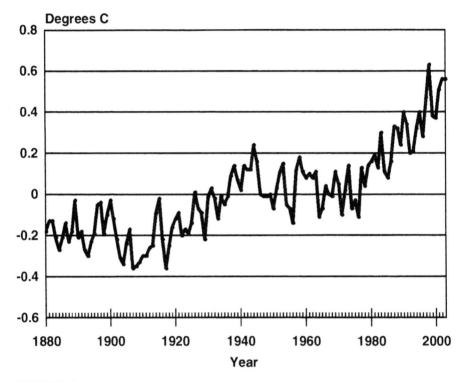

FIGURE 2.1
Departures of globally averaged surface temperatures from mean values. (Global land/sea temperature data available online at ftp://ftp.ncdc.noaa.gov/pub/data/anomalies/annual_land.and.ocean.ts.)

between 1957 and 1975, but warmed at a slower rate since that time. Experts suggest that much of these differences may be caused by changing atmospheric lapse rates with time, perhaps because of factors such as El Niño Southern Oscillations (ENSOs), volcanic eruptions, and global warming. Over longer timescales, these differences are expected to average out. There has also been considerable controversy about apparent differences between trends in lower atmospheric temperatures measured by satellite. However, recent studies suggest that the satellite MSU data have been contaminated by radiative effects of stratospheric cooling. When the MSU data are corrected for this bias, net warming in the lower atmosphere since 1979 appears to be very similar to that at the surface.[17-22]

Other parts of the global climate system are also beginning to show the effects of a global warming. Snowmelt, for example, has been occurring earlier across most of the Northern Hemisphere. Most glaciers and ice sheets in polar and alpine regions have been shrinking, particularly in Alaska and Europe. Many of the small glaciers are expected to completely disappear within decades. Likewise, some of the large ice shelves in Antarctica have been thinning.[23-29] Meanwhile, sea ice cover has been retreating dramatically across the Arctic.[30] The rate of heat uptake though these cryospheric melting processes is estimated to be similar to that occurring within the atmosphere. Borehole temperature measurements of the Earth's lithosphere indicate that that component of the climate system is also storing additional heat at similar rates.[31] More dramatically, waters within the upper 3 km of the world's oceans have increased their heat content at rates some ten times greater than this.[32]

Although the above results collectively indicate that the entire global climate system is heating up, the spatial and temporal patterns of this warming are varied and complex. Some regions have warmed much more rapidly than the global average and others much less so, or have even cooled. For example, the Antarctic Peninsula has warmed rapidly in recent decades, while other parts of Antarctica have cooled.[33–35] Likewise, the northwestern Arctic and much of Siberia have warmed by up to 3°C over the past 50 years, while the North Atlantic, the North Pacific and the northeastern U.S. have all cooled slightly.[36,37] In general, winter and spring seasons have warmed more than summer and fall seasons. These complex spatial and temporal patterns reflect shifts in global atmospheric circulation patterns that are occurring concurrently with the gradual rise in average temperatures. While such circulation changes have always been a contributor to normal climate variability, there are indications that recent changes may be at least partially attributable to warmer global climates.[38,39]

2.2.3.2 Precipitation Trends

Precipitation data records are much less representative of global trends than are those for temperature, since precipitation by its very nature is far less homogeneous. Furthermore, there is scant precipitation data for the Earth's ocean areas, which represent 70% of its surface. However, available records suggest a recent 0.5 to 1% increase/decade in annual average precipitation over most land areas in the mid to high latitudes of the Northern Hemisphere. Increases have been somewhat more modest over the tropics. There also appears to be a corresponding upward trend in both cloud cover and tropospheric water vapor content over much of the Northern Hemisphere. Water content in the comparatively dry stratosphere has also been increasing by about 1%/year. In contrast, there has been a modest decline (about 0.3%/decade) in precipitation over the Northern Hemisphere's sub-tropics. There are no clear indications of precipitation trends in the Southern Hemisphere, although some regions within South America and Africa show decreases. A number of countries have also experienced an increase in the number of wet days, and an increased proportion of total precipitation as heavy rain. As a result, most of the large watershed basins of the world have experienced a significant shift toward higher frequency of extreme hydrological floods during the 20th century.[40–45]

2.2.3.3 Other Climate-Related Trends

A broad range of indicators show that global ecosystems are already responding to recent changes in climate. About 80% of recent changes in behavior of more than 1500 biological species examined in various studies appear to be consistent with that expected due to regional changes in climate. On average, species have shifted their distributions poleward by some 6 km/decade and advanced the onset of their spring activities by 2 to 5 days/decade. Tropical ocean corals have also undergone massive bleaching in recent years. If such ecological responses to changes in climate differ significantly among species, this could effectively tear ecosystem communities apart.[46–49]

There are significant trends in climate extremes as well. For example, warm summer nights have become more frequent over the past few decades, particularly in mid-latitude and sub-tropic regions. This has contributed to a reduction in the number of frost days and in the intra-annual extreme temperature range. There has also been an increase in some regions in the extreme amount of precipitation derived from wet spells, in the number of heavy rainfall events, and/or in the frequency of drought. Hydrological data indicate that three quarters of 20th century extreme flooding events in major river basins of the world have occurred since 1953. This increase in extreme flood frequency appears to be very unusual, with an estimated 1.3% probability of being entirely due to natural variability. On the other hand, they are consistent with expected responses to warmer climates.[40,41]

Finally, changes in extreme weather behavior have also caused a global rise in related economic losses. In 2002, for example, losses due to record-setting floods in Europe and other weather-related disasters around the world resulted in economic losses in excess of U.S. $55 billion.[50,51]

2.3 CAUSES OF PAST CLIMATE CHANGE

The preceding discussion indicates that changes in the Earth's climate in recent decades are becoming increasingly unusual relative to that of the past several millennia. However, this evidence by itself does not help explain why these changes take place. To do so requires a more careful look at how the climate system works, how it responds to various external and internal forces that are exerted upon it over time, and how these responses might be modeled for use in climate simulations.

2.3.1 CLIMATE SYSTEM ENERGY BALANCE

In a very simple way, the Earth's climate system can be thought of as a giant heat engine, driven by incoming energy from the sun. As the solar energy passes through the engine, it warms the Earth and surrounding air, setting the atmospheric winds and the ocean currents into motion and driving the evaporation–precipitation processes of the water cycle. The result of these motions and processes is what we experience as weather and, when averaged over time, climate. The energy entering the climate system eventually leaves it, returning to space either as reflected shortwave solar radiation (unused by the climate system) or as emitted infrared radiation. As long as this outgoing energy leaves at the same rate as it enters, our atmospheric heat engine will be in balance and the Earth's average temperature will remain relatively constant. However, if some external factor causes an imbalance between the rates at which energy enters and leaves the climate system, global temperatures will change until the system responds and reaches a new equilibrium.

The flow of energy through the system is largely regulated by the Earth's atmosphere, although the radiative properties of the Earth's surface are also important factors. About 99% of the dry atmosphere is made up of nitrogen and oxygen, which are comparatively transparent to both incoming shortwave and outgoing infrared radiation. Hence they have little effect on the energy passing through the atmosphere. It is the variety of aerosols and gases that make up much of the remaining 1% of the dry atmosphere that, together with water vapor and clouds, function as the primary regulators of the crucial energy flows. They do so by reflecting, absorbing, and re-emitting significant amounts of both incoming solar radiation and outgoing heat energy.[52]

2.3.1.1 Incoming Solar Energy

Averaged around the Earth, the amount of sunlight entering the atmosphere is about 342 watts per square meter (W m^{-2}). However, approximately 31% of this incoming shortwave energy is reflected back to space by the atmosphere and the Earth's surface. The remaining 69% (about 235 W m^{-2}) is absorbed within the atmosphere and by the surface and thus provides the fuel that drives the global climate system. The amount of shortwave radiation returned to space by clouds and aerosols varies considerably with time and from one location to another. For example, major volcanic eruptions can abruptly produce large amounts of highly reflecting sulfate aerosols in the stratosphere that can remain there for several years before they settle out due to the forces of gravity. Alternatively, human emissions of sulfate aerosols into the lower atmosphere can significantly increase the reflection of incoming sunshine in industrialized regions compared to less-polluted areas of the world. Observational data indicate that, on average, clouds and aerosols currently reflect about 22.5% of incoming radiation back to space. Likewise, the amount of incoming energy reflected from the surface also depends on the time of year and the location. That is because snow and ice, which cover much of the Earth's mid- to high-latitude surfaces during winters, are highly reflective. On the other hand, ice-free ocean surfaces and bare soils are low reflectors. When averaged over time and space, the Earth's surface reflects almost 9% of the solar radiation entering the atmosphere back to space.

  In addition to reflecting and scattering incoming solar radiation, the atmosphere also absorbs almost 20% of it. About two thirds of this absorption is caused by water vapor. A second significant absorber is the ozone layer in the stratosphere, which absorbs much of the ultraviolet part of incoming solar energy. Thus this layer not only protects the Earth's ecosystems from the harmful effects of this radiation but also retains a portion of the sun's energy in the upper atmosphere. Another one tenth of the absorption can be attributed to clouds. Finally, a small fraction of the absorption is due to other absorbing gases and aerosols (particularly dark aerosols such as soot).

2.3.1.2 Outgoing Heat Radiation

The Earth's atmosphere and surface, heated by the sun's rays, eventually release all of this energy back to space again by giving off long-wave infrared radiation. When the climate system is in equilibrium, the total amount of energy released back to space by the climate system must, on average, be the same as that which it absorbs from the incoming sunlight — that is, 235 W m^{-2}. However, as the infrared radiation tries to escape to space, it encounters several major obstacles that can absorb much of it before it reaches the outer atmosphere — primarily clouds and absorbing gases. This absorbed energy is then reradiated in all directions, some back to the surface and some upward where other absorbing molecules at higher levels in the atmosphere are ready to absorb the energy again. Eventually, the absorbing molecules in the upper part of the atmosphere emit the energy directly to space. Hence, these gases make the atmosphere opaque to outgoing heat radiation, much as opaque glass will affect the transmission of visible light. Together with clouds, they provide an insulating blanket around the Earth, keeping it warm. Because they retain heat in somewhat the same way that glass does in a greenhouse, this phenomenon has been called the *greenhouse effect*, and the absorbing gases that cause it, *greenhouse gases*. Important naturally occurring greenhouse gases include water vapor, carbon dioxide, methane, ozone, and nitrous oxide.

The magnitude of the thermal insulating effect caused by greenhouse gases and clouds can be estimated fairly easily. Theoretically, the average radiating temperature required to release 235 W m^{-2} to space is −19°C. Yet we know from actual measurements that the Earth's average surface temperature is more like +14°C, some 33°C higher. This is enough to make the difference between a planet that is warm enough to support life and one that is not.

2.3.2   PAST CLIMATE FORCINGS

Primary causes for changes in the amount of energy entering or leaving the climate system (called climate forcings) involve alterations in the intensity of sunlight reaching the Earth's atmosphere, changes in the reflective properties of the Earth's surface, and/or variations in the concentrations of aerosols and greenhouse gases in the atmosphere. Studies of past climates indicate that such factors occur naturally and change constantly — on timescales varying from months to millions of years, and at spatial scales from local and regional to global. However, since the onset of human civilization some 8000 years ago, humans are also becoming an increasingly important factor.[53]

2.3.2.1   Natural Climate Forcing Factors

The most widely accepted hypothesis for explaining the largest variations in global temperatures during the past 420,000 years is that of solar forcing due to changes in the Earth's orbit around the sun. The 100,000-year glacial–interglacial cycle, for example, appears to be linked to the well-documented changes in eccentricity of the Earth's orbit around the sun. Similarly, changes in the obliquity and precession of the Earth's orbit likely contribute to climate variability at intervals of 41,000 and 22,000 years, respectively. These orbital changes affect both the total and the seasonal distribution of incoming sunlight across the Earth's surface. However, while the large glacial–interglacial cycles correlate well with changes in orbital eccentricity, the net annual solar forcing caused by those changes is far too weak to fully explain the amplitude of the climate cycles. Hence, various feedback processes appear to be significantly amplifying this forcing. Paleo studies indicate that changes in atmospheric greenhouse gas concentrations and altered surface albedos are two such important positive-feedback mechanisms. For example, analyses of Antarctic and Greenland ice cores indicate a strong correlation between past long-term changes in climate and the natural atmospheric concentrations of carbon dioxide (CO_2), methane (CH_4), and nitrous oxide (N_2O), all important greenhouse gases. The correspondence between atmospheric carbon dioxide, methane concentrations, and local Antarctic temperatures during the past 420,000 years has been remarkable. However, the various processes involved in such millennial scale changes in climate are very complex, and can differ between hemispheres. Furthermore, they may also differ from one cycle to the next, suggesting that past events may not be good analogues for the current interglacial. Past interglacials may also have been significantly longer than the 10,000 years previously thought.[54,55]

While orbital forcing factors may have been very important on millennial time-scales, their role in climate forcing on century and decadal timescales is quite minor. On these shorter timescales, aerosol emissions from volcanic eruptions and solar irradiance cycles appear to be far more important. In fact, many of the variations in climate over the past 300 years appear to be closely linked to changes in sunspot cycle behavior. However, the mechanisms by which relatively small change in solar irradiance can significantly affect climate are as yet not well understood.[56,57]

2.3.2.2 Human Interference with the Climate System

There is now clear evidence that another major forcing factor is at work on the climate system. Although humans may have started affecting regional climates many thousands of years ago, their role as agents of climate change on a global scale has escalated rapidly since the beginning of industrialization. This factor is now expected to dominate over all natural forcings and internal climate variations likely to occur over the next century and beyond. Human activities may, in fact, be ushering in a radically new stage in the Earth's climate that some are referring to as the "Anthropocene."[1] The following paragraphs describe two prominent aspects of the climate system that humans have altered in the past, and how these are likely to change in the future.

Regional Surface Albedo and Hydrology. Humans have been significantly transforming the Earth's regional landscape ever since the onset of agrarian human societies in Asia and Africa some 8000 years ago.[52] Over the millennia, they have changed vast areas of forested lands into agricultural fields (and, in some places, back again), dry lands into wetlands or wetlands into dry, and rural landscapes into city environments. Such changes in land use have altered the local albedo of the Earth's surface and hence have influenced how much sunlight is reflected back to space. In some cases, these changes have reduced surface reflection and caused a warming influence. In others, they have increased reflectivity and caused a local cooling. For example, studies indicate that deforestation in mid to high latitudes of the Northern Hemisphere caused winter season albedo to increase significantly because snow-covered fields are much more reflective than the trees they replaced. In fact, such changes from forest to agricultural landscapes may have caused a net global cooling effect of between 0.1 and 0.2°C over the past three centuries. In addition to changing surface albedo, land-use change also affects regional evaporation, evapo-transpiration, rainfall, atmospheric circulation, and cloud cover. However, the regional and seasonal complexities of this forcing factor are poorly understood and hence its importance in helping to explain past changes in climate is still difficult to quantify.[58-61]

Changing Atmospheric Composition. The other major way humans are interfering with the climate system is through emissions into the atmosphere of greenhouse gases and aerosols. These emissions change their abundance within the atmosphere, and thus gradually change the atmosphere's role in regulating the flow of energy into and out of the Earth's climate system.[62,63] For example:

- Over the past 150 years, humans have cumulatively emitted approximately 1500 billion tonnes of carbon dioxide into the atmosphere. About two thirds of these emissions were caused by the combustion of fossil fuels for energy, the remainder by deforestation. Emissions from the latter have been relatively stable in recent decades (at about 6 billion tonnes of CO_2 per year). However, those from fossil fuel use continue to rise quite rapidly, increasing from an average 20 billion tonnes/year in the 1980s to about 23 billion tonnes per year during the 1990s. Fortunately, natural processes are removing a significant fraction of these human emissions from the atmosphere through enhanced absorption in surface oceans and increased uptake by terrestrial vegetation. However, the amount of CO_2 in the atmosphere still increased at the rate of some 12 billion tonnes per year during the 1990s. Atmosphere concentrations, which were at a remarkably stable level of about 260 to 280 parts per million by volume (ppmv) throughout the Holocene, had increased to levels of about 374 ppmv by 2002 (Figure 2.2).This is an increase over pre-industrial levels of about 33%. There are indications, in fact, that current levels may be unprecedented in the past 20 million years. The net direct radiative forcing caused by this increase is estimated to be about 1.5 W m^{-2}.
- Human activities have contributed to dramatic increases in other greenhouse gases as well. Atmospheric methane concentrations have more than doubled over the past century, while those for nitrous oxide have increased by about 15%. Tropospheric ozone has also increased substantially over much of the industrialized world, and entirely new and powerful greenhouse gases such as halocarbons and sulfur hexafluoride are now being added in significant amounts. Collectively, these have added about 1 W m^{-2} to the positive forcing caused by carbon dioxide. Meanwhile, human emissions of halocarbons have also indirectly contributed to a decrease in ozone within the stratosphere, slightly offsetting the above forcings by between –0.1 and –0.2 W m^{-2}.
- Finally, there has also been a progressive increase in anthropogenic emissions of aerosols and their precursors into the atmosphere. While most of these aerosols have relatively short atmospheric lifetimes of

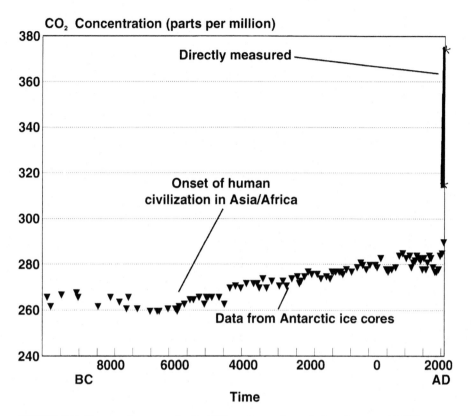

FIGURE 2.2 Changes in atmospheric CO2 concentrations during the past 11,000 years. (Vostok ice core data from Barnola, J.-M. et al., available on-line at http://cdiac.ornl.gov/trends/co2/vostok.htm; observed data for Mauna Loa from http://cdiac.ornl.gov/ftp/maunaloa-co2/maunaloa.co2.)

days to weeks, their continuous production in industrialized regions of the world have resulted in a large and sustained increase in their concentrations over and downwind of these regions. From a climate perspective, these aerosols play several important roles. First, they directly affect the amount of incoming sunlight that is reflected back to space or absorbed within the atmosphere. Second, fine aerosols also function as condensation nuclei and hence alter the amount and properties of cloud. They thus indirectly affect the absorption and reflection of incoming radiation through the role of these clouds. Finally, as these aerosols settle out of the atmosphere onto the Earth's surface, they can affect surface albedo. This is particularly true for soot on snow or ice. While there are large uncertainties associated with these effects, experts suggest that the net direct effect of increased aerosol concentrations over the past century is likely negative (on the order of -0.5 W m^{-2}), hence offsetting some of the warming effects of rising greenhouse gas concentrations. Their indirect effects through altered cloud conditions may be larger, but are even more uncertain.

2.3.3 SIMULATING CLIMATE FORCINGS UPON A DYNAMIC SYSTEM

Climate and climate variability at the Earth's surface are ultimately consequences of the way the atmosphere and the oceans redistribute and release heat energy that the Earth has absorbed from the sun. Because the intensity of the solar radiation changes with latitude, time of day, and time of year, all parts of the planet are not heated equally. Solar heating is greatest in the tropics, creating an excess of incoming relative to outgoing radiation. Temperatures here are subsequently much warmer than the global average, remaining consistently within a few degrees of 30°C during all seasons of the year. At the opposite extreme, the Earth's polar regions experience a net loss of energy to space. The magnitude of this polar energy loss varies significantly with time and space. On average, it is largest in winter seasons and weakest in summer. Land areas also experience higher winter heat losses than ocean areas. Some polar land areas that have dark surfaces can experience net gains in summer, when these surfaces are free of snow. Hence, temperatures in polar regions can vary from regional extremes of nearly 20°C in the Arctic during summer to regional lows of –60°C during Antarctic winters.

Much of the sun's energy absorbed at the Earth's surface is used to evaporate water from ocean and land surfaces and ecosystems. The more heat at the surface and the warmer the air temperature, the greater the amount of water vapor that can be evaporated from the surface and retained within the atmosphere. However, once air becomes saturated with water vapor, it condenses again into tiny water droplets or ice crystals that form clouds. When the conditions are right, these droplets or crystals fall to the ground as precipitation. Where, when, how much, and what type of precipitation falls will depend on the characteristics of a range of local atmospheric and surface factors. Furthermore, since atmospheric moisture is also transported horizontally by air currents, the precipitation patterns that emerge around the Earth are also influenced by the global atmospheric circulation patterns. As a result the distribution of precipitation around the planet presents an even more complex pattern than that for atmospheric circulation. Some areas receive large surpluses of rainfall that support very lush, rich ecosystems, while others do not receive enough to nourish vegetation and so become deserts.

Numerous other factors also affect the Earth's climate. These include the daily and seasonal cycles in solar forcing, the effects of clouds and of snow and ice, the influence of topography, and the impact of processes and activities within the biosphere. All of these elements are interconnected, interacting parts of the climate system. If a change in one of these parts upsets the balance of that system, it is likely to initiate complex reactions in some or all of the other parts, until a new equilibrium is established. Some reactions occur very rapidly, while others occur very, very slowly. Furthermore, some may increase the initial change (a process known as *positive feedback*), while others may oppose and partially offset it (a *negative feedback*).

The proxy data and climate observations discussed in Section 2.2 of this chapter provide invaluable information on *what* has happened to the Earth's climate in the past and *how* it is behaving today. However, because of the complexities and non-linearity of the dynamic global climate system, these data by themselves are limited in their usefulness to explain *why* the climate system changes with time and in space. Hence they also cannot be used with credibility to project how future climates might change. To develop such predictive capability, the complex feedbacks that inextricably couple together the various components of the climate system must first be carefully studied with the help of simple models and approximated as complex mathematical equations that capture the fundamental physical, chemical, and biological processes involved. These can then be integrated within a complex model of the entire dynamic global system. Such climate system models have been under development for some four decades, and are now sufficiently advanced to include a dynamic circulating atmosphere coupled to a circulating global ocean system with a responsive ice cover. Some modelers have now also coupled the dynamics of global ecosystems within their climate system models, thus including a representation of the biogeochemical feedbacks between the climate system and the biosphere. Although these models, generally referred to as coupled climate models, are intricate approximations of how the Earth's climate system functions and require the largest and most advanced supercomputers available to run, they continue to be limited by computing capacity, by inadequate observational data for parameterization, calibration, and verification, and by gaps in scientific knowledge of feedbacks within the climate system. Hence, in many respects, they still represent quite crude reproductions of the behavior of the real climate system. Nevertheless, they have become valuable tools for adding to the scientific understanding of why climate changes, and how it may change in the future.

2.3.4 ATTRIBUTING RECENT CLIMATE CHANGE

In recent decades, climate modelers have collaborated closely with climate forcing experts to examine how natural and human climate forcings have changed over the past century and how these changes may have affected the global climate systems. While differing in temporal and spatial detail and in the types of models and methodologies used, related studies indicate that combined natural and human forcings to date explain the global surface temperature trends of the past century quite well. Similar results are achieved in analyses of the observed rise in heat storage in the upper oceans. The relative balance between natural and human forcings, however, varies significantly during the century. During the first half of the century, the average intensity of solar irradiance slowly increased, while a decline in the frequency of major volcanic eruptions resulted in a decrease in average volcanic dust loading in the atmosphere. The net result was a gradual positive trend in natural climate forcing (and hence a warming influence). On the other hand, net human forcings during that period were relatively neutral, partly because the cooling effects of rising human aerosol emissions partially offset the modest increases in greenhouse gas concentrations. Hence, much of the observed warming between 1900 and 1950 can be explained by natural causes. However, a major shift in these forcing factors occurred in mid-century. During the past 50 years, volcanic eruptions again became more frequent (increasing average atmospheric concentrations of volcanic aerosols) and the average solar forcing stabilized. The net effect was to create a net negative natural forcing during that period.

In fact, when the two periods are combined, the net natural forcing over the entire century is near zero. In contrast, human influences have escalated rapidly during the past half century. Model studies suggest that the observed increase in concentrations of greenhouse gases over the past century would be enough to already warm the Earth's surface by 0.9°C. However, this warming influence has been partially offset by a net cooling effect due to increased aerosol concentrations and by stratospheric ozone depletion. Although the magnitude of this offset is poorly understood, it is estimated to be on the order of 0.4°C. Hence the net human-induced warming since 1900 may be somewhere in the range of 0.5°C, which is very close to the 0.6°C warming observed in the climate records. This signal is stronger in the Southern Hemisphere than the Northern Hemisphere, since the much lower concentrations of anthropogenic aerosols generate a much weaker masking of the signal caused by the enhanced greenhouse gas concentrations.[64–66]

Further supporting the model evidence for the significance of the human role is the evidence from the paleo data. These data indicate that the 20th century warming is likely without precedent during at least the past 2000 years, and hence difficult to explain on the basis of natural factors alone. Hence, experts agree that most of the observed warming over the past 50 years can be attributed to human activities.[7,66–70]

2.4 PROJECTED CLIMATE CHANGE FOR THE NEXT CENTURY

2.4.1 FUTURE CLIMATE FORCING SCENARIOS

The Earth's surface albedo and hydrology — and hence climate — will be significantly affected by future changes in vegetation cover. One important factor in such changes will be human behavior with respect to land use, which is difficult to predict. However, experts suggest that changes in land use over the next century will have finite limits that range between a continuation of current deforestation rates until most virgin forests have been depleted (the worst-case scenario) and a gradual increase in global forest cover due to the combined effects of a decline in deforestation activities and increased reforestation efforts (best-case scenario).[71] Another important factor affecting vegetation distribution is climate change itself. Such climate effects will vary substantially with time and space. In polar and sub-polar regions, for example, forest cover is expected to increase as treelines respond to warmer temperatures. In other areas, changing climates may increase desertification or cause forest landscapes to change to grasslands through natural successional processes. Modelers are now beginning to include related albedo and hydrological feedbacks into their advanced climate system models.[72–77]

Experts have also provided a range of plausible futures for greenhouse gas and aerosol emissions. These suggest that, unless deliberate measures are taken to curtail future emissions, another 3600 to 8000 billion tonnes of carbon dioxide will be emitted into the atmosphere over the next century as a result of fossil fuel combustion and land-use change. Allowing for terrestrial and ocean removal processes, they project that this will almost certainly result in a doubling of pre-industrial atmospheric CO_2 concentrations by 2100, and that a tripling could be possible. Concentrations of other greenhouse gases are also expected to increase, although not as significantly as carbon dioxide. In contrast, measures to curtail their emissions of aerosols for local environmental reasons are expected to help eventually cause these to decline relative to current levels. These changes in greenhouse gases and aerosols are likely to increase in global radiative forcing by 2100 by between about 4 and 9 W m^{-2} (relative to pre-industrial levels).[78,79]

2.4.2 CLIMATE MODEL PROJECTIONS

2.4.2.1 Temperature

Climate models are the primary tools used to project how future changes in radiative forcings might affect the climates of the next century. In its *Third Assessment Report*, the IPCC noted that, when allowing for uncertainties in future changes in greenhouse and aerosol concentrations as well as for the range of plausible climate sensitivity to such changes, expected warming of average surface temperatures will be between 1 and 2.5°C by 2050, increasing to between 1.4 and 5.8°C by 2100, relative to mean climate conditions of the past few decades.[80] Delayed effects from changes in radiative forcings to date already commit the world to 0.5°C of that warming, even if all emissions of greenhouse gases were to stop immediately.[81] Some experts have attempted to put some probabilities on these projections. For example, they suggest a 95% probability that global climates will warm by at least 1.7°C by 2100, and a 5% probability that it could warm by more than 4.9°C.[82,83] As illustrated in Figure 2.3, at the lower end of this range, changes will be unprecedented in the history of human civilization, while at the upper end they are comparable to the magnitude of change during the last deglaciation, but take place at about 50 times the rate.[10,78,84]

While the above results indicate that there is considerable agreement between model results on the significance of global-scale changes in temperature, there is much less agreement with respect to regional changes. Despite these differences, there are a number of common features.[78,85] For example:

- *Land areas warm more than ocean surfaces*. This is primarily a consequence of the thermal inertia of oceans. This ocean inertia is also the principal reason for a significant delay of decades in approaching a new equilibrium climate in response to an enhanced forcing regime.
- *The Arctic polar region warms more than the tropics*. The primary reason for this polar amplification is the reduction in the extent and duration of snow and ice cover on the surface, which thus reduces surface albedo, causing a positive feedback. Although such amplification is also expected to eventually occur in the Antarctic region, model experiments suggest a delayed cryospheric response to global warming in that region, relative to that in the Arctic.
- Nighttime temperatures will, on average, warm more than daytime temperatures, thus reducing the daily temperature range.

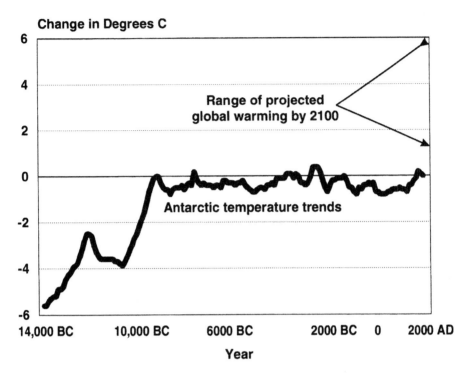

FIGURE 2.3 Projected changes in average Antarctic temperatures to 2100, in comparison with Antarctic trends of the past 15,000 years. (Ice core trends were adapted from data available online at http://cdiac.ornl.gov/trends/temp/vostok/jouz_tem.htm; future projections are based on multimodel regional simulations for several SRES scenarios, available on the IPCC data center web site http://ipcc-ddc.cru.uea.ac.uk.)

- *Ocean circulation is expected to slow down.* The turnover of the global oceans is largely determined by thermohaline processes that affect surface water densities. Generally, model studies agree that enhanced precipitation in the high latitudes of the Northern Hemisphere is likely to decrease the rate of deep-water formation in the North Atlantic and hence weaken the thermohaline circulation system. Melting of sea ice will add to this freshwater input. A weaker ocean circulation will also influence ocean heat transport mechanisms and may thus cause some surface ocean regions, like areas of the North Atlantic, to actually cool while the rest of the world warms.
- *Natural oscillations in the climate system will be superimposed on the projected upward trends in temperatures, and hence will modulate both the temporal and spatial response of climates to enhanced radiative forcing.* This adds significantly to the temporal and spatial uncertainty of the model projections, particularly at the regional scale.
- *There is also evidence that the pattern of future warming will increasingly be like that of current El Niño years, with enhanced warming in the central and eastern tropical Pacific relative to the western Pacific.* This, in turn, causes global atmospheric circulation patterns to change.

2.4.2.2 Projected Changes in Precipitation

Warmer surface temperatures will increase surface evaporation and hence enhance the global hydrological cycle. Various model projections suggest that, by mid-century, global precipitation response to projected warming could be anywhere between a small decrease of –0.2% and an increase of 5.6%. The atmospheric holding capacity for water vapor should also increase by about 7%/°C of warming. This means increased surface water loss through evaporation processes, averaged over the Earth's surface, will likely exceed increases in precipitation. Since the additional release of latent heat when water vapor condenses within the atmosphere will further invigorate storms, there will be a greater increase in rate of heavy precipitation than of average precipitation. That is, precipitation events (including snow storms) are likely to become fewer but more intense. The fraction of precipitation falling as snow will decrease, as will the duration of the snow season.

However, concurrent shifts in atmospheric circulation and hence precipitation patterns suggest that these changes will vary significantly from region to region and season to season. In general, the El Niño–like pattern indicated in many of the models suggests an eastward shift of precipitation in the tropical Pacific. Various models also suggest an enhanced positive Arctic Oscillation under warmer climates. Most high-latitude areas are expected to experience significant increases in precipitation, while much of the sub-tropical regions are likely to experience decreases. In the tropics, projected changes are smaller and less consistent among models. On average, the intensity of rainfall increases. Most models also project an increase in droughtiness of mid-continental regions in summer. By the time of CO_2 doubling, the Sahara Desert will have become significantly hotter and drier and its margins may have shifted northward by about 0.55° of latitude.[45,75,85–90]

2.4.2.3 Permafrost

The permafrost that underlies two thirds of the Arctic land mass is expected to undergo extensive thawing. For example, by 2100, the thickness of the active layer overtop most continuous permafrost zones is projected to increase by 30 to 40%. Much of the permafrost would completely disappear in discontinuous zones. Since snow acts as an insulator that slows the penetration of winter cold temperatures into the ground, any decrease in snow cover under warmer climates could help offset some of expected permafrost loss. However, losses would still cause significant ground settlement in most affected regions, and catastrophic slope movement in some locations. While much of the melting ground ice may evaporate, there will also be a pronounced effect on the permeability of soils, the melting of surface ice, and the seasonality of stream flows. These changes, in turn, will affect ecosystems. There is a remote risk that permafrost degradation could cause the release of large amounts of methane gas from natural gas hydrates, perhaps explosively.[91–94]

2.4.2.4 Severe Weather

Much of the potential risk of danger associated with warmer climates relates to the frequency of severe weather events that exceed the tolerance levels of ecosystems and/or socioeconomic systems. These events are often regional and local in scale and by definition occur infrequently. Hence they are much more difficult to simulate than larger-scale climate phenomena. However, various studies have helped provide some useful clues regarding how the behavior of weather extremes might change under warmer climates.[41,95-98] For example:

- Temperature extremes. An increase in average global temperatures will very likely cause a disproportionately larger increase in the frequency and intensity of extreme hot days and a decrease in the probability of extreme cold days. Because of the added discomfort of higher humidity, periods of intense heat stress become even more frequent.
- Precipitation extremes increase more than their means, and intense rainfall events become almost twice as frequent in many regions. For example, within the next century, extreme 1-in-20-year U.K. rainfall events could occur once every 3 to 5 years. Over central and northern Europe, 1-in-40-year winter precipitation extremes may increase to 1 in 8 years by the time of a transient CO_2 doubling. A similar increase occurs for summer monsoonal rainfall extremes over much of southern Asia. However, such events are projected to become less frequent over the Mediterranean and northern Africa. Generally, the lengths of wet and dry spells are both also expected to increase.
- Some studies suggest that the total number of mid-latitude winter storms is likely to decrease, but that the number of intense storms will rise. Dominant storm tracks may also be displaced poleward.
- Intense El Niño and La Niña events, which can be very disruptive to normal weather patterns, may become more frequent.
- While it is uncertain whether the frequency of tropical storms will rise, warmer ocean surfaces imply that the potential upper threshold for tropical storm intensity will increase.
- Changes in precipitation extremes, together with changes in timing of snowmelt, will have significant impacts on river flows and flood risks. In one study, 6 of 14 large river basins assessed are projected to experience a higher probability of floods, while only 2 show a decrease. Another study projects that, for a quadrupled CO_2 climate, some extra-tropical river basins, particularly in Russia, could see current 1-in-100-year flood events increase in frequency to once every 2 to 3 years. For Canadian rivers, the return period of such extreme floods could become once every 4 years for the Fraser River, once every 8 years for the St. Lawrence River, and once every 11 years for the Nelson River (Figure 2.4). However, most high-latitude rivers will have lower severe spring flood risks because of less snowmelt.

2.4.2.5 Risks of Large-Scale Abrupt Changes in Climate

In general, climate models suggest that global climates will evolve gradually in response to enhanced radiative forcings. However, there is considerable evidence

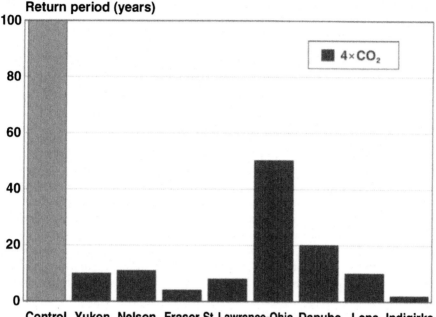

FIGURE 2.4 Return periods for the current 100-year flood for selected North American and Russian River basins under a 4X CO2 scenario. The control bar is the current 100-year flood event. (From Milly, P.C.D. et al., Nature, 415, 514-517, 2002.)

from paleo-climatological data that the Earth's climate can change abruptly from one relatively stable mode to another. Such abrupt changes would be far more catastrophic in nature, since the changes would be large and very rapid, and hence the potential for progressive adaptation of ecosystems and socioeconomic infrastructures would be minimal. Experts have identified three possible processes by which enhanced radiative forcing could cause such abrupt discontinuities in climates over the next decade.[99–101]

The first is the risk of a complete shutdown of the global ocean thermohaline circulation system. Since this system is the driver for the Gulf Stream that keeps Western Europe some 10°C warmer than it would be in its absence, its cessation could cause a calamitous cooling in that region even as the rest of the world warms. It would also dramatically alter the global atmospheric circulation patterns and hence the distribution of rainfall.

The second is the possible collapse of the West Antarctic Ice Sheet. This ice sheet is considered to be unstable and its demise through glacial surging (rather than *in situ* melting) could be triggered by 2100. It stores enough water to raise global sea levels by 5 to 6 m. A panel of experts has noted that, over the next 200 years, there is a 30% chance that the rapid decay of this ice sheet alone would contribute 20 cm/century to global sea levels, and a 5% risk of a 1 m/century contribution.

The third risk relates to the potential release of large amounts of methane contained within the solid hydrates found in the Arctic permafrost zone, particularly under the ocean. Paleo studies indicate that such large releases from the ocean bottom likely occurred during an abrupt climate warming some 55 million years ago. Since methane is a potent greenhouse gas, a repeat of such an event could abruptly launch the Earth into a super-greenhouse effect era.

2.5 SUMMARY AND CONCLUSIONS

Local climate is a major determinant of both the composition and behavior of a region's ecosystems and of the infrastructure and culture of human society residing within it. Hence, climatic statistics are an important factor in ecosystem management and planning for socioeconomic development. Although decision makers often treat these statistics as a constant, there is clear evidence that climate has changed, is changing, and will continue to change. Such changes occur naturally and, during the geologic past, have at times been quite great. However, global climates have been remarkably stable during the current interglacial, resulting in natural climate fluctuations that have been relatively modest. Human interference with the climate system, primarily through land-use change and changes in atmospheric composition, is now adding an unprecedented and increasingly dominant force for change. There is convincing evidence that this interference has already caused a substantial increase in global mean temperatures over the past 50 years. Projected changes for the next century will likely exceed anything yet experienced in human history and could rival the magnitude of very large changes during the past million years, but at a much more rapid rate. Such rapid change will have dramatic implications for, *inter alia*, the global hydrological cycle, ecosystem composition, and the frequency and severity of extreme weather events.

ACKNOWLEDGMENTS

In preparing this overview of climate and climate change science, the author has relied heavily on the assessments provided by the Intergovernmental Panel on Climate Change, particularly the Third Assessment completed in 2001. That assessment can be accessed on line at www.ipcc.ch. The author also wishes to acknowledge the valuable comments provided by Mike Price, Jag Bhatti, and two anonymous reviews of the document.

REFERENCES

1. Berger, A. and Loutre, M.F., An exceptionally long interglacial ahead? *Science*, 297, 1287–1288, 2002.
2. World Meteorological Organization, *Report of the International Conference on the Assessment of the Role of Carbon Dioxide and Other Greenhouse Gases in Climate Variations and Associated Impacts*, October 1985, Villach, Austria, WMO-No. 661, 1986.

3. Beltrami, H., Earth's long-term memory, *Science,* 297, 206–207, 2002.
4. Mann, M.E., The value of multiple proxies, *Science,* 297, 1481–1482, 2002.
5. Shrag, D.P. and Linsley, B.K., Corals, chemistry and climate, *Science,* 296, 277–278, 2002.
6. Folland, C.K. et al., Observed climate variability and change, in *Climate Change 2001: The Scientific Basis Contribution of Working Group 1 to the Third Assessment Report of the Intergovernmental Panel on Climate Change*, J.T. Houghton et al., Eds. Cambridge University Press, Cambridge, 2001, 99–181.
7. Folland, C.K. et al., Global temperature change and its uncertainties since 1861, *Geophys. Res. Lett.,* 28, 2621–2624, 2001.
8. Bush, A.B. and Philander, S.G.H., The climate of the Last Glacial Maximum: results from a coupled Atmosphere-Ocean General Circulation Model, *J. Geophys. Res.,* 104, 24509–24525, 1999.
9. Weaver, A.J., Eby, M., Fanning, A.F., and Wiebe, E.C., Simulated influence of carbon dioxide, orbital forcing and ice sheets on the climate of the Last Glacial Maximum, *Nature,* 394, 847–853, 1998.
10. CDIAC on line at http://cdiac.ornl.gov/trends/temp/vostok/jouz_tem.htm.
11. Adams, J.M., Global Land Environments since the Last Interglacial. Oak Ridge National Laboratory, TN. Unpublished report available online at http://www.esd.ornl.gov /ern/qen/nerc.html, 1997.
12. Gajewski, K. et al., The climate of North America and adjacent ocean waters *ca* 6ka, *Can. J. Earth Sci.,* 37, 661–681, 2000.
13. Hewitt, C.D. and Mitchell, J.F.B., A fully coupled GCM simulation of the climate of the mid-Holocene, *Geophys. Res. Lett.,* 25, 361–364.
14. Jones, P.D. and Moberg, A. Hemispheric and large-scale surface air temperature variations: an extensive revision and an update to 2001, *J. Climate,* 16, 206–223, 2003.
15. Stone, D.A. and Weaver, A.J., Daily maximum and minimum temperature trends in a climate model, *Geophys. Res. Lett.,* 29(9), 10.1029/2001GL014556, 2002.
16. Mann, M.E. and Jones, P.D., Global surface temperatures over the past two millennia. *Geophys. Res. Lett.,* 30(15), 1820,doi:10.1029/2003GL017814, 2003.
17. Christy, J.R., Parker, D.E., Stendel, M., and Norris, W.B., Differential trends in tropical sea surface and atmospheric temperatures since 1979, *Geophys. Res. Lett.,* 28, 183–186, 2001.
18. Gaffen, D.J., Sargent, M.A., Habermann, R.E., and Lanzante, J.R., Sensitivity of tropospheric and stratospheric temperature trends to radiosonde data quality, *J. Climate,* 13, 1776–1795, 2000.
19. Hegerl, G.C. and Wallace, J.M. Influence of patterns of climate variability on the difference between satellite and surface temperature trends, *J. Climate,* 15, 2412–2428, 2002.
20. Jin, M. and Dickinson, R.E., New observational evidence for global warming from satellite, *Geophys. Res. Lett.,* 29, doi:10.1029/2001GL013833, 2002.
21. Vinnikov, K.Y. and Grody, N., Global warming trend of mean tropospheric temperature observed by satellites, *Science,* 302, 269–272, 2003.
22. Fu, Q., Johanson, C.M., Warren, S.G., and Seidel, D.J., Contribution of stratospheric cooling to satellite-inferred tropospheric temperature trends, *Nature,* 429, 55–58, 2004.
23. Arendt, A.A., Echelmeyer, K.A., Harrison, W.D., Lingle, C.S., and Valentine, V.B., Rapid wastage of Alaska glaciers and their contribution to rising sea level, *Science,* 297, 382–386, 2002.

24. Meier, M.F., Dyurgerov, M.B., and McCabe, G.J., The health of glaciers: recent changes in glacier regime, *Climatic Change* 59, 123–135, 2003.

25. Reichert, B.K., Bengtsson, L., and Oerlemans, J., Recent glacier retreat exceeds internal variability, *J. Climate*, 15, 3069–3081, 2002.

26. Rignot, E. and Thomas, R.H., Mass balance of polar ice sheets, *Science*, 297, 1502–1506, 2002.

27. Shepherd, A., Wingham, D.J., Payne, T., and Slvarca, P., Larsen Ice Shelf has progressively thinned, *Science*, 302(5646), 856–859, 2003.

28. Stone, R.S., Dutton, E.G., Harris, M., and Longenecker, D., Earlier spring snowmelt in northern Alaska as an indicator of climate change, *J. Geophys. Res.*, 107(D10), 10.1029/2000JD000286, 2002.

29. Thompson, L.G. et al., Kilimanjaro ice core records: evidence of Holocene climate change in tropical Africa, *Science*, 298, 589–593, 2002.

30. Comiso, J.C., A rapidly declining perennial sea ice cover in the Arctic, *Geophys. Res. Lett.*, 29, 1956 doi:10.1029/2002GL015650, 2002.

31. Beltrami, H., Smerdon, J.E., Pollack, H.N., and Huang, S., Continental heat gain in the global climate system, *Geopys. Res. Lett.*, 29, doi:10.1029/2001GL014310, 2002.

32. Levitus, S., Antonov, J., Wang, J., Delworth, T., Dixon, K., and Broccoli, A., Anthropogenic warming of Earth's climate system, *Science*, 292, 267–270, 2001.

33. Doran, P.T., Priscu, J.C., and Lyons, W.B., Antarctic climate cooling and terrestrial ecosystem response, *Nature*, 415, 517–520, 2002.

34. Thompson, D.W.J. and Solomon, S., Interpretation of recent southern hemisphere climate change, *Science*, 296, 895–899, 2002.

35. Turner, J., King, J.C., Lachlan-Cope, T.A., and Jones, P.D., Recent temperature trends in the Antarctic, *Nature*, 418, 291–292, 2002.

36. Przyblak, R., Changes in seasonal and annual high-frequency air temperature variability in the Arctic from 1951–1990, *Int. J. Climatol.*, 22, 1017–1032, 2002.

37. Robinson, W.A., Reudy, R., and Hansen, J.E., General Circulation Model simulations of recent cooling in the east-central United States, *J. Geophys. Res.*, 107, 4748, doi:10.1029/2001JD001577, 2002.

38. Mortiz, R.E., Bitz, C.M., and Steig, E.J., Dynamics of recent climate change in the Arctic, *Science*, 297, 1497–1502, 2002.

39. Visbeck, M.H. Hurrell, J.W., Polvani, L. and Cullen, H.M., The North Atlantic Oscillation: past, present and future, *Proc. Natl. Acad. Sci. U.S.A.*, 98, 12876–12877, 2001.

40. Frich, P., Alexander, L.V., Della-Marta, P., Gleason, B., Haylock, M., Klein Tank, M.G. and Peterson, T., Observed coherent changes in climatic extremes during the second half of the twentieth century, *Climate Res.*, 19, 193–212, 2002.

41. Milly, P.C.D., Wetherald, R.T., Dunne, K.A., and Delworth, T.L., Increasing risk of great floods in a changing climate, *Nature*, 415, 514–517, 2002.

42. New, M., Todd, M., Hulme, M., and Jones, P., Precipitation measurements and trends in the twentieth century, *Int. J. Climatol.*, 21, 1899–1922, 2001.

43. Rosenlof, K.H. et al., Stratospheric water vapor increases over the past half-century, *Geophys. Res. Lett.*, 28, 1195–1198, 2001.

44. Ross, R.J. and Elliott, W.P., Radiosonde-based Northern Hemisphere tropospheric water vapor trends, *J. Climate*, 14, 1602–1612, 2001.

45. Trenberth, K.E., Dai, A., Rasmussen, R.M., and Parsons, D.B., The changing character of precipitation, *Bull. Am. Meteorol. Soc.*, 84, 1205–1215, 2003.

46. Parmesan, C. and Yohe, G., A globally coherent fingerprint of climate change impacts across natural systems, *Nature*, 421, 37–42, 2003.

47. Penuelas, J. and Filella, I., Responses to a warming world, *Science*, 294, 793–795, 2001.
48. Root, T.L. Price, J.T., Hall, K.R., Schneider, S.H., Rosenzweig, C., and Pounds, J.A., Fingerprints of global warming on wild animals and plants, *Nature*, 421, 57–60, 2003.
49. Walther, G.-R. et al., Ecological responses to recent climate change, *Nature*, 416, 389–395, 2002.
50. Schiermeier, Q., Insurers left reeling by disaster year, *Nature*, 421, 6619:99, 2003.
51. Schnur, R., Climate science — the investment forecast, *Nature*, 415, 483–484, 2002.
52. Baede, A.P.D. et al., The climate system: an overview, in *Climate Change 2001: The Scientific Basis Contribution of Working Group I to the Third Assessment Report of the Intergovernmental Panel on Climate Change*, J.T. Houghton et al., Eds., Cambridge University Press, Cambridge, 2001, chap. 1, 85–98.
53. Ruddiman, W.F., The anthropogenic greenhouse era began thousands of years ago, *Climatic Change*, 61, 261–293, 2003.
54. Crowley, T.J., Cycles, cycles everywhere, *Science*, 295, 1473–1474, 2002.
55. Pepin, L., Raynaud, D., Barnola, J.M., and Loutre, M.F., Hemispheric roles of climate forcings during glacial-interglacial transitions as deduced from the Vostok record and LLN-2D model experiments, *J. Geophys. Res.*, 106, 31885–319189, 2001.
56. Rind, D., The Sun's role in climate variations, *Science*, 296, 673–678, 2002.
57. Shindell, D.T., Schmidt, G.A., Miller, R., and Mann, M.E., Volcanic and solar forcing of climate change during the pre-industrial era, *J. Climate*, 16, 4094–4107, 2003.
58. Chase, T.N., Pielke, R.A., Sr., Kittel, T.G.F., Zhao, M., Pitman, A.J., Running , S.W., and Nemani, R.R., The relative climatic effects of land-cover change and elevated carbon dioxide combined with aerosols: a comparison of model results and observations, *J. Geophys. Res.*, 106, 31685–31691, 2001.
59. Heck, P., Lüthi, D., Wernli, H., and Schär, C., Climate impacts of European-scale anthropogenic vegetation changes: a sensitivity study using a Regional Climate Model, *J. Geophys. Res.*, 106, 7817–7835, 2001.
60. Matthews, H.D., Weaver, A.J., Eby, M., and Meissner, K.J., Radiative forcing of climate by historical land cover change, *Geophys. Res. Lett.*, 30(2), 1055, DOI:10.1029/2002GL016098, 2003.
61. Zhao, M. and Pitman, A.J., The impact of land cover change and increasing carbon dioxide on the extreme and frequency of maximum temperature and convective precipitation, *Geophys. Res. Lett.*, 28, 10.1029/2001GL013476, 2002.
62. Climate Monitoring and Diagnostic Laboratory, Carbon cycle, in *CMDL Summary Report #26*. Available online at http://www.cmdl.noaa.gov/publications/annrpt26/index.html, 2002.
63. Prentice, J.C. et al., The Carbon cycle and atmospheric carbon dioxide, in *Climate Change 2001: The Scientific Basis Contribution of Working Group I to the Third Assessment Report of the Intergovernmental Panel on Climate Change*, J.T. Houghton et al., Eds., Cambridge University Press, Cambridge, 2001, 183–237.
64. Gillett, N.P., Zwiers, F.W., Weaver, A.J., Hegerl, G.C., Allen, M.R., and Stott. P.A., Detecting anthropogenic influence with a multi-model ensemble, *Geophys. Res. Lett.*, 29(20), 1970, doi:10.1029/2002GL015836, 2002.
65. Tett, S.F.B. et al., Estimation of natural and anthropogenic contributions to twentieth century temperature change, *J. Geophys. Res.*, 107(D16), 10.1029/2002JD000028, 2002.
66. Thorne, P.W. et al., Assessing the robustness of zonal mean climate change detection, *Geophys. Res. Lett.*, 29(19), 1920, doi:10.1029/2002GL015717, 2002.

67. Mann, M.E., Rutherford, S., Bradley, R.S., Hughes, M.K., and Keimig, F.T., Optimal surface temperature reconstruction using terrestrial borehole data, *J. Geophys. Res.* 108(D7), 4203, doi:10.1029/2002JD002532, 2003.

68. Mitchell, J.F.B. et al., Detection of climate change and attribution of causes, in *Climate Change 2001: The Scientific Basis Contribution of Working Group I to the Third Assessment Report of the Intergovernmental Panel on Climate Change*, J.T. Houghton et al., Eds. Cambridge University Press, Cambridge, 2001, 697–738.

69. Ramaswamy, V. et al., Radiative forcing of climate change, in *Climate Change 2001: The Scientific Basis. Contribution of Working Group I to the Third Assessment Report of the Intergovernmental Panel on Climate Change*, J.T. Houghton et al., Eds., Cambridge University Press, Cambridge, 2001, 349–416.

70. Risbey, J.S. and Kandlikar, M., Expert assessment of uncertainties in detection and attribution of climate change, *Bull. Am. Meteorol. Soc.*, 83, 1317–1325.

71. Watson, R.T. et al., eds., *Land Use, Land-Use Change and Forestry*, Intergovernmental Panel on Climate Change, Cambridge University Press, Cambridge, 2000.

72. Bergengren, J.C., Thompson, S.L., Pollard, D., and Deconto, R.M., Modeling global climate-vegetation interactions in a doubled CO_2 world, *Climatic Change*, 50, 31–75, 2001.

73. Betts, R.A., Offset of the potential carbon sink from boreal forestation by decreases in surface albedo, *Nature*, 408, 187–190, 2000.

74. Cox, P.M., Betts, R.A., Jones, C.D., Spall, S.A., and Totterdell, I.J., Acceleration of global warming due to carbon cycle feedbacks in a Coupled Climate Model, *Nature*, 408, 184–187, 2000.

75. Douville, H., Planton, S., Royer, J.-F., Stephenson, D.B., Tyteca, S., Kergoat, L., Lafont, S., and Betts R., Importance of vegetation feedbacks in doubled-CO_2 climate experiments, *J. Geophys. Res.*, 105, 14841–14861, 2000.

76. Foley, J.A., Levis, S., Costa, M.H., Cramer, W., and Pollard, D., Incorporating dynamic vegetation cover within Global Climate Models, *Ecol. Appl.*, 10, 1620–1630, 2000.

77. Harding, R., Kuhry, P., Christensen, T.R., Sykes, M.T., Dankers, R., and van der Linden, S., Climate feedbacks at the tundra–taiga interface, *Ambio Special Report*, 12, 47–55, 2002.

78. Cubasch, U. et al., Projections of future climate change, in *Climate Change 2001: The Scientific Basis Contribution of Working Group I to the Third Assessment Report of the Intergovernmental Panel on Climate Change*, J.T. Houghton et al., Eds., Cambridge University Press, Cambridge, 2001, 525–582.

79. Nakicenovic, N. and Swart, R., Eds., *Emission Scenarios 2000*. Special Report of the Intergovernmental Panel on Climate Change, Cambridge University Press, Cambridge, 2000, 599 pp.

80. Houghton, J.T. et al., Eds., *Climate Change 2001: The Scientific Basis, Contribution of Working Group I to the Third Assessment Report of the Intergovernmental Panel on Climate Change*, Cambridge University Press, Cambridge, 2001, 881 pp.

81. Hansen, J. et al., Climate forcings in Goddard Institute for Space Studies SI2000 simulations, *J. Geophys. Res.*, 107, 4347–4384, 2002.

82. Reilly, J., Stone, P.H., Forest, C.E., Webster, M.D., Jacoby, H.D., and Prinn, R.G., Uncertainty and climate change assessments, *Science*, 293, 430, 2001.

83. Webster, M.D. et al., Uncertainty Analysis of Climate Change and Policy Response, *Climatic Change*, 61, 295–320, 2003.

84. IPCC Data Centre on line at http://ipcc-ddc.cru.uea.ac.uk.

85. Boer, G.J., Yu, B., Kim, S.J., and Flato, G.M., Is there observational support for an El Nino-like pattern of future global warming? *Geophys. Res. Lett.*, 31, L06201, doi:10.1029/2003GL018722, 2004.

86. Yang, F., Kumar, A., Schlesinger, M.E., and Wang, W., Intensity of hydrological cycles in warmer climates, *J. Climate*, 16, 2419–2423, 2003.

87. Gillett, N.P., Zwiers, F.W., Weaver, A.J., and Stott, P.A., Detection of human influence on sea-level pressure, *Nature*, 422, 292–294, 2003.

88. Liu, P., Meehl, G.A., and Wu, G., Multi-model trends in the Sahara induced by increasing CO_2, *Geophys. Res. Lett.*, 29(18), 1881, doi:10.1029/2002GL015923, 2002.

89. Räisänen, J., CO_2-induced changes in interannual temperature and precipitation variability in 19 CMIP2 experiments, *J. Climate*, 15, 2395–2411, 2002.

90. Wilby, R.L. and Wigley, T.M.L., Future changes in the distribution of daily precipitation totals across North America, *Geophys. Res. Lett.*, 29(7), doi:10.1029/ 2001GL013048, 2002.

91. Hecht, J., Earth's ancient heat wave gives a taste of things to come, *N. Sci.*, 173(2372), 21, 2002.

92. Nelson, F.E., Anisimov, O.A., and Shiklomanov, N.I., Climate change and hazard zonation in the circum-Arctic permafrost regions, *Nat. Hazards*, 26, 203–225, 2002.

93. Stendel, M. and Christensen, J.H., Impact of global warming on permafrost conditions in a coupled GCM, *Geophys. Res. Lett.*, 29(13), doi:10.1029/2001GL014345, 2002.

94. Stieglitz, M., Dery, S.J., Romanovsky, V.E., and Osterkamp, T.E., The role of snow cover in the warming of Arctic permafrost, *Geophys. Res. Lett.*, 30(13), 10.1029/2003GL017337, 2003.

95. Herbert, J.M. and Dixon, R.W., Is the ENSO phenomenon changing as a result of global warming? *Phys. Geogr.*, 23, 196–211, 2003.

96. Huntingford, C., Jones, R.G., Prudhomme, C., Lamb, R., Gash, J.H.C., and Jones, D.A., Regional Climate Model predictions of extreme rainfall for a changing climate. *Q. J. R. Meteorol. Soc.*, 129, 1607–1621, 2003.

97. Palmer, T.N. and Räisänen, J., Quantifying the risk of extreme seasonal precipitation events in a changing climate, *Nature*, 415, 512–514, 2002.

98. Timmermann, A., Changes of ENSO stability due to greenhouse warming, *Geophys. Res. Lett.*, 28, 2061–2064, 2001.

99. Alley, R.B. et al., Abrupt climate change, *Science*, 299, 2005–2010, 2003.

100. Smith, J.B. et al., Vulnerability to climate change and reasons for concern: a synthesis, in *Climate Change 2001. Third Assessment Report. WG II. Climate Change 2001: Impacts, Adaptation and Vulnerability,* J.J. McCarthy et al., Eds., Cambridge University Press, Cambridge, 2001, 911–967.

101. Vaughan, D.G. and Spouge, J.R., Risk estimation of collapse of the West Antarctic Ice Sheet, *Climatic Change*, 52, 65–91, 2002.

3 Impact of Climate Change on Agriculture, Forestry, and Wetlands

L.D. Mortsch

CONTENTS

3.1 INTRODUCTION

Climate — moisture, light, heat, and cold — has an important influence on agriculture, forest, and wetland systems. A human-caused increase of greenhouse gases in the atmosphere is projected to alter the global climate. These changes are described by Hengeveld in Chapter 2. This chapter provides an overview, for the Canadian context, of implications of a changing climate for agriculture, forestry, and wetland ecosystems. The topic is broad with a rapidly expanding body of literature; this review is not intended to be comprehensive but serves as an introduction to a limited number of biophysical impacts affected by temperature increases, precipitation variability, and hydrologic changes. Potential adaptation measures to deal with the impacts are listed. The review focuses primarily on Canadian literature but draws on sources from other regions when there are no Canadian examples.

3.2 ADAPTING TO A CHANGING CLIMATE

To date, much of the focus for dealing with human-caused climate change has been mitigation — reducing emissions and increasing sinks of greenhouse gases — to prevent or slow climate change. The Kyoto Protocol, while reducing emissions of greenhouse gases, only delays doubling of carbon dioxide (CO_2) in the atmosphere as concentrations of greenhouse gases in the atmosphere continue to increase.[1] Doubling of CO_2 in the atmosphere is a distinct possibility by 2100 and a tripling or quadrupling by that time may occur depending on the various scenarios of economic development, population growth, and associated greenhouse gas emissions.[1] The long residence time of some greenhouse gases means that their radiative forcing effects may be felt far into the future.[1] Impacts of human-caused climate change are likely and adaptation — responding to the impacts of climate change — will need to be undertaken.

Adaptation, as defined in the Third Assessment Report of the Intergovernmental Panel on Climate Change (IPCC), is "any adjustment that takes place in natural or human systems in response to actual or expected impacts of climate change, aimed at moderating harm or exploiting beneficial opportunities."[2] Adaptation has been described in a number of ways including the timing of the adaptation, its method of implementation, and who is adapting. Contrasting adjectives — reactive or anticipatory, autonomous or planned, and private or public — help further conceptualize adaptation for this chapter. Reactive adaptation describes the actions taken to cope with actual damages concurrent with or shortly after an impact, often an extreme event. Meanwhile, anticipatory adaptation is longer range in perspective. It involves assessing potential future changes and developing responses to cope with or lessen the impacts whether they happen or not. Autonomous adaptation is synonymous with natural systems where there are automatic reactions to the stresses or opportunities of changing environmental cues and conditions. Individuals and their systems can respond in a similar manner to economic, social, and environmental cues. Planned adaptation is undertaken with an awareness that climate is changing or is about to change and purposeful action is needed to return to, maintain, or achieve a desired state. To further that end, climate change information is explicitly incorporated into the policy-making process. Private and public adaptation distinguishes between actions that are undertaken by individuals and firms and those that are government sponsored. Most of the adaptation responses presented in this chapter are anticipatory and planned and can be undertaken both privately and publicly at the individual, sectoral, or regional scale.

Although this chapter lists potential adaptations to deal with climate change impacts, it does not explore the capacity to undertake these adaptations and the barriers to successful adaptation. For example, the amount, rate, and scale of climate change may be greater than what has been experienced in the past. The rate of climate change affects the ability to adapt. Slow, gradual changes allow for the evolution of impacts, the formulation of plans, and assembly of resources to adapt. Extreme, rapid changes challenge the capability of natural and human systems to respond.

Adaptation can be implemented through behavioral, technical, institutional, financial, or regulatory changes but is reliant on the capacity to adapt. Adaptive capacity affects the perception of changes in climate and attendant risk(s), determination of the degree and types of response options, and the implementation of change. The propensity to adapt is affected by the sociocultural, economic, political, technological, and institutional context that enhances or impedes adoption of strategies.[3-5]

3.3 PROJECTED CLIMATE CHANGES FOR CANADA

Rising concentrations of greenhouse gases in the atmosphere are anticipated to lead to a suite of changes in climate.[1] The most certain outcome is an increase in global mean air temperature. Over the 21st century, the mean annual temperature for North America could warm 1 to 3°C with a low greenhouse gas emission scenario and 3.5 to 7.5°C for a high emissions scenario.[6] Warming is projected to occur in all seasons and all regions of Canada, but the greatest warming is expected in winter and in the Arctic and southern and central Prairies.[7,8] In temperate regions, the growing season is expected to lengthen by 5 to 10 days for a mean annual temperature increase of 1°C.[9,10] Daily minimum temperatures are likely to increase more than daily maximum temperatures.[11] Globally, average annual precipitation is projected to increase with the greatest increase taking place in the mid and high latitudes.[1,12] Many areas in Canada can expect an increase in precipitation on an annual basis. Although there may be changes in the seasonality of precipitation, especially during a critical period such as summer when some areas may experience a decrease in precipitation, precipitation in winter is expected to increase.[7,8,12]

Much of Canada experiences periods of frozen ground, snow cover, and frozen rivers and lakes; these elements of the cryosphere are sensitive to climate change. With warming, winter snow depth and spring snow cover, particularly in southern Canada, are likely to decline.[13] Similarly, the duration and thickness of ice cover on freshwater lakes and rivers are expected to decrease. An earlier breakup of ice influences evaporation, discharge, and ecology.[13] Southern margins of the sporadic and discontinuous permafrost regions are vulnerable to warming of ground temperatures causing permafrost thaw; the zone of permafrost is expected to move northward.[14,15] The onset, rate, and magnitude of this melt vary due to regional differences in climate, snow cover, and surface properties. Warming and melt generally deepen the active layer thickness as well as enhance the potential for evaporation.

Key changes in hydrology are expected. A prime consideration is that evaporation and evapotranspiration are enhanced by warmer air temperatures, which decrease available soil moisture content and cause drying.[8,16] Most of Canada is expected to experience soil moisture decreases but the southern Prairies may have serious summer deficits by 2090.[8] Also, changes in snow climatology are likely to influence the amount and timing of the snowmelt, peak runoff, and groundwater, river, and lake recharge in snow-dominated watersheds.[13,17] Where precipitation decreases or does not offset evaporative demands, surface water resources decrease; a key concern is summer and fall flows could be much lower. Recharge of groundwater is also influenced by evaporation and precipitation and will be reflected in changes to groundwater levels and discharge to rivers and lakes.[18-22] In areas where precipitation increases significantly, there may be concerns for flooding associated with winter rain on snow and frozen ground or extreme precipitation events.

Extreme climatic events (rare in frequency of occurrence or intensity) are projected to increase.[1,11,12] With warming, the number of extremely warm days and associated heat waves are expected to increase. For example, a 1-in-80-year temperature extreme today may occur with 1-in-10-year probability by 2050.[11] More intense precipitation events are also projected; the probability of extreme precipitation events is enhanced by a warm atmosphere's higher water-holding capacity and greater instability due to convective activity.[8,11,12] Floods, from more intense precipitation events, are projected to increase while at the same time droughts are expected to increase in frequency and duration due to higher moisture stress.[23]

3.4 OBSERVED TRENDS IN CANADA

Significant trends in climate have been observed and some are consistent with projections of climatic change due to an enhanced greenhouse effect. Trends relevant to agriculture, forestry, and wetland ecosystems are described in order to highlight where current changes might already be influencing their functioning and require adaptation.

Significant increases in average annual temperature ranging from 0.5 to 2.0°C have been detected for southern Canada (south of 60° latitude) from 1900 to 1998. On average, Canada has warmed 0.9°C during this period.[24] Winter and spring temperatures exhibited the greatest warming; summer has less warming while fall has shown little change.[24,25] For the period 1950 to 1998, the northwest (Yukon and the Northwest Territories) and British Columbia have had significant warming but cooling has been observed in Labrador and Newfoundland.[24] Daily maximum temperatures have warmed (0.5 to 1.5°C) across southern Canada during the 20th century. Yet, daily minimum temperatures have warmed more (1.0 to 2.0°C).[26] A significant trend for an earlier spring (advance of 0°C-isotherm) occurred in western Canada.[27] Observations of a longer frost-free period and longer growing season correspond to early spring warming.[26] More frequent fluctuations of temperatures around 0°C due to late winter and early spring warming increased the frequency of thaw–freeze cycles.[28]

Most of Canada south of 60° latitude has experienced a significant increase in precipitation from 5 to 30% during the period 1900 to 1998.[24] Warming, with temperatures rising above freezing, can change the proportion of precipitation falling as rain vs. snow; the trend is significant in the most southerly regions in spring.[24,29,30] Winter and early-spring snow depths in Canada decreased during the period 1946 to 1995 coincident with an increase in air temperature.[30] Similarly, the areal extent of snow cover has diminished. Snow cover depth and duration have become more unreliable with the snow cover season ending earlier and reaching an earlier date of maximum snow depth.[31]

Warming affects regional hydrology by shifting the magnitude and timing of events, particularly if they are tied to snow accumulation and melt. For example, the freshet, a period of high flow during spring, is linked to above-freezing temperatures and melting of snow pack. An assessment of 84 unregulated river basins extending from northwestern Ontario to Alberta detected an earlier spring snowmelt runoff with more northerly rivers displaying the greatest response.[32] Similarly, Whitfield and Cannon[33] reported an earlier onset of runoff and Zhang et al.[34] mapped a significant trend toward earlier occurrence of the freshet season across Canada.

Melting of lake and river ice and evaporation from water bodies are influenced by warming. In the Northern Hemisphere for the period 1846 to 1995, ice breakup on small lakes advanced 6.5 days and freeze dates occurred 5.8 days later.[35] In the Experimental Lakes Area of northwestern Ontario, average annual evaporation increased approximately 50% (average of 9 mm/year) associated with a mean annual air temperature increase of 1.6°C and a precipitation decrease (approximately 60% of highest years) over the period 1970 to 1990. Evaporation increased by an average of 35 mm for each 1°C increase in annual air temperature but was more pronounced with summer warming (68 mm/1°C).[36] The net effect was decreased stream flow, with annual runoff declining significantly from about 400 mm/year to less than 150 mm/year in the late 1980s.[36]

Plant phenology — timing of life cycle stages such as budding, leaf emergence, and flowering of many plant species — is influenced by temperature (cold and heat accumulation above a threshold) and/or photoperiod.[37–39] Satellite-based observations from 1982 to 1999, using normalized difference vegetation index (NDVI) as a surrogate for plant photosynthetic activity and hence growth, have detected changes. There have been increases in growing season NDVI throughout Eurasia from 40°N to 70°N but a more patchy response in North America.[40] However, parts of boreal Canada, Alaska, and northeastern Asia had observed decreases in NDVI that may be caused by temperature-induced drought.[40] Ground-based measurements also point to a trend of earlier phenological events. In Wisconsin, phenological events started earlier; the response was greater for events earlier in the year than later.[38] First-bloom date and first-leaf date for lilacs in North America have advanced and may be linked to spring warming.[41] Similarly, Cayan et al.[42] reported that the date of first budding of honeysuckle as well as lilacs advanced since the 1970s due to warming. Flowering of the trembling aspen in Edmonton, Alberta was 26 days earlier in 1997 than in 1900.[37]

Fire activity (number of fires and area burned) has varied significantly from year to year. Yet, a significant increase in fire activity was determined at three scales of analysis: Canada, Ontario, and northwestern Ontario for the period 1917 to 2000 that was not solely due to an expanding fire protection area but could be attributed to climate change and changes in land-use patterns.[43] Area burned by wild-land fires in Canada over the past four decades of the 20th century has had a pronounced upward trend coincident with warming during the fire season. Results suggest that an increase in Canadian forest fire occurrence projected due to human-caused climate change has already been detected.[44]

3.5 PROJECTED IMPACTS OF CLIMATE CHANGE AND POTENTIAL ADAPTATION STRATEGIES

3.5.1 FORESTRY

Climate change, as reflected by changes in temperature and precipitation and other climatic elements, can influence the occurrence, extent, frequency, intensity, timing, and duration of forest disturbances. Disturbances — drought, fire, and insects — influence the composition, structure, and functioning of forests. Forests at the margins of their temperature or moisture range or those already under stress due to pollution or diseases are likely to be more vulnerable to a changing climate.

3.5.1.1 Moisture Stress and Drought

Future changes in temperature and precipitation will affect moisture availability in forests. Warmer air temperatures increase water losses due to evaporation and evapotranspiration causing drier conditions and also reduce the water use efficiency of plants.[7,46] A longer, warmer growing season enhances these effects and amplifies the potential for moisture stress and drought especially if precipitation does not offset drying. In the future, forests may experience moisture stress more frequently and for longer periods.

Responses to moisture stress and drought, including reduction in growth and decline in health of trees, are modulated by forest characteristics, age-class structure, and soil type and depth. Young, small plants such as seedlings and saplings are susceptible. Large trees with more stored nutrients and carbohydrates are less sensitive to drought and are only affected by more severe conditions.[45] Trees growing in shallow and stony soils (e.g., over the Canadian Shield or till deposits in northwestern Ontario) are more likely to be affected by water deficit in the future.[46] Similarly, shallow-rooted trees such as black spruce, jack pine, and red pine are vulnerable to drought while deep-rooted trees can absorb water from greater soil depths and are not as predisposed to stress.

Moisture stress and drought enhance forest susceptibility to disturbances such as insect and disease infestation and forest fires. Climate conditions such as above-average temperatures leading to soil drying have caused fine root death in sugar maples and white birch.[46] Root diseases and stem decay and decline require a stressed host before infection or disease expression can occur. Large areas of dead or damaged trees due to drought provide a large fuel source when fires are initiated. Drought also reduces the moisture content of fuel (dead trees, undergrowth, and organic matter) on the forest floor and predisposes the forests to fire.

Adaptation:

- Plant deeper-rooted and drought-tolerant species.[46]
- Thin stands to reduce crowding and stress and to increase vigor.[46]
- Establish mixed conifer–hardwood forests to encourage buildup of forest litter and humus in order to improve soil moisture retention.
- Undertake assisted regeneration to preserve forest sustainability.[7]
- Maintain forest health and biodiversity by building on sustainable forest management initiatives.[7]

3.5.1.2 Insects

Insect outbreaks such as spruce bud worm, gypsy moth, and mountain pine beetle are major disturbance factors in Canadian forests that influence forest stand structure, composition, and functioning. Insect disturbance patterns are expected to change dramatically due to warmer temperatures, increased frequency of droughts, changes in host resistance, the removal of natural controls (parasites, predators, or pathogens) on some insects, and higher carbon dioxide concentrations. Although it is difficult to predict with certainty which trees species will form the future forests and which insects will become major disturbance factors, changes in the type, frequency, duration, and severity of insect outbreaks are expected due to climate change.

Changes in temperature (and precipitation) can influence the development, survival, reproduction, dispersal, and geographic distribution of insects. Earlier, warmer spring temperatures and later, warmer fall temperatures create a longer period in which insects can successfully grow and mature. The current distribution of insects may expand northward and to higher altitudes because temperature is no longer limiting. In temperate and boreal forests, an increase in the summer temperature may accelerate the rate of development in insects and their reproductive potential could increase. Warmer winter temperatures may result in better over-winter survival of larvae.[46–48]

The mountain pine beetle, a significant pest of mature pine forests in western North America, provides a good case study. The extent of the mountain pine beetle is limited by temperature, not the distribution of the host trees (e.g., lodgepole pine). The minimum winter temperature of –40°C exerts a strong influence in defining the range of the beetle. In warmer areas, the beetles can successfully hatch, develop, and mature before the onset of winter while in cooler areas they cannot mature as rapidly and many die over winter.[49] To date, outbreaks of the beetle have been restricted to southern British Columbia. Yet, warming associated with climate change, particularly warmer winters, may make conditions more favorable allowing expansion northward and eastward as well as upward into higher elevations. Carroll et al.[50] showed that the northern boundary of climatic suitability for the beetle could shift 7° latitude north with a 2.5°C increase in temperature. Uninfested forests in northern British Columbia, northwestern Alberta, and southern Yukon and the

Northwest Territories could be vulnerable as well as high-elevation pine forests (e.g., whitebark and limber pine forests) in southeastern British Columbia. Moreover, within the mountain pine beetle's current range, infestations and outbreaks may occur more regularly and with greater severity as warmer conditions enhance larval growth, reproduction, and winter survival.

Temperature increases and drought stress can cause changes in plant physiology and resistance that benefit insect pests. For example, in a drought-stressed balsam fir, sucrose content almost tripled; this sucrose stimulates the feeding of certain stages of spruce budworm and accelerates their growth.[51] Also, Mattson et al.[52] observed greater growth of spruce budworm larvae on trees experiencing dry conditions. Drought-stressed trees can have temperatures that are 2 to 4°C warmer, which can be beneficial to insects, for example, increasing spruce bud worm fecundity and survival.[53]

Higher concentrations of carbon dioxide in the atmosphere may increase the carbon–nitrogen ratio of leaves.[54] This reduces the nutritional value of the vegetation and requires increased feeding by insects to meet nutritional needs (nitrogen).[46,55]

Adaptation:

- Monitor for insect outbreaks and damage.[45]
- Undertake sanitation cutting of infested trees to reduce area affected and encourage healthy stands.[45]
- Shorten stand rotations to reduce the period of vulnerability and increase vigor.
- Develop increased resistance to insect pests through breeding.
- Control competing vegetation, through thinning or weed control, in order to reduce stress in regenerating trees and encourage desired species composition of forest.[46]
- Use insecticides selectively to control insects and reduce timber volume losses.[46]

3.5.1.3 Forest Fires

Forest fire activity (number of fires and area burned) is influenced by weather, ignition sources, fuels, and human activity. However, in Canada, weather (and climate) is the most important natural factor influencing forest fires as it affects drought frequency, forest health, fuel availability and dryness, lightning, and wind.[56] In a changing climate, forests are expected to be exposed to more fire activity as fuel loads accumulate, fire season lengthens, and more extreme fire weather conditions occur. Fires may become larger and more frequent, causing loss of forest resources and creating the requirement for more investment in fire management.

The Canadian forest fire season typically extends from late April to the end of August.[57] An assessment using the Forest Fire Danger Rating System criteria and $2 \times CO_2$ scenarios showed that the length of the forest fire season increased 30 days on average (22%) for forested areas of Canada.[58] An increase in temperature (shorter winters) was the prime factor in extending the fire season. Although the potential for drier fuels due to precipitation decrease and higher evaporative losses was also an important determinant.

Temperature, precipitation conditions, and evaporative losses influence the moisture content of the surface and organic layers of the forest floor, and forest fuels help to create the conditions that promote fire development and spread should a fire be initiated by people or lightning. Projected changes in the moisture content of the forest floor using daily General Circulation Model (GCM)-based scenarios resulted in an increase in the projected number of human-caused fires (18% increase for 2050 and 50% increase for 2100) in Ontario during summer.[59] Lightning frequency in North America is also projected to increase due to climate change.[60] Lightning fire occurrence increased 30% in the U.S. using a $2 \times CO_2$ scenario.[61] Forest fire management agencies may have the challenge of dealing with more fires in the future.

Changes in the potential for fire activity in Canada have been explored using the Canadian forest Fire Weather Index (FWI) and numerous climate change scenarios. These assessments have all demonstrated a significant increase in the severity of fire weather conditions in the future (a surrogate for increased fire activity).[62-64] The mean seasonal fire weather severity index increased, and the area under high to extreme fire weather danger expanded. Yet there was a contrast between eastern and western Canada reflecting different climate regimes; increases in fire weather conditions were greatest in the continental interior of west-central Canada.[63] June and July remained the months with the greatest area of extreme fire weather danger but the high to extreme fire severity season started earlier and ended later than current conditions. An assessment linking FWI components with area burned and two climate change scenarios projected significant increases in area burned even with large regional variation in fire activity. Area burned in Canadian ecozones increased by 74 to 118% by 2100 using $3 \times CO_2$ scenarios.[56]

Major forest fire episodes are often due to short-term extreme fire weather conditions where a large number of fires overwhelm forest fire management resources.[63] An increase in future forest fire activity necessitates an examination of forest fire management policy and investment required to provide an acceptable level of fire protection. The level of fire detection and suppression effort may need to be guided by what is at risk.[63] For example, fires near major population centers, recreation areas, and forest industry would be actively detected and suppressed while remote northern fires would burn unsuppressed unless specific resources were at risk.[46]

Adaptation:

- Monitor weather and forest fire conditions.
- Develop forecasts of fire season severity.[65]
- Continue public education on causes and prevention of forest fires.[59]
- Continue public communication programs during dry and dangerous forest fire conditions.[59]
- Develop policy to establish restricted forest access and use during dangerous fire conditions.[59]
- Develop community forest fire emergency plans.

- Undertake salvage logging of fire-killed trees in accessible areas.[66]
- Develop "fire-smart landscapes" into long-term forest management planning, which incorporates harvesting, regeneration, and stand tending to reduce spread and intensity of fires and its impacts.[7]

3.5.2 AGRICULTURE

A warmer climate and a longer growing season due to climate change may benefit crop and livestock productivity, although there may be drought, soil erosion, water shortages, and pest outbreaks that have negative consequences. These factors will interplay at different levels to result in both advantages and disadvantages to Canadian agriculture.

3.5.2.1 Crop Growth and Yields

An increase in atmospheric CO_2 improves plant water-use efficiency and increases plant growth and yields, although the nutritional quality of crops may not increase as much as the yield volume.[67,68] Certain crops, C_4 plants of tropical origin, such as maize, millet, sorghum, and sugarcane, receive modest benefits. Most crop plants are C_3 (temperate and boreal origin) and have demonstrated significant increases in yield with elevated atmospheric CO_2 unless they become overheated or lack certain nutrients and moisture. Where areas become hot and dry, C_4 species generally have a competitive advantage over C_3 species because they have better water-use efficiencies and can tolerate hotter, drier conditions. However, due to elevated CO_2 the C_3 species will also have improved photosynthesis and water-use efficiencies.[69]

Food crops have critical low and high temperature thresholds as well as optimum temperatures that influence performance and distribution. Lobell and Asner[70] related decadal-scale climate changes to large-scale crop production in the U.S. from 1982 to 1998. Gradual temperature changes during the period had a measurable impact on trends in crop yield; corn and soybean yield showed a 17% relative decrease for each 1°C increase in growing season temperature. When wheat, rice, and maize experienced temperatures near the critical maximum threshold, effects included reversal of vernalization, spiklet sterility, and loss of pollen viability, respectively.[70] Warmer winter temperatures increased the risk of reversing vernalization in winter wheat; in certain areas a shift to growing spring wheat was recommended to maintain production.[70] Rising nighttime temperatures were usually considered a benefit to crops; yet, recently researchers demonstrated that rice grain yield declined 10% for each 1°C increase in growing season minimum temperature (usually nighttime temperature) during the dry season.[71]

The magnitude of temperature increase affects the severity of the climate change impacts on crop production. In the Third Assessment Report of the IPCC a 2.5°C increase was identified as a threshold for global agricultural impacts.[72] With temperature increases below this threshold, there were both gains and losses from a global perspective, but above this threshold effects were primarily deleterious. Initial increases in air temperatures exacerbated moisture stress and negatively affected growth and yield of some crops in warm and drought-prone areas while warming and a longer growing season benefited crops in northern areas and at high elevations.

In Canada, a considerable body of research has assessed the impacts of climate change scenarios and elevated levels of CO_2 on crop production usually focusing on changes in agroclimatic resources (thermal and moisture conditions) and biophysical yields.[4,5] The results indicate that impacts vary across regions and commodity. De Jong et al.[73] developed GCM-based temperature and precipitation scenarios to determine effects on future crop yields in a number of agricultural regions across Canada. The scenarios projected warmer and slightly wetter climate conditions, which contributed to an advancement of planting dates: 1 to 2 weeks for eastern and central Canada and approximately 3 weeks in the west. Yields for winter wheat, soybeans, and potatoes were projected to increase substantially while yields for barley, wheat, and canola showed no significant change. Similarly, changes in length of growing season and growing degree-days projected by means of a $2 \times CO_2$ climate change scenario were used to determine crop yield changes for 12 key agricultural regions of Quebec. Results varied by region and crop type. The C_4 cereal, oleaginous, and specialty crops (potato, tobacco, and sugar beet) experienced yield increases up to 40%. However, legumes, C_3 cereals, and vegetables (onions, tomatoes, and cabbage) suffered yield declines. The declines were caused by acceleration of maturation rate due to higher temperatures exacerbated by moisture stress.[74] In Atlantic Canada, projected corn heat units (CHU) increased substantially using a CGCM1 scenario. Yields for grains and soybeans increased with little change indicated for barley.[75] An assessment of fruit production in British Columbia found that apple and grape production benefited from a 1 month longer growing season; winter damage was also reduced. However, benefits were offset by increasing potential for extreme heat and pests surviving milder winters.[76]

Estimates of changes in future yields due to increasing CO_2 and improved agroclimatic conditions have been based on simple models of very complex interactions. Although the modeling has incorporated factors such as longer growing seasons, more heat units, and longer frost-free periods, there are other climate-sensitive biophysical factors that have not been considered (changes in soil fertility, abundance and types of pests and diseases, crop nutrient requirements, and crop and weed interactions).[77–79] Adaptation and key technological and management changes of critical importance to productivity improvements also have not been incorporated into the assessments.[80] In the MINK (Missouri, Iowa, Nebraska, and Kansas) regional assessment in the U.S., McKenney et al.[81] demonstrated the importance of incorporating the potential adaptation responses of farmers. The adaptations included technological improvements in crop breeding, pest control, and harvesting techniques as well as crop substitutions, alterations in planting dates, and more efficient irrigation. With these adaptations, projections for three of four major crops (soybean, wheat, and sorghum) showed enhanced performance under the 2030 climate change scenario while corn yields declined. All yields declined in projections without adaptation.

Adaptation:

- Shift timing of planting and harvesting to take advantage of longer growing seasons. Earlier seeding options are easy to implement but other factors such as temperature and moisture stress could suppress yields.[4,5]
- Switch from spring wheat to higher yielding winter wheat in Prairies.[69] Conversion would be beneficial in southern sites making more effective use of early spring moisture.[4]
- Shift to longer-maturing, higher-yield cultivars (e.g., corn, soybeans, and phaseolous beans).[5,74]
- Change irrigation and tillage practices. Irrigation appears to be a very effective response but may not be sustainable.[4]

3.5.2.2 Livestock and Forage Production

Air temperature is a critical bioclimatic element for livestock production. For example, summer thermal stress has a demonstrated effect on reproduction, weight gain, milk output, feeding behavior, feed conversion efficiency, and meat quality.[68,82] In severe conditions, heat stress has caused death. In 1995, more than 5000 head of cattle died in Nebraska–Iowa feedlots during a 3-day heat wave where high air temperatures combined with clear skies, high solar radiation, and low wind speeds. The critical factor was that nighttime temperatures remained high and cattle had no opportunity for cool-off and recovery.[83] Significant numbers of feedlot cattle were lost during an August 1992 heat wave because the relatively cool summer had not allowed acclimatization to high temperatures.[83] Poultry operations have also exhibited vulnerability to high temperatures. More than 500,000 poultry were lost in Quebec during the July 2002 heat wave because temperatures exceeded the critical threshold of 35°C inside enclosures.[84] Current sensitivity to extreme temperatures illustrates the vulnerability of operations to an increase in the frequency, severity, and duration of extreme temperature events due to climate change.

Only a few studies have assessed the direct impacts of climate change on livestock production while most have explored the indirect effects of changes in forage. In warm regions, livestock productivity decreases due to warmer summer temperatures. Higher temperatures during warm periods affected animal weight gain, dairy production, and feed conversion efficiency.[68,85–87] Warmer fall and winter temperatures, particularly in northern areas, benefited livestock production due to reduced feed requirements, better survival of young, and lower energy requirements.[67,87,88]

Yield of forages (stems and leaves) often increased due to CO_2 enrichment but nutritional quality (e.g., minerals and higher carbon and lower protein content) decreased.[89,90] Animal behavior was affected by climate, which also affected production; for example, high daytime temperatures influenced grazing time of livestock.[91] The stress of higher temperatures combined with poorer quality of forage affected productivity in a cow/calf operation. While forage production in this operation increased on average, greater variability from year to year increased uncertainty for grazing management. Fewer cattle were grazed in order to ensure good animal growth and vigor; overall production decreased.[86]

Adaptation:

- Reduce heat stress by providing shade, using sprinklers, or lessening crowding.[82]
- Shift to heat-resistant breeds or replace cattle with sheep.
- Increase stocking density (often double normal level) during the first half of the growing season followed by no grazing through the remainder of the season. Cattle can be sent to feedlots early or finished at pasture with supplements.[80]
- Lengthen grazing period with fewer numbers of cattle stocked.[80,86]
- Move herds to where pasture and feed are more plentiful.[80]

3.5.2.3 Drought

A longer, warmer frost-free season combined, in many instances, with increases in precipitation during the growing season points to a more favorable climate for agriculture in many regions across Canada. However, the gains due to more precipitation could be offset by substantial increases in evapotranspiration with warmer air temperatures. Larger seasonal moisture deficits and an increased frequency and intensity of droughts could affect crop production.[69,72,92] Vulnerable regions include drought-prone areas such as the Prairies and the interior of British Columbia as well as those areas that currently have little experience coping with drought.[69] Some areas will experience changes in agricultural suitability. For example, south-central Saskatchewan and southeast Alberta currently produce the highest quality spring wheat in Canada. The area is projected to become more arid, affecting its suitability for wheat production and making it more appropriate for livestock grazing.[69]

Two impact assessments in the Prairies illustrate how scenarios of precipitation and temperature changes can affect soil moisture and provide opportunities or impacts for agricultural productivity. Nyirfa and Harron[93] assessed moisture availability over the Prairies for 2040–2069 and, although precipitation was projected to increase, it was not sufficient to offset the moisture losses from warmer temperatures and higher evaporation rates (Canadian CGCM1 scenario). Spring-seeded small grain crops were vulnerable. Conversely, where there were sufficient precipitation increases, there were opportunities for agriculture. McGinn et al.[94] found that moisture remained the same or higher in the top 120 cm of the soil profile under future climate scenarios than present-day conditions in many areas of the Prairies. Also, warmer temperatures allowed earlier seeding dates for spring wheat (advanced by 18 to 26 days), which meant that crops could be harvested before the onset of arid conditions in late summer. Yet the study demonstrated variability in benefits; agricultural productivity in southeastern Saskatchewan and southern Manitoba was vulnerable due to precipitation decreases.

The impacts of the 1988 drought in the Prairies illustrate the current sensitivities of livestock production that could be exacerbated in the future. During the drought, livestock production was compromised by poor pastureland, dust storms, and poor germination and growth affecting the availability and quality of feed grain with associated price increases.[80] Water shortages in dugouts and local streams affected the availability of water for livestock. Water quality issues caused by the occurrence of sulfur and blue-green algae created serious problems for cattle operations.[95]

Adaptation:

- Enhance water conservation and water use efficiency (e.g., improve irrigation efficiency).[96]
- Change current agricultural practices (use drought-resistant cultivars, change timing of planting, switch to dryland farming).[5]
- Continue drought monitoring and seasonal climate predictions.[96]
- Develop drought risk management plans that enhance the capacity to cope outside current ranges.[4]
- Purchase crop insurance.[5,96]

3.5.2.4 Agricultural Water Supply

Of particular concern is a sufficient, secure supply of water (quantity and quality) to meet needs of crops, livestock, and farming communities in a changing climate. Agricultural water supplies dependent on wells, dugouts, reservoirs, and glacier-fed and snowmelt rivers are vulnerable to changes in climate. Warmer temperatures cause higher rates of evaporation that increase losses in surface waters including reservoir storage as well as reduce surface runoff and groundwater recharge. Similarly, warming reduces winter snow cover, which contributes to the spring melt that is so critical to recharging groundwater, rivers, lakes, and reservoirs. More than 1300 glaciers on the eastern slopes of the Rockies are now 25 to 75% smaller than in 1850, and retreat is projected to continue with warming.[28] At present, the melting and retreat provide runoff to the glacier-fed rivers of the Prairies; flow is highest in summer when demand for drinking water, aquatic habitat, and irrigation is greatest. However, when many of the glaciers disappear, flow will significantly decline, particularly in summer with the potential for critical supply–demand mismatches. The western Canadian prairies are highly dependent on flow from glacier-fed rivers for irrigation; 10% of the base flow for irrigation comes from glacier runoff.[97] At present, there already is a reduction in runoff to these rivers and less water for irrigation.[98]

Adaptation:

- Enhance water conservation (e.g., snow management to increase storage).[7,96]
- Improve irrigation efficiency.[96]
- Improve on-farm water infrastructure and management (e.g., wells, pasture pipelines, constructed ponds, and small reservoirs).[7]

3.5.2.5 Soil Erosion

Soil erosion rates are expected to increase with climate change.[99] The intensity of precipitation events are expected to increase; for example, a 1-in-40-year rainfall event today may occur with 1-in-20-year probability by 2090.[11] Precipitation intensity has a greater influence on runoff and soil erosion than frequency of precipitation although they are both important factors.[100,101] As the intensity of precipitation increases, soil erosion is enhanced, and entrainment and delivery of sediments, nutrients, and pesticides to surface waters grows; non-point-source water pollution increases. Soil type, slope, and vegetation cover affect soil erosion. In agricultural areas, timing of planting, harvesting, and tillage practices led to different vulnerabilities for soil erosion during the seasons. Spring can be a high rainfall and runoff period with little vegetative cover to prevent erosion. Winter rain on bare soil also enhances runoff and erosion. Higher-intensity precipitation events contribute to soil degradation; once soil has been destabilized by an "extreme erosion episode," it becomes more vulnerable to subsequent smaller events, leading to even greater negative results.[101]

Adaptation:

- Undertake more conservation tillage practices (no-till and reduced tillage) that provide benefits of storing more soil water, enhancing soil carbon and reducing the potential for soil erosion. Although these practices can have a negative effect on early season soil temperature, seed germination, and seedling emergence with a small overall effect on final crop yield, warmer temperatures may offset the effects.[102]

3.5.3 Wetland Ecosystems

Wetlands are an interface between the terrestrial and aquatic environments and their development and ecological viability depend on water saturation for at least part of the year. They are particularly sensitive to changes in hydrology as air temperature, regional precipitation, surface runoff, snow cover, length of the freezing season, permafrost, groundwater storage, and evapotranspiration are altered due to climate change. Changes in the water balance in marshes, fens, bogs, swamps, and peatlands influence their structure, functioning, productivity, area, and distribution, as well as affect wetland-dependent fish, wildlife, and activities.

3.5.3.1 Evaporation Exceeds Precipitation

A key driver of water levels in a wetland is the interplay between input from precipitation, surface runoff, and groundwater balancing losses due to evaporation and transpiration. Warmer air temperatures, shorter, warmer winters, and longer summers (with higher potential for evapotranspiration) are projected to enhance evaporative losses and reduce water levels. Evaporation increased by 2 to 3% for each 1°C increase in temperature.[16,36] Prairie pothole wetlands occur in a semi-arid region; their areal extent, vegetation communities, and primary productivity respond to phases of drying, regenerating, and flooding.[103] The number of these potholes with standing water can vary dramatically during wet and dry periods.[104] During dry periods, these wetlands are particularly vulnerable to encroachment and conversion to agricultural land.

Adaptation:

- Maintain vegetation buffer strips around wetlands.
- Develop land-use policies and incentives to protect wetlands from encroachment.

3.5.3.2 Altered Water Level Regime

Climate change could affect shoreline wetlands through changes in the mean level, annual range, and seasonal cycle as well as the timing, amplitude, and duration of water levels. Water levels are critical in altering vegetation communities, areal extent of wetlands, and suitability of habitat for wetland-dependent birds, animals, and fish.

Two possible climate change scenarios, an increased frequency and duration of low water levels and a changed temporal distribution and amplitude of seasonal water levels, were used to assess the implications for Great Lakes shoreline wetlands. Most climate change impact assessments of the Great Lakes water resources project lower net basin supplies and a reduction in water levels.[105,106] Low water levels may have a negative impact on wetland areal extent and productivity due to drying of landward wetland areas and the uncertain success of colonizing vegetation on exposed mud flats.[107] Certain wetlands are more vulnerable to water level decline due to their geomorphic form and dominant vegetation communities. Marshes can adapt to lower levels if they are not prevented from migrating by a barrier beach. Precambrian Shield marshes, located in irregular slope and rocky areas, have fewer sites for colonization. Swamps are less adaptable as their characteristic vegetation, trees, cannot regenerate and colonize quickly. The timing and degree of flooding in a wetland affects wetland vegetation; for example, high winter water levels can cause emergent vegetation die-off.

The Long Point wetland complex on Lake Erie was used as a case study to understand wetland vegetation community relationships with historical water level fluctuations and from that to infer how the wetland might be affected by projected lake level declines.[108] Key responses to lower water levels included expansion of tall dense dry emergent and meadow vegetation; shifts to meadow, treed, upland vegetation in landward and high elevation portions of the wetland; a decrease in open water, submergent, and floating leaved-vegetation communities; and a decrease in complexity due to less interspersion and larger patches of vegetation communities.

Adaptation:

- Dyke wetlands to manage and maintain water levels.
- Develop land-use zoning policies to prevent encroachment onto exposed shorelines and hardening of shorelines.

3.5.3.3 Permafrost Melt

The highest abundance of wetlands is found in the mid to high latitudes (between 50 and 70° north latitude) but this is also the area that is expected to experience the greatest warming due to climate change. Approximately 50% of peatlands in Canada are found in continuous or discontinuous permafrost areas.[109] Peatlands are important net sinks of carbon because their rates of plant production are greater than their rates of organic matter decomposition.[110] Important controls on the carbon budget of peatlands include plant communities, peat quality, temperature (air and peat), and hydrology, particularly the water table position. Anaerobic conditions in a peat profile usually develop due to waterlogged conditions; high water levels and warm temperatures stimulate the production of methane (CH_4). Emissions of CO_2 increase under dry, aerobic conditions. Climate change is expected to affect these controls and influence the role of wetlands as a carbon sink.

Most studies suggest that peatlands in areas of continuous permafrost may become net sources of carbon dioxide to the atmosphere due to climate warming.[109] Warming, permafrost degradation, and lower water levels deepen the aerobic layer within the peatland profile and poorly decomposed peat, currently frozen, becomes available for decomposition.[111] The response in the discontinuous permafrost zone may be more complex since vegetation patterns and hydrology affect peat land–permafrost relationships and there may be sections within a peatland with and without permafrost and where permafrost is accumulating or degrading. The lower water tables projected using climate change scenarios suggests that most peatland sites should have a decrease in methane emissions; however, increases may occur in discontinuous permafrost areas where the peat formations (palsa and plateau) are currently dry and produce lower methane emissions but permafrost melting creates collapsed bogs and fens with higher water tables and methane emissions. Methane emissions from Canadian wetlands is expected to decrease in southern regions but increase in northern regions due to warmer peat temperatures, longer growing season, and permafrost collapse features.[110] Although climatic change will likely increase net productivity, increases in CO_2 emissions from lowered water table and warmer temperatures suggest that carbon storage in northern peatlands will decrease. Many peatlands may become a source rather than a sink for carbon due to a lower water table.[110]

3.6 SUMMARY AND CONCLUSION

A number of climate change sensitivities and vulnerabilities were described for agriculture, forestry, and wetlands in this chapter. The magnitude and rate of climate change affect the ability of natural and managed systems to adapt to a changing climate and influence the severity of impacts. Much research is required to understand the sensitivities, impacts, and adaptation response strategies. The research needs to link natural and social science expertise as well as consider the implications of a changing climate in the context of other stresses such as globalization, land-use change, and pollution.

Proactive adaptation strategies in the managed agriculture and forestry sectors can be directed at behavioral, institutional, technological, regulatory, and economic changes. In unmanaged ecosystems such as wetlands, the options for adaptation are more limited and it is not certain how much humans can "manage" the response of these ecosystems. Where adaptation is possible, long-term and short-term strategies will need to be developed and implemented. Adaptation requires significant long-term investment decisions and planning for lags in implementation.

REFERENCES

1. IPCC, *Climate Change 2001: The Scientific Basis*, McCarthy, J.J. et al., Eds., Cambridge University Press, New York, 2001.
2. Smit, B. and Pilifosova, O., Adaptation to climate change in the context of sustainable development and equity, in *Climate Change 2001: Impacts, Adaptation and Vulnerability*, McCarthy, J.J. et al., Eds., Cambridge University Press, New York, 2001, chap. 18.
3. Bryant, C.R. et al., Adaptation in Canadian agriculture to climatic variability and change, *Climatic Change*, 45, 181, 2000.
4. Wall, E., Smit, B., and Wandel, J., Canadian Agri-food Sector Adaptation to Risks and Opportunities: Position Paper on Climate Change, Impacts and Adaptation in Canadian Agriculture, C-CIARN, Guelph, Canada, 2004.
5. Brklacich, M. et al., Implications of global climatic change for Canadian agriculture: a review and appraisal of research from 1984 to 1997, in *The Canada Country Study Impacts and Adaptation, Vol. VII: National Sectoral Volume*, Koshida, G. and Avis, W., Eds., Environment Canada, Downsview, Ontario, 1997, chap. 4.
6. Cohen, S. et al., North America, in *Climate Change 2001: Impacts, Adaptation and Vulnerability*, McCarthy, J.J. et al., Eds., Cambridge University Press, New York, 2001, chap. 15.
7. Lemmen, D.S. and Warren, F.J., Eds., *Climate Change Impacts and Adaptation: A Canadian Perspective*, Climate Change Impacts and Adaptation Directorate, Natural Resources Canada, Ottawa, Ontario, 2004.
8. Hengeveld, H., Projections for Canada's Climate Future — A Discussion of Recent Simulations with the Canadian Global Climate Model, Report CCD 00-01 Special Edition, Environment Canada, Downsview, Ontario, 2000.
9. Matsumoto, K. et al., Climate change and extension of the *Ginkgo biloba* L. growing season in Japan, *Glob. Change Biol.*, 9, 1634, 2003.
10. Norby, R.J., Hartz-Rubin, J.S., and Verbrugge, M.J., Phenological responses in maple to experimental atmospheric warming and CO_2 enrichment, *Glob. Change Biol.*, 9, 1792, 2003.
11. Kharin, V.V. and Zwiers, F., Changes in the extremes in an ensemble of transient climate simulations with a coupled atmosphere-ocean GCM, *J. Climate*, 13, 3760, 2000.
12. Francis, D. and Hengeveld, H., *Extreme Weather and Climate Change*, Environment Canada, Downsview, Ontario, 1998.
13. Brown, R.D. et al., Climate variability and change — cryosphere, in *Threats to Water Availability in Canada*, National Water Research Institute Scientific Assessment Report Series No. 3 and ACSD Science Assessment Series No. 1, National Water Research Institute, Burlington, Ontario, 2004, 107.

14. Camill, P., Permafrost thaw accelerates in boreal peatlands during late 20th century climate warming, *Climatic Change*, 68, 135, 2005.
15. Stendel, M. and Christensen, J.H., Impact of global warming on permafrost conditions in a coupled GCM, *Geophys. Res. Lett*, 29, doi:10.1029/2001GL014345, 2002.
16. Lockwood, J.G., Is potential evapotranspiration and its relationship with actual evapotranspiration sensitive to elevated atmospheric CO_2 levels? *Climatic Change*, 41, 193, 1999.
17. Whitfield, P., Reynolds, C.J., and Cannon, J., Modelling streamflow in present and future climates: examples from the Georgia Basin, British Columbia, *Can. Water Resour. J.*, 27, 427, 2002.
18. Allen, D.M., Mackie, D.C., and Wei, M., Groundwater and climate change: a sensitivity analysis for the Grand Forks aquifer, southern British Columbia, Canada, *Hydrogeol. J.*, 12, 270, 2004.
19. Chen, C.C., McCarl, B.A., and Schimmelpfennig, D.E., Yield variability as influenced by climate: a statistical investigation, *Climatic Change*, 66, 239, 2004.
20. Piggott, A. et al., Exploring the dynamics of groundwater and climate interaction, in Proceedings of the 54th Canadian Geotechnical Conference and 2nd Joint IAH-CNC and CGS Groundwater Specialty Conference. Canadian Geotechnical Society and the Canadian National Chapter of the International Association of Hydrogeologists, Mahmound, M., van Everdingen, R., and Carss, J., Eds., Calgary, Alberta, 2001, 401.
21. Rivera, A., Allen, D.M., and Maathuis, H., Climate variability and change — groundwater resources, in *Threats to Water Availability in Canada*, National Water Research Institute Scientific Assessment Report Series No. 3 and ACSD Science Assessment Series No. 1, National Water Research Institute, Burlington, Ontario, 2004, 77.
22. Croley, T.E. and Luukkonen, C.L., Potential effects of climate change on ground water in Lansing, Michigan, *J. Am. Water Resour. Assoc.*, 39, 149, 2003.
23. Whetton, P.A. et al., Implications of climate change due to the enhanced greenhouse effect on floods and droughts in Australia, *Climatic Change*, 25, 289, 1993.
24. Zhang, X. et al., Temperature and precipitation trends in Canada during the 20th century, *Atmos. Ocean*, 38, 395, 2000.
25. Vincent, L. A., Zhang, X., and Hogg, W. D., Maximum and minimum temperature trends in Canada for 1895–1995 and 1945–1995, *Preprints, 10th Symposium on Global Change Studies*, American Meteorological Society, Dallas, TX, 1999, 95.
26. Bonsal, B.R. et al., Characteristics of daily and extreme temperatures over Canada, *J. Climate*, 14, 1959, 2001.
27. Bonsal, B.R. and Prowse, T.D., Trends and variability in spring and autumn 0°C-isotherm dates over Canada, *Climatic Change*, 57, 341, 2003.
28. Canadian Council of Ministers of the Environment, Climate, Nature, People: Indicators of Canada's Changing Climate, Canadian Council of Ministers of the Environment, Winnipeg, Manitoba, 2003.
29. Brown, R.D. and Goodison, B.E., Interannual variability in reconstructed Canadian snow cover, 1915–1992, *J. Climate*, 9, 1299, 1996.
30. Brown, R.D. and Braaten, R.O., Spatial and temporal variability of Canadian monthly snow depths, 1946–1995, *Atmos. Ocean*, 36, 37, 1998.
31. Brown, R.D., Northern hemisphere snow cover variability and change, 1915–97, *J. Climate*, 13, 2339, 2000.
32. Burn, D., Hydrologic effects of climatic change in west-central Canada, *J. Hydrol.*, 160, 53, 1994.
33. Whitfield, P.H. and Cannon, A.J., Recent variations in climate and hydrology in Canada, *Can. Water Resour. J.*, 25, 19, 2000.

34. Zhang, X. et al., Trends in Canadian streamflow, *Water Resour. Res.*, 37, 987, 2001.
35. Magnuson, J. et al., Historical trends in lake and river ice cover in the Northern Hemisphere, *Science*, 289, 1743, 2000.
36. Schindler, D.W. et al., The effects of climatic warming on the properties of boreal lakes and streams at the Experimental Lakes Area, Northwestern Ontario. *Limnol. Oceanogr.*, 41, 1004, 1996.
37. Beaubien, E.G. and Freeland, H.J., Spring phenology trends in Alberta, Canada: links to ocean temperature, *Int. J. Biometeorol.*, 44, 53, 2000.
38. Bradley, N.L. et al., Phenological changes reflect climate change in Wisconsin, *Proc. Natl. Acad. Sci. U.S.A.*, 96, 9701, 1999.
39. Penuelas, J. and Filella, I., Responses to a warming world, *Science*, 294, 793, 2001.
40. Zhou, F. et al., Variation in northern vegetation activity inferred from satellite data of vegetation index during 1981–1999, *J. Geophys. Res. Atmos.*, 106, 20069, 2001.
41. Schwartz, M.D. and Reiter, B.E., Changes in North American spring, *Int. J. Climatology*, 20, 929, 2000.
42. Cayan, D.R. et al., Changes in the onset of spring in the Western United States, *Bull. Am. Meteor. Soc.*, 82, 399, 2001.
43. Podur, J., Martell, D.L., and Knight, K., Statistical quality control analysis of forest fire activity in Canada, *Can. J. For. Res.*, 32, 195, 2002.
44. Gillett, N.P. et al., Detecting the effect of climate change on Canadian forest fires, *Geophys. Res. Lett.*, 31, L18211, 2004.
45. Dale, V.H. et al., Climate change and forest disturbances, *BioScience*, 51, 723, 2001.
46. Colombo, S.J. et al., The Impacts of Climate Change on Ontario's Forests, Forest Research Information Paper No. 143, Ministry of Natural Resources, Ontario, 1998.
47. Ayres, M.P. and Lombardero, M.J., Assessing the consequences of global change for forest disturbance from herbivores and pathogens, *Sci. Total Environ.*, 262, 263, 2000.
48. Volney, W.J.A. and Fleming, R.A., Climate change and impacts of boreal forest insects, *Agric. Ecosyst Environ.*, 82, 283, 2000.
49. BC Ministry of Water, Land and Air Protection, Indicators of Climate Change for British Columbia, 2002, Ministry of Water, Land and Air Protection, Water, Air and Climate Change Branch, Victoria, BC, 2002.
50. Carroll, A.L., Taylor, S.W., and Régnière, J., Effects of climate change on range expansion by the mountain pine beetle in British Columbia, BC Ministry of Water, Land and Air Protection, 2004.
51. Mattson, W.J. and Haack, R.A., The role of drought in outbreaks of plant-eating insects, *BioScience,* 37, 110, 1987.
52. Mattson, W.J., Slocum, S.S., and Koller, C.N., Spruce Budworm performance in relation to foliar chemistry of its host plants, USDA For. Ser., Gen. Tech. Rep. NE-85, 55, 1983.
53. Fleming, R.A. and Candau, J.N., Influences of climatic change of some ecological processes of an insect outbreak system in Canada's boreal forests and the implication for biodiversity, *Environ. Monit. Assess.*, 49, 235, 1998.
54. Fajer, E.D., Bower, M.D., and Bazzaz, F.A., The effects of enriched carbon dioxide atmospheres on plant-insect herbivore interactions, *Science*, 243, 1198, 1989.
55. Bazzaz., F.A. and Fajer, E.D., Plant life in a carbon dioxide rich world, *Sci. Am.*, 266, 68, 1992.
56. Flannigan, M.D. et. al., Future area burned in Canada, *Climatic Change*, 72, 1, 2004.
57. Stocks, B.J. et al., Large forest fires in Canada, 1959–1997, *J. Geophys. Res.*, 107, 8149, doi:10.1029/2001JD000484, 2002.

58. Wotton, B.M. and Flannigan, M.D., Length of the fire season in a changing climate, *For. Chron.*, 69, 187, 1993.
59. Wotton, B.M., Martell, D.L., and Logan, K.A., Climate change and people-caused forest fire occurrence in Ontario, *Climatic Change*, 60, 275, 2003.
60. Fosberg, M.A. et al., Global change: effects on forest ecosystems and wildfire severity, in *Fire in the Tropical Biota: Ecosystems Process and Global Challenges*, Goldammer, J.G., Ed., Ecological Studies 84, Springer-Verlag, Berlin, 1990, 483.
61. Price, C. and Rind, D., The impact of a $2 \times CO_2$ climate on lightning-caused fires, *J. Climate*, 7, 1484, 1994.
62. Flannigan. M.D. et al., Future wildfire in circumboreal forests in relation to global warming, *J. Veg. Sci.*, 9, 469, 1998.
63. Stocks, B.J. et al., Climate change and forest fire potential in Russian and Canadian boreal forest, *Climatic Change*, 38, 1, 1998.
64. Wotton, B.M. et al., Estimating current and future fire climates in the boreal forest of Canada using a regional climate model, in *Proceedings of the 3rd International Conference on Forest Fire Research and 14th Conference on Fire and Forest Meteorology*, Coimbra, Portugal, 1998, 1207.
65. Brown, T.J., Hall, B.L., and Westerling, A L., The impact of the twenty-first century climate change on wildland fire danger in the western United States: an applications perspective, *Climatic Change*, 62, 365, 2004.
66. Thorman, M.K. et al., Land-use practices and changes — forestry, in Threats to Water Availability in Canada, National Water Research Institute Scientific Assessment Report Series No. 3 and ACSD Science Assessment Series No. 1, National Water Research Institute, Burlington, Ontario, 2004, 57.
67. Schimmelpfennig, D. et al., Agricultural Adaptation to Climate Change: Issues of Long Run Sustainability, Agricultural Economic Report No. 740, Economic Research Service, U.S. Department of Agriculture, Washington, D.C., 1996.
68. Rötter, R. and van de Geijn, S.C., Climate change effects on plant growth, crop yield and livestock, *Climatic Change*, 43, 651, 1999.
69. Smith, D.L. and Almaraz, J.J., Climate change and crop production: contributions, impacts and adaptations, *Can. J. Plant Pathol.*, 26, 253, 2004.
70. Lobell, D.B. and Asner, G.P., Climate and management contributions to recent trends in U.S. agricultural yields, *Science*, 299, 1032, 2003.
71. Peng, S., Rice yields decline with higher night temperature from global warming, *Proc. Natl. Acad. Sci. U.S.A.*, 101, 9971, 2004.
72. IPCC, *Climate Change 2001: Impacts, Adaptation and Vulnerability*, McCarthy, J.J. et al., Eds., Cambridge University Press, New York, 2001.
73. de Jong, R. et al., Crop Yield Variability under Climate Change and Adaptive Crop Management Scenarios, final project report submitted to the Climate Change Action Fund, Government of Canada, 1999.
74. Singh, B. et al., Impacts of a GHG-induced climate change on crop yields: effects of acceleration in maturation, moisture stress and optimal temperature, *Climatic Change*, 38, 51, 1998.
75. Bootsma, A. et al., Adaptation of Agricultural Production to Climate Change in Atlantic Canada, final report submitted to the Climate Change Action Fund, Government of Canada, 2001.
76. Neilsen, D. et al., Impact of Climate Change on Crop Water Demand and Crop Suitability in the Okanagan Valley, British Columbia, final report submitted to the Climate Change Action Fund, Government of Canada, 2002.

77. Bertrand, A. and Castonguay, Y., Plant adaptations to overwintering stresses and implications of climate change, *Can. J. Bot.*, 81, 1145, 2003.
78. Patterson, D.T. et al., Weeds, insects and diseases, *Climatic Change*, 43, 711, 1999.
79. Boland, G.J. et al., Climate change and plant diseases in Ontario, *Can. J. Plant Pathol.*, 26, 335, 2004.
80. Cohen, R.D.H. et al., Evaluation of the Effects of Climate Change on Forage and Livestock Production and Assessment of Adaptation Strategies on the Canadian Prairies, Saskatchewan Research Council Publication 11363-1E02, final report submitted to the Prairie Adaptation Research Collaborative, Climate Change Action Fund, Saskatoon, Saskatchewan, 2002.
81. McKenney, M.S., Easterling, W.E., and Rosenberg, N.J., Simulation of crop productivity and responses to climate change in the year 2030: the role of future technologies, adjustments and adaptations. *Agric. For. Meteorol.*, 59, 103, 1992.
82. Hahn, G.L. and Mader, J.A., Heat waves in relation to thermoregulation, feeding behavior, and mortality of feedlot cattle, in *Proceedings of the 5th International Livestock Environment Symposium*, Minneapolis, MN, 1997, 563.
83. Mader, T.L., Hu, Q.S., and Harrington, J.A., *Evaluating Models Predicting Livestock Output Due to Climate Change*, Great Plains Regional Center of the National Institute for Global Environmental Change, University of Nebraska, Lincoln, 2003.
84. DesJarlais, C. et al., *Adapting to Climate Change*, Ouranos, Montreal, Quebec, 2004.
85. Klinedinst, P.L. et al., The potential effects of climate change on summer season dairy cattle milk production and reproduction, *Climatic Change*, 23, 21, 1993.
86. Hanson, J.D. and Baker, B.B., Comparison of the effects of different climate change scenarios on rangeland livestock production, *Agric. Syst.*, 41, 487, 1993.
87. Erkert, J.B. et al., The impact of global warming on local incomes from range livestock systems, *Agric. Syst.*, 48, 87, 1995.
88. Rotter, R. and van de Geijn, S.C., Climate change effects on plant growth, crop yield and livestock, *Climatic Change*, 43, 651, 1999.
89. Mooney, H.A. and Koch, W., The impact of rising CO_2 concentrations on the terrestrial biosphere. *Ambio*, 23, 74, 1994.
90. Jablonski, L.M., Wang, X., and Curtis, P.S., Plant reproduction under elevated CO_2 conditions: a meta-analysis of reports on 79 crop and wild species, *New Phytol.*, 156, 9, 2002.
91. Ownesby, C., Cochran, R., and Auen, L., Effects of elevated carbon dioxide on forage quality for ruminants, in *Carbon Dioxide, Populations, and Communities*, Koerner, C. and Bazzaz, F., Eds., Academic Press, San Diego, CA, 1996, 363.
92. Motha, R.P. and Baier, W., Impacts of present and future climate change and climate variability on agriculture in the temperate regions: North America, *Climatic Change*, 70, 137, 2005.
93. Nyirfa, W.N. and Harron, B., Assessment of climate change on the agricultural resources of the Canadian Prairies, unpublished report submitted to the Prairie Adaptation Research Collaborative (PARC), Canada, 2002.
94. McGinn, S.M., Shepherd, A., and Akinremi, O., Assessment of climate change and impacts on soil moisture and drought on the Prairies, unpublished report submitted to Climate Change Action Fund, Government of Canada, 2001.
95. PFRA (Prairie Farm Rehabilitation Authority), *Watch for Sulphates and Blue-Green Algae in Cattle Water Supplies.* Available online at http://www.agr.gc.ca/pfra/drought/article_e.htm, 2003.

96. Kurukulasuriya, P. and Rosenthal, S., Climate Change and Agriculture: A Review of Impacts and Adaptations, Paper No. 91, World Bank Environment Department, Washington, D.C., 2003.

97. O'Neill, D., Threats to water availability in Canada — a perspective, in Threats to Water Availability in Canada, National Water Research Institute Scientific Assessment Report Series No. 3 and ACSD Science Assessment Series No. 1, National Water Research Institute, Burlington, Ontario, 2004, xi.

98. Demuth, M.N. and Pietroniro, A., The Impact of Climate Change on the Glaciers of the Canadian Rocky Mountain Eastern Slopes and Implications for Water Resource-Related Adaptation in the Canadian Prairies, Phase I, Headwaters of the North Saskatchewan River Basin, Climate Change Action Fund–Prairie Adaptation Research Collaborative, PARC Project P55, Canada, 2003.

99. Pruski, F.F. and Nearing M.A., Climate-induced changes in erosion during the 21st century for eight U.S. locations, *Water Resour. Res.*, 38, 34-1, 2002.

100. Phillips, D., White, D., and Johnson, B., Implications of climate change scenarios for soil erosion potential in the USA, *Land Degrad. Rehabil.*, 4, 61, 1993.

101. Soil and Water Conservation Society, *Conservation Implications of Climate Change: Soil Erosion and Runoff from Cropland*, Soil and Water Conservation Society, Ankeny, IA, 2003.

102. Mehti, B.B., Madramootoo, C.A., and Mehuys, G.R., Yield and nitrogen content of corn under different tillage practices, *Agron. J.*, 91, 631, 1999.

103. van der Valk, A.G. and Davis, C.B., Primary production of Prairie glacial marshes, in *Freshwater Wetlands: Ecological Processes and Management Potential*, Good, R.E. and Wingham, D.F., Eds., Academic Press, New York, 1978, 21.

104. Conly, M.F. and van der Kamp, G., Monitoring the hydrology of Canadian prairie wetlands to detect the effects of climate change and land use changes, *Environ. Monit. Assess.*, 67, 195, 2001.

105. Mortsch, L. et al., Climate change impacts on the hydrology of the Great Lakes–St. Lawrence system, *Can. Water Resour. J.*, 25, 153, 2000.

106. Lofgren, B.M. et al., Evaluation of the potential impacts on Great Lakes water resources based on climate scenarios of two GCMs, *J. Great Lakes Res.*, 28, 537, 2002.

107. Mortsch, L.D., Assessing the impact of climate change on the Great Lakes shoreline wetlands, *Climatic Change*, 40, 391, 1998.

108. Hebb, A.J., Implementation of a GIS to assess the effects of water level fluctuations on the wetland complex at Long Point, Ontario, Masters of Environmental Studies Thesis, University of Waterloo, Waterloo, Ontario, 2003.

109. Robinson, S.D. and Moore T.R., The influence of permafrost and fire upon carbon accumulation in high boreal peatlands, Northwest Territories, Canada, *Arct. Antarct. Alp. Res.*, 32, 155, 2000.

110. Moore, T.R., Roulet, N.T., and Waddington, J.M., Uncertainty in predicting the effect of climatic change on the carbon cycling of Canadian peatlands, *Climatic Change*, 40, 229, 1998.

111. Halsey, L.A., Vitt, D.H., and Zoltai, S.C., Disequilibrium of permafrost in boreal continental western Canada to climate change, *Climatic Change*, 30, 57, 1995.

Part II

Managed Ecosystems —
State of Knowledge

4 Anthropogenic Changes and the Global Carbon Cycle

J.S. Bhatti, M.J. Apps, and R. Lal

CONTENTS

4.1 INTRODUCTION

Although climatic fluctuations have occurred often over the past 420,000 years, the rates of increase in temperature in the last 100 years are unprecedented in both magnitude and cause. Similarly, the rates of increase in atmospheric greenhouse gas (GHG) concentrations over the 20th century do not appear in the paleo record, and are causally linked with the recent changes in global temperature. Significantly, human industrial development is clearly linked to for the changes in GHG concentrations. Over much of the preceding half million years, the fluctuations of atmospheric GHGs and global average temperature remained in a relatively narrow, correlated band (Chapter 2), implying a natural balance in the exchange of GHGs between the atmosphere and planetary surface.[1]

The 19th century, however, witnessed the start of a dramatic change in this balance which to date has already recorded a 32% increase in CO_2 relative to the average of the past 420,000 years, a change whose rate is still accelerating.[2] These changes have been driven by human perturbations to the global carbon (C) cycle — changes that have been both *direct*, introducing new C to the active cycle through fossil fuel use and land-use change (LUC), and *indirect*, affecting the biospheric portion of the active C cycle through environmental stresses and perturbations to other global biogeochemical cycles. The observed response of the global climate system to this change during the 20th century, expressed in terms of global mean temperature, is modest (+0.6°C) but has already led to detectable impacts.[3] The predicted changes in climate for the 21st century and beyond are now more certain and predicted to be higher, and faster, than previously estimated — perhaps +6°C or more by 2100.[2] Although terrestrial and oceanic ecosystems currently absorb an amount equal to about 60% of the direct anthropogenic emissions of CO_2 to the atmosphere, the natural physiological mechanisms that are thought to be responsible for this increased uptake are not expected to function as effectively in the future (Chapter 9). Thus, in the absence of purposeful mitigation strategies, the terrestrial CO_2 sink will likely decrease and could even become a source during the 21st century,[4] accelerating the changes in climate. Changes in the global C budget, dominated by CO_2 although CH_4 is also important, play a vital role in determining global climate.

4.2 GLOBAL CARBON CYCLE

Understanding the mechanisms that regulate the global C cycle and the exchange of C between the atmosphere and various natural and anthropogenic components (illustrated in Figure 4.1) is central to finding ways to mitigate or adapt to global climate change.

4.2.1 CARBON POOLS

Five principal global C reservoirs can be identified: atmosphere, vegetation, soils, oceans, and fossil fuels. In 1999, the atmosphere contained about 767 gigatons of C (Gt C) in the form of CO_2.[2] This C corresponds to an average atmospheric concentration of CO_2 of 365 parts per million by volume (ppm), although the actual CO_2 concentration varies slightly from place to place and from season to season.[2] Notably, concentrations and seasonal variations are somewhat higher in the Northern than in the Southern Hemisphere because the main anthropogenic sources of CO_2 are located north of the Equator and because there are larger biospheric exchanges over land surfaces (which is greater in the Northern Hemisphere) than oceans (which is greater in the Southern Hemisphere). During the 1990s, the average concentration of CO_2 increased by 1.5 ppm/year,[5] and is continuing to rise in the first decade of the 21st century at an even higher rate. The 5 ppmv increase during 2001–2003 was the highest ever recorded.[6] Thus, by 2005 the atmosphere was estimated to contain 807 Gt C in the form of CO_2, an increase of 42 Gt C since 1999.

The terrestrial C pool, the third largest pool, comprises reservoirs in soil and vegetation. The vegetation pool, made up of all vegetation types but dominated in mass by trees, is estimated at 610 Gt C.[7] The soil C pool is made up of two components: the soil organic C (SOC) pool estimated at 1550 Gt C and the soil inorganic carbon (SIC) pool estimated at 950 Gt C.[8,9] Thus, the soil C pool of 2500 Gt C is about four times the size of the vegetation pool and about three times the

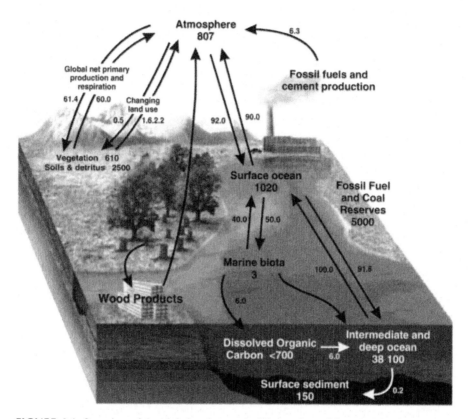

FIGURE 4.1 Overview of the global carbon cycle. The stocks and fluxes of C (Gt) between various components and the atmosphere. (Modified from Bhatti et al.[76])

atmospheric pool. The total terrestrial C pool is about 3060 Gt C. The amount of C stored in the geological formations as fossil fuel is considerably larger — on the order of 5000 Gt C — of which the vast majority is in the form of coal (4000 Gt C) and the rest as oil and gas (500 Gt C each). In comparison, the terrestrial C pool of 3060 Gt C is about 61% of the estimated fossil fuel pool and about four times the atmospheric pool. The fossil fuel parts of the geological pools as reported here include only that carbon that originated from biological processes in the far distant past, and does not include carbonates in sedimentary rocks (1×10^6 Gt C), which were primarily formed through abiotic chemical and physical processes. These latter deposits contain about 1700 Gt C and occur primarily in arid and semiarid regions.[10]

The oceans contain the largest C pool at 38,000 Gt C — but most of these vast stores are effectively held out of circulation in the form of dissolved bicarbonate in the intermediate and deep ocean.[7]

4.2.2 CARBON EXCHANGE

All these C reservoirs are interconnected by biotic, abiotic, and anthropogenic processes. For example, 60 Gt C is exchanged in each direction between vegetation and the atmosphere each year through the biological processes of photosynthesis (uptake) and respiration (release). Similarly, 90 Gt C is emitted and 92 Gt C absorbed by the ocean each year[7,11] through a combination of physical exchange and biological activity at the ocean surface. In comparison, only 6.3 Gt C/year is emitted by human combustion of fossil fuel and another 1.6 to 2.0 Gt C/year by land-use change, but these fluxes are emissions only, with no compensating uptake directly associated with them. Clearly, were it possible to enhance photosynthetic uptake and avoid the re-emission through decomposition (i.e., sequestering), even 5% of the photosynthetic C in terrestrial ecosystems would drastically offset the industrial emissions. Over short timescales (a few years), this is not difficult: the challenge, however, is whether such sequestration can be carried out in a sustainable way. Additional issues at hand are how much do each of the four terrestrial nongeological pools contribute to the enrichment of CO_2 concentration in the atmosphere, which pools are potential sinks of atmospheric CO_2, and can they be managed in some way to ensure this sink?

On the source side of the sink-source balance, combustion of fossil fuels and depletion of the geological pool is an obvious and readily quantifiable term. Another obvious but not easily quantifiable source is deforestation and the attendant biomass burning that occurs largely, but not exclusively, in the tropics. Yet another important but neither obvious nor easily quantifiable source is the emission of CO_2 and other GHGs through soil degradation. Each year, soils globally release about 4% of their pool (60 Gt C) into the atmosphere — about ten times the fossil fuel combustion. Although most of this is associated with the natural processes of decay, decomposition, and combustion that form part of the balanced carbon cycle, additional releases are associated with human land-use practices and changes in land use. The exact magnitude of the loss is not known, and may in fact be greater than 60 Gt C because of anthropogenic perturbations to ecosystems leading to degradation. On the other hand, the so-called "missing C" (the amount required to close the balance between estimates of total sink, total source, and atmospheric C increase; see Chapter 9) may also be associated with uptake by soils and other terrestrial ecosystems. These issues can be resolved only when the mechanisms that underlie all major fluxes of the global C cycle are understood.

Boreal forests and their associated peatlands represent the largest terrestrial reservoir of C,[2] as well as being located in a region especially sensitive to climate change. The boreal biome, therefore, plays a critical role in the global C cycle and has the capacity to either accelerate or slow climate change to some degree, depending on whether the forest ecosystems act as a net source or a net sink of C. This source or sink status is, however, not a static characteristic of the ecosystem, but changes over time as a result of alterations to forest age-class structure, disturbance regime, and resource use.[12,13]

Currently, about 78% of the direct human perturbations to the global C cycle are due to fossil fuel combustion, emissions of which now exceed 6 Gt C/year and continue to increase rapidly. (To put this global emission in perspective for a single year, it is equivalent to the total incineration of half of all trees in Canada — with no residues, charcoal, or shoot left behind. Alternatively, to offset the fossil emissions by growing forests, it would be necessary to create a forest biomass equal to half that in Canadian forests every year.) In addition, since the mid 19th century, LUC has resulted in the cumulative emission of ~156 Gt C of anthropogenic CO_2 to the atmosphere. This LUC flux is about 56% of that from fossil fuel use (~280 Gt C) and continues to be an important anthropogenic emission (2.2 Gt C/year).[14] Human land-use practices, therefore, play a significant role in the contemporary C cycle.

Of the 7.6 ± 0.8 Gt C/year of CO_2 added to the atmosphere by human activities during the period 1980 to 1995, less than half (3.2 ± 1.0 Gt C/year) remains there, with the rest taken up about equally by the oceans and by terrestrial ecosystems.[15] Earth's biosphere thus actively removes some of the new C that humans have added to the atmosphere and into the active C cycle. Terrestrial ecosystems, in particular, appear to have sequestered (taken up and retained) 2.3 ± 0.9 Gt C/year, even after accounting for the loss of between 2.0 and 2.2 Gt C/year from deforestation.[14] Likewise, the world's oceans sequester a similar amount of the new C added to the active cycle by human activities.

The biosphere thus appears to be attempting to restore the balance that prevailed for the previous 420,000 years. But it is losing the battle: atmospheric CO_2 concentrations are already at unprecedented levels and rising at a rate never before seen in the geological record (Chapter 9). Moreover, it is unclear whether the biosphere can continue to function as a net sink into the future. At the present, scientific know-how required to explain and predict changes in the mechanisms responsible for the present net biospheric uptake is severely limited. More specifically:

- Will these mechanisms continue to offset the direct anthropogenic emissions? Or will the mechanisms decline in strength, or even fail entirely as the C cycle–climate system moves into a new mode of operation,[16] as several terrestrial and ocean model simulations alarmingly suggest?[4,17]
- Are the changes in the C balance of Canada's forest associated with an altered natural disturbance regime,[12] a warning that the putative sink is already disappearing?

Although it is not possible to address these questions with full certainty at this time, they are of obvious importance to humanity. Whether forests and agriculture ecosystems can continue to provide both the goods (e.g., food and fiber) and services (e.g., recreation, spiritual, and social) that humans have come to depend on is a question that remains to be answered. There is an urgent need to assess the impact of human activities on the terrestrial biosphere and its contribution to the global C cycle. Climate change affects both the distribution and character of the landscape through changes in temperature, precipitation, and natural disturbance patterns. These impacts are not entirely separable from the effects of other global changes such as increases in CO_2, NO_x, and O_3 levels, and anthropogenic pressures which may be exacerbated by climate change. Figure 4.2 illustrates the interactions among climate, vegetation, disturbance regimes, and C pools. The following sections (Land Use and Land-Use Change; Land Degradation and Soil Erosion; CO_2 Fertilization; Drainage; and NO_2 Fertilization) deal with the impacts of various anthropogenic agents on ecosystems and their contributions to the C cycle.

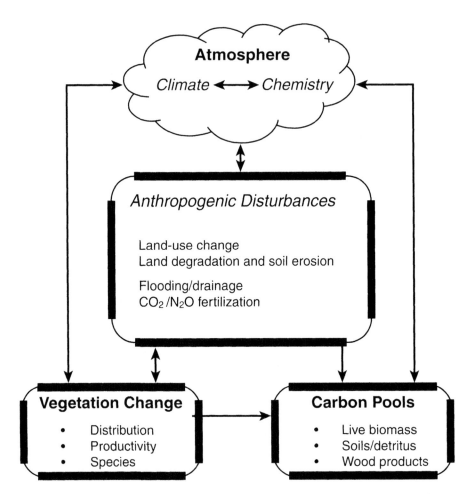

FIGURE 4.2 Feedbacks between the atmosphere and various components of the boreal forest. (Modified from Bhatti et al., 2002.)

4.3 LAND USE AND LAND-USE CHANGE

Loss of forested areas is a major conservation issue with important implications for climate change. The proportion of land surface covered with agriculture is relatively small (7%) in Canada compared to that under forest (50%).[18] With increases in population and food demand over the last century, large forested areas of the boreal region are being converted to agricultural use.[15] However, the rate of forest loss and fragmentation of different ecosystems in the boreal biome, and the associated anthropogenic factors that influence these rates, is not well established.

   Forested lands are influenced by natural and anthropogenic causes, including harvesting, degradation, large-scale wildfire, fire control, pest and disease outbreaks, and conversion to nonforest use, particularly agriculture and pastures. These disturbances often cause forests to become sources of CO_2 because the rate of net primary productivity is exceeded by total respiration or oxidation of plants, soil, and dead organic matter — net ecosystem production (NEP) < 0.[20]

Between 1975 and 2001, 18.7 million hectare (Mha) of forest was harvested in Canada and 15.2 Mha successfully regenerated.[18] The harvest techniques (site preparation, planting and spacing, and thinning) as well as harvest methods (clear-cutting or partial cutting) and factors that affect how much and what type of material is removed from the site have a significant influence on the C balance. After harvesting, a forest stand's net C balance is a function of the photosynthetic uptake minus the autotrophic and heterotrophic respiration that occurs. While a stand is young, the losses through decomposition outweigh the gains through photosynthesis, resulting in a net source (Chapter 9).

Increasing the C uptake can be accomplished through techniques that reduce the time for stand establishment (such as site preparation, planting, and weed control), increase available nutrients for growth, or through the selection of species that are more productive for a particular area. Decreasing the losses can be accomplished through modification of harvesting practices such as engaging in lower-impact harvesting (to reduce soil disturbance and damage to residual trees), increasing efficiency (and hence reducing logging residue), and managing residues to leave C on site[21] (Chapter 9).

Rapid expansion of agriculture along its southern boreal has been recognized as at risk for more than 50 years.[22] The conversion of native upland and lowland into agriculture and urban lands has escalated, resulting in the contemporary patchwork of ecosystem types.[19] Losses of C include both the initial depletion associated with the removal of natural vegetation and the subsequent losses from soil through mineralization, erosion, and leaching in the perturbed ecosystems. In the prairie provinces of Canada alone, it is estimated that there was a net deforestation of 12.5 Mha between 1869 and 1992.[23] Using the Canadian Land Inventory Database to examine changes between 1966 and 1994, Hobbs[24] estimated that forests of the southern boreal plains of Saskatchewan declined from 1.8 Mha in 1966 to 1.35 Mha by 1994,[24] an overall conversion of 24% of the boreal transition zone to agriculture since 1966. A more recent study suggests that forestland is being converted into agriculture, industrial, and urban development at the rate of 1215 ha/year along the southern boreal zone of Canada.[25] This rate is approximately three times the world average: the loss of boreal forests and wetlands is equal to, and in some regions greater than, that occurring in tropical rainforests. These estimates suggest that all the wetland and forested areas in the boreal transitional zone will be lost by 2050 unless purposeful action is taken to reverse the present trend.

Conversion of natural to agricultural ecosystems causes a net emission of CO_2 and other GHGs into the atmosphere. In addition to decomposition of biomass with the attendant release of CO_2, agricultural activities also deplete the soil C pool through reduction of biomass inputs and changes in temperature and moisture regimes, which further accelerate decomposition. Soil drainage aimed at managing water table depth and soil cultivation (to control weeds and prepare seedbeds) also accelerates soil erosion and mineralization of the SOC pool. Most agricultural soils in the North America have lost 30 to 50% (30 to 40 Mg C/ha) of the preexisting carbon pool following conversion from natural to agricultural ecosystems. Thus,

SOC pools in most agricultural soils are well below their potential capacity by an amount equal to the historic C loss since conversion to agricultural ecosystems.

The above discussion has focused on CO_2, but similar conclusions can be drawn for other GHGs, such as CH_4 and N_2O.[2] For example, N_2O emissions are influenced by the timing and amount of fertilizer applications and hence, intensity of management. Changes in land cover also alter the uptake of CH_4 by soils, and different agricultural practices differ in their CH_4 emission profiles. Increases in animal populations have also contributed to the increase in atmospheric CH_4. Enteric fermentation, the digestion process in ruminant animals such as cattle, sheep, and goats, adds an estimated 100 Gt of CH_4 per year to the atmosphere.

Virtually all these emissions also vary with alterations in climatic and ecological conditions, leading to a heterogeneous spatial and temporal pattern of GHG emissions from the terrestrial biosphere that is strongly influenced by physical, biogeochemical, socioeconomic, and technical factors. Actual land use and the resulting land cover are important controls on these emissions, and when mitigation policies are evaluated, aggregated assessments using global averages to calculate the emissions are no longer valid. State-of-the-art assessments must be dynamic, geographical and regionally explicit, and include the most important aspects of the physical subsystem, the biogeochemical subsystem, and land use and changes therein.

Farm operations also incur hidden C costs. The average emission (calculated in carbon equivalent units) per hectare is 15 kg C for moldboard plowing,[1] 11 kg C for sub-soiling, 8 kg C for heavy tandem disking, 8.0 kg C for chiseling, 6.0 kg C for standard disking, 4.0 kg C for cultivation, and 2.0 kg C for rotary hoeing.[26] Thus, emissions are 35 kg C/ha for complete conventional tillage operations compared with 6.0 kg C/ha for disking only, and none for no-till farming. Emissions associated with pump irrigation are 150 to 285 kg C/ha/year depending on the source of energy and depth of the water table.[27,28]

Other agricultural activities also led to emission of GHGs, especially CO_2 and N_2O (Figure 4.3). In addition, there are hidden C costs for application of nitrogenous fertilizers and pesticides.[26] Estimates of emissions (given in equivalent C units) for production, transportation, and packaging of fertilizer are 1 to 3 kg C/kg for N, 0.2 kg C/kg for P, 0.15 kg C/kg for K, and 0.16 kg C/kg for lime.[26] The hidden C costs are even higher for pesticides and range from 6.3 kg C/kg for herbicides, 5.1 kg C/kg for insecticides and 3.9 kg C/kg for fungicides.[26]

Enhancing the use efficiency of agricultural chemicals and irrigation water can have beneficial C implications. The use efficiency of N is generally low, and fertilizer use is a significant cause of increased N_2O emission.[29] It is thus important to minimize losses of fertilizers (especially nitrogenous fertilizers) by erosion, leaching, and volatilization.[30,31] Integrated nutrient management and integrated pest management can be valuable strategies for reducing emissions. While increasing N stocks through incorporation of cover crops in the rotation cycle is a useful strategy, N_2O emission and leaching of NO_3 into the groundwater can also occur when the N is biologically fixed. Sustainable management must seek to enhance the use efficiency of C-based inputs while simultaneously decreasing losses of these fertilizers, thereby achieving both environmental and economic benefits.

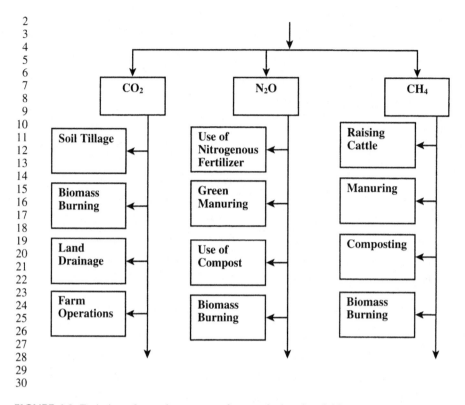

FIGURE 4.3 Emission of greenhouse gases from agricultural activities.

4.4 CO₂ FERTILIZATION

CO_2 fertilization, discussed in Chapters 5, 9, and 16, theoretically has the potential to increase photosynthetic uptake of CO_2 in terrestrial plants by up to 33%.[32] The CO_2 fertilization effect may be expected to enhance the growth of some tree species and forest ecosystems, allowing them to absorb more C from the atmosphere (Chapter 16). Whether the enhancement of photosynthesis by elevated CO_2 actually results in net removal of CO_2 from the atmosphere at the ecosystem level, however, is a subject of intense debate (e.g., Reference 33). Notably, forest inventory data indicate that the net effect on C-stocks is less than the enhancement of gross photosynthesis alone would suggest, and may account for less than a few percent increase in accumulated C in forest vegetation.[34]

Many of the experimental studies on elevated CO_2 response have been conducted on tree seedlings, often in growth chambers, under conditions not otherwise limiting plant growth.[32,35] Several field experiments are currently under way that employ free air CO_2 enrichment (FACE) technology by which the CO_2 (and other gases) around growing plants may be modified to simulate future levels of these gases under climate change.[33,36] These experiments, however, have not been conducted for long enough to determine what the long-term effects of elevated CO_2 levels might be once canopy closure is reached.[37] While the response of mature forests to increases in atmospheric CO_2 concentration has not been demonstrated experimentally, it will likely be different from that of individual trees and young forests (see References 34 and 38 and Chapter 15).

Chen et al.[39] and others have hypothesized that Canada's forest net primary productivity (NPP) may be increasing, and that this increase may be due in part to CO_2 fertilization.[39] This disagrees with the inventory measurements reported for U.S. forests over the past century.[34] Forest age-class dynamics, LUC, and alterations in natural disturbance patterns appear to have a much larger influence than CO_2 fertilization on forest growth in North American forests.[12,34] There is a growing consensus in the scientific community that CO_2 fertilization effects, to the extent that they exist, can be expected to saturate (that is, their contribution to continued net CO_2 removals will go to zero) over the next 100 years or so,[37,40] or even reverse. This occurs because increases in CO_2 levels stimulate increases in gross photosynthesis at a diminishing rate, while increases in temperature stimulate increases in respiration at an exponential rate thereby reducing the net photosynthetic uptake.[37] Additional increases in decomposition further reduce the net sink and may even result in a source.[41]

The difference in the response of C_3 and C_4 plants to increasing CO_2 concentrations is also well documented, and different biomes have significantly different proportions of C_3 and C_4 plants.[42] Based on this factor alone, temperate and boreal forests would be expected to be more sensitive to CO_2 fertilization than grasslands. Even within a biome, between plant species or even genotypes there is a marked differential response to CO_2 fertilization. A managed temperate forest planted with a highly sensitive species may store larger amounts of C than an otherwise equivalent forest planted with less sensitive species, or a comparable tract of old forest. Therefore, the CO_2 fertilization effect is quite heterogeneous over time and space.

4.5 NO_x FERTILIZATION AND OZONE

The concentration of N_2O in the atmosphere increased about 0.25% per year during the 1990s, and has increased about 13% since pre-industrial times (from 275 to 312 ppbv).[2] The primary sources of N_2O are the combustion of fossil fuels, use of fertilizers, livestock, and burning of biomass. Because of the widespread use of anhydrous ammonia, it is estimated that about 5% of the N in fertilizer applied to fields in Ontario, Canada is converted to N_2O and about 11% to NO_x.

In boreal forest ecosystems, N is a limiting factor to vegetation growth because most of it occurs in forms that cannot be readily used by most plants. Human activities have increased the supply of N in some regions of the eastern boreal forest. It has been suggested that increased N deposition (due to NO_x atmospheric pollution) may temporarily enhance forest C sequestration in N-limited ecosystems, leading to a short-term C gain in net primary productivity (NPP).[43]

Different forest ecosystem types vary greatly in their potential for C sequestration. Woody tissues typically have C:N ratios >300 and lifetimes >100 years. Hence, it might be expected that if higher wood production with excess N can be obtained, it would result in large removal of C from the atmosphere over long time periods.

Even if the high C:N ratios are maintained, however, the positive effect on forest growth from N deposition in boreal forests will likely be negated in the medium term as other factors, such as other nutrients and water,[44] become limiting to their growth. In North America, N deposition has not appreciably affected C accumulation rates at the landscape level.[34] On the contrary, the evidence is that excess deposition has harmful effects at the stand level, on both forested and aquatic ecosystems.[45]

Nevertheless, lack of available N is a limiting factor in most of boreal ecosystems. In addition to the direct effects of nutrient addition in stimulating NPP, enhanced N supply operates synergistically with CO_2 fertilization, and may also increase the soil C storage capacity. The net effect of these increases with the increased oxidation and microbial decomposition at the projected higher future temperatures on both biomass production and soil humification, however, is difficult to predict with present data and understanding.

Annual mean ground-level ozone (O_3) concentrations in Canada are increasing, particularly in urban areas.[46] At least 2 Mha of Canada's productive eastern forest is exposed annually to damaging levels.[47] Exposure of western forests to O_3 is difficult to estimate with the present lack of ground-level monitoring data, but some southeastern forest ecosystems are likely to be more exposed because of significant industrial expansion in these areas. O_3 can adversely affect forest ecosystems by impairing tree physiology, in particular by decreasing the rate of photosynthesis in some species and altering carbohydrate allocation patterns in others.[48] With respect to the latter, C transfer is commonly increased to the shoots, but decreased to the roots. While this gives an apparent increase in their growth rates, it also makes trees more vulnerable to drought, nutrient deficiencies, and winter damage.

FACE study results indicate that the growth rate increases due to CO_2 fertilization observed in some tree species are often negated by the effects of tropospheric O_3.[49] For example, Isebrands et al.[49] reported negative responses in aspen (*Populus tremuloides*) and birch (*Betula papyrifera*) aboveground estimated stem volumes relative to the controls after 3 years of fumigation with O_3 and O_3 + CO_2. A stimulation of 20 to 30% increase with CO_2 alone was also completely offset by O_3. While experimental studies have shown reduced growth rates in some forest species exposed to O_3,[49] there is no evidence that changes in O_3 levels result in any significant changes (either negative or positive) in forest growth rates at the landscape level. Any effects are likely to be region specific and occur against a longer-term background of climate and forest change.

Carbon is only one of several important constituents of soil organic matter. Even if C is supplied to the soil through application of crop residues and other biomass, it may not be converted into humus if there is not enough N and other essential elements (e.g., P, S, Ca, Mg). The C:N ratio in crop residues is often 80:1 or 100:1. In contrast, the C:N ratio in humus is typically between 10:1 and 12:1. The implication is that humification of C in crop residues and other biosolids is limited by availability of N. It is estimated that sequestration of 1000 kg of C in humus requires 83.3 kg of N, 20 kg of P, and 14.3 kg of S.[50] Because of this requirement for N, more C is sequestered in croplands that are fertilized.[51] Counteracting the GHG benefits of this uptake, however, is the fact that addition of N fertilizers for crop growth and humification of biosolids also increases the quantity of mineralized N,[52] with an attendant emission of N_2O and leaching of NO_3 into the groundwater.

4.6 LAND DEGRADATION

Land degradation can occur through either degradation of the vegetation cover or the underlying soil but ultimately results in reduced performance of both parts of the ecosystem. Degradation of soil occurs as a result of excessive utilization, environmental changes, and/or careless management of agricultural areas or lands used for pasture or forestry. Degradation can span the range of vegetation cover reduction to severe soil erosion. Soil degradation may be physical, chemical, or biological (Figure 4.4). These degradation processes adversely affect NPP both directly and indirectly, and reduce the amount of biological material returned to the soil. Consequently, the C input into the system is lower than the C out of the system, resulting in depletion of the soil C pool and an atmospheric source.

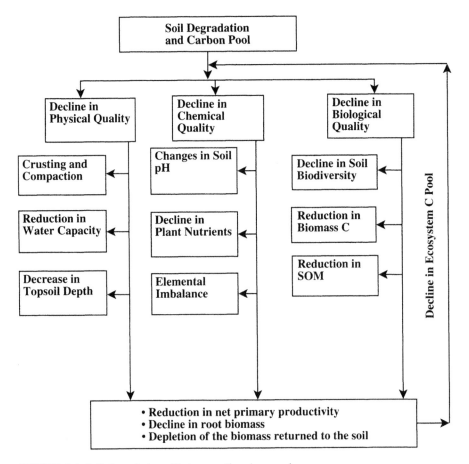

FIGURE 4.4 Soil degradation effects on soil carbon pool.

LUC associated with a loss of vegetation typically has an initial rapid loss of C and nutrients that is generally ascribed directly to the LUC itself; it can, however, result in a longer-term degradation that generates additional depletion. The loss of belowground C is especially significant because this loss can be much more rapid than the rate of its formation or replacement; these pools are usually regarded as long-term storage. The magnitude of these degradation processes is relevant because they occur in at least 70% of drylands[53] that occupy almost 40% of the land surface. They are also an important factor in the carbon balance in humid areas, such as rain forests and tundra. Is there a CO_2 fertilization effect on these degraded lands? If antidesertification or land management measures are taken, will the C stocks return to their original values, or will they be higher or lower?

These environmental stresses, combined with other factors such as increased fire frequency, and the introduction of exogenous plant and animal species, pests, and diseases to natural and managed biomes contribute to further magnify the uncertainty in predictions of the future of the terrestrial sink and the effect of CO_2 fertilization and climate change.

Soil degradation and desertification are serious global issues affecting both large areas and diverse ecosystems (Table 4.1). Soil degradation and desertification are severe in warm and arid climates. These are biophysical processes, but they are driven by socioeconomic and political forces. The problem of soil degradation is exacerbated by overexploitation of natural resources through deforestation of steep slopes, excessive grazing, cultivation of steep or marginal lands, and mining of soil fertility by low-input- or no-input-based extractive agricultural practices.

Soil degradation adversely affects soil quality and depletes the SOC pool. Biological soil degradation, in comparison with physical and chemical degradation processes, is directly related to the depletion of the SOC pool and the reduction in soil biodiversity. Soil degradation has positive feedback elements due to the

TABLE 4.1
Global Extent of Soil Degradation

Land Degradation Process	Global Areas* Affected (10^6 ha)
1. Soil degradation	
i. Water erosion	749
ii. Wind erosion	280
iii. Chemical degradation	146
iv. Physical degradation	39
2. Desertification	
i. Irrigated cropland	43
ii. Rainfed cropland	216
iii. Rangeland	3334

*These estimates include moderate, severe, and extreme forms of soil degradation.
Source: Adapted from Lal.[30]

interactions among the difference processes involved, so that, once triggered, the degradation may accelerate over time in both magnitude and rate.

As a result of these feedback effects, most degraded soils are severely depleted of their SOC pool. The historic loss of the SOC pool in degraded soils and ecosystems, however, has created the potential for creating a C sink: restoring degraded soils and ecosystems provides an opportunity to restore some of the depleted SOC.

4.7 SOIL EROSION

Accelerated soil erosion is the most widespread form of soil degradation. Globally, the total land area affected is estimated at more than 1 billion hectares — about 750 Mha through water erosion and 290 Mha by wind erosion (Table 4.2). Regionally, hot spots of erosion include the Himalayan-Tibetan ecosystem in South Asia, the Loess Plateau in China, sub-Saharan Africa and the East African highlands, the highlands of Central America, the Andean region, Haiti and the Caribbean.[31] The annual sediment transport into the ocean by the world's rivers is estimated at 15 to 20 Gt C.[54] Global transport of C in rivers is estimated at 0.74 Gt C/year.[55]

Soil erosion leads to a preferential removal of soil organic material, because it is a light fraction and is concentrated in the vicinity of the soil surface. The enrichment ratio (i.e., the concentration of organic carbon in the eroded soil relative to that in the non-eroded soil) of the sediments is generally more than 1, and often as high as 5 to 10. Consequently, the SOC pool of eroded soils is severely depleted, often by as much as 30 to 45 Mg C/ha. The fate of C displaced by erosion is obviously important but a highly debated topic. Some sedimentologists[56,57] argue that C transported to and buried under the aquatic ecosystems and ocean is permanently sequestered. Others[31] suggest that a sizable proportion of the displaced C is

TABLE 4.2
Global Extent of Moderate, Severe, and Extreme
Forms of Soil Erosion by Water and Wind

Region	Area Affected (106 ha)	
	Water Erosion	Wind Erosion
Africa	169	98
Asia	317	90
South America	77	16
Central America	45	5
North America	46	32
Europe	93	39
Oceanic	4	16
World	751	296

Note: Figures in Tables 4.1 and 4.2 differ because of different sources of data and in some cases a slight form of erosion is also included in Table 4.2.
Source: Adapted from Lal.[31]

emitted to the atmosphere as CO_2 prior to its burial in aquatic ecosystems and floodplains. Lal[31] estimated that on the global scale, water erosion translocates about 4.0 to 6.0 Gt C/year. Of this, 2.8 to 4.2 Gt C/year is redistributed over the landscape and transferred to local depressional sites, 0.4 to 0.6 Gt C/year is transported into the ocean by world rivers, and 0.8 to 1.2 Gt C/year is emitted to the atmosphere as CO_2. Thus, adoption of conservation-effective measures can drastically reduce the erosion-induced emission of soil into the atmosphere.

Restoration of eroded and degraded soils has a large potential to sequester C and offset a fraction of the anthropogenic emissions. The data in Table 4.3 show the potential of soil C sequestration through restoration of eroded soils and degraded ecosystems. The total potential of C sequestration in these ecosystems is 0.1 to 0.2 Gt C/year over a 50-year period.[30] Suggested cropping practices that may restore some of the depleted SOC in eroded agricultural soils include reduction in tillage, growing of perennial forages cover, and application of organic amendments to the soil. An almost 70% increase in total C over a 5-year period has been observed in the Canadian Prairies with continuous legume/cereal production, reduced tillage, and nutrient additions via fertilizer or composted manure applications.[58]

4.8 WETLAND DRAINAGE

In Canada, most peatlands occur in the boreal zone and are generally unaffected by agricultural, urban, ports/harbors, and industrial development. Flooding forested and wetland areas in the boreal zone for hydroelectric reservoirs generates massive fluxes of dissolved organic carbon (DOC) into the water, accelerates peat decomposition, and increases methane and CO_2 fluxes to the atmosphere.[46,59,60] For example, Kelly et al.[59] experimentally flooded a boreal wetland in Ontario, causing the carbon dynamics of the site to change from a sink of 6.6 g C/m^2/year to a source of 130 g C/m^2/year. Turetsky et al.[60] estimated that 0.8 ± 0.2 Gt C/year is released from approximately 780 km^2 of hydroelectric reservoirs in peatlands across western boreal Canada.

TABLE 4.3
Potential of Soil Carbon Sequestration in Eroded and Degraded Soils of the World

Degraded Soil	Land Area That Can Be Restored (10^6 ha)	Mean Rate of SOC Sequestration (kg/ha/yr)	Annual Rate of Sequestration (Gt C/yr)
Water and wind erosion	500	200–400	0.1–0.2
Physical degradation	75	100–200	0.01–0.02
Chemical degradation	20	100–200	0.002–0.004
Total			0.112–0.224

Source: Adapted from Lal.[30]

In the Canadian Prairie/parkland, drainage is still a current practice, although the major peat deposits lie farther north in the boreal forest. Approximately 17% of farmers whose lands supported wetlands had drained one or more of them between 1990 and 1992.[61] Drainage has serious consequences because it changes the habitat entirely and lowers the water table, altering both the processes of photosynthesis (C uptake) and decomposition (C release). Since 1950, the U.S. has lost 87 Mha of its original wetlands, primarily due to expansion of agriculture.[53] In Canada, agriculture alone accounts for an estimated loss of 20 Mha of the pre-settlement wetlands.[62] Losses in Europe are along the same lines — 67% in France (1990–1993), 57% in Germany (1950–1985), and 60% in Spain (1948–1990).[63] Peatlands/wetlands have been drained for the growth of crops and trees, production of fuel, and the harvesting of horticultural moss.[64] The emission of CO_2 increases when northern peatlands are drained or degraded.[65,67] Drained and cut-over boreal peatlands remain an ongoing source of CO_2 emissions for a long time, even in wet years.[68] Globally, the long-term drainage of peatlands has resulted in the emission of about 0.0085 Gt C/year of CO_2, while burning of fuel peat adds an additional 0.026 Gt C/year.[69] In Sweden, farmed organic soils represent less than 10% of total arable land, but contribute as much as 10% of the total national anthropogenic CO_2 emission.[70] In Canada, it has been suggested that draining an additional 5% of Canadian peatlands would be sufficient to offset the putative existing peatland carbon sink of the country.[68] Draining of peatlands also significantly increases nitrous oxide emissions.[71,72] In Finland, N_2O emissions from farmed organic soils amount to 25% of total anthropogenic N_2O emissions.[70]

In western Canada, it has been estimated that at least 75% of prairie/parkland wetlands have already been lost through agricultural drainage, many of which are subsequently only marginally productive under crop management.[73] The loss of SOC from such wetlands when converted to agricultural usage may be as much as 50%.[74,75] Another key element is the degradation of riparian zones and associated uplands via cultivation or overgrazing. This has the direct impact of reducing the amount of vegetated habitat available to sequester C, but also has a negative impact on the remaining wetland through nutrient loading associated with transport of fertilizer and pesticides in runoff with sediments.

4.9 CONCLUSION

This chapter centers on a key question: Will the present source–sink relationship of the terrestrial biosphere be maintained? More specifically, will the currently observed sequestration by terrestrial ecosystems decrease with time, or can it be maintained and even increased over the next 100 years?

To answer these questions, a reliable projection of the C budget for 50 to 100 years into the future is needed. However, such predictions are difficult to make with certainty given the present state of knowledge. In Chapters 5, 6, 7, 9, 10, and 11 some of these uncertainties are laid out in more detail. In Chapters 15, 16, and 17 the research that appears to be needed to improve the reliability of the predictions is examined. With present knowledge and data, however, it is possible to predict the likely trends in the ecosystems — whether there will be increase or decrease in the relative importance of different components of the terrestrial ecosystems — so that a general trend in the C budget can be projected. Based on the analysis presented above, the following trends appear to be likely:

- Emissions from land-use or land cover change from forest and wetland/peatlands to agriculture will almost surely increase given the sustained increase in food and fiber demand over the next 50 years.
- The emission of CO_2 from the soils (agricultural and forest) as climate warms will become an increasingly important source through the 21st century.
- Increased CO_2 fertilization coupled with N deposition may partially offset these expected increases in emissions with increased plant productivity, but the magnitude of these offsets remains highly debated.
- Risks of soil erosion and other degradation processes, with attendant emission of CO_2 and other GHGs, are likely to increase with high demographic pressure and warming climate.
- The frequency and severity of disturbances (fire, dieback due to insects) are expected to change, with the expected increases over the short term in northern ecosystems leading to increase in mortality and shifts in the age-class structure of the existing forests.
- Where increases in the disturbance regime (fire, insects, disease, wind throw) occur due to climate change, forest carbon stocks will be reduced, with large emissions of carbon to the atmosphere.
- In some locales these emissions may be offset by increased uptake stimulated by better growing conditions.
- Methane emissions from wetlands and peatlands are expected to decrease in southern regions but these GHG emission reductions will to some extent be offset by increased CO_2 releases. Moreover, there may be an increase in CH_4 emissions in northern regions associated with higher temperature, longer growing seasons, and permafrost melting.

Over the medium term (one to two centuries), shifts in the distribution of vegetation will lead to a reduction in tree cover as regeneration processes greatly lag those of mortality. During this period, the terrestrial ecosystems will tend to lose C and act as sources. In the longer term (longer than one to two centuries) if a stable, warmer, and wetter climate prevails, the terrestrial ecosystems will favor higher vegetation densities and hence provide the potential for increased C stocks, acting as sinks while this transition takes place. Whether such potential can be achieved is critically dependent on human actions.

It is possible that the current C source–sink relationships for the terrestrial biosphere may remain approximately in balance, especially if the current sinks could be increased. However, a more likely outcome for the foreseeable future is that the biosphere as a whole will become a net source. This will have important implications for the development of strategies to stabilize the concentration of GHGs in the atmosphere.

REFERENCES

1. Apps, M.J., Special Paper: Forests, the global carbon cycle and climate change, in *Forests for the Planet, Proceedings of XII World Forestry Congress*, Quebec City, 21–28 September 2003, 2003, 139.
2. IPCC (Intergovernmental Panel on Climate Change), *Climate Change 2001: The Scientific Basis, Contribution of Working Group I to the Third Assessment Report of the Intergovernmental Panel on Climate Change*, Houghton, J.T., Ding, Y., Griggs, D.J., Noguer, M., van der Linden, P.J., Dai, X., Maskell, K., and Johnson, C.A., Eds., Cambridge University Press, New York, 2001.
3. Reilly, J. et al., Uncertainty and climate change assessments, *Science*, 293, 430, 2001.
4. Cox, P.M. et al., Acceleration of global warming due to carbon-cycle feedbacks in a coupled climate model, *Nature*, 409, 184, 2000.
5. Zwiers, F.W., The 20-year forecast, *Nature*, 416, 690, 2002.
6. Keeling, C.D. and Whorf, T.P., Atmospheric CO_2 records from sites in the SIO air sampling network, in Trends: A Compendium of Data on Global Change, Carbon Dioxide Information Analysis Center, Oak Ridge National Laboratory, U.S. Department of Energy, Oak Ridge, TN, 2004.
7. Schimel, D.S., Terrestrial ecosystems and the carbon cycle, *Global Change Biol.*, 1, 77, 1995.
8. Batjes, N.H., A world data set of derived soil properties by FAO-UNESCO soil unit for global modelling, *Soil Use Manage.*, 13, 9, 1997.
9. Eswaran, H., ven den Berg, E., and Reoch, P., Organic carbon in soils of the world, *Soil Sci. Soc. Am. J.*, 57, 192, 1993.
10. Eswaran, H., Van den Berg, E., and Reich, P., Organic carbon in soils of the world. *Soil Sci. Soc. Am. J.*, 57, 192, 1993
11. Raich, P.B. and Schlesinger, W.H., The global carbon dioxide flux in soil respiration and its relationship to vegetation and climate, *Tellus*, 44B, 81, 1992.
12. Kurz, W.A. and Apps, M.J., A 70-year retrospective analysis of carbon fluxes in the Canadian forest sector, *Ecol. Appl.*, 9, 526, 1999.
13. Kauppi, P.E. et al., Technical and economic potential of options to enhance, maintain and manage biological carbon reservoirs and geo-engineering, in *Climate Change 2001: Mitigation. Contribution of Working Group III to the Third Assessment Report of the Intergovernmental Panel on Climate Change*, Metz, B. Davidson, O., Swart, R., and Pan, J. Eds., Cambridge University Press, Cambridge, 2001, chap. 4.
14. Houghton, R.A., Revised estimates of the annual net flux of carbon to the atmosphere from changes in land use and land management 1850–2000, *Tellus*, 55B(2), 378, 2003.
15. Houghton, R.A., Interannual variability in the global carbon cycle, *J. Geophys. Res.*, 105, 20121, 2000.
16. Falkowski, P. et al., The global carbon cycle: a test of our knowledge of Earth as a system, *Science*, 290, 291, 2000.
17. Sarmiento, J.L. et al., Air-sea CO_2 fluxes and carbon transport: a comparison of three ocean general circulation models, *Global Biogeochem. Cycles*, 14, 1267, 2000.
18. Natural Resources Canada, The State of Canada's Forests 2003–2004, Natural Resources Canada, Ottawa, 2004.
19. Houghton, R.A., 2000. Emissions of carbon from land-use change, in *The Carbon Cycle*, T.M.L. Wigley and D.S. Schimel, Eds., Cambridge University Press, New York, 2000, 63–76.
20. IGBP Terrestrial Carbon Working Group, The terrestrial carbon cycle: implications for the Kyoto Protocol, *Science*, 280, 1393, 1998.

21. Binkley, C.S. et al., Sequestering carbon in natural forests, in Economics of Carbon Sequestration in Forestry, Sedjo, R.A., Sampson, R.N., and Wisniewski, J., Eds., *Crit. Rev. Environ. Sci. Technol.*, 27, 23, 1997.

22. Davidson, C. 1998. Issues in measuring landscape fragmentation. *Wildl. Soc. Bull.*, 26, 32, 1998.

23. Ramankutty, N. and Foley, J.A., Estimating historical changes in land cover: North American croplands from 1850 to 1992, *Global Ecol. Biogeogr.*, 8, 381, 1999.

24. Hobbs, N.T. and Theobald, D.M., Effects of landscape change on wildlife habitat: applying ecological principles and guidelines in the western United States, in *Applying Ecological Principles to Land Management*, V.H. Dale and R.A. Haeuber, Eds., Springer Verlag, New York, 2001, 374.

25. Fitzsimmons, M., Effects of deforestation and reforestation on landscape spatial structure in boreal Saskatchewan, Canada, *Can. J. For. Res.*, 32, 843, 2002.

26. Lal, R., Carbon emission from farm operations, *Environ. Int.*, 30, 981, 2004.

27. Follett, R.F., Soil management concepts and carbon sequestration in cropland soils, *Soil Tillage Res.*, 61, 77, 2001.

28. West, T.O. and Marland, G.A., A synthesis of carbon sequestration, carbon emissions, and net carbon flux in agriculture: comparing tillage practices in the U.S., *Agric. Econ. Environ.*, 91, 217, 2002.

29. Sitaula, B.K., Bakken, L.R., and Abrahamsen, G., N-fertilization and soil acidification effects on N_2O and CO_2 emission from temperate pine forest soil, *Soil Biol. Biochem.*, 27, 1401, 1995.

30. Lal, R., Off-setting global CO_2 emissions by restoration of degraded soils and intensification of world agriculture and forestry, *Land Degradation Dev.*, 14, 309, 2003.

31. Lal, R., Soil erosion and global carbon budget, *Environ. Int.*, 29, 437, 2003.

32. Norby, R.J. et al., Tree responses to rising CO_2 in field experiments: implications for the future forest, *Plant Cell Environ.*, 22, 683, 1999.

33. Luo, Y. et al., Sustainability of terrestrial carbon sequestration: a case study in Duke Forest with inversion approach, *Global Biogeochem. Cycles*, 17, 1021, 2003.

34. Caspersen, J. P. et al., Contributions of land-use history to carbon accumulation in U.S. forests, *Science*, 290, 1148, 2000.

35. Curtis, P.S. and Wang, X.Z., A meta-analysis of elevated CO_2 effects on woody plant mass, form, and physiology, *Oecologia*, 113, 299, 1998.

36. Curtis, P.S., Vogel, C.S., Wang, X., Pregitzer, K.S., Zak, D.R., Lussenhop, J., Kubiske, M., and Teeri, J.A., Gas exchange, leaf nitrogen, and growth efficiency of *Populus tremuloides* in a CO_2-enriched atmosphere, *Ecol. Appl.*, 10, 3, 2000.

37. IPCC (Intergovernmental Panel on Climate Change), *Land Use, Land-Use Change, and Forestry*, Watson, R.T., Novel, I.R., Bolin, N.H., Ravindranath, N.H., Verardo, D.J., and Dokken, D.J., Eds., Cambridge University Press, New York, 2000.

38. Körner, C. and Bazzaz, F.A., Eds., *Community, Population, Evolutionary Responses to Elevated CO_2*, Academic Press, San Diego, CA, 1996.

39. Chen, W., Chen, J., and Cihlar, J., An integrated terrestrial ecosystem carbon-budget model based on changes in disturbance, climate, and atmospheric chemistry, *Ecol. Model.*, 135, 55, 2000.

40. Schimel, D.S. et al. Recent patterns and mechanisms of carbon exchange by terrestrial ecosystems, *Nature*, 414, 169, 2001.

41. Knorr, W., Prentice, I.C., House, J.I., and Holland, E.A., Long-term sensitivity of soil carbon turnover to warming, *Nature*, 433, 298, 2005.

42. Wolfe, D.W. and Erickson, J.D., Carbon dioxide effects on plants: uncertainties and implications for modeling crop response to climate change, in *Agricultural Dimension*, Kaiser, H.K. and Drennen, T.E., Eds., St. Lucie Press, Australia, 1993, 153.

43. Nadelhoffer, K.J. et al., Nitrogen deposition makes a minor contribution to carbon sequestration in temperate forests, *Nature*, 398, 145, 1999.

44. Oren, R. et al., Soil fertility limits carbon sequestration by forest ecosystem in a CO_2-enriched atmosphere, *Nature*, 411, 469, 2001.

45. Schindler, D.W., A dim future for boreal waters and landscapes, *BioScience*, 48, 157, 1998.

46. Munn, R.E. and Maarouf, A.R., Atmospheric issues in Canada, *Sci. Total Environ.*, 203, 1, 1997.

47. McLaughlin, S. and Percy, K., Forest health in North America: some perspectives on actual and potential roles of climate and air pollution, *Water Air Soil Pollut.*, 116, 151, 1999.

48. Percy, K. et al., State of science and knowledge gaps with respect to air pollution impacts on forests: reports from concurrent IUFRO 7.04.00 working party sessions, *Water Air Soil Pollut.*, 116, 443, 1999.

49. Isebrands, J.G. et al., Interacting effects of multiple stresses on growth and physiological processes in northern forests, in *Responses of Northern U.S. Forests to Environmental Change*, Ecological Studies 139, Mickler, R.E., Birdsey, R.A., and. Hom, J., Eds., Springer-Verlag, Berlin, 2000, 149.

50. Himes, F.L., Nitrogen, sulfur and phosphorus and the sequestering of carbon, in *Soil Processes and the Carbon Cycle*, Lal, R., Kimble, J.M., Follett, R.F., and Stewart, B.A., Eds., CRC Press, Boca Raton, FL, 1998, 315.

51. Jenkinson, D.S., Adams, D.E., and Wild, A., Model estimate of CO_2 emissions from soil in response to global warming, *Nature*, 351, 304, 1991.

52. Glendining, M.T. and Powlson, D.S., The effect of long continued application of inorganic nitrogen fertilizer on soil organic nitrogen — a review, in *Soil Management — Experimental Basis for Sustainability and Environment Quality*, Lal, R. and Stewart, B.A., Eds., CRC Press, Boca Raton, FL, 1995, 385.

53. Waleing, D.E. and Webb, B.W., Erosion and sediment yield: a global overview, in *Erosion and Sediment Yield: Global and Regional Perspectives, Proc. Exeter Symp.*, July 1996, 1996.

54. Degens, E.T., Kempe, S., and Richey, J.E., Biochemistry of major world rivers, in *Biochemistry of Major World Rivers*, Degens, E.T., Kempe, S., and Richey, J.E., Eds., John Wiley & Sons, Chichester, U.K., 1991, 323.

55. Stallard, R.F., Terrestrial sedimentation and carbon cycle: coupling weathering and erosion to carbon burial, *Global Biogeochem. Cycles*, 12, 231, 1998.

56. Smith, S.V. et al., Budgets of soil erosion and deposition for sedimentary organic carbon across the conterminous U.S., *Global Biogeochem. Cycles*, 15, 697, 2001.

57. Janzen, H.H. et al., Management effects on soil C storage on the Canadian prairies, *Soil Till. Res.*, 47, 181, 1998.

58. Duchemin, E. et al., Production of the greenhouse gases CH_4 and CO_2 by hydroelectric reservoirs in the boreal region, *Global Biogeochem. Cycles*, 9, 529, 1995.

59. Kelly, C.A. et al., Increases in fluxes of greenhouse gases and methyl mercury following flooding of an experimental reservoir, *Environ. Sci. Technol.*, 31, 1334, 1997.

60. Turetsky, M.R. et al., Current disturbance and the diminishing peatland C sink, *Geophys. Res. Lett.*, 29, 1526, 2002.

61. Canadian Wetlands Conservation Task Force, Wetlands: A Celebration of Life, Issue Paper no. 193-1, North American Wetlands Conservation Council (Canada), Ottawa, Ontario, 1993.

62. Lal, R.T. et al., *Soil Degradation in the U.S.*, CRC/Lewis Publishers, Boca Raton, FL, 2003, 209.

63. Rubec, C.D.A., The status of peatland resources in Canada, in *Global Peat Resources*, Lappalainen, E., Ed., International Peat Society, Saarijärvi, Finland, 1996, 243.

64. Armentango, T.W. and Menges, E.S., Patterns of change in the carbon balance of organic soil-wetlands of the temperate zone, *J. Ecol.*, 74, 755, 1986.

65. Roulet, N. et al., Methane flux from drained northern peatlands: effect of a persistent water table lowering on flux, *Global Biogeochem. Cycles*, 7, 749, 1993.

66. Funk, D.W. et al., Influence of water table on carbon dioxide, carbon monoxide, and methane fluxes from taiga bog microcosms, *Global Biogeochem. Cycles*, 8, 271, 1994.

67. Alm, J. et al., Carbon balance of a boreal bog during a year with an exceptionally dry summer, *Ecology*, 80, 161, 1999.

68. Waddington, J.M, Warner, K.D., and Kennedy, G.W., Cut-over peatlands: a persistent source of atmospheric CO_2, *Global Biogeochem. Cycles*, 16, 10.1029/2001G B001398, 2002.

69. Gorham, E., Northern peatlands: role in the carbon cycle and probable responses to climatic warming, *Ecol. Appl.*, 1, 182, 1991.

70. Kasimir-Klemedtsson, E. et al., Greenhouse gas emissions from farmed organic soils, a review, *Soil Use Manage.*, 13, 2245, 1997.

71. Freeman, C., Lock, M.A., and Reynolds, B., Fluxes of CO_2, CH_4, and N_2O from a Welsh peatland following simulation of water table draw-down: potential feedback to climate change, *Biogeochemistry*, 19, 51, 1993.

72. Regina, K. et al., Emissions of N_2O and NO and net nitrogen mineralization in a boreal forested peatland treated with different nitrogen compounds, *Can. J. For. Res.*, 28, 132, 1998.

73. Environment Canada, Wetlands in Canada: A Valuable Resource, Fact Sheet 86-4, Lands Directorate, Ottawa, Ontario, 1986.

74. Schlesinger, W.H., *Biogeochemistry: An Analysis of Global Change*, 2nd ed., Academic Press, New York, 1997, 588.

75. Euliss, N.H., Jr., Olness, A., and Gleason, R.A., Organic carbon in soils of prairie wetlands in the United States, presented at The Carbon Sequestration Workshop, Oak Hammock Marsh, Manitoba, April 19–20, 1999.

76. Bhatti, J.S., van Kooten, G.C., Apps, M.J., Laird, L.D., Campbell, I.D., Campbell, C., Turetsky, M.R., Yu, Z., and Banfield, E., Carbon balance and climate change in boreal forests, in *Towards Sustainable Management of the Boreal Forest*, Burton, P.J., Messier, C., Smith, D.W., and Adamowicz, W.L., Eds., NRC Research Press, National Research Council of Canada, Ottawa, 2003, 799–855.

5 Plant/Soil Interface and Climate Change: Carbon Sequestration from the Production Perspective

G. Hoogenboom

CONTENTS

5.1 INTRODUCTION

Agricultural production systems are very complex and have to deal with the dynamic interaction of living organisms that are controlled by their inherent genetics and both the edaphic and atmospheric environment. In addition, the human component of the agricultural production system has the potential to manage crops and livestock at various levels. A rangeland system with free roaming animals does not require the intensive management that is required in a greenhouse production system, where vegetables and flowers are raised with both the edaphic and atmospheric environment controlled. It is this range of components of the agricultural production system that is exposed to climate change and climate variability and where the managers of these productions systems have to handle decisions for mitigation, adaptation, and reductions in risks and uncertainty.

With respect to climate change, agriculture is considered both to be the cause of climate change and to be affected by climate change.[1] Even for low-input systems, such as the rangeland system mentioned previously, agricultural production, including both crop and livestock systems, requires inputs. Inputs for both the extensive and intensive systems include fertilizer, irrigation, and chemicals for crop production, and shelter, feed, and water for animals. Most of these inputs require energy during their production process, such as oil and other resources that are used for the production of fertilizers and chemicals, for transportation from the factory to the farm, and during the application process, such as the operation of the pump for irrigation applications or the use of a tractor for the application of fertilizers and pesticides. In all these cases the use of energy in the form of fossil fuels causes the release of CO_2 and other pollutants into the atmosphere. In addition, because agriculture involves natural processes, there is also release of other trace gases, such as nitrous oxides (NO_x) that are part of the natural soil nitrogen transformation processes,[2-4] or methane (CH_4) emission from flooded rice production systems.[5-7] The former is discussed in Chapter 4, while the latter is not really an issue for the temperate climate of Canada, which does not allow for the production of tropical crops, except under controlled conditions. The trace gases that are produced or released by livestock systems are discussed in Chapters 12 and 13.

Animals play an important role in the agricultural system. They are a critical component of the food chain in the form of meat, eggs, and milk, and other processed animal products. As a source of food for humans, animals require feed as either raw or processed plant material. In addition, animals can play a critical role for animal traction and they are considered as capital in developing countries. The proper handling of animal manure is an issue that is a concern for both developed and developing countries, specifically with respect to climate change, due to the volatile nature of some of manure compounds and the release of trace gases that affect the atmosphere[8-10] and with respect to water quality where nitrogen (N), phosphorus (P), and microbial contamination are of concern.[11] However, from the cropping system perspective animal manure is considered to be beneficial, as it adds valuable organic matter to the soil and improves overall soil quality. These issues are discussed in other chapters while this chapter mainly addresses the interaction of the crop with the atmosphere, the impact of climate change on crop production, and the potential role of crops for carbon sequestration.

5.2 SOIL–PLANT–ATMOSPHERE AND CLIMATE CHANGE

5.2.1 PRECIPITATION

The plant as a living organism is extremely vulnerable to its environment. The plant uses the soil as its main source for water to replace the water that is lost through transpiration, a process required for evaporative cooling of the plant due to the absorption of radiation during the daytime. More than 90% of the plant consists of water. Although plant tissue has some buffering capacity, wilting can occur rather quickly if the water lost through transpiration is not rapidly replaced with water uptake by the root system. Any form of drought stress will affect most of the growth and development processes in the plant, such as elongation and expansion, and will cause stomatal closure, resulting in a reduction in photosynthesis. Water uptake also allows the plant to extract nutrients required for growth of plant tissue, including the production of proteins, lipids, organic acids, and other components. The roots provide the plant with an anchor system to support its canopy for optimal exposure to solar radiation and to protect against wind damage and other atmospheric processes.[12]

The atmospheric component of the soil–plant–atmosphere system is the main cause of the vulnerability of plants to local weather conditions. Most of the agricultural production systems across the world, including Canada and the U.S., are rainfed systems. Precipitation, including rainfall and snow, is extremely variable, both temporarily from day to day and from one year to the next, as well as spatially from one location to another location, sometimes even within a farmer's field.[13,14] Climate normals are based on the average of 30 years of daily weather data and normally do not show much change.[15] However, both the temporal and spatial variability of precipitation are of major concern to farmers and producers. Most of the variability in crop production for rainfed systems can be explained by the variability in rainfall.[16,17]

One issue that in some cases is not extensively addressed in climate change deliberations is precipitation. As stated earlier, most of the agricultural production systems across the world are rainfed systems, with precipitation as the only source of water for growing a crop. Even if both the CO_2 and the local temperature increase are beneficial to the growth and development of a crop, but water is not available due to changes in the climate or weather and climate variability, then the ultimate impact can be crop failure and an economic loss to the farmer. Although climatologists normally refer to total annual precipitation, what is critical for optimal crop growth is an even distribution of rainfall during the entire growing season in amounts that replace the water lost by soil evaporation and transpiration on a regular basis. It is expected that climate change will cause alterations in the duration of the rainy seasons, the occurrence and frequency of drought spells, both short term and long term, and other extreme events,[18] which all potentially can have a negative impact on overall crop growth and development and ultimately crop yield.[19] However, these predictions for future climate vary, depending on the climate change scenario and the particular model that is used.

5.2.2 TEMPERATURE

Climate zones are characterized by local precipitation and temperature conditions, ranging from arid to humid with respect to precipitation, and artic to tropical, with respect to temperature. Although water is a necessary requirement for all plant growth, it is the temperature that determines the main crops or species that can be grown in a region. All crops have a typical temperature response that defines the minimum and maximum temperatures that limit plant growth as well as an optimum temperature for maximum growth. Although, in general, all plants have similar biochemical processes that define photosynthesis, respiration, partitioning, growth, development, water uptake, and transpiration, each process has a unique temperature response that shows the adaptation of a plant to its environment.[20] For instance, citrus crops normally do not grow in Canada, as the temperatures during the winter months are too low. Rapeseed or canola grows very well in Canada but is normally not grown in other regions of North America. Some horticultural crops in the southeastern U.S. are planted at staggered planting dates, with the earliest planting in Florida, followed by Georgia, South Carolina, North Carolina, etc. In this case the growers are trying to benefit from the optimum temperatures during a special period of the spring season that provides the best growth and development.

Development is a key component of crop growth, defining how quickly a plant moves from one reproductive phase to the next phase, and it ultimately determines the total length of the growing season from planting to harvest. For example, temperature is the main factor that determines the number of days to flowering and the number of days to physiological and harvest maturity. The former can affect the time required for total canopy closure that is needed for optimum biomass production, while the latter determines the total grain filling duration required to obtain maximum yield. For certain crops, such as winter wheat and fruits, temperature can also affect early development through vernalization. This process basically prohibits the plant from developing too fast if it is exposed to favorable conditions early during the growing season, such as a fall planting for wheat. Although a longer growing season, in general, increases yield potential, there are certain risks associated with long growing seasons, such as early frost in temperate climates, the start of the dry season in semi-arid environments, or adverse weather conditions such as hail, hurricanes, tornadoes, and drought. Most crops have a critical or base temperature below which no development occurs. When the temperature increases above this temperature, the crop's development rate is normally a function of the difference between the current temperature and the base temperature, sometimes referred to as degree-days. Most crops also have an optimum or cardinal temperature, above which the development rate does not increase further. Again, this optimum temperature and its range vary from species to species. It has also been found that at very high temperatures development might actually slow, mainly due to the adverse effect on most of the plant's biochemical processes. The high temperatures that are predicted as a consequence of climate change for some of the subtropical and tropical regions are of concern, especially if they are in the range that can have a negative impact on crop growth and development.

5.2.3 SOLAR RADIATION

The sun is the ultimate energy source for all atmospheric processes.[21,22] Solar radiation is also the main energy input factor that ultimately determines plant growth and biomass production. The photosynthesis process creates carbohydrates that are distributed to the various plant components, resulting in the growth of leaf, stem, root, and reproductive components, such as ears, heads, and pods. Most crops show an asymptotic response to solar radiation that reaches a plateau at high light levels due to certain limitations of the biochemical processes that are associated with photosynthesis.

Solar radiation is a combination of intensity and duration due to the dynamic nature of the solar system. Sunrise and sunset slowly change each day, depending on the season and location, and determine the duration of daylight hours. At solar noon the plant is normally exposed to the highest amount of solar radiation, especially under clear skies, but this period normally lasts only for a few hours at most. As the sun moves through the sky, the plant adapts to this change in solar radiation intensity and, in some cases, leaves track the sun to optimize the reception of direct sunlight. The combination of total daylight hours and instantaneous light intensity determines the total amount of solar radiation that a plant is exposed to on a daily basis and determines the daily amount of carbohydrates produced by the photosynthesis process.

In addition to the total solar energy and light intensity, plants also respond to day length through their vegetative and reproductive development processes. Day length is normally defined as the period from sunrise to sunset, although plants can also be sensitive to the twilight period prior to sunrise and after sunset. Crops can be characterized as short-day, long-day, or day-neutral plants. Short-day plants show a delay in reproductive development when the day length exceeds a certain threshold, normally around 12 hours, while long-day plants show a delay in development when the day length drops below the threshold day length, also normally around 12 hours. In general, day-neutral plants will flower under any day length condition. Plants that are photoperiod sensitive cannot necessarily be moved to a different region where temperatures are more favorable, as the change in photoperiod could adversely affect vegetative and reproductive development. For example, some varieties of barley are very sensitive to long day lengths. When grown under long days of the north they reach maturity very quickly and have poor yields, while grown under the short-day conditions of an Australian winter these varieties remain vegetative for a long period and are high yielding.

The impact of climate change on solar radiation is rarely discussed.[19] Any changes in precipitation will also directly affect solar radiation because of changes in cloud cover. For certain regions it is expected or predicted that precipitation might increase, causing a decrease in solar radiation. Depending on the timing during the growing season and the location, this could also affect potential photosynthesis and biomass production, especially for the higher latitudes where solar radiation is sometimes limiting.

5.2.4 CARBON DIOXIDE

Carbon dioxide is the main atmospheric component that is absorbed by the plant as part of the photosynthesis process and forms the basic building block for the production of carbohydrates. Crops are categorized as either C_3 or C_4 crops, depending on the biochemical pathways of the photosynthesis process. Some of the tropical grasses and cereals, including maize, sorghum, and millet, are considered C_4 crops, while the more temperate crops, including wheat, barley, and soybean, are considered C_3 crops. In general, C_3 plants are more responsive to an increase in CO_2 levels than C_4 crops.

It is a well-known fact that the CO_2 concentration in the atmosphere has slowly increased from 320 ppm in 1960 to 380 ppm in 2004, as recorded at the Mauna Loa Observatory in Hawaii.[23] The increase in CO_2 in itself is beneficial to agriculture, as it acts like a fertilizer and enhances photosynthesis and plant growth. Some of the increases in yield that have been observed by national agricultural statistic services are partially due to the increase in CO_2, in addition to advances in agricultural technology.[24]

5.2.5 INTERACTION

Why is it important to understand these basic processes that undergird plant growth and development? Climate change is expected to affect local weather conditions and especially their variability. Any modification of the weather conditions will directly affect plant growth and development and ultimately agricultural production. In most cases when farmers state that they had either a good or bad year, this is mainly due to the weather conditions that were different during the past growing season when compared to previous growing seasons, e.g., the season was dryer than normal, or colder than normal, or the temperature was near optimal for growth and development. Some of the changes in weather conditions can have a positive effect on plant growth and development, while others can be negative. The overall impact is a function of when these weather conditions occur during the life cycle of a plant and the intensity of these conditions. Because of the dynamic nature of plants, they will immediately respond to any changes in weather conditions, caused either by the natural temporal and spatial variability in weather conditions or by the more permanent changes in weather conditions caused by climate change. However, plants are more affected by changes in extremes than changes in average conditions, as most of the processes that control plant growth and development are nonlinear. Exceptions include disasters, such as changes in the timing of the first or last frost date, which can immediately destroy a crop, a hail storm, or changes in the frequency and intensity of precipitation, which will also affect plant growth.

5.3 CARBON SEQUESTRATION

5.3.1 PHOTOSYNTHESIS

As a consequence of the inherent nature of the photosynthesis process in which ambient CO_2 is used to create sugars and carbohydrates, plants sequester carbon.

Because of these unique characteristics, plants are the main living organisms on Earth that have the capacity to mitigate the increase in CO_2 concentration in the atmosphere. One should also remember that plants are the main source of oxygen, as it is one of the products of photosynthesis. Humans and animals need oxygen on a continuous basis in order to survive. Plants that can potentially contribute to carbon sequestration through photosynthesis are associated with most of the eco-systems that can be found around the world, including the plankton that lives in the ocean, the natural vegetation of all undisturbed ecosystems, the crops that we grow as part of our agricultural production systems, and the trees of pristine and managed forests.[25] During the growth process of any of these organisms, carbon is being sequestered. One could then pose the question: why not grow more crops or grow more trees to mitigate climate change through carbon sequestration? Unfortunately this solution is not that simple. Trees normally grow very slowly. Although the potential to sequester carbon is fairly large, the actual carbon seques-tration rate on an annual basis is very small, especially for the temperate climate found in Canada and similar climatic zones. Chapter 9 discusses the impact of climate change on forestry in more detail. Unfortunately in some areas of the world the reverse of carbon sequestration is currently occurring through defores-tation. Trees are being removed and burned to create land for agricultural produc-tion, such as in the Brazilian Amazon. During this burning process CO_2 that was originally sequestered by the trees during their photosynthesis and growth process is released back into the atmosphere.[26]

5.3.2 CROP BIOMASS

Agricultural crops grow much faster than trees. However, due to their inherent role in the food chain, most of the biomass that is produced does not contribute to permanent carbon sequestration. For most of the agronomic crops the economic yield consists of grains. The grains are either processed as feed for consumption by livestock or as food products for human consumption. As soon as these products are consumed, most of the CO_2 is released due to the animal and human digestion and respiration processes. The remaining carbohydrates and other by-products are released as human and animal excreta in the form of urine and feces. In many ancient Asian societies the human excreta were considered a valuable resource and human waste was recycled into cropland as organic fertilizer, sometimes referred to as night soil. In most modern societies waste is treated in sewage plants. During the treatment of human waste in sewage plants the potential carbon sequestration of crops ends, as all the CO_2 that was originally sequestered by the crop is released again. Human consumption of crop products, therefore, does not add much to the potential for carbon sequestration. One could potentially consider the carbon that is sequestered in the human population growth in general and especially of overweight people, but this is relatively minor. Most of the food that we eat is lost again through our metabolic processes. However, there is scope to capture the gases that are released during the composting and sewage process and to use the biogas as an alternative energy source, thereby mitigating the effect of CO_2 released into the atmosphere by burning of the traditional fossil fuels.[27,28]

In addition to the seeds or grains, plants also produce large amounts of vegetative biomass that mainly consists of carbohydrates and related components. There are various options that farmers have for using this biomass. The by-products can be harvested in the form of straw or fodder, which basically means that the plant biomass is removed from the field, or they can be kept on the field to help improve the overall soil quality. If the straw or fodder is harvested, it has an economic value and can be used as feed for animals, as a source for more permanent products, such as paper and carton, as a source for biofuels, and various other applications. As feed for livestock plant biomass basically follows the same transformation process as the use of grains for animal feed. Upon consumption of biomass by the animals, some CO_2 is released into the atmosphere during the digestion process, while the remaining carbon is lost through manure. If the manure is ultimately returned to the fields that are being used for crop production, there is potential benefit for soil improvement and carbon sequestration through soil organic matter, which can be a relatively large sink for carbon.[29,30] The use of crop biomass for other products also leads to short- and long-term carbon sequestration, although the potential benefits are still unclear. In pasture systems all biomass is either directly consumed by livestock or harvested as hay and provided to the animals as feed at a later date. The process of carbon transformations is similar to the one described previously for crop biomass of grain cereals and other agricultural crops. Chapter 8 discusses some of the issues associated with the impact of climate change on pasture systems.

Some might state that the use of biofuels is ultimately beneficial to the environment. However, one needs to carefully analyze the complete production system and the impact on the total environment, not just the positive impact on air pollution due to a reduction in the burning of fossil fuels. The use of biofuels is indeed a cleaner technology when compared to the use of fossil fuels. In addition, there are also some strong political and economic benefits. It is important to note that the production of crops such as maize or sugarcane for biofuels does require inputs, especially fertilizers. In most cases inorganic fertilizers are being used, which in turn require fossil fuels during their production process. The expected net gain in carbon sequestration and energy use could actually be a net loss, depending on the quantity and quality of the inputs and outputs of the overall system. In addition, there is a significant negative impact on the overall edaphic system, as all biomass, except for the roots, is removed from the field and could cause potential soil degradation through erosion if not managed well by the farmer. Chapter 11 discusses additional issues associated with biomass and energy.

5.3.3 Roots

One potential plant component that is often ignored in the topic of carbon sequestration is the root system and other associated belowground components of the plant such as the nodules of grain legumes. It was stated previously how important plant roots are for water uptake and nutrient supply for overall plant growth and plant health. Crops can partition a relatively large part of their biomass to the root system to support these activities. For most crops the belowground components are not harvested, except for a few root and tuber crops such as potato, cassava, and aroids.

Upon harvest of the aboveground components, the roots are left in the soil and thereby become a potential source for carbon sequestration that can be up to 10 to 25% of the total aboveground biomass. Bolinder et al.[31] estimated for winter wheat that 17% of the biomass was in the roots, for oats 29% of the biomass was in the roots, and for barley 33% of the biomass was in the roots.

Any plant material that is left on the field or in the soil after final harvest, including roots, leaves, stems, and other plant components, becomes part of the organic residue material of the soil surface and soil profile system. In addition, animal manure can be returned to the field, adding to the total organic material that is available as organic fertilizer. Through the microbiological processes this material is slowly decomposed into different components, including NO_3 and NH_4. Depending on the rate of these transformation processes, which are not only controlled by environmental conditions such as soil temperature, soil moisture, oxygen, and pH, but also by the presence and composition of the microbes, some carbon is permanently stored through carbon sequestration while the remainder is released back into the atmosphere as CO_2 or CH_4. These processes are discussed in detail in Chapter 12 on ruminant contributions to methane and global warming. However, these dynamic organic matter transformation processes ultimately determine the potential for carbon sequestration of the agricultural production system. A detailed review of the potential of U.S. cropland and grazing lands to sequester carbon and mitigate the greenhouse effect is provided by Lal et al.[32] and Follett et al.[33]

5.4 UNCERTAINTY IN MEASUREMENT OF CLIMATE CHANGE EFFECTS

The issue of climate change is, in some cases, still somewhat controversial. Many people, especially the popular press, associate climate change with global warming. In 2003, the *Daily Telegraph* (London) referred to feast and famine as global warming scorched farms across Europe. Some of the weather changes that we have experienced during the last few years are due to climate variability and some changes are due to climate change.

The change in temperature, sometimes referred to as "global warming," needs to be analyzed carefully, including, for instance, the changes that have been observed for many locations in Canada.[34-36] A recent study found some interesting differences between the weather experienced in Quebec between 1742 and 1756 and the current climate.[37] The summers and winters appeared to have been milder than most of the 20th century, except for a few periods, while the springs and autumns were cooler. This resulted in shorter growing seasons when compared to the 20th century. Many reporting weather stations have recorded a long-term increase in temperature, while others have reported a long-term decrease in temperature. For example, in the southeastern U.S. it is well known that the temperature has decreased during the last century, rather than increased.[38] Although it is indeed true that the temperatures at most of the main reporting weather stations have increased, one should carefully study the environment where these observations have been recorded. Many of these stations are located at airports where buildings, runways, and the tarmac have greatly affected the local environment. In addition, the heat island effect of major cities is well known, as buildings hold heat better than the surrounding environment. In the U.S., the National Weather Service has found that many of the weather stations of the Cooperative Weather Network have siting problems due to changes in the local environment, especially trees and shrubs. Many of the long-term temperature and rainfall records, which sometimes span more than a century, are based on these stations. In many cases this change in local conditions is unknown or not reported in the meta-data of each station.[39,40] One should keep in mind that for some of the temperate climates, such as for Canada, a 1° decrease in temperature can have a much more devastating impact on agriculture than a 1° increase.

As a consequence of the interest of many government agencies and nongovernmental organizations in the potential impact of climate change on the various economic sectors, including agriculture and management ecosystems, the issue of climate change has been studied extensively.[41–48] A quick literature search on the Internet located hundreds of scientific papers published during the last 10 to 15 years on the impact of climate change on agriculture and water resources, as well as on carbon sequestration. However, determining the impact of climate change on agriculture in general or more specifically on a particular crop or livestock system is somewhat difficult due to the uncertainty associated with climate change, especially the predictions and future projections of the General Circulation Models or Global Climate Models (GCMs). There is even more uncertainty for the predictions at a regional scale, which are very important for agricultural impact studies.[49,50]

In traditional agronomic research, experiments are based on a set of fixed changes to inputs and associated factors, such as planting date, fertilizer application rate or date, and variety or cultivar. These factors are varied at different levels and the response of the crop to these changes is determined through improvement in yield and yield components. The combination of input factors that provides the highest yield or, more appropriately, the highest gross margin or economic return, is normally recommended to the farmer and disseminated through agrotechnology transfer. Unfortunately, climate change predictions by the current GCMs cover a wide range.[49,51] In most cases an ensemble of predictions is used, rather than single predictions to deal with the uncertainty in these predictions.[52–55] As the GCMs improve with scientific advancements, the predictions should also change and one hopes improve to provide a more realistic climate prediction that can be used for impact assessment studies.

5.4.1 CONTROLLED ENVIRONMENTS

Climate change deals with uncertainty in changes in weather and climate, including CO_2 concentration, temperature, precipitation, and solar radiation. It is rather difficult to impose these conditions under normal field experiments, as it requires a modification to the local environment. Traditionally agriculture has modified the environment to optimize plant growth and development and increase yield, including both the soil and aerial environment.[56] In the past most of the temperature impact studies have been conducted in greenhouses and growth chambers. However, some of the limitations of these environmental conditions are that the soil system is artificial and that most of the plants are grown in pots, causing them to become root bound.[57] Growth chambers do have an advantage in that temperature, light, CO_2, and in some cases humidity and dewpoint temperature can be tightly controlled. In addition, one can conduct studies that determine the interactive effects of changes in temperature, CO_2, and other atmospheric factors if an adequate number of growth chambers are available.

5.4.2 Sunlit Chambers

Sunlit chambers have been developed for growing plants outdoors to circumvent some of the issues associated with growing plants in containers. For many of these sunlit chambers one or more factors can be controlled, including temperature and the ambient CO_2 concentration, but the heating and cooling requirements as well as the control systems are quite elaborate. One of the main objectives of these chambers is to be able to grow plants outdoors in the local soil to allow the roots to grow naturally. They are therefore referred to as Soil-Plant-Atmosphere Research (SPAR) units.[58,59] Unfortunately, even the SPAR units do not provide much control of the belowground environment, a factor often ignored in climate change studies. However, a well-designed SPAR unit that is airtight does have the capability to measure the net fluxes of CO_2[60] and determine the potential of carbon sequestration for the soil–plant system, as one can determine the exact amount of carbon that has been sequestered by the plants in either aboveground biomass or the roots. Biosphere 2 is an example of a large-scale controlled environment system.[61–64] Unfortunately, the operational costs were too high to maintain it as either a research or commercial facility.

An example of a SPAR unit is shown in Figure 5.1. This system is part of the Georgia Envirotron facility.[65] These are large SPAR units, measuring 2×2 m, and they provide control of air temperature, dewpoint temperature or relative humidity, and CO_2 levels.[66] The control of temperature and humidity in these chambers is better and more uniform than in indoor chambers, despite the rapid changes caused by the external variation in sunlight and temperature.[67] The units were designed to be portable in order to be able to measure the impact of climate change in farmers' fields. Similar units, although not portable, are also in operation at the University of Florida, Mississippi State University, and other locations across the world.[68,69] The SPAR units have been used to study the impact of climate change, especially increases in temperature and CO_2, on a wide range of crops, including cotton, rice, and soybean.[70–72] These units are able to measure gas exchange, including net photosynthesis and evapotranspiration, and can be used to determine a complete mass balance for both water and carbon. However, observations in SPAR units are restricted to nondestructive measurements, such as vegetative and reproductive development, canopy height, number of leaves, and reproductive structures. SPAR units require a large amount of resources for operation, including both capital as well as human resources.

Open top chambers are an alternative to SPAR units. However, they provide less control of the atmospheric environment, especially temperature, relative humidity, and solar radiation, but they can be used to expose small plot-grown plants to

FIGURE 5.1 Sunlit growth chamber with complete control of air temperature, relative humidity or dewpoint temperature, and CO_2 concentrations above ambient.

different levels of CO_2 and other trace gases.[73-78] Especially ozone (O_3), a trace gas associated with climate change due to anthropogenic changes and air pollution, is known to have a negative impact on plant growth and development, leading ultimately to a decrease in crop production.[75,79-82]

5.4.3 FREE-AIR CO_2 ENRICHMENT

A research facility that has been developed to specifically determine the impact of the increase in ambient CO_2 on crops under field conditions is the Free-Air CO_2 Enrichment (FACE) facility.[83-85] Plants are grown outdoors in a regular field, normally under less than ideal conditions such as those found in a farmer's field, and artificial CO_2 enrichment is applied to determine the "true" interaction between the soil–plant–atmosphere system and the increase in CO_2 concentration. One of the first FACE facilities for agriculture was developed at the research facility of the USDA-ARS in Phoenix, AZ. Crops that have been studied include cotton, sorghum, and wheat.[86-88] One FACE facility has recently been developed at the University of Illinois to study the interaction of changes in both ambient CO_2 and O_3 concentrations.

5.4.4 EXPERIMENTAL CASE STUDY

An example of an experimental climate change impact study conducted in the Georgia Envirotron is shown in Figure 5.2.[65] The main goal of this experiment was to determine the impact of an increase in ambient CO_2 concentration and temperature on biomass production for maize, a C_4 crop, and soybean and peanut, C_3 crops. To

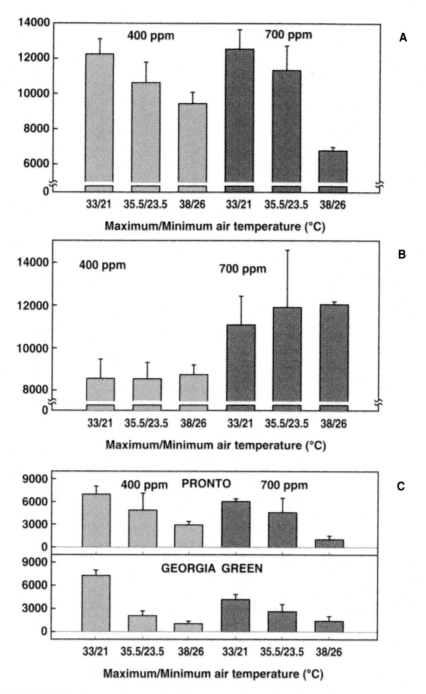

FIGURE 5.2 The impact of an increase in temperature and ambient CO_2 concentration on total aboveground dry matter for maize at beginning of grain filling at 63 days after sowing (A), for soybean at beginning of pod at 68 days after sowing (B), and for final pod yield of peanut for the cultivars Pronto and Georgia Green at harvest maturity (C).

define the base temperature, we used the typical summer weather data from Camilla, GA, which was 33°C for the maximum temperature and 21°C for the minimum temperature. Ambient CO_2 was set at 400 ppm. We then increased both the maximum and minimum temperature by 2.5 and 5°C and the CO_2 level to 700 ppm, resulting in a total of six different treatments, i.e., three temperature levels and two CO_2 levels. It is interesting to see the difference in response of these three crops to the increased temperature treatments. Total biomass for maize at the start of grain filling, which was observed at 63 days after sowing, decreased with an increase in temperature, while the impact of the increase in CO_2 was minimal. Soybean did not show any significant differences between the three temperature combinations at the start of pod filling, which was observed at 68 days after sowing, although the total biomass at +5°C was slightly higher than the other two combinations. However, there was a significant increase in total biomass when the CO_2 concentration increased from 400 to 700 ppm. It is important to note that these results only show the impact on potential carbon sequestration by soybean and maize for these conditions, not the impact on yield and associated harvest factors.

The impact on final pod yield of two peanut cultivars, e.g., Pronto and Georgia Green, is shown in Figure 5.2C. Peanut showed a high sensitivity to the high temperature combinations that were used in this study, as shown by a more than 50% decrease in pod yield for the temperature combination 38°C/25°C. Surprisingly, pod yield for the 700 ppm treatment of both cultivars was lower than for the ambient concentration of 400 ppm for the control temperature combination of 33°C/21°C. Yield was higher for Georgia Green for the +2.5 and +5°C temperature and 700 ppm treatments and the same for Pronto for the +2.5°C temperature, but less for the +5°C temperature and 700 ppm treatment. Any increase in temperature in Georgia due to climate change could reduce potential peanut yield, even if the ambient CO_2 concentration continues to increase.

5.4.5 CROP SIMULATION MODELS

As a result of the fairly artificial nature of experimental studies of climate change and the impact on crop growth, development, and yield, a more comprehensive approach is needed. Crop simulation models integrate the current scientific knowledge of many different disciplines, including not only crop physiology, but also plant breeding, agronomy, agrometeorology, soil physics, soil chemistry, soil microbiology, plant pathology, entomology, economics, and various others.[89] A computer model is a mathematical representation of a real-world system. Crop simulation models can, therefore, predict growth, development, and yield of many different crops as a function of soil and weather conditions, crop management, and genetic coefficients (Figure 5.3). Simulation models have been developed for most of the major agronomic crops, including wheat, rice, maize, sorghum, millet, soybean, peanut, and cotton.[90] The Decision Support System for Agrotechnology Transfer (DSSAT) Version 4.0 includes computer models for more than 20 different crops.[91,92] Other well-known models include the Erosion Productivity Impact Calculator (EPIC[93–96]), the Agricultural Production Systems sIMulator (APSIM[97,98]), Simulateur mulTIdisciplinaire pour les Cultures Standard (STICS[99,100]), and *ecosys*.[101–103] Sim-

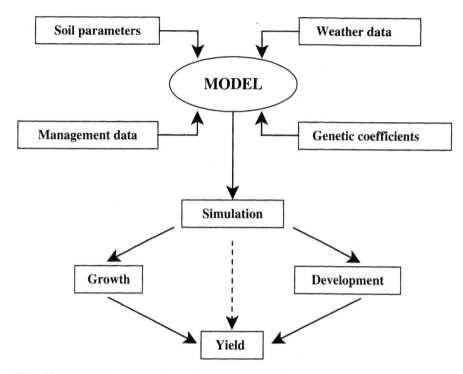

FIGURE 5.3 The importance of weather parameters, soil conditions, crop management, and genetic coefficients on the simulation of crop growth and development.

ulation models have also been developed for rangeland and pasture systems.[104–106] Most crop simulation models operate on a daily time step and simulate processes such as vegetative and reproductive development, photosynthesis, respiration, and biomass partitioning, soil evaporation, transpiration, and root water uptake, and the soil and plant nitrogen processes.[90,91,107–111] The crop models use daily weather data, including solar radiation, precipitation, and maximum and minimum temperature, as input in order to be able to simulate crop responses to local weather and climate conditions.[112,113]

The potential impact of climate change on crop production can only be determined with crop simulation models due to the uncertainty associated with climate change, especially the long-term implications of changes in our local climate.[114] The crop models can use the estimates for the changes in atmospheric conditions and how these changes influence temperature, precipitation, and other local weather variables provided by the GCMs as input.[115–118] Crop simulation models also allow for the evaluation of different "What-If" type scenarios for agricultural management practices, such as crop and cultivar selection, optimum planting dates, and fertilizer and irrigation management,[119–122] as well the interaction with local weather conditions.[17] Recent improvements in crop simulation models have allowed for a more accurate simulation of the soil carbon balance, a key issue when studying carbon sequestration.[102,123–125] Ultimately the models can be used to determine potential strategies for adaptation and mitigation.[126–132]

When studying climate change, carbon sequestration, and policies for mitigating climate change, it is important to consider the socioeconomic aspects of the agricultural system, especially the local farmer. Farmers have had a long history of coping with the variability in local weather conditions and the economic risks associated with their management decisions. The early climate change studies did not explicitly deal with adaptations that farmers might apply due to climate change;[133] sometimes these are referred to as the "dumb farmer" studies. Although farmers traditionally are risk averse, they adapt to changes in their local environment and modify their cropping practice when needed, such as crop or cultivar selection, planting date, and other management decisions, if they think that it can improve their overall operation and long-term economic sustainability.[134] In some cases farmers have been ahead with respect to the adoption of new technologies that cope with changes in the environment when compared to researchers and their scientific advancements. An example is the adoption of yield monitors as part of precision farming technologies.[135]

5.5 CLIMATE CHANGE IMPACT

In the early 1990s the U.S. Environmental Protection Agency commissioned one of the first studies to determine the impact of climate change on global agriculture.[136] The basic methodology that was used included a suite of crop simulation models that encompasses DSSAT.[92] The outputs of three different GCMs were used to modify the local long-term historical weather conditions, and yield estimates were obtained for wheat, rice, soybean, and maize. This same methodology was used by scientists representing more than 20 countries.[136] Assuming a fixed increase in temperature of 2°C, soybean yield was predicted to increase by 15%, wheat by 13%, rice by 9%, and maize by 8%. However, a temperature increase of 4°C caused a 7% decrease in rice yield, a 4% decrease for soybean, a 1.5% decrease for maize, and a 1% decrease for wheat. When the outputs of the GCMs were applied to the local long-term historical weather conditions, there was a more drastic impact on agricultural production. For example, for wheat in Canada, the decrease in yield ranged from 10 to 38% while the average decrease in yield at the global level ranged from 16 to 33%. Overall, this study found that crop yields in the mid- and high-latitude regions, such as Canada, were less adversely affected than yields in the low-latitude regions. It was also found that farm-level adaptations in the temperate regions can generally offset the potential detrimental effects of climate change.[136]

The results of these impact studies, in general, are inconsistent due to the various scenarios that can be used and the uncertainties associated with the outcomes of the GCMs.[49–51,137] McGinn et al.[138] found that crop yields in Alberta increased by 21 to 124% when outputs of the Canadian Climate Centre GCMs were used. In some cases not only the scenarios predicted by the GCMs, but also the crop simulation models that are used can affect the outcome of the predictions and impact assessments.[139,140] The Global Change and Terrestrial Ecosystems (GCTE) Focus group 3 project of the International Geosphere-Biosphere Programme (IGBP) developed networks for different crops to study the impact of global change on managed ecosystems, particularly the impact on crop yield. Report 2 lists 19 different models for simulating growth, development, and yield for wheat.[141] Unfortunately, these types of inventory and comparison studies are rare. In most cases it is very difficult for model users to decide which of these models would be most appropriate to determine the impact of climate change on yield and carbon sequestration. One of the most extensive model comparisons was conducted for potato by Kabat et al.,[142] with a detailed analysis and comparison of eight different potato models. These studies should not necessarily be considered as a model competition, but more an evaluation of the advantages and disadvantages of the various modeling approaches. A few crop simulation model comparisons have been conducted for climate change applications, including wheat.[143–146]

Key to how these models respond to temperature is the internal temperature response curves. Traditionally a degree-day approach is used, which defines a base temperature for development and a threshold value to reach the various developmental stages, such as anthesis and maturity. However, it is easier to compare the impact of temperature using a development or growth rate, as shown in Figure 5.4. The most conservative degree-day approach would show a proportional increase in the development rate for each degree increase in temperature above the base temperature. The base temperature for wheat and barley are normally considered to be 0°C, while the base temperature for maize is 8°C. Most crops also have an optimum temperature, above which there is no further increase in the rate of development. This is shown by the optimum temperature response depicted in Figure 5.4A. The optimum temperature for wheat and barley are considered to be 15°C, while the optimum temperature for maize is 34°C. However, there are different interpretations of these cardinal temperatures as well as different implementations of the temperature response curves, such as the curve linear response curve shown in Figure 5.4B.[147,148] The calculated growth or development rate will be different depending on the type of equation that has been implemented, especially when the temperatures are above the optimum temperature. Unfortunately these equations are extremely critical in modeling the impact of climate change on crop growth, yield, and carbon sequestration.[149]

5.5.1 MODELING CASE STUDY

As an example we modeled wheat growth, development, and yield for Swift Current, Saskatchewan. The model we used was CSM-CEREALS-Wheat[91] as implemented in DSSAT Version 4.0.[92] The crop management information was based on a spring wheat experiment conducted by Campbell et al.[150–152] in 1975. This data set has been used as one of the experimental data sets for evaluation of the wheat simulation model. After model evaluation we selected one treatment, specifically, rainfed, and one application of nitrogen at 164 kg N/ha prior to planting. We increased the daily maximum and minimum temperature with 0.5°C increments until we reached an increase of 5°C and kept all other conditions the same, including the ambient CO_2 concentration. The model response showed that total aboveground biomass decreased linearly with an increase in temperature. Grain yield seemed to be highest at a temperature increase of 2.0°C (Figure 5.5). This response can partially be explained by changes in development. The number of days from planting to anthesis

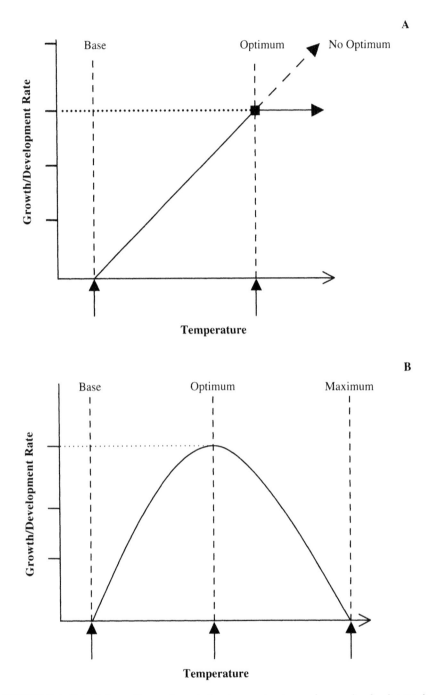

FIGURE 5.4 Calculation of growth and development rates, using a simple degree-day approach (A) and a more complex temperature response curve (B).

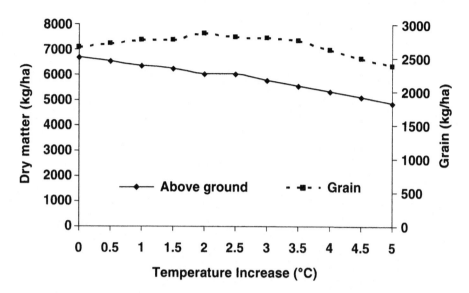

FIGURE 5.5 The impact of temperature increase on aboveground dry matter and grain yield for wheat one scenario, i.e., rainfed spring wheat grown in Swift Current, Saskatchewan using weather conditions for 1975.

decreased from 57 to 51 days and the number of days from anthesis to maturity decreased from 31 to 22 days. Therefore, the total growing season was reduced by 15 days from 88 to 73 days.

Although this example can be used to help explain differences in response due to changes in single environmental variables,[153] it cannot be used for impact assessment studies. First, one should include at least 30 years of historical weather data to account for the seasonal weather variability. In addition, if these trends are consistent with a decrease in the number of days to maturity, it is highly likely that a farmer would plant a longer-season wheat variety to fully benefit from this change in weather conditions. Due to the shorter season duration, the amount of nitrogen fertilizer was more than sufficient with the original wheat variety, but could have changed if a different variety had been used, such as a longer-season variety. One should keep in mind that management inputs have to be adjusted if significant changes are predicted in growing season duration. This will ultimately affect the predictions of potential yield, biomass production, and carbon sequestration.

5.6 ISSUES AND FUTURE DIRECTIONS

5.6.1 MANAGEMENT DECISIONS AND POTENTIAL IMPACT

Decisions are made on a continuous basis in the agricultural production system by stakeholders, policy makers, agribusinesses, and many others that are directly or indirectly affected by agriculture, including consumers. There are tactical decisions

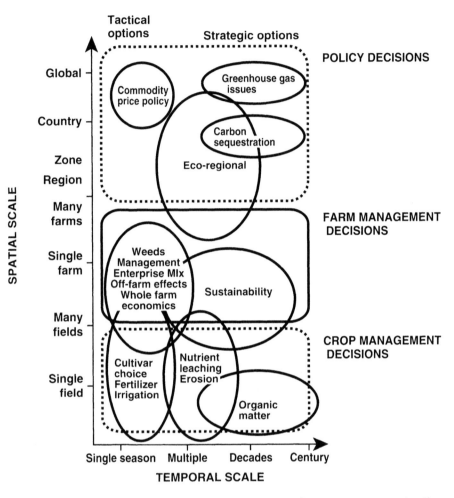

FIGURE 5.6 The complex interaction of crop management, farm management, and policy decisions with agricultural production at different spatial and temporal scales and the environment. (Modified from Meinke.[172])

made by the farmer or grower on a continuous basis, such as irrigation and fertilizer applications, as well as long-term strategic and planning decisions. The outcomes of these decisions affect not only crop performance in a single field, but also the farm, a region, or a country (Figure 5.6). An application of nitrogen fertilizer applied today could ultimately affect carbon sequestration 10 or 20 years from now. It is very important to carefully evaluate all these options and the potential impact of a decision on this complex system when studying carbon sequestration at the soil–plant interface in the context of agricultural production and agronomic adaptation to climate change.[130]

Agriculture's primary role is to provide food, feed, and fiber for all humankind. The world population continues to increase and therefore the demand for food also continues to increase. As the standard of living increases, there will also be a shift in basic food requirements from the traditional cereal staples to meat, requiring a change from production for human consumption to production for animal feed.[154] This increases the pressure on the limited natural resources that are available for farming, especially in developing countries.[155] If land is being used for food and fiber production, it reduces the potential for carbon sequestration by smallholder farmers. As stated previously, agriculture is really a "risky" business and weather variability is one of the main factors that control agricultural production.[156] It is expected that climate change will affect climate variability, including a change in the frequency and intensity of extreme events. It is extremely difficult for a farmer to prepare for extreme events, unless they can be predicted ahead of time. Even for many of these cases, such as hurricanes and tornadoes, farmers do not have much flexibility to protect their crop. However, farmers have options to adapt to climate change by modifying their management practices. It will be up to the research community to provide farmers with new technologies, such as varieties that are adapted to changes in the local temperature.[157]

5.6.2 Uncertainty in Benefits

The already established increase in ambient CO_2 concentration associated with climate change is a positive factor for agriculture and has resulted in an increase in biomass production and yield. This can be beneficial to both the farmer and the local environment. A well-balanced management of the biomass at harvest could result in both short-term and long-term carbon sequestration through the soil environment. In general, a potential increase in temperature could also be beneficial to agriculture, especially for the mid and higher latitudes, such as Canada. It would extend the potential crop growing season by, for example, delaying the first frost date and improving the temperature conditions during the growing season to provide optimum growth and development. However, it could be detrimental to the lower latitudes where the temperatures are already high and an increase would have a negative impact on plant growth and development.

It is still unclear what will happen to rainfall and snow under a changing climate. A decrease in rainfall would be detrimental to agriculture, unless natural resources are available for supplemental irrigation. An increase in rainfall could be detrimental if it occurs in the form of high-intensity rainfall events, flash floods, or regular floods. Even if rainfall would not be affected, an increase in temperature would normally also cause an increase in evapotranspiration. For prairie conditions in Canada, where there is normally some water deficit, this would reduce the water availability, resulting in more drought stress and a potential reduction in biomass growth and yield. However, the most recent climate change simulations with the Canadian Climate Centre GCM predicted an increase in temperature for the Canadian Prairie, resulting in earlier planting dates due to spring warming.[158] Some scenarios predicted soil moisture to increase, while others showed a decrease, especially in Alberta. The impact and potential for adoptive strategies that are selected are very much a function of the scenarios that were selected.

5.6.3 Research Gaps

Although agricultural research has studied many aspects of plants and their interaction with the biotic and abiotic environment in detail, much is still unknown, from plant genetics to field-level physiology.[159,160] A better understanding is needed, for example, of the impact of both high and low temperature on crop growth and development as well as the interactive effects of CO_2 and temperature.[161] A better understanding is also needed of the interactive effects of other factors associated with climate change, air pollution, and other atmospheric conditions, such as ultraviolet-B (UV-B) radiation and O_3 and their impact on crop growth and development and ultimately carbon sequestration. Finally, changes in pests and disease dynamics and the interaction with the local crops under climate change should also be considered.[162–164] A 1°C rise in temperature could see significant changes in local insect, disease, weed, and other pest populations, some of which could be unfamiliar to both farmers and researchers. This might require changes in applications in pesticides, if available and labeled appropriately, and breeding for resistance against pests that were previously unknown. For example, hurricane Ivan affected the southeastern U.S. during the fall of 2004 and carried Asian soybean rust from South America to several soybean growing states in the U.S. It will take up to 7 years to develop a soybean variety that is resistant to this strain of rust. A high correlation has been found between the occurrences of these hurricanes and climate variability, such as the El Niño and La Niña events.[51,165] If the frequency of hurricanes will increase, we could see occurrences of new pests and diseases in regions that were previously not exposed to them.

Another issue that has not really been studied extensively is the potential change in the soil environment due to climate change. One would expect that if the air temperature increases, the temperature of the soil surface and subsequent layers will also increase. This in turn could have a significant impact on the microbiological processes that occur in the soil, including possibly higher turnover rates of some of the organic carbon pools. Changes in precipitation will affect soil moisture conditions and could also influence these microbiological processes and microbe populations. Any changes in the soil environment, both biotic and abiotic, will ultimately affect crop growth and development and biomass production and yield, as well as carbon sequestration.

To be able to comprehend the response of the agricultural system to climate change, comprehensive simulation models will be needed that integrate the state of science. However, one needs to keep in mind that crop simulation and other agronomic models are only a mathematical representation of the cropping system and are never perfect. Improved data collection procedures and additional experimental data are also needed for model improvement and model evaluation. This will establish the credibility of these models to simulate and predict local crop production and to allow for scenario analysis and the development of information that can be used for decisions associated with climate change mitigation and carbon sequestration. The direct use of models to study the potential impact of climate change has a somewhat limited value for the local farmer, but could have a strong policy value if the correct decisions are made and implemented.

5.6.4 Stakeholders

When developing agricultural policies that mitigate the potential impact of climate change and address issues associated with carbon sequestration, it is important to involve the stakeholders and to keep in mind that the livelihood and long-term economic sustainability of farmers are at stake.[166] Traditionally, farmers have been the shepherds of the land and it has been to their benefit to take care of the precious natural resources of planet Earth. This includes management practices such as no-tillage that adds some of the crop organic material back into the soil, improves soil quality, and ultimately establishes the potential for carbon sequestration.[167–169] A recent study with the Canadian Economic and Emissions Model for Agriculture showed that changing tillage practices from conventional tillage to zero tillage had the greatest potential for carbon sequestration and net reduction in greenhouse gas emissions.[170] Other studies have shown that a conversion from cultivated land to grassland could also increase the potential for carbon sequestration.[171] However, what impact do these changes have on our farming communities? A significant consideration in all decisions that are recommended and policy changes that are implemented should be the socioeconomic impact, including the net return for farmers and the potential reduction in risk that farmers have to cope with. Unfortunately, a large part of this risk is associated with weather variability that is driven by climate change and climate variability.

5.7 SUMMARY AND CONCLUSIONS

Over the centuries farmers have adapted their crop management strategies to adjust to the local changes in weather conditions due to climate change and climate variability in order to reduce their risk and vulnerability and to obtain an optimum crop yield. Many of these changes were based on research outcomes of studies conducted at the plant/soil interface. The climate is changing, but the nature of these changes for the future climate is unclear. To be able to determine the impact of climate change on the plant/soil interface and especially crop production, we must partially rely on research outcomes of other disciplines, especially the oceanic and atmospheric sciences. As a consequence of the uncertainty and variability of the current climate predictions and projections, it is rather difficult to determine the potential impact of climate change on crop growth and development. Research advances have provided us with several state-of-the-art research facilities for determining climate change impact, such as indoor growth chambers, outdoor sunlit and SPAR units, and the FACE facility. In these facilities plants can be grown under controlled atmospheric conditions, including ambient CO_2, temperature, relative humidity or dewpoint temperature, and light or solar radiation. Research advances have also provided us with state-of-the-art computer simulation models that can predict growth, development, and yield of many different crops, including rangelands and pastures. Both approaches have allowed us to determine the potential impact of climate change on crop yield and provided us with possible management scenarios for mitigation. However, there are still many unknowns and research gaps in these studies, such as the interaction with insects, diseases, weeds, and other pests, the impact of high temperatures on plant growth, or the impact of the interactions of various trace gases, UV-B, and other atmospheric factors. The socioeconomic component of the farming system is also important, including the changes in management practices that can be made by producers to mitigate the potential negative impact of climate change, and policy decisions at regional or national levels. Agriculture is directly benefiting from the increase in ambient CO_2 concentration as it has allowed plants to sequester carbon at a higher rate, resulting in an increase in biomass production and crop yield. The impact of the predicted changes in temperature is unclear and depends on the magnitude of these changes. Current predictions that integrate the outcomes of the GCMS with crop simulation models show that the tropics and subtropics will be negatively affected, while cooler regions such as Canada might benefit from higher temperatures, resulting in longer growing seasons. However, the potential benefits of these longer growing seasons are closely associated with the predicted changes in precipitation. An increase in precipitation amounts, in general, will benefit most regions, while a decrease could cause a potential reduction in yield, even if temperatures were more favorable. Changes in the occurrence of precipitation should also be considered, especially if they affect the duration of drought spells or the rainy season, or extreme events, such as hurricanes and flooding. As the climate continues to change we could see significant changes at the plant/soil interface with respect to crop growth and yield and the potential for carbon sequestration. Especially the most vulnerable regions across the world should be closely monitored to avoid any unexpected surprises with respect to crop failures, which would affect our potential capacity to provide food, feed, and fiber to human kind.

REFERENCES

1. Desjardins, R.L., Kulshreshtha, S.N., Junkins, B., Smith, W., Grant, B., and Boehm, M. Canadian greenhouse gas mitigation options in agriculture. *Nutrient Cycling Agroecosyst.* 60(1–3):317–326, 2001.
2. Colbourn, P. Denitrification and N_2O production in pasture soil: the influence of nitrogen supply and moisture. *Agric. Ecosyst. Environ.* 39:267–278, 1992.
3. Mosier, A.R., Duxbury, J.M., Freney, J.R., Heinemeyer, O., and Minami, K. Assessing and mitigating N_2O emissions from agricultural soils. *Climatic Change* 40:7–38, 1998.
4. Conen, F., Dobbie, K.E., and Smith, K.A. Prediction N_2O emissions from agriculture land through related soil parameters. *Global Change Biol.* 6:417–426, 2000.
5. Neue, H.U., Wassmann, R., Lantin, R.S., Alberto, M.C.R., Aduna, J.B., and Javellana, A.M. Factors affecting methane emission from rice fields. *Atmos. Environ.* 30:1751–1754, 1996.
6. Kulshreshtha, S.N., Junkins, B., and Desjardins, R. Prioritizing greenhouse gas emission mitigation measures for agriculture. *Agric. Syst.* 66(3):145–166, 2000.
7. Allen, L.H., Jr., Albrecht, S.L., Colon-Guasp, W., Covell, S.A., Baker, J.T., Pan, D., and Boote, K.J. Methane emissions of rice increased by elevated carbon dioxide and temperature. *J. Environ. Qual.* 3:1978–1991, 2003.
8. Martins, O. and Dewes, T. Loss of nitrogenous compounds during composting of animal wastes. *Bioresour. Technol.* 42(2):103–111, 1992.

9. Luo, J., Kulasegarampillai, M., Bolan, N., and Donnison, A. Control of gaseous emissions of ammonia and hydrogen sulphide from cow manure by use of natural materials. *N.Z. J. Agric. Res.* 47(4):545–556, 2004.

10. Thompson, R.B. and Meisinger, J.J. Gaseous nitrogen losses and ammonia volatilization measurement following land application of cattle slurry in the mid-Atlantic region of the USA. *Plant Soil* 266(1–2):231–246, 2004.

11. Elmi, A., Madani, A., Gordon, R., Macdonald, P., and Stratton, G.W. Nitrate nitrogen in the soil profile and drainage water as influenced by manure and mineral fertilizer application in a barley-carrot production system. *Water, Air Soil Poll.* 160(1–4):119–132, 2005.

12. Waisel, Y., Eshel, A., and Kafkafi, U. *Plant Roots: The Hidden Half.* Marcel Dekker, New York, 2002.

13. Krajewski, W.F., Ciach, G.J., and Habib, E. An analysis of small-scale rainfall variability in different climatic regimes. *J. Sci. Hydrol.* 48(2):151–162, 2003.

14. Nijbroek, R., Hoogenboom, G., and Jones, J.W. Optimal irrigation strategy for a spatially variable soybean field: a modeling approach. *Agric. Syst.* 76(1):359–377, 2003.

15. Mitchell, M. and Kienholz, J. A climatological analysis of the Koppen Dfa/Dfb boundary in eastern North-America, 1901–1930. *Ohio J. Sci.* 97(3):53–58, 1997.

16. Gommes, R. Current climate and population constraints on world agriculture, in *Agricultural Dimensions of Global Climate Change,* H.M. Kaiser and T.E. Drennen, Eds. St. Lucie Press, Boca Raton, FL, 2003, 67–86.

17. Hoogenboom, G. Contribution of agrometeorology to the simulation of crop production and its applications. *Agric. For. Meteorol.* 103(1–2):137–157, 2000.

18. Changnon, D. and Changnon, S.A., Jr. Evaluation of weather catastrophe data for use in climate change investigations. *Climatic Change* 38:435–445, 1998.

19. Cutforth, H.W. Climate change in the semiarid prairie of southwestern Saskatchewan: Temperature, precipitation, wind, and incoming solar energy. *Can. J. Soil Sci.* 80(2):375–385, 2000.

20. Hall, A.E. *Crop Responses to Environment.* CRC Press, Boca Raton, FL, 2001.

21. Monteith, J.L. *Principles of Environmental Physics.* Arnold, London, 2003.

22. Jones, H.G. *Plants and Microclimate. A Quantitative Approach to Environmental Plant Physiology.* 2nd ed. Cambridge University Press, Cambridge, 1992.

23. Keeling, C.D. and Whorf, T.P. Atmospheric CO_2 records from sites in the SIO air sampling network. In Trends: A Compendium of Data on Global Change. Carbon Dioxide Information Analysis Center, Oak Ridge National Laboratory, U.S. Department of Energy, Oak Ridge, TN, 2004. Available online at http://cdiac.ornl.gov/trends/co2/sio-mlo.htm [URL Verified 03/02/2005].

24. Stooksbury, D. Is climate change already giving us greater maize yields? *N. Sci.* 23:48, 1991.

25. Garcia-Oliva, F. and Masera, O.R. Assessment and measurement issues related to soil carbon sequestration in land-use, land-use change, and forestry (LULUCF) projects under the Kyoto protocol. *Climatic Change* 65:347–364, 2004.

26. Davidson, E.A. and Artaxo, P. Globally significant changes in biological processes of the Amazon basin: results of the large-scale biosphere-atmosphere experiment. *Global Change Biol.* 10(5):519–529, 2004.

27. Eymontt, A. and Romaniuk, W. Opportunities and obstacles for implementation of agricultural and domestic waste treatment systems. *Microbiology* 66(5):574–577, 1997.

28. Kalia, A.K. and Singh, S.P. Development of a biogas plant. *Energ. Sourc.* 26(8):707–714, 2004.
29. VandenBygaart, A.J., Gregorich, E.G., and Angers, D.A. Influence of agricultural management on soil organic carbon: A compendium and assessment of Canadian studies. *Can. J. Soil Sci.* 83(4):363–380, 2003.
30. Hannam, I. International and national aspects of a legislative framework to manage soil carbon sequestration. *Climatic Change* 65:365–387, 2004.
31. Bolinder, M.A., Angers, D.A., and Dubuc, J.P. Estimating shoot to root ratios and annual carbon inputs in soils for cereal crops. *Agric. Ecosyst. Environ.* 63(1):61–66, 1997.
32. Lal, R., Kimble, J.M., Follett, R.F., and Cole, C.V., Eds. *The Potential of U.S. Cropland to Sequester Carbon and Mitigate the Greenhouse Effect.* Lewis Publishers, Boca Raton, FL, 1999.
33. Follett, R.F., Kimble, J.M., and Lal, R., Eds. *The Potential of U.S. Grazing Lands to Sequester Carbon and Mitigate the Greenhouse Effect.* Lewis Publishers, Boca Raton, FL, 2001.
34. Shabbar, A. and Bonsal, B. An assessment of changes in winter cold and warm spells over Canada. *Nat. Hazards* 29(2):173–188, 2003.
35. Bonsal, B.R. and Prowse, T.D. Trends and variability in spring and autumn 0 degrees C-isotherm dates over Canada. *Climatic Change* 57(3):341–358, 2003.
36. Zhang, X.B., Vincent, L.A., Hogg, W.D., and Niitsoo, A. Temperature and precipitation trends in Canada during the 20th century. *Atmos. Ocean* 38(3):395–429, 2000.
37. Slonosky, V.C. The meteorological observations of Jean-Francois Gaultier, Quebec, Canada: 1742–56. *J. Climate* 16(13):2232–2247, 2003.
38. Alexandrov, V.A. and Hoogenboom, G. 2001. Climate variation and crop production in Georgia, USA, during the 20th century. *Climate Res.* 17(1):33–43, 2001.
39. Quayle, R.G., Easterling, D.R., Karl, T.R., and Hughes, P.Y. Effects of recent thermometer changes in the Cooperative Station Network. *Bull. Am. Meteorol. Soc.* 72:1718–1723, 1991.
40. Guttman, N.B. and Baker, C.B. Exploratory analysis of the difference between temperature observations recorded by ASOS and conventional methods. *Bull. Am. Meteorol. Soc.* 77(12):2865–2873, 1996.
41. Kimball, B.A., Rosenberg, N.J., Allen, L.H., Jr., Heichel, G.H., Stuber, C.W., Kissel, D.E., and Ernst, S., Eds. *Impact of Carbon Dioxide, Trace Gases, and Climate Change on Global Agriculture.* ASA Special Publication No. 53, Madison, WI, 1990.
42. Downing, T.E., Ed. *Climate Change and World Food Security. Proceedings of the NATO Advanced Research Workshop "Climate Change and World Food Security."* Series I: Global Environmental Change 37. Springer, Berlin, 1996.
43. Kaiser, H.M. and Drennen, T.E., Eds. *Agricultural Dimensions of Global Climate Change.* St. Lucie Press, Boca Raton, FL, 2003.
44. Rosenberg, N.J., Ed. *Towards an Integrated Impact Assessment of Climate Change: The MINK Study.* Kluwer Academic Publishers, Dordrecht, the Netherlands, 1993.
45. Rosenzweig, C. and Hillel, D. *Climate Change and the Global Harvest.* Oxford University Press, New York, 1998.
46. Reddy, K.R. and Hodges, H.F., Eds. *Climate Change and Global Crop Productivity.* CABI Publishing, Wallingford, Oxon, U.K., 2000.
47. Mearns, L.O., Ed. *Issues in the Impacts of Climate Variability and Change on Agriculture.* Kluwer Academic Publishers, Dordrecht, the Netherlands, 2003.
48. Hitz, S. and Smith, J. Estimating global impacts from climate change. *Global Environ. Change* 14:201–218, 2004.

49. Hengeveld, H.G. *Projections for Canada's Climate Future. A Discussion of Recent Simulations with the Canadian Global Climate Model.* Climate Change Digest CCD 00-01, Special Edition. Meteorological Service of Canada, Downsview, Ontario, Canada, 2000.

50. Mearns, L.O., Giorgi, F., McDaniel, L., and Shields, C. Climate change scenarios for the southeastern US based on GCM and regional model simulations. *Climatic Change* 60(1–2):7–35, 2003.

51. National Assessment Synthesis Team (NAST). Climate Change Impacts on the United States: The Potential Consequences of Climate Variability and Change. U.S. Global Change Research Program, Washington, D.C., 2000.

52. Collins, M. Climate predictability on interannual to decadal time scales: the initial value problem. *Climate Dynam.* 19(8):671–692, 2002.

53. Peng, P.T., Kumar, A., van den Dool, H., and Barnston, A.G. An analysis of multi-modal ensemble predictions for seasonal climate anomalies. *J. Geophys. Res. Atmos.* 107(D23):4710, 2002.

54. Hassan, A.S., Yang, X.Q., and Zao, S.S. Reproducibility of seasonal ensemble integrations with the ECMWF GCM and its association with ENSO. *Meteorol. Atmos. Phys.* 86(3–4):159–172, 2004.

55. Stainforth, D.A., Aina, T., Christensen, C., Collins, M., Faull, N., Frame, D.J., Kettleborough, J.A., Knight, S., Martin, A., Murphy, J.M., Piani, C., Sexton, D., Smith, L.A., Spicer, R.A., Thorpe, A.J., and Allen, M.R. Uncertainty in predictions of the climate response to rising levels of greenhouse gases. *Nature* 433:403–406, 2005.

56. Barfield, B.J. and Gerber, J.F., Eds. *Modification of the Aerial Environment of Plants.* American Society of Agricultural Engineers, St. Joseph, MI, 1979.

57. Ainsworth, E.A., Davey, P.A., Bernacchi, C.J., Dermody, O.C., Heaton, E.A., Moore, D.J., Morgan, P.B., Naidu, S.L., Yoo Ra, H., Zhu, X., Curtis, P.S., and Long, S.P. A meta-analysis of elevated CO_2 effects on soybean physiology, growth and yield. *Global Change Biol.* 8:695–709, 2002.

58. Liu, L., Hoogenboom, G., and Ingram, K.T. Controlled-environment sunlit plant growth chambers. *Crit. Rev. Plant Sci.* 19(4):347–375, 2000.

59. Kim, S., Reddy, V.R., Baker, J.T., Gitz, O.C., and Timlin, D.J. Quantification of photosynthetically active radiation inside sunlit growth chambers. *Agric. For. Meteorol.* 126:117–127, 2004.

60. Bernier, P.Y., Stewart, J.D., and Hogan, G.D. Quantifying the uncontrolled CO_2 dynamics of growth chambers. *J. Exp. Bot.* 45:1143–1146, 1994.

61. Allen, J., and Nelson, M. Biospherics and Biosphere 2, Mission One (1991–1993). *Ecol. Eng.* 13:15–29, 1999.

62. Dempster, W.F. Biosphere 2 engineering design. *Ecol. Eng.* 13:31–42, 1999.

63. Kang, D. Simulation of the water cycle in Biosphere 2. *Ecol. Eng.* 13:301–311, 1999.

64. Zabel, B., Hawes, P., Stuart, H., and Marino, B.V.D. Construction and engineering of created environment: overview of the Biosphere 2 closed system. *Ecol. Eng.* 13:43–63, 1999.

65. Ingram, K.T., Hoogenboom, G., and Liu, L. The Georgia Envirotron — a multidisciplinary research facility for the study of interacting environmental stresses on plants. ASAE Paper 98-4151, American Society of Agricultural Engineers, St. Joseph, MI, 1998.

66. Liu, L., Hoogenboom, G., Ingram, K.T., and Prussia, S.E.. Design of a prototype movable sunlit growth chamber for field research. ASAE Paper 97-4027, American Society of Agricultural Engineers, St. Joseph, MI, 1997.

67. Liu, L., Hoogenboom, G., and Ingram, K.T. Control and performance of a movable sunlit growth chamber. ASAE Paper 98-4153, American Society of Agricultural Engineers, St. Joseph, MI, 1998.

68. Jones, P., Jones, J.W., Allen, L.H., Jr., and Mishoe, J.W. Dynamic computer control of closed environmental plant growth chambers. Design and verification. *Trans. ASAE* 27:879–888, 1984.

69. Pickering, N.B., Allen, L.H., Jr., Albrecht, S.L., Jones, P., Jones, J.W., and Baker, J.T. Environmental plant chambers: control and measurement using CR-IOT dataloggers. In *Computers in Agriculture 1994*, D.G. Watson, F.S. Zazueta, and T.V. Harrison, Eds. American Society of Agricultural Engineers, St. Joseph, MI, 1994, 29–35.

70. Baker, J.T. and Allen, L.H., Jr. Effects of CO_2 and temperature on rice: a summary of five growing seasons. *J. Agric. Meteorol.* 48(5):575–582, 1993.

71. Reddy, K.R., Robana, R.R., Hodges, H.F., Liu, X.J., and McKinion, J.M. Interactions of CO_2 enrichment and temperature on cotton growth and leaf characteristics. *Environ. Exp. Bot.* 39:117–129, 1998.

72. Baker, J.T., Allen, L.H., Jr., Boote, K.J., Jones, P., and Jones, J.W. Response of soybean to air temperature and carbon dioxide concentration. *Crop Sci.* 29:98–105, 1989.

73. Heagle, A.S., Body, D.E., and Heck, W.W. An open-top field chamber to assess the impact of air pollution on plants. *J. Environ. Qual.* 2(3):365–368, 1973.

74. Heagle, A.S., Philbeck, R.B., Ferrell, R.E., and Heck, W.W. Design and performance of a large, field exposure chamber to measure effects of air quality on plants. *J. Environ. Qual.* 18:361–368, 1989.

75. Heagle, A.S., Miller, J.E., and Pursley, W.A. Atmospheric pollutants and trace gases: growth and yield responses of potato to mixtures of carbon dioxide and ozone. *J. Environ. Qual.* 32:1603–1610, 2003.

76. Allen, L.H., Jr., Drake, B.G., Rogers, H.H., and Shinn, J.H. Field techniques for exposure of plants and ecosystems to elevated CO_2 and other trace gases. *Crit. Rev. Plant Sci.* 11(2–3):85–119, 1992.

77. Kimball, B.A. Cost comparisons among free-air CO_2 enrichment, open-top chamber, and sunlit controlled-environment chamber methods of CO_2 exposure. *Crit. Rev. Plant Sci.* 11:265–270, 1992.

78. Weigel, H.-J., Mejer, G.J., and Jäger, H.J. Impact of climate change on agriculture: open-top chambers as a tool to investigate long-term effects of elevated CO_2 levels on plants. *Angew. Bot.* 66:135–142, 1992.

79. Heck, W.W., Cure, W.W., Rawlings, J.O., Zaragoza, L.J., Heagle, A.S., Heggestad, H.E., Kohut, R.J., Dress, L.W., and Temple, P.J. Assessing impacts of ozone on agricultural crops: II. Crop yield functions and alternative exposure statistics. *J. Air Pollut. Cont. Assess.* 34:810–817, 1984.

80. Allen, L.H., Jr. Plant responses to rising carbon dioxide and potential interactions with air pollutants. *J. Environ. Qual.* 19:15–34, 1990.

81. Fuhrer, J. Critical level for ozone to protect agricultural crops: Interaction with water availability. *Water Air Soil Pollut.* 85:1355–1360, 1995.

82. Fuhrer, J. Agroecosystem responses to combinations of elevated CO_2, ozone, and global climate change. *Agric. Ecosyst. Environ.* 97(1–3), 1–20, 2003.

83. Allen, L.H., Jr. Free-air CO_2 enrichment field experiments: An historical overview. *Crit. Rev. Plant Sci.* 11(2–3):121–134, 1992.

84. Pinter, P.J., Jr., Kimball, B.A., Wall, G.W., LaMorte, R.L., Hunsaker, D.J., Adamsen, F.J., Frumau, K.F.A., Vugts, H.F., Hendrey, G.R., Lewin, K.F., .Nagy, J., Johnson, H.B., Wechsunge, F., Leavitt, S.W., Thompson, T.L., Matthias, A.D., and Brooks, T.J. Free-air CO_2 enrichment (FACE): blower effects on wheat canopy microclimate and plant development. *Agric. For. Meteorol.* 103(4):319–333, 2000.

85. Ainsworth, E.A. and Long, S.P. What have we learned from 15 years of free-air CO_2 enrichment (FACE)? A meta-analytic review of the responses of photosynthesis, canopy. Properties and plant production to rising CO_2. *N. Phytol.* 165(2):351–371, 2005.

86. Kimball, B.A., LaMorte, R.L., Seay, R.S., Pinter, P.J., Jr., Rokey, R.R., Hunsaker, D.J., Dugas, W.A., Heuer, M.L., Mauney, J.R., Hendrey, G.R., Lewin, K.F., and Nagy, J. Effects of free-air CO_2 enrichment on energy balance and evapotranspiration of cotton. *Agric. For. Meteorol.* 70:259–268, 1994.

87. Kimball, B.A., Pinter, P.J., Jr., Garcia, R.L., LaMorte, R.L., Wall, G.W., Hunsaker, D.J., Wechsung, G., Wechsungs, F., and Kartschalls, T. 1995. Productivity and water use of wheat under free-air CO_2 enrichment. *Global Change Biol.* 1:429–442, 1995.

88. Pinter, P.J., Jr., Kimball, B.A., Garcia, R., Wall, G., Hunsaker, D.J., and LaMorte, R.L. 1996. Free-air CO_2 enrichment: responses of cotton and wheat crops. In *Carbon Dioxide and Terrestrial Ecosystems*, G.W. Koch and H.A. Mooney, Eds. Academic Press, San Diego, CA, 1996, 215–248.

89. Peart, R.M. and Curry, R.B., Eds. *Agricultural Systems Modeling and Simulation.* Marcel Dekker, New York, 1998.

90. Hoogenboom, G. 2003. Crop growth and development. In *Handbook of Processes and Modeling in the Soil–Plant System,* D.K. Bendi and R. Nieder, Eds. The Haworth Press, Binghamton, New York, 2003, 655–691.

91. Jones, J.W., Hoogenboom, G., Porter, C.H., Boote, K.J., Batchelor, W.D., Hunt, L.A., Wilkens, P.W., Singh, U., Gijsman, A.J., and Ritchie, J.T., 2003. DSSAT Cropping System Model. *Eur. J. Agron.* 18:235–265, 2003.

92. Hoogenboom, G., Jones, J.W., Wilkens, P.W., Porter, C.H., Batchelor, W.D., Hunt, L.A., Boote, K.J., Singh, U., Uryasev, O., Bowen, W.T., Gijsman, A.J., du Toit, A., White, J.W., and Tsuji, G.Y. Decision Support System for Agrotechnology Transfer Version 4.0 [CD-ROM]. University of Hawaii, Honolulu, 2004.

93. Williams, J.R., Jones, C.A., Kiniry, J.R., and Spanel, D.A. The EPIC crop growth model. *Trans. ASAE* 32(2):497–511, 1989.

94. Stockle, C.O., Williams, J.R., Rosenberg, N.J., and Jones, C.A. A method for estimating the direct and climatic effects of rising atmospheric carbon dioxide on growth and yield of crops: Part I. Modification of the EPIC model for climate change analysis. *Agric. Syst.* 38:225–238, 1992.

95. Stockle, C.O., Dyke, P.T., Williams, J.R., Jones, C.A., and Rosenberg, N.J. A method for estimating the direct and climatic effects of rising atmospheric carbon dioxide on growth and yield of crops: Part II. Sensitivity analysis at three sites in the Midwestern USA. *Agric. Syst.* 38:239–256, 1992.

96. Easterling, W.E., Chen, X., Hays, C., Brandle, J.R., and Zhang, H. 1996. Improving the validation of model-simulated crop yield response to climatic change: an application to the EPIC model. *Climate Res.* 6:263–273, 1996.

97. McCown, R.L., Hammer, G.L., Hargreaves, J.N.G., Holzworth, D.P., and Freebairn, D.M. APSIM: a novel software system for model development, model testing and simulation in agricultural systems research. *Agric. Syst.* 50:255–271, 1996.

98. Keating, B.A., Carberry, P.S., Hammer, G.L., Probert, M.E., Robertson, M.J., Holzworth, D., Huth, N.I., Hargreaves, J.N.G., Meinke, H., Hochman, Z., McLean, G., Verburg, K., Snow, V., Dimes, J.P., Silburn, M., Wang, E., Brown, S., Bristow, K.L., Asseng, S., Chapman, S., McCown, R.L., Freebairn, D.M., and Smith, C.J. An overview of APSIM, a model designed for farming systems simulation. *Eur. J. Agron.* 18:267–288, 2003.

99. Brisson, N., Ruget, F., Gate, P., Lorgeou, J., Nicoullaud, B., Tayot, X., Plenet, D., Jeuffroy, M.-H., Bouthier, A., Ripoche, D., Mary, B., and Justes, E. STICS: a generic model for simulating crops and their water and nitrogen balances. II. Model validation for wheat and maize. *Agronomie* 22:69–92, 2002.

100. Brisson, N., Gary, C., Justes, E., Roche, R., Mary, B., Ripoche, D., Zimmer, D., Sierra, J., Bertuzzi, P., Burger, P., Bussiere, R., Cabidoche, Y.M., Cellier, P., Debaeke, P., Gaudillere, J.P., Henault, C., Maraux, F., Sequin, B., and Sinoquet, H. An overview of the crop model STICS. *Eur. J. Agron.* 18:309–332, 2003.

101. Grant, R.F. Changes in soil organic matter under different tillage and rotation: mathematical modeling in *ecosys. Soil Sci. Soc. Am. J.* 61:1159–1175, 1997.

102. Grant, R.F., Wall, G.W., Kimball, B.A., Frumau, K.F.A., Pinter, P.J., Hunsaker, D.J., and La Morte, R.L. Crop water relations under different CO_2 and irrigation: testing of *ecosys* with the Free Air CO_2 Enrichment (FACE) experiment. *Agric. For. Meteorol.* 95:27–51, 1999.

103. Grant, R.F., Juma, N.G., Robertson, J.A., Izaurralde, R.C., and McGill, W.B. Long term changes in soil carbon under different fertilizer, manure, and rotation: testing mathematical model *ecosys* with data from the Breton Plots. *Soil Sci. Soc. Am. J.* 65:205–214, 2001.

104. Hanson, J.D., Baker, B.B., and Bourdon, R.M. Comparison of the effects of different climate change scenarios on rangeland livestock production. *Agric. Syst.* 41:487–502, 1993.

105. Hunt, H.W., Trlica, M.J., Redente, E.F., Moore, J.C., Detling, J.K., Kittel, T.G.F., Walter, D.E., Fowler, M.C., Klein, D.A., and Elliott, E.T. Simulation model for the effects of climate change on temperate grassland ecosystems. *Ecol. Modelling* 53:205–246, 1991.

106. Hunt, H.W., Morgan, J.A., and Read, J.J. Simulating growth and root-shoot partitioning in prairie grasses under elevated atmospheric CO_2 and water stress. *Ann. Biol.* 81:489–501, 1998.

107. Charles-Edwards, D.A., Doley, D., and Rimmington, G.M. *Modelling Plant Growth and Development.* Academic Press, Sydney, Australia, 1986.

108. Hanks, J. and Ritchie, J.T., Eds. *Modeling Plant and Soil Systems.* American Society of Agronomy, Madison, WI, 1991.

109. Goudriaan, J. and van Laar, H.H. *Modelling Potential Crop Growth Processes. Textbook with Exercises.* Kluwer Academic, Dordrecht, the Netherlands, 1994.

110. Hu, B.-G. and Jaeger, M, Eds. *Plant Modeling and Applications.* Tsinghua Academic Press, Beijing, China, 2003.

111. Yin, X. and van Laar, H.H. *Crop Systems Dynamics. An Ecophysiological Simulation Model for Genotype-by-Environment Interactions.* Wageningen Academic, Wageningen, the Netherlands, 2005.

112. Zalud, Z. and Stastna, M. 2000. Sensitivity of the CERES-Maize yield simulation to the selected weather data. *Rocnik* 4:53–61, 2000.

113. Heinemann, A.B., Hoogenboom, G., and Chojnicki, B. The impact of potential errors in rainfall observation on the simulation of crop growth, development, and yield. *Ecol. Modeling* 157(1):1–21, 2002.

114. Curry, R.B., Peart, R.M., Jones, J.W., Boote, K.J., and Allen, L.H., Jr. Simulation as a tool for analyzing crop response to climate change. *Trans. ASAE* 33(3):981–990, 1990.

115. Easterling, W.E., Rosenberg, N.J., McKenney, M.S., Jones, C.A., Dyke, P.T., and Williams, J.R. Preparing the erosion productivity impact calculator (EPIC) model to simulate crop response to climate change and the direct effects of CO_2. *Agric. For. Meteorol.* 59:17–34, 1992.

116. Easterling, W.E., Mearns, L.O., Hays, C.J., and Marx, D. Comparison of agricultural impacts of climate change calculated from high and low resolution climate change scenarios. *Climatic Change* 51:173–197, 2001.

117. Hoogenboom, G., Tsuji, G.Y., Jones, J.W., Singh, U., Godwin, D.C., Pickering, N.B., and Curry, R.B. Decision support system to study climate change impacts on crop production, in *Climate Change and Agriculture: Analysis of Potential International Impacts,* C. Rosenzweig, L.H. Allen, Jr., L.A. Harper, S.E. Hollinger, and J.W. Jones, Eds. ASA special publication 59. American Society of Agronomy, Madison, WI, 1995, 51–75.

118. Reddy, K.R., Doma, P.R., Mearns, L.O., Boone, M.Y.L., Hodges, H.F., Richardson, A.G., and Kakani, V.G. Simulating the impacts of climate change on cotton production in the Mississippi Delta. *Climate Res.* 22(3):271–281, 2002.

119. Whisler, F.D., Acock, B., Baker, D.N., Fye, R.E., Hodges, H.F., Lambert, J.R., Lemmon, H.E., McKinion, J.M., and Reddy, V.R. Crop simulation models in agronomic systems. *Adv. Agron.* 40:141–208, 1986.

120. Reddy, K.R., Hodges, H.F., and McKinion, J.M. Crop modeling and applications: a cotton example. *Adv. Agron.* 59:225–290, 1997.

121. Tsuji, G.Y., Hoogenboom, G., and Thornton, P.K., Eds. *Understanding Options for Agricultural Production.* Systems Approaches for Sustainable Agricultural Development. Kluwer Academic, Dordrecht, the Netherlands, 1998.

122. Ahuja, L., Ma, L., and Howell, T.A., Eds. *Agricultural System Models in Field Research and Technology Transfer.* Lewis Publishers, Boca Raton, FL, 2002.

123. Gijsman, A.J., Hoogenboom, G., Parton, W.J., and Kerridge, P.C. Modifying the DSSAT crop models for low-input agricultural systems, using a SOM/residue module from CENTURY. *Agron. J.* 94:462–474, 2002.

124. Zhang, Y., Li, C.S., Zhou, X.J., and Moore, B. A simulation model linking crop growth and soil biogeochemistry for sustainable agriculture. *Ecol. Modelling* 151:75–108, 2002.

125. Smith, P. 2004. Carbon sequestration in croplands: the potential in Europe and the global context. *Eur. J. Agron.* 20:229–236, 2004.

126. Tubiello, F.N., Rosenzweig, C., and Volk, T. Interactions of CO_2 temperature and management practices: simulations with a modified version of CERES-Wheat. *Agric. Syst.* 49:135–152, 1995.

127. Alexandrov, V.A. and Hoogenboom, G. The impact of climate variability and change on major crops in Bulgaria. *Agric. For. Meteorol.* 104(4):315–327, 2000.

128. Alexandrov, V.A. and Hoogenboom, G. Vulnerability and adaptation assessments of agricultural crops under climate change in the Southeastern USA. *Theor. Appl. Climatol.* 67:45–63, 2000.

129. Chipanshi, A.C., Chanda, R., and Totolo, O. Vulnerability assessment of the maize and sorghum crops to climate change in Botswana. *Climatic Change* 61:339–360, 2003.

130. Easterling, W.E., Chhetri, N., and Niu, X. Improving the realism of modeling agronomic adaptation to climate change: simulating technological substitution. *Climatic Change* 60:149–173, 2003.

131. Droogers, P. 2004. Adaptation to climate change to enhance food security and preserve environmental quality: example for southern Sri Lanka. *Agric. Water Manage.* 66(1):15–33, 2004.

132. Mall, R.K., Lal, M., Bhatia, V.S., Rathore, L.S., and Singh, R. Mitigating climate change impact on soybean productivity in India: a simulation study. *Agric. For. Meteorol.* 121:113–125, 2004.

133. Johnston, T. and Chiotti, O. Climate change and the adaptability of agriculture: a review. *J. Air Waste Manage. Assoc.* 50(4):563–569, 2000.

134. Smithers, J. and Blay-Palmer, A. Technology innovation as a strategy for climate adaptation in agriculture. *Appl. Geogr.* 21(2):175–197, 2001.

135. Batte, M.T. and Arnholt, M.W. Precision farming adoption and use in Ohio: case studies of six leading-edge adopters. *Comput. Electron. Agric.* 38(2):125–139, 2003.

136. Rosenzweig, C., Allen, L.H., Jr., Harper, L.A., Hollinger, S.E., and Jones, J.W., Eds. *Climate Change and Agriculture: Analysis of Potential International Impacts.* ASA Special Publication 59. American Society of Agronomy, Madison, WI, 1995.

137. Boer, G.J. 2004. Long time-scale potential predictability in an ensemble of coupled climate models. *Climate Dynam.* 23(1):29–44, 2004.

138. McGinn, S.M., Toure, A., Akinremi, O.O., Major, O.J., and Barr, A.G. Agroclimate and crop response to climate change in Alberta, Canada. *Outlook Agric.* 28(1):19–28, 1999.

139. Toure, A., Major, D.J., and Lindwall, C.W. Comparison of 5 wheat simulation-models in Southern Alberta. *Can. J. Plant Sci.* 75(1):61–68, 1995.

140. Toure, A., Major, D.J., and Lindwall, C.W. Sensitivity of 4 wheat simulation-models to climate-change. *Can. J. Plant Sci.* 75(1):69–74, 1995.

141. Global Change and Terrestrial Ecosystems (GCTE). *GCTE Focus 3 Wheat Network: 1996 Model and Experimental Metadata.* Report 2, 2nd ed. GCTE Focus 3 Office, Wallingford, U.K., 1996.

142. Kabat, P., Marshall, B., van den Broek, B.J., Vos, J., and van Keulen, H. *Modelling and Parameterization of the Soil-Plant-Atmosphere System. A Comparison of Potato Growth Models.* Wageningen Pers, Wageningen, the Netherlands, 1995.

143. Semenov, M.A., Wolf, J., Evans, L.G., Eckersten, H., and Iglesias, A. Comparison of wheat simulation models under climate change. II. Application of climate change scenarios. *Climate Res.* 7:271–281, 1996.

144. Wolf, J., Evans, L.G., Semenov, M.A., Eckersten, H., and Iglesias, A. Comparison of wheat simulation models under climate change. I. Model calibration and sensitivity analyses. *Climate Res.* 7:253–270, 1996.

145. Wolf, J. Comparison of two soya bean simulation models under climate change. II. Application of climate change scenarios. *Climate Res.* 20:71–81, 2002.

146. Mall, R.K. and Aggarwal, P.K. Climate change and rice yields in diverse agro-environments of India. I. Evaluation of impact assessment models. *Climatic Change* 52:315–330, 2002.

147. McMaster, G.S. and Wilhelm, W.W. Comparison of the equations for predicting the phyllochron of wheat. *Crop Sci.* 35:30–36, 1995.

148. McMaster, G.S. and Willhelm, W.W. Growing degree-days: one equation, two interpretations. *Agric. For. Meteorol.* 87:291–300, 1997.

149. White, J.W., Ed. *Modeling Temperature Response in Wheat and Maize. Proceedings of a Workshop.* CIMMYT, El Batan, Mexico, 2003.
150. Campbell, C.A., Davidson, H.R., and Warder, F.G. Effects of fertilizer N and soil moisture on yield, yield components, protein content and N accumulation in the aboveground parts of spring wheat. *Can. J. Soil Sci.* 57(3):311–327, 1977.
151. Campbell, C.A., Cameron, D.R., Nicholaichuk, W., and Davidson, H.R. Effects of fertilizer N and soil moisture on growth, N content, and moisture use by spring wheat. *Can. J. Soil Sci.* 57(3):289–310, 1977.
152. Campbell, C.A. and Paul, E.A. Effects of fertilizer N and soil moisture on mineralization, N recovery and A values, under spring wheat grown in small lysimeters. *Can. J. Soil Sci.* 58(1):39–51, 1978.
153. Bannayan, M., Hoogenboom, G., and Crout, N.M.J. Photothermal impact on maize performance: a simulation approach. *Ecol. Modeling* 180(2–3):277–290, 2004.
154. Smil, V. *Feeding the World. A Challenge for the Twenty-First Century.* MIT Press, Cambridge, MA, 2000.
155. Sanchez, P.A. Linking climate change research with food security and poverty reduction in the tropics. *Agric. Ecosyst. Environ.* 82:371–383, 2001.
156. Hammer, G.L., Nicholls, N., and Mitchell, C., Eds. *Applications of Seasonal Climate Forecasting in Agricultural and Natural Ecosystems. The Australian Experience.* Kluwer Academic, Dordrecht, the Netherlands, 2000.
157. Salinger, M.J., Stigter, C.J., and Das, H.P. Agrometeorological adaptation strategies to increasing climate variability and climate change. *Agric. For. Meteorol.* 103(1–2):167–184, 2000.
158. McGinn, S.M. and Shepherd, A. Impact of climate change scenarios on the agroclimate of the Canadian prairies. *Can. J. Soil Sci.* 83:623–630, 2003.
159. White, J.W., McMaster, G.S., and Edmeades, G.O. Physiology, genomics and crop response to global change. *Field Crops Res.* 90(1):1–3, 2004.
160. White, J.W., McMaster, G.S., and Edmeades, G.O. Genomics and crop response to global change: what have we learned? *Field Crops Res.* 90(1):165–169, 2004.
161. Crafts-Bradner, S.J. and Salvucci, M.E. Analyzing the impact of high temperature and CO_2 on net photosynthesis: biochemical mechanisms, models and genomics. *Field Crops Res.* 90(1):75–85, 2004.
162. Chakraborty, S., Tiedemann, A.V., and Teng, P.S. Climate change: potential impact on plant diseases. *Environ. Pollut.* 108:317–326, 2000.
163. Scherm, H., Sutherst, R.W., Harrington, R., and Ingram, J.S.I. Global networking for assessment of impacts of global change on plant pests. *Environ. Pollut.* 108:333–341, 2000.
164. Seem, R.C., Magarey, R.D., Zack, J.W., and Russo, J.M. Estimating disease risk at the whole plant level with General Circulation Models. *Environ. Pollut.* 108:389–395, 2000.
165. Tartaglione, C.A., Regional Effects of ENSO on US Hurricane Landfalls, M.S. thesis, Florida State University, Tallahasee, 45 pp.
166. Reilly, J. and Schimmelpfennig, D. Irreversibility, uncertainty, and learning: portraits of adaptation to long-term climate change. *Climatic Change* 45(1):253–278, 2000.
167. Uri, N.D. Conservation practices in US agriculture and their implication for global climate change. *Sci. Total Environ.* 256(1):23–38, 2000.
168. Rosenzweig, C. and Hillel, D. Soils and global climate change: challenges and opportunities. *Soil Sci.* 165(1):47–56, 2000.

169. Arnalds, A. Carbon sequestration and the restoration of land health. An example from Iceland. *Climatic Change* 65:333–346, 2004.

170. Boehm, M., Junkins, B., Desjardins, R., Kulshreshtha, S., and Lindwall, W. Sink potential of Canadian agricultural soils. *Climatic Change* 65:297–314, 2004.

171. Grant, B., Smith, W.N., Desjardins, R., Lemke, R., and Li, C. Estimated N_2O and CO_2 emissions as influenced by agricultural practices in Canada. *Climatic Change* 65:315–332, 2004.

172. Meinke, H. The current capability of simulating crop rotations and related issues, in *Rotation Models for Ecological Farming, CAMASE/PE Workshop Report*, M.C. Plentinger and F.W.T. Penning de Vries, Eds. Quantitative Approaches in Systems Analysis No. 10. DLO Research Institute for Agrobiology and Soil Fertility, Wageningen & C.T. de Wit Graduate School for Production Ecology, Wageningen, the Netherlands, 1997, 133–136.

6 Carbon Dynamics in Agricultural Soils

R. Lal

CONTENTS

6.1 INTRODUCTION

Soils are integral component of natural ecosystems. The functioning of ecosystems, including the cycling of elements and transfer of mass and energy, is moderated through soils. Yet, the focus of soil science research during the 19th and 20th centuries has mainly been on two principal soil functions: (1) the medium for plant (crops, pastures, and trees) growth, and (2) the foundation for civil structures (roads, buildings, hydraulic dams). The unprecedented growth in human population during the 20th century led to (1) widespread adoption of the "Green Revolution" technologies based on intensive management of plant nutrients and available water capacity of the soil in the effective root zone for enhancing crop yields, (2) rapid expansion of urban centers, which often necessitated use of fertile topsoil for brick making, along with development of highways, parking lots, airports, and recreational facilities, and dams to create water reservoirs and hydroelectric generation facilities. Consequently, the 20th century witnessed rapid advances in soil science with focus on soil as a medium for plant root growth (e.g., edaphology) with specific reference to water and nutrient movement through the soil and uptake by plants, elemental balance, aeration for root/microbial respiration, soil reaction, and optimization of these properties and processes through anthropic interventions involving tillage or lack of it, water table management, precision farming, fertigation, etc. Equally impressive advances were made in soil mechanics and soil hydrology to meet the requirements for engineering functions of soil.

The human society faces different challenges at the onset of the 21st century. While there is no cause for complacency in the pursuit of agricultural intensification for advancing food production to meet the needs of human population growing at 1.3%/annum and expected to reach 8.5 to 9.0 billion by 2050 and eventually stabilize at the 10 to 11 billion level by 2100. Yet there are other pressing demands on world soil resources that also require attention. In addition to the two usual functions, human needs in the 21st century demand prioritization of research on several other functions of soil. Important among these are (1) repository for industrial, urban, and nuclear waste, (2) storehouse of germ plasm and biodiversity, (3) archive of human and planetary history, (4) biomembrane to denature pollutant and purify water, and (5) moderate the climate. This chapter specifically focuses on the function of soil as moderator of the climate, through its influence on the global C cycle. World soils constitute the third largest global C pool comprising 1550 Pg of soil organic carbon (SOC) and 950 Pg of soil inorganic carbon (SIC). Indeed, the soil C pool of 2500 Pg is 3.3 times the atmospheric pool of 760 Pg and 4.0 times the biotic pool of 620 Pg.

Thus, the objective of this chapter is to illustrate the importance of world soils in the global carbon (C) cycle. The focus of the discussions is on the interaction between the soil and the atmospheric C pools under the changing climatic conditions, with specific reference on C dynamics in agricultural soils because of the need for their intense management to meet the needs for food, feed, fiber, and fuel production.

6.2 SOIL AS MODERATOR OF EARTH'S CLIMATE

Soil affects and is affected by the climate through numerous interactive processes (Figure 6.1). Soil affects climate by influencing outgoing (albedo and the long wave radiation) and incoming (insolation) radiation through its effects on air quality and the concentration of dust and other particulate materials, and changing amount and distribution of precipitation through its effect on relative humidity and temperature under local conditions. In turn, climate strongly affects soil properties[1-3] through its influences on rate and depth of weathering, intensity and severity of the cycles of erosion and deposition, quantity and quality of soil C pool and its stratification, and soil reaction (pH) and the attendant changes in elemental composition and cycling by flora and fauna (Figure 6.1). Strong interactive processes between soil and climate have long been recognized by the Dokuchaev School in Russia.[4] The interactions between soil and the atmosphere (climate) are closely linked to those between soil and biosphere, soil and lithosphere, and soil and hydrosphere (Figure 6.2). Soil's effects on climate are through its influence on the global C cycle. Indeed, the interactive processes between soil and the environment (e.g., biosphere, hydrosphere, lithosphere, and the atmosphere) are moderated through changes in soil processes including the global C cycle. Physical, chemical, and biological processes and properties are influenced by the climate (Table 6.1). Important soil processes that

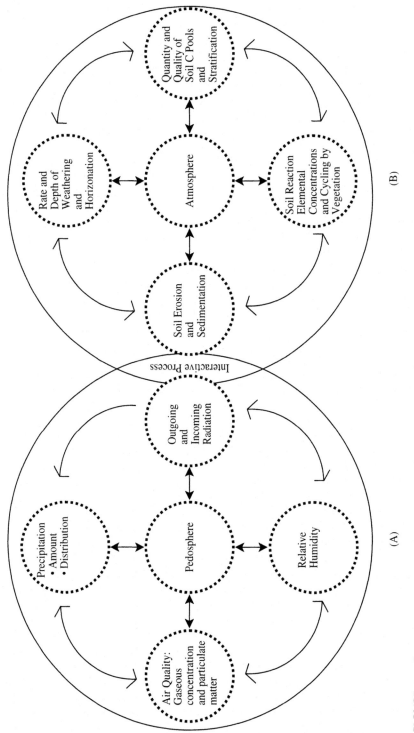

FIGURE 6.1 Effects of (A) soil on the atmosphere, (B) atmosphere on soil, and of the interactive processes on the climate.

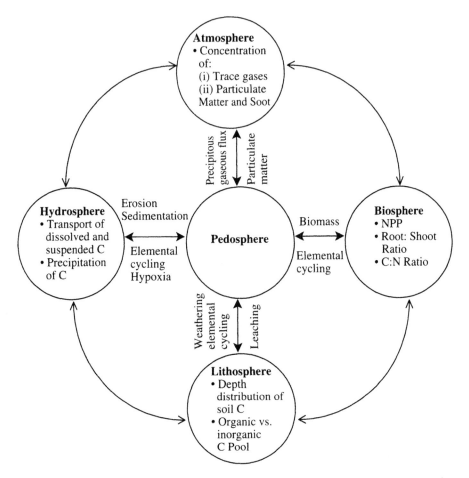

FIGURE 6.2 Interactive processes in soil with its environment with strong influence on the global C cycle.

are influenced by climate and have strong impact on the soil C cycles are soil aggregation and erosion, oxidation/mineralization of soil organic matter (SOM), and methanogenesis. In addition, climate also affects the N cycle through its impact on nitrification/denitrification processes and emission of N_2O into the atmosphere. Thus, natural or anthropogenic change in climate can drastically affect soil properties,[5,6] and the magnitude of change may depend on the antecedent conditions and the degree of climate change.

6.2.1 SOILS AND THE GLOBAL CARBON CYCLE

Soils provide numerous ecosystem services of value to humans and functioning of the biosphere.[7-9] In this regard, the importance of soil in moderating the global C cycle cannot be overemphasized. Historically, soils have been the source of atmospheric enrichment of CO_2 ever since the dawn of settled agriculture about 10,000

TABLE 6.1
Soil Processes and Properties That Are Affected by Climate and That Strongly Affect the Soil Carbon Pool

Characteristics	Soil Processes	Soil Properties
I. Physical	(a) Structure/aggregation	(i) Clay content and mineralogy
		(ii) Cementing agents (sesquoxides, carbonates, organic polymers)
	(b) Erosion	(i) Soil erodibility
		(ii) Rate of new soil formation
		(iii) Transportability and sedimentation
	(c) Water retention and transmission	(i) Plant available water capacity
		(ii) Least limiting water range
		(iii) Infiltration rate
		(iv) Deep percolation
	(d) Crusting, compaction, and hard setting	(i) Bulk density
		(ii) Porosity and pore size distribution
		(iii) Soil strength
II. Chemical	(a) Ion exchange	(i) Elemental concentration
		(ii) Ionic species
	(b) Leaching	(i) Soil reaction (pH)
		(ii) Charge density
	(c) Diffusion	(i) Concentration gradient
		(ii) Tortuosity
III. Biological	(a) Oxidation/mineralization	(i) Decomposition constant
		(ii) C:N ratio and lignin/suberin contents
	(b) Soil respiration	(i) Soil microbial biodiversity
		(ii) Biomass C
		(iii) Soil enzymes
	(c) Methanogenesis	(i) Methanogenic bacteria
		(ii) Substrate composition
	(d) Nitrification/denitrification	(i) Bacterial population
		(ii) NO_3 concentration

years ago.[10] Most agricultural soils have lost 25 to 75% of their antecedent SOC pool due to historic land use.[11] The historic loss of SOC pool is estimated at 66 to 90 Pg C, of which the loss due to accelerated erosion by water and wind is 19 to 32 Pg C.[12] Therefore, the SOC pool in most agricultural soils is drastically below their potential maximum determined by the pedologic and climatic factors. This deficit in SOC pool, which can be filled through conversion to a restorative land use and adoption of recommended land use and management practice, is also called the soil C sink capacity. The C sink capacity of agricultural soils is estimated to be about 35 to 40 Pg over a 50- to 100-year period.[13,14] Sequestration of 1 Pg of atmospheric C in soil is equivalent to reduction of atmospheric CO_2 by 0.47 ppm. Total SOC pool to 1-m depth is 1500 Pg compared to 760 Pg in the atmosphere.[11]

6.2.2 Soil Carbon Dynamics

The magnitude and rate of depletion of SOC pool depend on land use and soil/plant management practices (Table 6.2). Practices that lead to severe depletion of SOC pool include deforestation, conversion of natural to agricultural ecosystems, biomass burning and residue removal, soil tillage, and extractive or fertility mining practices. These practices set in motion those processes that exacerbate mineralization of SOM and increase the rate and cumulative amount of CO_2-C emission. Attendant changes in soil properties, with a positive feedback on emission of CO_2 and other greenhouse gases (GHGs), are reduction in the amount and stability of aggregates, increase in bulk density with a decrease in available water-holding capacity, and reduction in hydraulic conductivity and infiltration rate (Table 6.2).

The depletion in SOC pool due to conversion of natural to agricultural ecosystems and soil cultivation occurs due to (1) reduction in the amount of biomass returned to the soil, (2) increase in the rate of mineralization usually associated with the change in soil temperature and moisture regimes, and (3) increase in losses of SOC pool due to erosion and leaching. In sloping and plowed soils prone to erosion by water and/or tillage,[15] severe depletion occurs as a consequence of all three processes.

The SOC dynamics in agricultural soils can be described by static and dynamic models. The static model has been developed and used for five decades.[16-19] It states that SOC equals gains minus losses of SOC (Equation 6.1).

$$\frac{dC}{dt} = A - KC \qquad (6.1)$$

where C is the SOC pool, K is the decomposition constant, and A is the amount of C added to the soil through root biomass, crop residue, and other biosolids applied as amendments, and t is time. At steady state, when the addition of SOC by humification equals the loss by decomposition (and other processes), $dC/dt = 0$ and Equation 6.1 can be rewritten as

$$C = A/K \qquad (6.2)$$

The decomposition constant K is influenced by practices, processes, and properties outlined in Table 6.2. Sometimes Equation 6.2 is written in the form

$$C = hA/K \qquad (6.3)$$

where h is the humification efficiency, which is to 10 to 12% of the annual biomass addition in temperate climates.[20] In contrast to the static model, the dynamic exponential model is an improvement over the static model.[21-29] The two-component dynamic model is shown in Equation 6.4.

$$C_t = K_1 A/K_2 (1 - e^{-k_2 t}) + C_0 e^{-k_2 t} \qquad (6.4)$$

TABLE 6.2
Land Use and Soil/Plant Management Practices That Exacerbate the Emission of CO_2 from Soil to the Atmosphere

Practice	Processes Affected	Properties Altered
A. Deforestation	1. Energy balance	(i) Soil temperature
	2. Water balance	(ii) Soil moisture
	3. Compaction	(iii) Bulk density
	4. Erosion	(iv) Porosity
	5. Shift in vegetation	(v) SOC pool
	6. Nutrient cycling	(vi) Nutrient reserve
B. Biomass burning	1. Energy balance	(i) Soil temperature
	2. Water balance	(ii) Soil moisture
	3. Nutrient balance	(iii) Mineralization rate
	4. Runoff/leaching	(iv) Soil reaction (pH)
	5. Net primary productivity (NPP)	(v) Hydrophobicity
C. Biomass removal	1. C/elemental cycling	(i) SOC pool
	2. Crusting/compaction	(ii) Nutrient pool
	3. Activity of soil fauna	(iii) Bulk density
	4. Runoff/erosion	(iv) Infiltration rate
	5. NPP	(v) Soil temperature and moisture regimes
D. Soil tillage	1. Gaseous flux	(i) Bulk density
	2. Erosion/runoff	(ii) Infiltration rate
	3. Aggregation	(iii) Stability and amount of aggregation
	4. Compaction/crusting	(iv) Soil temperature and moisture regimes
	5. Diffusion	(v) Permeability
E. Extractive subsistence farming	1. SOC depletion	(i) Soil structure
	2. Nutrient depletion	(ii) SOC content
	3. Elemental cycling	(iii) Nutrient reserve
F. Drainage	1. Anaerobiosis	(i) Soil moisture and temperature regimes
	2. Methanogenesis	(ii) Rate of mineralization
	3. Nitrification	(iii) Leaching
	4. Denitrification	(iv) Soil reaction

where C_t is the SOC pool at time t, C_O is the antecedent SOC pool at time $t = O$, K_1 is the annual rate at which biomass is humified and added to the soil and is good for SOC sequestration, and K_2 is the annual rate of SOC loss by mineralization and erosion, and K_2 is bad for SOC sequestration. A is the accretion or annual addition of C to the soil as crop residue or other biosolids. Similar to the static model (Equation 6.1), the first term $[K_1A/K_2(1 - e^{-k_2 t})]$ is an estimate of the addition to the SOC pool through crop residue, etc., and the second term $(C_0 e^{-k_2 t})$ is an estimate of the decomposition of C_0. The difference between the two terms is the net amount of C_t at any time. Taking the derivative of Equation 6.4 with respect to t leads to Equation 6.5:

$$\frac{dC}{dt} = K_1 A - K_2 C \qquad (6.5)$$

which at equilibrium, when $dC/dt = 0$, gives Equation 6.6:

$$C = K_1 A/K_2 \qquad (6.6)$$

Similar to the K in Equation 6.3, K_1 is strongly influenced by the quality of crop residue (e.g., C:N ratio, lignin and suberin contents). In contrast, K_2 is influenced by soil properties, climatic factors, and management practices. Good and bad farming/land-use practices affecting the magnitude of constants K_1 and K_2 are outlined in Table 6.3.

Models described in Equations 6.1 through. 6.6 are based on several assumptions:[30]

1. The rate of mineralization depends on the amount of SOC at time t.
2. The rate of mineralization is not limited by lack of other elements (e.g., N).
3. The decomposition constants (K_1 and K_2) do not change over time.
4. All components of the SOC pool are equally susceptible to mineralization.

The objective of soil and crop management is to maximize C by moderating K_1, K_2, and A through tillage methods, residue management, integrated nutrient management, use of compost and biosolids, and cropping systems based on complex rotations and use of cover crops.

6.3   SOIL CARBON SEQUESTRATION

Soil C sequestration implies transfer of a fraction of atmospheric CO_2 into soil C pool through conversion of pant residue into humus, and retention of humus-C in soil for a long time. Enhancing the SOC pool of agricultural soils has numerous advantages. In comparison with engineering techniques (e.g., geologic sequestration, mineralization), SOC enhancement is a natural process, has no adverse ecological impacts, is cost-effective, and improves soil quality. Restoration of soil quality through SOC enhancement improves biomass/agronomic productivity, improves water quality by reducing erosion and sedimentation and non-point-source pollution, improves air quality by reducing wind erosion, and mitigates global warming by reducing the net rate of enrichment of atmospheric CO_2. In some cases, however, herbicide effectiveness may be decreased in soils containing high SOC concentration.[31,32]

Several important mechanisms of protection of SOC sequestered in soil include physical, chemical, and biological processes,[33–38] some of which are described below.

TABLE 6.3
Factors Affecting the Value of Constants K_1 and K_2

Parameters	Factors Increasing the Value of Constants	
	K_1 Representing Humification Efficiency (good practices)	K_2 Representing Loss by Erosion and Mineralization (bad practices)
1. Climate		
(i) Rainfall	High	Low
(ii) Temperature	Low	High
(iii) Type	Temperate, boreal, tundra, taiga	Tropics, subtropics
2. Soil		
(i) Clay	High	Low
(ii) Minerology	2:1, high-activity clays	1:1, low-activity clays
(iii) Water retention	High	Low
(iv) Type	Heavy texture, poorly drained	Light texture, excessively drained
3. Soil Management		
(i) Tillage	No-till, conservation tillage	Plow tillage
(ii) Residue	Surface mulch	Incorporation, removal, burning
(iii) Fertility	Integrated nutrient management	Nutrient deficit, fertility mining
4. Crop Management		
(i) Rotations	Complex	Simple
(ii) Cover crops	Winter cover crops	Continuous cropping
(iii) Agroforestry	With tree-based systems	Without tree-based systems
(iv) Farming systems	With animal and ley farming	Without animal
5. Landscapes		
(i) Slope gradient	Gentle to none	Undulating to steep
(ii) Position	Foot slopes	Summit and shoulder slopes
(iii) Shape	Concave/depositional	Convex
(iv) Drainage density	Low	High
(v) Aspect	North facing	South facing

6.3.1 AGGREGATION

Physical protection of SOC is an important mechanism of increasing the residence time of C in soil, and it involves its encapsulation within a stable aggregate. Humic compounds, comprising long-chain polymers, stabilize micro-aggregates against disruptive forces including chemical, mechanical, and biological processes. Several models have been proposed suggesting the role of SOC in stabilization of soil aggregates.[39]

The classical model of Edwards and Bremmer[34] illustrates the mechanism of physical protection of SOC through stabilization of micro-aggregates (Equation 6.7):

$$\text{Micro-aggregate} = [(Cl–P–OM)_x]_y \qquad (6.7)$$

where Cl is clay particle, P is polyvalent cation (e.g., Fe^{3+}, Al^{3+}, Ca^{2+}, Mg^{2+}), OM is organic molecule, and x and y are the number of these units bonded together by cementing agents to form a secondary particle or a microaggregate. The OM thus sequestered is physically protected against microbial processes and is not mineralized. The strong bonding agents (e.g., polyvalent cations, long-chain organic polymers) stabilize the aggregate while weak bonds (e.g., Na^+) disperse/slake the aggregate (Equation 6.8).

$$[(Cl-P-OM)_x]_y \underset{Aggregation}{\overset{Dispersion}{\rightleftharpoons}} y(Cl-P-OM)_x \underset{Aggregation}{\overset{Dispersion}{\rightleftharpoons}} xy(Cl-P-OM)$$

(6.8)

Dispersion or breakdown of micro-aggregates (such as by raindrop impact or by the shearing effect of flowing water) exposes the OM to microbial processes leading to emission of CO_2 into the atmosphere. Indeed, accelerated soil erosion enhances emission of CO_2 from soil to the atmosphere.[40] Predominant processes are water runoff, soil erosion, gaseous diffusion, crusting and compaction, anaerobiosis, and depletion of SOC and nutrient pools. Decline in soil quality, caused by a range of degradative processes, exacerbates depletion of the SOC pool and emission of CO_2.

6.3.2 ILLUVIATION

Deep transfer of SOC into the subsoil, away from the surface zone prone to natural and anthropogenic perturbations, is another strategy of increasing the residence time of C in soil. The SOC buried deep in the subsoil is protected against erosion by water and disruption by plowing and animal/vehicle traffic. The rate of mineralization is also lower in the subsoil than in the surface soil.

Illuviation of SOC occurs with bioturbation (e.g., earthworms) and movement with percolating water from surface into the subsoil either as dissolved organic carbon (DOC) or suspended colloid along with the clay particles. Reprecipitation of DOC in the subsoil following reaction with silica and other compounds and deposition of clay-humus colloids in the deeper layers is another mechanism of transfer of SOC from surface into the subsoil.

6.3.3 SECONDARY CARBONATES

The soil C pool comprises two components: SOC and soil inorganic carbon (SIC) subpools. Agricultural soils in arid and semi-arid regions also have the potential of sequestering SIC. The SIC subpool contains primary carbonates (e.g., calcite, dolomite, aragonite, and siderite). These primary carbonates are of lithogenic origin and occur in soil due to weathering of the parent material. In contrast, there are also secondary carbonates that occur in soil due to some pedogenic processes. There are two mechanisms of SIC sequestration: (1) formation of secondary carbonates, and (2) leaching of bicarbonates into the ground water. Visible accumulation of secondary carbonates is a common occurrence in soils of arid and semi-arid climates.[41] Secondary carbonates occur as carbonate films, threads, concretions, and pendants.[42] They may also occur as laminar caps, caliche, and calrete.[43] In gravelly soils, secondary carbonates occur as pebble coatings on lower surfaces of pebbles. Secondary carbonates are more stable than SOC.

There are four principal mechanisms of formation of secondary carbonates:[44,45] (1) dissolution of carbonates in the surface soil followed by translocation and reprecipitation into the subsoil,[46] (2) capillary rise of Ca^+ from shallow groundwater and its reprecipitation in the surface layer,[47] (3) carbonate dissolution and reprecipitation *in situ,*[48] and (4) formation of carbonates by activity of soil fauna such as termites.[45] Chemical reactions leading to formation of secondary carbonates are outlined in Equations 6.9 through 6.11.[41,44]

$$CO_2 + H_2O \rightleftarrows H_2CO_3 \tag{6.9}$$

$$H_2CO_3 \rightleftarrows H^+ + HCO_3^- \tag{6.10}$$

$$Ca^{+2} + 2H\,CO_3^- \rightleftarrows CaCo_3 + H_2O + CO_2 \tag{6.11}$$

Decreasing water content, decreasing partial pressure of CO_2 in soil air, and increasing the products of Ca^{2+} or HCO^-_3 in soil favor the precipitation of secondary carbonates (Equation 6.11). These reactions normally occur at a pH range of 7.3 to 8.5, and an adequate amount of Ca^{2+} (brought in from outside the ecosystem by deposition, liming or other amendment) must be present in the soil.

The rate of formation of secondary carbonates is generally low, and may range from 2 to 10 kg C/ha/year.[49,50] The rate of formation of secondary carbonates in semi-arid Saskatchewan, Canada, was 9.9 to 13.4 kg C/ha/year,[44,51,52] and is generally high in calcareous soils.[53] Monger and Gallegos[54] reported that the rate of formation of secondary carbonates in the southwestern U.S. ranges from 1 to 14 kg C/ha/year.

Leaching of bicarbonates into the groundwater, especially in irrigated soils and in humid environments, is another mechanism of SIC sequestration. Nordt et al.[55] demonstrated that flux of HCO^-_3 into the groundwater and rivers is a major factor of conveying carbonates to the ocean (Equation 6.12).

atm CO_2

$\downarrow$

Plant C   *Emission*

$\downarrow$ $\uparrow$ (6.12)

SOC $\longrightarrow$ *Soil* CO_2 $\longrightarrow$ HCO_3^- $\longrightarrow$ *Secondary Carbonates*

$\downarrow$ $\downarrow$

| *groundwater* | $\longleftarrow$ HCO_3

In soils of strongly calcareous parent materials, the rate of leaching of HCO^-_3 in groundwater (and eventually into the ocean) may be 0.25 to 1.0 Mg C/ha/year.[56] Leaching of bicarbonates is also an important mechanism in irrigated soils, especially when water used for irrigation is not saturated with bicarbonates.

6.4 TECHNOLOGICAL OPTIONS OF CARBON SEQUESTRATION IN AGRICULTURAL SOILS

There are numerous options for C sequestration. The choice of an appropriate strategy depends on economic and environmental considerations.[57] Three strategies of SOC sequestration in agricultural soils are (1) restoration of degraded soils and ecosystems,[58] (2) conversion of agriculturally marginal soils to pastures[59] or forests,[60] and (3) agricultural intensification on prime agricultural soils through adoption of recommended management practices (RMPs), which enhance agronomic/biomass productivity.[61] Benefits of SOC sequestration upon conversion from plow tillage to no-till are widely recognized.[62-68]

There are three factors that reduce CO_2 emissions by elimination of plowing and conversion to no-till farming. (1) Fossil fuel combustion in plow-based systems of 30 to 40 kg C/ha/year[77] is avoided by adoption of no-till farming. (2) Conversion to no-till reduces the spontaneous flux of CO_2 observed immediately after plowing and continuing for several weeks.[78-80] (3) Conversion to no-till reduces soil erosion and avoids erosion-induced emission of CO_2 from SOC transported and redistributed over the landscape.[40]

Soil fertility management is an important strategy of SOC sequestration. Extractive farming practices based on mining soil fertility deplete SOC pool. The term "mining soil fertility" implies using plant nutrients (N, P, K, Zn, etc.) contained in humus through its mineralization rather than adding nutrients through chemical fertilizers and organic amendments (e.g., manure, compost). In contrast, agronomic practices that maintain or enhance soil fertility also enhance the SOC pool. In this regard, judicious use of fertilizer, especially N, is important to SOC sequestration.

Incorporation of cover crops in the rotation cycle is another important strategy for improving soil fertility and enhancing the SOC pool in cropland soils.[81-83] Cover crops enhance SOC sequestration by increasing input of biomass C during critical periods.[84-87]

The frequency of growing a cover crop in the rotation cycle, with the main objective of enhancing or maintaining SOC pool, can be determined through the use of Equation 6.1.[12] The duration of cropping (D_c) and cover cropping (D_{cc}) can be adjusted to attain the desired level of SOC, thereby decreasing the decomposition constant K (Equation 6.1). Assuming the equilibrium SOC in soils as C_e (thus $dC/dt = 0$), Equation 6.1 can be written as Equation 6.13:

$$(-K_c C_e + A_c)D_c + (K_{cc}C_e + A_{cc})D_{cc} = 0 \qquad (6.13)$$

where K_c and K_{cc} refer to decomposition constants, A_c and A_{cc} are accretion constants, and D_c and D_{cc} are duration of cropping and cover cropping periods, respectively.[12]

Reorganization of Equation 6.13 allows estimation of the ratio of cultivation to the cover cropping period for the desired level of C_e (Equation 6.14).

$$D_c = \frac{D_{cc}(A_{cc} - K_{cc}C_e)}{(K_cC_e - A_c}$$ (6.14)

In Equation 6.14, C_e can be fixed for a recommended land use and soil/crop management practice. Thus, Equation 6.14 is a useful management tool for identifying suitable cropping systems for the region with the objectives of SOC sequestration.[12]

6.5 RATES OF SOIL CARBON SEQUESTRATION

Rates of soil C sequestration reported in the literature are highly variable, and depend on (1) soil, climate, land use, and management, and (2) methods of assessment. The rate of SOC sequestration ranges from 300 to 500 kg C/ha/year upon conversion from plow tillage to no-till,[61–71] and 100 to 400 kg C/ha/year with input of N and integrated nutrient management.[72–76] These rates are average rates measured on long-term experiments of 10 to 20 years, and rates can be maintained for 25 to 50 years. While the impact of these practices is not necessarily additive, rates as high as 1 Mg C/ha/year have been reported in agricultural soils of North America.[11] In contrast, the rate of formation of secondary carbonates is 5 o 15 kg C/ha/year and that of leaching of bicarbonates (in irrigated soils or humid regions) is 0.25 to 1.0 Mg C/ha/year.[11]

6.5.1 MEASUREMENT ISSUES RELATED TO SOIL CARBON STORAGE

Apparent variability in measured rates of soil C sequestration also depends on the sampling protocol and analytical procedures. The importance of accurate measurement of soil bulk density cannot be overemphasized.[88] The SOC pool can be measured on equal depth basis or equal mass basis,[89,90] and the rate of SOC sequestration differs depending on the choice of the method used. The SOC concentration in soil may also differ among the laboratory analytical technique used,[91] especially with the calibration of the method used. There are several new methods being developed including the remote sensing–based techniques,[92] inelastic neutron scattering method,[93,94] and laser-induced breakdown spectroscopy (LIBS)[95–97] and infrared spectroscopy.[98–100] Remote sensing–based techniques involve monitoring equipment installed on an airplane.[92] Similarly, nondestructive and *in situ* measuring techniques reduce the time and labor involved in sampling and preparation of samples, and are cost-effective. Once developed and usable, results obtained by these new methods may be more credible, less variable, and provide rapid and economic assessment of the SOC pool and its change over time.

Whatever technique is used, the data obtained must be corrected for coarse fragments, charcoal carbon, fossil carbon (especially in soils in vicinity of coal mines and in reclaimed mine soils), and carbonates.[91] The choice of the reference point or the baseline is also very important to estimating the rate of SOC sequestration. For example, the rate of SOC sequestration using the paired plot technique (e.g., no-till vs. plow tillage) may differ depending on whether the baseline SOC pool in plow till is taken when a part of it was converted to no-till or when soil samples were obtained at the end of the experiment. The rate of SOC sequestration may also differ if SOC pools in both plow till and no-till are compared with reference to a natural ecosystem.

Then, there is a question of the hidden C costs of input used. Hidden C costs refer to the amount of C emitted through fossil fuel combustion in manufacture, packaging, transport and application of chemical or amendments, lift and application of irrigation water, and in diesel use for soil tillage and other farming operations. The data in Table 6.4 show that hidden C costs are high for plow tillage, with inputs of nitrogenous fertilizers and pesticides.[77,101] Some argue that hidden C costs of input must be deducted from the gross SOC sequestration rate to assess the net rate.[102,103] In contrast, others believe that inputs are used for food production, and SOC seques-

---

TABLE 6.4
Hidden Carbon Costs of Agricultural Operations

Operation	Equivalent Carbon Emission
1. Tillage (kg C/ha)	
(i) Moldboard plowing	15.2
(ii) Chisel plowing	7.9
(iii) Disc plowing	8.3
(iv) Cultivation	4.0
2. Fertilizer (kg C/kg)	
(i) Nitrogen	1.3
(ii) Phosphorus	0.2
(iii) Potassium	0.15
(iv) Lime	0.16
3. Pesticides (kg C/kg)	
(i) Herbicides	6.3
(ii) Insecticides	5.1
(iii) Fungicides	3.9
4. Irrigation (kg C/ha)	
(i) Sprinkler	121.0
5. Farm operations (kg C/ha)	
(i) Planting/sowing	3.8
(ii) Fertilizer spreading	7.6
(iii) Corn silage	19.6
(iv) Corn harvesting (combine)	10.0
(v) Spraying	1.4

Source: Modified from Lal.[77]

tration is merely a by-product of the necessity of adopting recommended management practices.[11] However, the difference in hidden C costs of RMPs vs. traditional practices must be accounted for. Full cost accounting is extremely important.

6.6 CONCLUSIONS

Carbon sequestration in agricultural soils is an important strategy of reducing the net emission of CO_2 into the atmosphere while restoring degraded soils, improving soil and water quality, and increasing agronomic productivity and farm income. This is a truly win–win situation because of numerous ancillary benefits of SOC sequestration. Soil C sequestration can be achieved through adoption of conservation tillage based on crop residue mulch and use of cover crops, use of nitrogenous fertilizers and organic amendments, erosion management, and restoration of degraded soils. These practices increase rate of input of biomass carbon and decrease losses by erosion and mineralization. There are several new methods of soil carbon determination including the remote-sensing techniques, *in situ* or nondestructive methods based on assessment under field conditions. When available, use of these methods will make determination of soil carbon pool and its temporal changes credible, less variable, rapid, routine, and economic. A large database also exists on the hidden carbon costs of recommended input and on the economics of soil carbon sequestration.[68,77,102,103] While several promising technologies for SOC sequestration and the underpinning science are known, widespread adoption of recommended practices depends on a range of social and economic factors and the issues pertaining to human dimensions. Commodification of soil carbon, through trading carbon credits in domestic and international markets, with or without the Kyoto Protocol, may enhance the adoption rate of recommended technologies. Providing incentives to farmers, through government-sponsored programs, may be another option. Creating public awareness about the potential of soil carbon sequestration to enhance soil quality and mitigate the climate change through appropriate outreach programs may be an effective strategy of practical significance. Soil scientists and agronomists need to communicate with land managers, policy makers, and the public at large about the ecological, economic, and climatic benefits of enhancing/restoring the soil carbon pool. Soil carbon is an important natural resource, and there are numerous reasons for its improvement and restoration in agricultural soils.

REFERENCES

1. Jenny, H. 1943. *Factors of Soil Formation.* McGraw Hill, New York, 281 pp.
2. Jenny, H. 1980. *The Soil Resource: Origin and Behavior.* Springer-Verlag, New York, 377 pp.
3. Jenny, H. and C.D. Leonard 1934. Functional relationships between soil properties and rainfall. *Soil Sci.* 38: 363–381.
4. Neustruev, S.S. 1927. Genesis of Soils. *Russ. Pedol. Invest. USSR Acad. Sci.*, 5 pp.
5. Heal, O.W. 2001. Potential response of natural terrestrial ecosystems to Arctic climate change. *Buvsindi* 14: 3–16.

6. Nisbet, T. 2002. Implications of climate change: soil and water. *For. Comm. Bull.* 125: 53–67.

7. Costanza, R., R. d'Ange, R. de Groots et al. 1997. The value of the world's ecosystem services and natural capital. *Nature* 387: 253–260.

8. Daily, G.C., P.A. Matson, and P.M. Vitousek. 1997. Ecosystem services supplied by soil. In *Nature's Services: Societal Dependence on Natural Ecosystems*, G.C. Daily, Ed. Island Press, Washington, D.C., 113–132.

9. Odum, E. 1969. The strategy of ecosystem development. *Science* 164: 262–270.

10. Ruddiman, W. 2003. The anthropogenic greenhouse era began thousands of years ago. *Climatic Change* 6: 261–293.

11. Lal, R. 2004. Soil carbon sequestration impacts on global climate change and food security. *Science* 304: 1623–1627.

12. Lal, R. 1999. Soil management and restoration for C sequestration to mitigate the accelerated greenhouse effect. *Prof. Environ. Sci.* 1: 307–326.

13. Brown, S., J. Sathaye, M. Cannel, and P. Kauppi. 1996. Management of forests for mitigation of greenhouse gas emissions. In *Climate Change 1995: Impacts, Adaptations and Mitigation of Climate Change: Scientific and Technical Analyses. Contributions of Working Group II to the Second Assessment Report of IPCC*, R.T. Watson, M.C. Zingowers, and R.H. Moss, Eds. Cambridge University Press, Cambridge, 773–787.

14. Cole, C.V., J. Duxbury, J. Freney, O. Heinemeyer, K. Minami, A. Mosser, K. Paustian, N. Rosenberg, N. Sampson, D. Sauerbeck, and Q. Ahao. 1997. Global estimates of potential mitigation of greenhouse gas emissions by agriculture. *Nutr. Cycling Agroecosyst.* 49: 221–228.

15. Mech, S.J. and G.A. Free. 1942. Movement of soil during tillage operations. *Agric. Eng.* 23: 379–332.

16. Woodruff, C.M. 1949. Estimating the nitrogen delivery of soil from the organic matter determination as reflected by Sanborn field. *Soil Sci. Soc. Am. Proc.* 14: 208–212.

17. Greenland, D.J. 1971. Changes in the nitrogen status and physical conditions of soils under pasture, with special reference to the maintenance of the fertility of Australian soils used for growing wheat. *Soil Fertil.* 34: 237–251.

18. Jenkinson, D.S. and J.H. Raynor. 1977. The turnover of soil organic matter in some of the Rothamsted classical experiments. *Soil Sci.* 123: 298–305.

19. Stevenson, F.J. 1994. *Humus Chemistry: Genesis, Composition, Reaction.* John Wiley & Sons, New York.

20. Puget, P., R. Lal, C. Izaurralde, M. Post, and L. Owens. 2005. Stock and distribution of total and corn-derived soil organic carbon in aggregate and primary particle fractions for different land use and soil management practices. *Soil Sci.* 170: 256–279.

21. Dalal, R.C. and R.J. Mayor. 1986. Long-term trends in fertility of soils under continuous cultivation and cereal cropping in southern Queensland. II. Total organic carbon and its rate of loss from the soil profile. *Aust. J. Soil. Res.* 24: 281–292.

22. Schimel, D.S., D.C. Coleman, and K.A. Horton. 1985. Soil organic matter dynamics in aired rangeland and cropland toposequences in North Dakota. *Geoderma* 36: 201–214.

23. Parton, W.J., D.S. Schimel, C.V. Cole, and D.S. Ojima. 1987. Analysis of factors controlling soil organic matter levels in Great Plains grasslands. *Soil Sci. Soc. Am. J.* 51: 1173–1179.

24. Balesdent, J. and M. Balabane. 1992. Maize root-derived soil organic matter estimated by natural ^{13}C abundance. *Soil Biol. Biochem.* 24: 97–101.

25. Jenkinson, D.S. 1988. Soil organic matter and its dynamics. In *Russel's Soil Conditions and Plant Growth*, A. Wild, Ed. Longman, London, 564–607.

26. Huggins, D.R., G.A. Buyanovsky, G.H. Wagner, J.R. Brown, R.G. Darmody, T.R. Peck, G.W. Lesoing, M.B. Vanotti, and L.G. Bundy. 1998. Soil organic carbon in the tall grass prairies-derived region of the maize belt: effects of the long-term crop management. *Soil Tillage Res.* 47: 219–234.

27. Bayer, C., J. Mielniczuk, and L. Martin Neto. 2000. Efeito de sistemas de preparo e de culture na dinamica de matema organica e na mitigacao de emissoes de CO_2. *R. Bros. Ci solo* 24: 599–607.

28. Gregorich, E.G., B.H. Ellert, and C.M. Monreal. 1995. Turnover of soil organic matter and storage of corn residue carbon estimated from natural ^{13}C abundance. *Can. J. Soil Sci.* 75: 161–168.

29. Jenkinson, D.S., D.E. Adams, and A. Wild. 1991. Global warming and soil organic matter. *Nature* 351: 304–306.

30. Lal, R. 2000. World cropland soils as a source or sink for atmospheric CO_2. *Adv. Agron.* 71: 145–191.

31. Dielman, J.A., D.A. Mortenson, D.D. Buhler, C.A. Cambadella, and T.B. Moormann, 2000. Identifying associations among site properties and weed species abundance. I. Multivariate analyses. *Weed Sci.* 48: 567–575.

32. Burton, M.G., D.A. Mortenson, D.B. Mark, and T.L. Lindquist. 2004. Factors affecting the realized niches of common sunflower (*Helianthus annuus*) in ridge tillage corn: seed germination, emergence and survival. *Weed Sci.* 52: 779–787.

33. Emmerson, W.W. 1959. The determination of the stability of soil crumbs. *J. Soil Sci.* 5: 233.

34. Edwards, A.P. and J.M. Bremner. 1967. Microaggregates in soils. *J. Soil Sci.* 18: 65–73.

35. Tisdall, J.M. and J.M. Oades. 1982. Organic matter and water stable aggregates in soils. *J. Soil Sci.* 33: 141–163.

36. Golchin, A., J.M. Oades, J.O. Skjemstad, and P. Clark. 1994. Soil structure and carbon cycling. *Aust. J. Soil Res.* 32: 1043–1068.

37. Laird, D.A., D.A. Martens, and W.L. Kingery. 2001. Nature of clay-humic complexes in an agricultural soil: I. Chemical, biochemical, and spectroscopic analyses. *Soil Sci. Soc. Am. J.* 65: 1413–1418.

38. Six, J., R.T. Conant, E.A. Paul, and K. Paustian. 2002. Stabilization mechanisms of soil organic matter: implications for C-saturation of soils. *Plant Soil* 241: 155–176.

39. Blanco-Canqui, H. and R. Lal. 2004. Mechanisms of carbon sequestration in soil aggregates. *Crit. Rev. Plant Sci.* 23: 481–504.

40. Lal, R. 2003. Soil erosion and the global carbon budget. *Environ. Int.* 29: 437–450.

41. Lal, R., J.M. Kimble, and B.A. Stewart, Eds. 2000. *Global Climate Change and Pedogenic Carbonates*. Lewis/CRC Press, Boca Raton, FL, 378 pp.

42. Wilding, L.P., L.T. West, and L.R. Drees. 1990. Field and laboratory identification of calcic and petrocalcic horizons. In *Proc. Int. Soil Correlation Meeting (ISCOM IV): Characterization, Classification and Utilization of Aridisols*, J.M. Kimble and W.D. Nettleton, Eds., 3–17 October 1987, USDA-SCS, Lincoln, NE, 79–92.

43. Gile, L.H. 1993. Carbonate stages in sandy soils of the Leasburg Surface, southern New Mexico. *Soil Sci.* 156: 101–110.

44. Mermut, A.R. and A. Landi. 2004. Secondary/Pedogenic Carbonates. In *Encyclopedia of Soil Science*, R. Lal, Ed. Marcel Dekker, New York.

45. Monger, H.C. 2002. Pedogenic carbonate: links between biotic and abiotic $CaCO_3$, presented at 17th World Congress of Soil Science, 14–21 August 2002, Thailand, Symposium 20, Oral presentation 891.

46. Marion, G.M., W.H. Schlesinger, and P.J. Fonteyn. 1985. CALDEP: a regional model for soil $CaCO_3$ (caliche) deposition in southwestern deserts. *Soil Sci.* 139: 468–481.

47. Sobecki, T.M. and L.P. Wilding. 1983. Formation of calcic and argillic horizons in selected soils of the Texas Coast Prairie. *Soil Sci. Soc. Am. J.* 47: 707–715.

48. Rabenhorst, M.C. and L.P. Wilding. 1986. Pedogenesis on the Edwards Plateau, Texas. III. A new model for the formation of petrocalcic horizons. *Soil Sci. Soc. Am. J.* 50: 693–699.

49. Machette, M.N. 1985. Calcic soils of the southwestern United States. In *Soils and Quarternary Geomorphology of Southwestern United States*; Weide, D.L., Ed. Geological Society of America, Vol. 203, 1–21.

50. Schlesinger, W.H. 1985. The formation of caliche in soils of Mojave Desert, California. *Geochim. Cosmochim. Acta* 49, 57–66.

51. Landi, A. 2002. Carbon Balance in the Major Soil Zones of Saskatchewan., Ph.D. thesis, University of Saskatchewan, Canada.

52. Landi, A., A.R. Mermut, and D.W. Anderson. 2003. Origins and rate of pedogenic carbonate accumulation in Saskatchewan soils, Canada. *Geoderma* 17: 143–156.

53. Schlesinger, W.H., G.M. Marion, and P.J. Fonteyn. 1989. Stable isotope ratios and the dynamics of caliche in desert soils. In *Stable Isotopes in Ecological Research*, P.W. Rundel, J.R. Eghlringer, and K. Nagy, Eds. Springer-Verlag, Berlin, 309–317.

54. Monger, H.C. and R.A. Gallegos. 2000. Biotic and abiotic processes and rates of pedogenic carbonate accumulation in the southwestern United States — relationship to atmospheric CO_2 sequestration. In *Global Climate Change and Pedogenic Carbonates*, R. Lal, J.M. Kimble, and B.A. Stewart, Eds. CRC/Lewis Publishers, Boca Raton, FL, 273–290.

55. Nordt, L.C., L.P. Wilding, and L.R. Drees. 2000. Pedogenic carbonate transformations in leaching soil systems. In *Global Climate Change and Pedogenic Carbonates,* R. Lal, J.M. Kimble, and B.A. Stewart, Eds. CRC/Lewis Publishers, Boca Raton, FL, 43–64.

56. Wilding, L.P. 1999. Comments on paper by R. Lal. In *Carbon Sequestration in Soils: Science, Monitoring and Beyond*, N. Rosenberg, R. Cesar Izaurralde, and E.L. Malone, Eds. Battelle Press, Columbus, OH, 146–149.

57. Wigley, T.M.L., R. Richels, and J.A. Edmonds. 1996. Economic and environmental choices in the stabilization of atmospheric CO_2. *Nature* 379: 250–243.

58. Daily, G.C. 1995. Restoring value to the world's degraded lands. *Science* 269: 350–354.

59. Conant, R.T., K. Paustian, and E.T. Elliott. 2001. Grassland management and conversion into grassland: effects on soil carbon. *Ecol. Appl.* 11: 343–355.

60. Harrison, K., W. Broecker, and G. Bonani. 1993. The effect of changing land use on soil radiocarbon. *Science* 262: 725–726.

61. Lal, R., J.M. Kimble, R.F. Follett, and V.C. Cole. 1998. *Potential of U.S. Cropland for C Sequestration and Greenhouse Effect Mitigation*. NRCS, Washington, D.C.

62. Campbell, C.A. and R.P. Zentner. 1998. Crop production and soil organic matter in long-term crop rotations in the semi-arid northern Great Plains of Canada. In *Soil Organic Matter in Temperate Agroecosystems: Long-Term Experiments in North America,* E.A. Paul, K. Paustian, E.T. Elliott, and C.V. Cole, Eds. CRC Press/Lewis Publishers, Boca Raton, FL, 317–333.

63. Janzen, H.H., C.A. Campbell, E.G. Gregorich, and B.H. Ellert. 1998. Soil carbon dynamics in Canadian agroecosystems. In *Soil Processes and Carbon Cycles*, Lal, R., J.M. Kimble, R.F. Follett, and B.A. Stewart, Eds. CRC Press, Boca Raton, FL, 57–80.

64. Nyborg, M, M. Molina-Ayala, F.D. Solberg, R.C. Izaurralde, S.S. Malhi, and H.H. Janzen. 1998. Carbon storage in grassland soils as related to N and S fertilizer. In *Management of Carbon Sequestration in Soil*, Lal, R., J.M. Kimble, R.F. Follett, and B.A. Stewart, Eds. CRC Press, Boca Raton, FL, 421–432.

65. Paul, E.A., K. Paustian, E.T. Elliott, and C. V. Cole, Eds. 1997. *Soil Organic Matter in Temperate Agroecosystems: Long-Term Experiments in North America.* CRC Press, Boca Raton, FL.

66. Lal, R. 2000. World cropland soils as a source of sink for atmospheric CO_2. *Adv. Agron.* 71: 145–191.

67. Izaurralde, R.C., W.B. McGill, J.A. Robertson, N.G. Juma, and J.J. Thurston. 2001. Carbon balance of the Breton classical plots over half a century. *Soil Sci. Soc. Am. J.* 65: 431–441.

68. Marland, G., T.O. West, B. Schlamadinger, and L. Canella. 2003. Managing soil organic carbon in agriculture: the net effect on greenhouse gas emissions. *Tellus Ser. B Chem. Phys. Meteorol.* 55: 613–621.

69. Bruce, J.P., M. Frome, E. Haites, H. Janzen, R. Lal, and K. Paustian. 1999. Carbon sequestration in soils. *J. Soil Water Cons.* 54: 382–389.

70. Post, W.M. and K.C. Kwon. 2000. Soil carbon sequestration and land-use change: process and potential. *Global Change Biol.* 6: 317–327.

71. West, T.O. and W.M. Post. 2002. Soil organic carbon sequestration rates by tillage and crop rotation: a global data analysis. *Soil Sci. Soc. Am. J.* 66: 1930–1946.

72. Janzen, H.H. 1987. Effect of fertilizer on soil productivity in long-term spring wheat rotations. *Can. J. Plant Sci.* 67: 165–174.

73. Halvorson, A.D., C.A. Reule, and R.F. Follett. 1999. Nitrogen fertilization effects on soil carbon nitrogen in dryland cropping system. *Soil Sci. Soc. Am. J.* 63:912–917.

74. Halvorson, A.D., B.J. Wienhold, and A.L. Black. 2002. Tillage, nitrogen, and cropping system effects on soil carbon sequestration. *Soil Sci. Soc. Am. J.* 66: 906–912.

75. Gregorich, E.G., B.H. Ellert, C.F. Drury, and B.C. Liang. 1996. Fertilization effects in soil organic matter turnover and corn residue C storage. *Soil Sci. Soc. Am. J.* 60: 372–476.

76. Belanger, G., J.E. Richards, and D.A. Angers. 1999. Long term fertilization effects on soil carbon under permanent swards. *Can. J. Soil Sci.* 79: 99–102.

77. Lal, R. 2004. Carbon emission from farm operations. *Environ. Int.* 30: 981–990.

78. Reicosky, D.C. and M.J. Lindstron. 1993. Fall tillage method: effect on short-term carbon dioxide flux from soil. *Agron. J.* 85: 1237–1243.

79. Prior, S.A., D.C. Reicosky, D.W. Reeves, G.B. Runion, and R.L. Raper. 2000. Residue and tillage effects on planting implement-induced short-term CO_2 and water loss from a loamy sand soil in Alabama. *Soil Tillage Res.* 54: 197–199.

80. Rochette, P. and D.A. Angers. 1999. Soil surface carbon dioxide fluxes induced by spring, summer and fall moldboard plowing in a sandy loam. *Soil Sci. Soc. Am. J.* 63: 621–628.

81. Kuo, S., M.U. Sainju, and E. Jellum. 1997. Winter cover crop effects on soil organic carbon and carbohydrate in soil. *Soil Sci. Soc. Am. J.* 61: 145–152.

82. Nyakatawa, E.Z., K.C. Reddy, and K.R. Sistani. 2001. Tillage, cover cropping, and poultry litter effects on selected soil chemical properties. *Soil Tillage Res.* 58: 69–79.

83. Sainju, U.M., B.P. Singh, and W.F. Whitehead. 2002. Long-term effects of tillage, cover crops, and nitrogen fertilization on organic carbon and nitrogen concentrations in sandy loam soils in Georgia, USA. *Soil Tillage Res.* 63: 167–179.

84. Franzluebbers, A.J., F.M. Hons, and D.A. Zuberer. 1994. Long-term changes in soil carbon and nitrogen pools in wheat management system. *Soil Sci. Soc. Am. J.* 58: 1639–1645.

85. Bowman, R.A., R.S. Vigil, D.C. Nielsen, and R.L. Anderson. 1999. Soil organic matter changes in intensively cropped dryland systems. *Soil Sci. Am. J.* 63: 186–191.

86. Follett, R.F. 2001. Soil management concepts and carbon sequestration in cropland soils. *Soil Tillage Res.* 61: 77–92.

87. Gregorich, E.G., C.F. Drury, and J.A. Baldock. 2001. Changes in soil carbon under long-term maize in monoculture and legume-based rotation. *Can. J. Soil Sci.* 81: 21–31.

88. Lal, R. and J.M. Kimble. 2000. Importance of soil bulk density and methods of measurement. In *Global Climate Change and Pedogenic Carbonates*, R. Lal, J.M. Kimble, and B.A. Stewart, Eds. CRC/Lewis Publishers, Boca Raton, FL, 31–44.

89. Ellert, B.H. and J.R. Bethany. 1995. Calculation of organic matter and nutrients stored in soils under contrasting management regimes. *Can. J. Soil Sci.* 75: 525–538.

90. Ellert, B.H., H.H. Janzen, and B. McConkey. 2001. Measuring and comparing soil carbon storage. In *Assessment Methods for Soil Carbon*, R. Lal, J.M. Kimble, R.F. Follett, and B.A. Stewart, Eds. CRC/Lewis Publishers, Boca Raton, FL, 131–146.

91. Kimble, J.M., R. Lal, and R.F. Follett. 2001. Methods for assessing soil C pool. In *Assessment Methods for Soil Carbon*, R. Lal, J.M. Kimble, R.F. Follett, and B.A. Stewart, Eds. CRC/Lewis Publishers, Boca Raton, FL, 3–12.

92. Shepherd, K. and M.G. Walsh. 2002. Development of reflectance spectral libraries for characterization of soil properties. *Soil Sci. Soc. Am. J.* 66: 988–998.

93. Wielpolski, L., I. Orion, G. Hendrey, and H. Rogers. 2000. Soil carbon measurements using inelastic neutron scattering. *IEEE Trans. Nucl. Sci.* 47: 914–917.

94. Wielpolski, L., S. Mitra, G. Hendrey, Orion, S., Prior, H., Rogers, B., Runion, and A. Tonbert. 2004. Non-destructible Soil Carbon Anaylzer (NS-SCA). 2004 BNL Report 72200-2004.

95. Cremers, D.A., M.H. Ebinger, D.D. Breshears, P.J. Unkefer, S.A. Kammerdiener, M.J. Ferris, K.M. Catlett, and J.R. Brown. 2001. Measuring total soil carbon with laser-induced breakdown spectroscopy (LIBS). *J. Environ. Qual.* 30: 2202–2206.

96. Ebinger, M.H., M.L. Norfleet, D.D. Breshears, D.A. Cremers, M.J. Ferris, P.U. Unkefer, M.S. Lamb, K.L. Goddard, and C.W. Meyer. 2003. Extending the applicability of laser-induced breakdown spectroscopy for total soil carbon measurement. *Soil Sci. Soc. Am. J.* 67:1616–1619.

97. Harris, R.D., D.A. Cremers, M.H. Ebinger, and B.K. Bluhm. 2004. Determination of nitrogen in sand using laser-induced breakdown spectroscopy. *Appl. Spectrosc.* 58: 770–775.

98. Reeves, J.B., III and G.W. McCarty. 2001. Quantitative analysis of agricultural soils using near infrared reflectance spectroscopy and fiber-optic probe. *J. Near Infrared Spectrosc.* 9: 25–42.

99. Reeves, J.B., III, G.W. McCarty, and J.J. Meisenger. 1999. Near infrared reflectance spectroscopy for the analysis of agricultural soils. *J. Near Infrared Spectrosc.* 7: 179–193.

100. Reeves, J.B., III, G.W. McCarty, and V.B. Reeves. 2001. Mid-infrared diffuse reflectance spectroscopy for the quantitative analysis of agricultural soils. *J. Agric. Food Chem.* 49: 766–772.

101. West, T.O. and G. Marland. 2002. A synthesis of carbon sequestration, carbon emissions, and net carbon flux in agriculture, comparing tillage practices in the United States. *Agric. Ecosyst. Environ.* 91: 217–232.
102. Robertson, G.P., E.A. Paul, and R.R. Harwood. 2000. Greenhouse gases in intensive agriculture: contributions of individual gases to the relative forcing of the atmosphere. *Science* 289: 1922–1925.
103. Schlesinger, W.H. 1999. Carbon sequestration in soil. *Science* 284: 2095.

7 Plant Species Diversity: Management Implications for Temperate Pasture Production

M.A. Sanderson

CONTENTS

7.1 INTRODUCTION

Biodiversity refers to the broad array of genetic material, species, and ecosystems that make up the natural world including their variability and interactions. On a global scale, Earth's biodiversity encompasses 1.75 million described species from viruses to vertebrates and the habitats they occupy.[1] Conservation of the Earth's biodiversity has important implications for ecosystem functions (habitat, biological, or system properties or processes of ecosystems[2]) and the goods and services humans derive from them.

Some ecological research indicates that increased plant biodiversity benefits ecosystem functions such as primary productivity, nutrient retention, and resistance to weed invasions in experimental grasslands. These results and concepts have been extrapolated to management of forage and pastureland.[3,4] It is not clear, however, whether the results and concepts of basic ecological biodiversity studies apply to managed forage and grazing lands.

Early research on pasture management seemed to advocate relatively complex (diverse) mixtures of grasses and legumes.[5] During the 1950s, however, the emphasis of pasture management shifted to monocultures maintained by fertilizers and other inputs or simple mixtures of grasses and legumes (e.g., one of each) and management of those mixtures to maintain the legume component.[6,7] Since that time, the management focus has remained on monocultures or simple forage mixtures.

Pasture management in temperate regions is moving beyond the traditional concerns of optimizing the quality and quantity of herbage for animal production. New challenges in pasture management include such cross-cutting issues as sustainability, reducing inputs of fertilizers and pesticides, soil protection, C sequestration, resistance to invasion by alien plants and insects, and the aesthetic value of the landscape.[8,9] It is within this context that increased biodiversity may play an important role.

My objectives are to (1) discuss the role of biodiversity in temperate pasturelands, (2) consider the evidence for ecosystem benefits of increased plant species diversity in temperate grazing lands, and (3) consider whether it is worth managing for increased plant species diversity.

7.2 PASTURE BIODIVERSITY

Pastures contain a broad array of insects, invertebrates, fungi, animals, and plants. For example, grasslands in Northern Ireland vary greatly in plant and insect species depending on the level of management and inputs (Table 7.1). Generally, biodiversity decreased as management intensity increased. Semi-natural pastures in Sweden had a broad range of plants, insects, and birds (Figure 7.1). Nearly 300 species of herbaceous plants, 275 species of beetles, and several species of soil macroinvertebrates have been identified in pastures of the northeastern U.S.[10-14]

TABLE 7.1
Plant and Insect Biodiversity in Northern Ireland Grasslands[47]

Grassland Type	Plants	Beetles	Spiders
	No. of Species per Habitat		
Wet grassland	110	28	30
Unimproved grassland	104	27	47
Hay meadow	91	28	30
Limestone grassland	84	26	33
Heather moorland	70	34	42
Woodland	58	25	38
Improved grassland	42	22	18

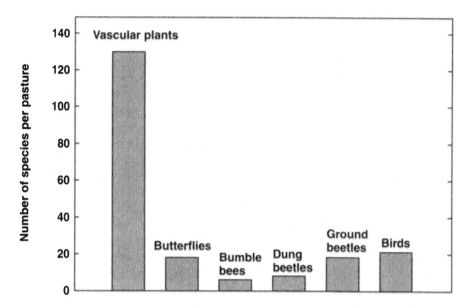

FIGURE 7.1 Biodiversity in semi-natural pastures of south-central Sweden.[48]

Pastures can be very diverse ecosystems, but many components of this biodiversity cannot be easily managed or directly manipulated. Management practices can influence insect and soil animal abundance and diversity,[15,16] but plant species diversity may be the component of biodiversity most amenable to management. This chapter focuses on plant species diversity because of its central role in primary production in the pasture ecosystem.

7.2.1 PLANT SPECIES DIVERSITY IN PASTURES

Plant species diversity refers to the number of species (species richness) and their relative abundance (species evenness) in a defined area. Ecologists use various indices that combine these two measures to describe diversity in plant communities. Spatial scale strongly influences plant species diversity in that (1) species richness increases with the area sampled, and (2) small-scale (alpha or within-community) diversity varies independently from large-scale (beta or across-communities) diversity. Evaluating species richness without taking into account evenness and spatial scale effects could underestimate the importance of diversity in pasture ecosystems.

On the surface, pasturelands may appear uniform with a homogeneous mixture of plant species. Closer examination, however, reveals a complex temporal and spatial structure of both species and species richness in pastures.[17] Plant species richness in traditionally managed grasslands (i.e., species-rich ancient grazing lands such as chalk grassland and heathland) in northwest Europe ranged from 50 to 60 species 100/m^2, whereas more intensively managed grasslands contained 10 to 20 species 100/m^2.[18]

7.2.2 PLANT DIVERSITY AND PASTURE ECOSYSTEM FUNCTION

The principal ecosystem functions (defined as biological system properties or processes of ecosystems) include primary productivity, nutrient cycling, and decomposition. In grasslands, many of these ecosystem functions provide ecosystem services of value to humans such as control of soil erosion, nutrient cycling, food and fiber production, along with recreational and cultural values.[2] Although we might be focused primarily on productive output in managed grasslands, the ecosystem goods and services provided by grazing ecosystems must be considered as well.

Reported benefits of plant diversity in grasslands include increased primary production; greater ecosystem stability in response to disturbance; reduced invasion by exotic species; and greater nutrient cycling and retention.[19] Although there is some consensus that plant diversity benefits grassland ecosystem function, there are reports that indicate no general benefit of increased plant diversity, and highly productive agricultural systems often rely on low plant species diversity.[20,21] The studies indicating benefits for plant diversity suggest that managing for increased plant species diversity on pasturelands could increase forage yield, improve yield stability, and reduce soil nutrient losses.

7.3 EVIDENCE FOR DIVERSITY EFFECTS IN PASTURELAND

Many of the studies on diversity effects in grasslands were done with several non-agronomic species in small plots that were clipped and not grazed. Thus, the results from these types of experiments are difficult to extrapolate to pastures. In this section, I discuss some of the evidence for plant diversity effects on primary and secondary productivity, ecosystem stability, resistance to invasion, and nutrient cycling in pastures. I then discuss some of the mechanisms involved.

7.3.1 COMPLEX FORAGE MIXTURES AND PRIMARY PRODUCTIVITY

One of the principal benefits ascribed to increased plant diversity in grassland systems has been increased primary productivity. In early applied research on complex forage mixtures, studies in Connecticut in the U.S. found no significant trend in yield with increasing seeded species richness (combinations of one to seven forage and legume species[22]). Research with combinations of cool-season grasses and legumes containing one to seven species in clipped and grazed plots under irrigation in Utah, however, indicated a positive relationship between herbage yield and seeded species richness.[23]

Research in Ontario, Canada concluded that mixture complexity per se was not as important as the use of strategically selected and appropriately managed complexity.[24] In New Zealand, pastures seeded with a mixture of 10 to 23 species of cool-season grasses and pasture herbs and grazed by sheep yielded more herbage than simple grass–legume mixtures.[25,26] Greater herbage production of the complex mixtures resulted from summer growth of legume and forb components. In southern England, a comparison of what was considered a species-poor mixture (6 to 17

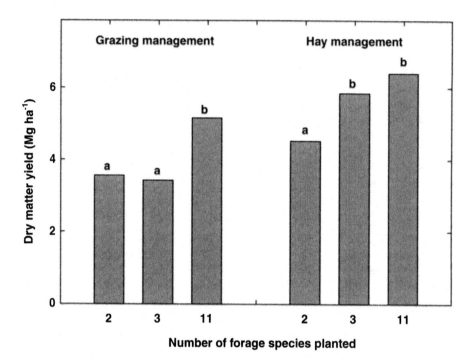

FIGURE 7.2 Dry matter yields of three forage mixtures under grazing in southeastern Pennsylvania. Two 0.4-ha pastures of a 2-species (orchardgrass-white clover), 3-species (orchardgrass, alfalfa, chicory), and an 11-species mixture [orchardgrass, alfalfa, chicory, white clover, red clover, birdsfoot trefoil, prairiegrass (*Bromus wildenowii*), meadow bromegrass (*Bromus biebersteinii* Roemer & J.A. Schultes), reed canarygrass (*Phalaris arundinacea* L.), perennial ryegrass, timothy (*Phleum pratense* L.), and tall fescue] were planted in the autumn of 1997. The soil was a Weikert — Loamy-skeletal, mixed, active, mesic Lithic (Typic) Dystrudepts. Pastures were grazed by Holstein dairy heifers during 1999 to 2002. Paddocks were stocked with 45 to 60 Holstein dairy heifers for a 1- to 2-day period of stay on a 30- to 45-day rotation interval. Grazing started in late April and ended the first week of October each year. Paddocks were cut for hay once in June of 1999 and 2000. Data are averages of two replicate pastures and 4 years. Bars with similar letters do not differ ($P < 0.05$).[51]

species of grasses, forbs, and a legume) with a species-rich mixture (25 to 41 species) under hay management at six sites over 4 years showed no difference in herbage yield between the mixtures in year 1. The species-rich mixture, however, yielded up to 42% more herbage than the species-poor mixture during years 2 to 4.[27]

We compared three forage mixtures (2-, 3-, or 11-species) on-farm in replicated 0.4-ha pastures grazed by dairy heifers or managed under a three-cut hay system for 4 years. The complex mixture yielded more forage dry matter than the two-species mixture, but this difference was due to the inclusion of a few highly productive forage species (Figure 7.2). The primary advantage of the 3- and 11-species mixtures resulted from the inclusion of chicory (*Cichorium intybus* L.) and alfalfa, both deep taproot species, on the drought-prone soil. A disadvantage was that nearly half of the planted species did not persist beyond 4 years in the complex forage

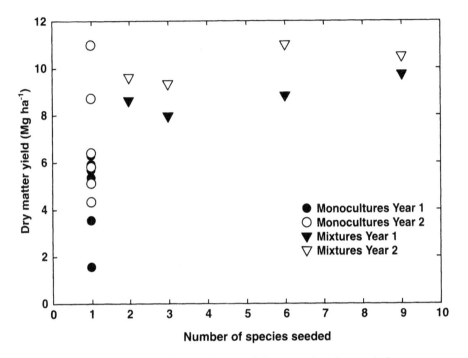

FIGURE 7.3 Dry matter yields of monocultures and forage species mixtures during two years at State College, PA.[53]

mixture. Similar results were obtained in a grazed small-plot trial that compared 13 forage mixtures (combinations of 2, 3, 6, or 9 species) under management-intensive grazing. The 6- and 9-species mixtures yielded more forage than the 2- and 3-species mixtures; however, the principal cause for greater yields was the inclusion of red clover (Figure 7.3).

Other field-plot studies have shown no benefit to forage production from highly complex forage mixtures.[28,29] Several studies in the New Zealand hill country reported inconsistent evidence of production responses to forage species richness.[30–32] In those studies, the environment (site, fertility, slope) influenced herbage yield more than species diversity.[31]

The contrasting results from studies conducted with different forage species, environments, and management conditions should not be surprising. Taken together, these studies suggest that in a stable environment with few limitations to production and good management herbage yield might be maximized from a low-diversity stand composed of species well adapted to that environment. As environmental heterogeneity and functional expectations increase, pastureland sustainability might be maximized from more complex mixtures.

7.3.2 GRAZING ANIMAL PRODUCTIVITY ON DIVERSE PASTURE MIXTURES

There are few studies on how biodiversity affects animal performance even though grazing animals play a key role in affecting plant species diversity in grazing lands.[33]

TABLE 7.2
Milk Production of Dairy Cows Grazing N-Fertilized Grass or Two Grass–Legume Mixtures in Minnesota[49]

Treatment	Carrying Capacity	Milk Production	
	Animal days/ha	kg/cow/day	kg/ha
Grass+N[a]	325	17.1	4733
Simple mixture[b]	300	16.8	4233
Complex mixture[c]	301	15.8	3789

[a] Smooth bromegrass and orchardgrass received 450 kg N/ha/year in three applications during year 1 and 235 kg/ha/year in two applications during year 2.
[b] Alfalfa, white clover, smooth bromegrass, and orchardgrass.
[c] Alfalfa, red clover, alsike clover, white clover, smooth bromegrass, orchardgrass, timothy, meadow fescue, and reed canarygrass.

TABLE 7.3
Milk Production and Dry Matter Intake of Dairy Cows Grazing Four Different Species Mixtures[50,52]

Forage Mixture	Milk Yield[a] (kg/cow/day)	Herbage Intake[a]	Dry Matter Yield (kg/ha)	
			2002	2003
Two species[b]	34.1	12.9	4800	9000
Three species	35.3	12.1	7400	9900
Six species	34.4	12.1	7900	11300
Nine species	34.3	11.6	7500	9000

[a] Data are means of four grazing periods in each of two years.
[b] Two-species mixture = orchardgrass and white clover; three-species mixture = orchardgrass, white clover, and chicory; six species mixture = orchardgrass, red clover, chicory, tall fescue, Kentucky bluegrass, and birdsfoot trefoil; nine-species mixture = orchardgrass, red clover, chicory, tall fescue, Kentucky bluegrass, birdsfoot trefoil, perennial ryegrass, alfalfa, and white clover.

On New Zealand high-country grazing lands species richness and evenness were weakly associated with sheep-carrying capacity.[34] Grazing research with lactating dairy cows in the mid-1960s indicated that there was no benefit to planting a complex mixture of grasses and legumes for grazing (Table 7.2).

More recent dairy grazing research demonstrated that milk production per cow did not differ among simple (orchardgrass–white clover) and complex swards (three to nine species of grasses, legumes, and chicory; Table 7.3). Forage production per hectare (and by extrapolation, animal production per hectare) was greater on complex forage mixtures compared to the simple grass-legume mixture during a dry year (2002) but not during a wet year (2003). Similar to the on-farm work, less than half of the species planted in the complex mixtures were present by the third year, indicating that species presence was not very stable in these mixtures. Future research in this area should focus on grazing trials that measure animal productivity, behavior, and selection on a range of species mixtures so that practical recommendations can be made for grazing management.

7.3.3 Ecosystem Stability

Another tenet of plant biodiversity theory is that increased diversity contributes to the stability of ecosystems. In a small-plot study, mixtures of up to 15 species of legumes, forbs, and grasses did not improve forage yield or yield stability.[30] Most of the mixtures decreased in species number during the 3-year study and became dominated by perennial grasses.

Research on New Zealand high-country grazing lands showed that species richness and evenness were weakly associated with the stability of sheep production (coefficient of variation in annual carrying capacity[34]). Stability (measured as the coefficient of variation in seasonal herbage production) of temperate grazing lands in southern Australia was not related to species richness.[35]

New Zealand researchers reported a high coefficient of variation (CV) for low numbers of species, and a decreasing CV as species number increased, evidence of reduced risk from species-rich grasslands.[30] Thus, the evidence is lacking for a clear-cut effect of plant species diversity on the stability of managed grazing lands.

7.3.4 Invasion by Exotic Species (Weeds and Pests)

Greater plant diversity in grassland ecosystems may contribute to resistance to invasion by weeds and pests. Weed abundance decreased in experimental pasture mixtures as the evenness of forage species increased.[36] In addition, species composition of the mixture affected weed abundance: mixtures based on tall fescue (*Festuca arundinacea* Schreb.) had fewer weeds in the soil seed bank and aboveground vegetation than did mixtures based on smooth bromegrass. Similar results were found in a series of greenhouse, field, and survey experiments with cool-season pasture species in the northeast.[37]

7.3.5 Diversity and Nutrient Cycling in Forage Plant Communities

Aboveground diversity influences nutrient cycling through microbial decomposition of plant litter. In complex pasture plant communities, the litter of different plant species intermingles and decomposes together. Research results concerning plant diversity effects on litter decomposition in experimental systems have been conflicting (e.g., positive effects,[38] no effects,[39] or mixed results[40]).

Increasing the diversity of grassland plant communities may increase nutrient retention. Soil nitrate levels, both within and below the rooting zone, were reduced as the number of plant species increased in growth chamber studies and in tallgrass prairie communities.[41,42] For example, increasing the diversity of nonleguminous species grown with legumes in grassland mixtures could help reduce nitrate leaching species while still benefiting production through N fertilization.[43]

7.3.6 Mechanisms to Explain Diversity Effects in Forage Plant Communities

Mechanisms proposed to explain the observed responses to plant diversity in experimental grasslands include (1) the "sampling effect," (2) facilitation, (3) niche differentiation/niche separation, and (4) the "insurance effect."[19]

The sampling effect occurs when mixtures overyield simply because of the greater probability of including an adapted, high-yielding plant species in a randomly assembled species-rich mixture. In some studies, the dominant influences on herbage accumulation were related to the presence of particular species rather than the number of species.[30] Generally, when higher-yielding species were included in species-rich mixtures, yields increased. The sampling effect may explain the greater herbage yields for complex forage mixtures in Table 7.3, Figure 7.2, and Figure 7.3. In those studies, it was the inclusion of specific forage species (chicory for the study in Table 7.3; chicory and alfalfa in Figure 7.2; and red clover in Figure 7.3) that accounted for most of the yield increase. Thus, sustainable herbage production would depend on the persistence of those species.

Facilitation involves plant species interactions that alter the environment in a way that benefits a neighboring species. For example, neighbors near species that lift water hydraulically from deep in the soil profile can use a significant proportion of that water resource, effectively ameliorating the adverse effects of drought for shallow-rooted species. Growing deep and shallow rooted grassland species in mixture may result in greater nutrient extraction from deeper soil layers by the deep-rooted species than would normally be observed in monoculture.[44] White clover growing in a complex mixture with chicory had improved leaf water relations and greater relative growth rate than similar white clover grown only with bluegrass.[45] It was hypothesized that hydraulic lift occurred, whereby chicory may have redistributed water from deep in the soil to the surface layer making it available to shallow-rooted species such as white clover. Research is under way to test the hydraulic lift hypothesis. The main conclusion was that the presence of a deep-rooted species was more important than simply species richness in affecting herbage productivity in a stressful environment. Neighboring plants can also favorably alter other environmental conditions. Shading by larger plants can lower soil temperature, reducing heat stress effects while also reducing evapotranspiration leading to improved leaf water relations of smaller neighboring species. Another common example of facilitation in forage agriculture is the use of a nurse crop to aid the establishment of perennial grasses and legumes.

Niche differentiation refers to different plant species coexisting by exploiting resources differently, either in time or space. For example, different plant species may obtain soil nutrients through different rooting depths or strategies, as occurs in grass–legume mixtures where the grass exploits the N symbiotically fixed by the legume. Or, species in mixtures may differ in phenology and have different periods of peak growth, which would distribute production more uniformly throughout the growing season.

Diversity theory suggests that greater plant diversity buffers or insures plant communities against environmental extremes. By having some species that are tolerant of different stresses, complex mixtures act to ensure and thereby stabilize productivity. For example, the productivity of grazing lands during summer drought may be improved by sowing a percentage of pastures to warm-season grasses[45] or by planting multispecies mixtures that include some of the more drought-resistant cool-season grasses and forbs.[46] Interactions among species within complex mixtures may also improve the ability of normally drought-sensitive species to maintain production under stressful conditions.

7.4 CONCLUSIONS AND RECOMMENDATIONS

Clearly, biodiversity plays a role in the proper functioning of temperate pasture ecosystems. However, can biodiversity or components of biodiversity (e.g., plant species diversity) be managed to influence ecosystem functions to benefit the output of pasture ecosystem goods and services? Limited research suggests that greater plant species diversity benefits herbage productivity and resistance to weed invasion. More research is needed to address key questions on how diversity affects secondary productivity (e.g., grazing animal performance). Full-scale grazing trials are needed along with systems research on the strategic placement of species within the farm landscape. Furthermore, we need to know more about grazing animal behavior and selectivity in diverse mixtures. It must also be remembered that management for increased plant species diversity is not simply a numbers game. Species identity, abundance, and spatial distribution across the landscape are critical features in pasturelands. Managing for high forage species diversity may not be appropriate for a highly productive, stable environment where the main objective is maximum forage production. Most temperate pasturelands, however, are highly variable in soils, landscapes, and climate and often fulfill multiple functions for producers (e.g., animal production, resource protection, aesthetic and social values). It is in these situations where greater plant diversity may be most beneficial.

REFERENCES

1. Alonso, A., Dallmeier, F., Granek, E., and Raven, P. *Biodiversity: Connecting with the Tapestry of Life*. Smithsonian Institution/Monitoring and Assessment of Biodiversity Program and President's Committee of Advisors on Science and Technology. Washington, D.C., 2001.
2. Costanza, R., d'Arge, R., de Groot, R., Farber, S., Grasso, M., Hannon, B., Limburg, K., Naeem, S., O'Neill, R.V., Paruelo, J., Raskin, R.G., Sutton, P., and van den Belt, M. The value of the world's ecosystem services and natural capital, *Nature* 387, 253, 1997.
3. Tilman, D., Duvick, D.N., Brush, S.B., Cook, R.J., Daily, G.C., Heal, G.M., Naeem, S., and Notter, D.R. Benefits of Biodiversity. Task force report 133. Council for Agricultural Science and Technology, Ames, IA, 1999.
4. Minns, A., Finn, J., Hector, A., Caldeira, M., Joshi, J., Palmborg, C., Schmid, B., Scherer-Lorenzen, M., Spehn, E., and Troubis, A. The functioning of European grassland ecosystems: potential benefits of biodiversity to agriculture. *Outlook Agric.*, 30, 179, 2001.

5. Foster, L. Herbs in pastures. Development and research in Britain, 1850–1984. *Biol. Agric. Hort.*, 5, 97, 1988.

6. Blaser, R.E., Skrdla, W.H., and Taylor, T.H. Ecological and physiological factors in compounding seed mixtures, *Adv. Agron.*, 4, 179, 1952.

7. Donald, C.M. Competition among crop and pasture plants. *Adv. Agron.*, 15(1), 114, 1963.

8. West, N.E. Biodiversity of rangelands. *J. Range. Manage.*, 46, 2, 1993.

9. Krueger, W.C., Sanderson, M.A., Cropper, J., Miller-Goodman, M., Kelley, C.E., Pieper, R.D., Shaver, P., and Trlica, M.J. Environmental impacts of livestock on U.S. grazing lands. Council for Agricultural Science and Technology issue paper 22. CAST, Ames, IA, 2002.

10. Adler, P.R., Sanderson, M.A., and Goslee, S.C. 2003. Survey of CRP and other grasslands in the Northeast U.S. Agron. Abstr. CD-ROM. American Society of Agronomy, Madison, WI.

11. Byers, R.A., Barker, G.M., Davidson, R.L., Hoebeke, E.R., and Sanderson, M.A. Richness and abundance of Carabidae and Stahylinidae (Coleoptera) in Pennsylvania dairy pastures under intensive grazing, *Great Lakes Entomol.*, 33, 81, 2001.

12. Byers, R.A. and Barker, G.M. Soil dwelling macroinvertebrates in intensively managed grazed dairy pastures in Pennsylvania, New York, and Vermont, *Grass Forage Sci.*, 55, 253, 2000.

13. Tracy, B.F. and Sanderson, M.A. Seedbank diversity in grazing lands of the Northeast United States, *J. Range Manage.*, 53, 114, 1999.

14. Tracy, B.F. and Sanderson, M.A. Patterns of plant species richness in pasture lands of the Northeast United States, *Plant Ecol.*, 149, 169, 2000.

15. Bardgett, R.D. and Cook, R. Functional aspects of soil animal diversity in agricultural grasslands, *Appl. Soil Ecol.*, 10, 263, 1998.

16. Kruess, A. and Tscharntke, T. Contrasting responses of plant and insect diversity to variation in grazing intensity, *Biol. Conserv.* 106, 293, 2002.

17. Parsons, A.J. and Dumont, B. Spatial heterogeneity and grazing processes, *Anim. Res.* 52, 161, 2003.

18. Peeters, A. and Janssens, F. Species-rich grasslands: diagnostics, restoration, and use in intensive livestock production systems, *Grassland Sci. Eur.*, 3, 375, 1998.

19. Fridley, J.D. The influence of species diversity on ecosystem productivity: How, where, why? *Oikos* 93, 514, 2001.

20. Huston, M.A., Aarssen, L.W., and Austin, M.P. No consistent effect of plant diversity on productivity, *Science* (Washington, D.C.), 289, 1255, 2000.

21. Wardle, D.A., Huston, M.A., and Grime, J.P. Biodiversity and ecosystem function: an issue in ecology, *Bull. Ecol. Soc. Am.*, 81, 235, 2000.

22. Brown, B.R. and Munsell, R.I. Species and varieties of grasses and legumes for pastures, Bull. 208. Storrs Agric. Exp. Sta. Storrs, CT, 1936.

23. Bateman, G.O. and Keller, W. Grass-legume mixtures for irrigated pastures for dairy cattle, Bull. 382. Utah Agric. Exp. Sta. Logan, UT, 1956.

24. Clark, E.A. Diversity and stability in humid temperate pastures. In *Competition and Succession in Pastures*, P.G. Tow and A. Lazenby, Eds. CAB International, New York, 2001, 103–118.

25. Ruz-Jerez, B.E., Ball, P.R., White, R.E., and Gregg, P.E.H. Comparison of a herbal ley with a ryegrass-white clover pasture and pure ryegrass sward receiving fertilizer nitrogen, *Proc. N.Z. Grassl. Assoc.*, 53, 225, 1998.

26. Daly, M.J., Hunter, R.M., Green, G.N., and Hunt, L. A comparison of multi-species pasture with ryegrass-white clover pastures under dryland conditions, *Proc. N.Z. Grassl. Assoc.*, 58, 53, 1996.

27. Bullock, J.M., Pywell, R.F., Burke, M.J.W., and Walker, K.J. Restoration of biodiversity enhances agricultural production, *Ecol. Lett.*, 4, 185, 2001.

28. Zannone, L., Assemat, L., Rotili, P., and Jacquard, P. An experimental study of intraspecfic competition within several forage crops, *Agronomie*, 3, 451, 1983.

29. Tracy, B.F. and Sanderson, M.A. Productivity and stability relationships in clipped pasture communities of varying diversity, *Crop Sci.*, 44, 2180, 2004.

30. Nicholas, P.K., Kemp, P.D., Barker, D.J., Brock, J.L., and Grant, D.A. Production, stability and biodiversity of North Island New Zealand hill pastures. Pages 21-9 to 21-10 in J.G. Buchanan-Smith et al., Eds. Proc. Intl. Grassl. Congr., 18th, Winnipeg, MB, Canada. 8–17 June 1997. Association Management Centre, Calgary, AB, Canada, 1997.

31. Dodd, M.B, Barker, D.J. and Wedderburn, M.E. Are there benefits of pasture species diversity in hill country, *Proc. N.Z. Grassl. Assoc.*, 65, 127, 2003.

32. White, T.A., Barker, D.J., and Moore, K.J. Vegetation diversity, growth, quality and decomposition in managed grasslands, *Agric. Ecosyst. Environ.*, 101, 73, 2004.

33. Rook, A.J. and Tallowin, J.R.B. Grazing and pasture management for biodiversity management, *Anim. Res.* 52, 181, 2003.

34. Scott, D. Sustainability of New Zealand high-country pastures under contrasting development inputs. 7. Environmental gradients, plant species selection, and diversity, *N.Z. J. Agric. Res.* 44, 59, 2001.

35. Kemp, D.R., King, W. McG., Gilmour, A.R., Lodge, G.M., Murphy, S.R., Quigley, P.E., Sanford, P., and Andrew, M.H. SGS Biodiversity theme: impact of plant biodiversity on the productivity and stability of grazing systems across southern Australia, *Aust. J. Exp. Agric.*, 43, 961, 2003.

36. Tracy, B.F., Renne, I.J., Gerrish, J.R., and Sanderson, M.A. Forage diversity and weed abundance relationships in grazed pasture communities, *Basic Appl. Ecol.*, 5, 543, 2004 (in press).

37. Tracy, B.F. and Sanderson, M.A. Relationships between forage plant diversity and weed invasion in pasture communities, *Agric. Ecosyst. Environ.* 102, 175, 2004.

38. Bardgett. R.D. and Shine, A. Linkages between plant diversity, soil microbial biomass, and ecosystem function in temperate grasslands, *Soil Biol. Biochem.*, 31, 317, 1999.

39. Wardle, D.A., Bonner, K.I., and Nicholson, K.S. Biodiversity and plant litter: experimental evidence which does not support the view that enhanced species richness improves ecosystem function, *Oikos*, 79, 247, 1997.

40. Hector, A., Beale, A.J., Minns, A., Otway, S.J., and Lawton, J.H. Consequences of the reduction of plant diversity for litter decomposition: effects through litter quality and microenvironment, *Oikos*, 90, 357, 2000.

41. Naeem S., Thompson, L.J., Lawler, S.P., Lawton, J.H., and Woodfin, R.M. Declining biodiversity can alter the performance of ecosystems, *Nature*, 368, 734, 1994.

42. Tilman, D., Wedin, D., and Knops, J. Productivity and sustainability influenced by biodiversity in grassland ecosystems, *Nature,* 379. 718, 1996.

43. Scherer-Lorenzen, M., Palmborg, C., Prinz, A., and Shulze, E.D. The role of plant diversity and composition for nitrate leaching in grasslands, *Ecology,* 84, 1539, 2003.

44. Berendse, F. 1982. Competition between plant populations with differing rooting depths. III. Field experiments, *Oecologia*, 53, 50, 1982.

45. Skinner, R.H., Gustine, D.L., and Sanderson, M.A. Growth, water relations, and nutritive value of pasture species mixtures under moisture stress, *Crop Sci.*, 44, 1361, 2004.

46. Lucero, D.W., Grieu, P., and Guckert, A. Effects of water deficits and plant interaction on morphological growth parameters and yield of white clover (*Trifolium repens* L.) mixtures. *Eur. J. Agron.*, 11, 167, 1999.

47. McAdam, J.H., Hoppe, G., Millsopp, C.A., Cameron, A., and Mulholland, F. Environmentally sensitive areas in Northern Ireland, re-monitoring of the west Fermanagh and Erne Lakeland ESA. Department of Agriculture for Northern Ireland, 1998.

48. Soderstrom, B., Svensson, G., Vessby, K., and Glimskar, A. Plants, insects, and birds in semi-natural pastures in relation to local habitat and landscape features, *Biodiversity Conserv.*, 10, 1839, 2001.

49. Wedin, W.F., Donker, J.D., and Marten, G.C. An evaluation of nitrogen fertilization in legume-grass and all-grass pasture. *Agron. J.*, 58, 185, 1965.

50. Soder, K.J., Sanderson, M.A., and Muller, L.D., Effects of forage diversity on intake and productivity of grazing lactating dairy cows, *Am. Assoc. Dairy Sci. Abstr.*, p. 93, 2004.

51. Sanderson, M.A., Skinner, R.H., and Tracy, B.F. Productivity of simple and complex mixtures of forages compared in on-farm pastures, *Proc. Am. Forage Grassl. Conf.*, 13, 429, 2004.

52. Sanderson, M.A., Taube, F., Tracy, B.F., and Wachendorf, M. Plant species diversity influences on forage production and performance of dairy cattle on pasture, *Grassl. Sci. Eur.*, 9, 632, 2004.

53. Deak, A., Hall, M.H., and Sanderson, M.A. Forage production and species diversity in pastures, *Proc. Am. Forage Grassl. Conf.*, 13, 220, 2004.

8 Net Ecosystem Carbon Dioxide Exchange over a Temperate, Short-Season Grassland: Transition from Cereal to Perennial Forage

V.S. Baron, D.G. Young, W.A. Dugas, P.C. Mielnick, C. La Bine, R.H. Skinner, and J. Casson

CONTENTS

8.1 INTRODUCTION

North American grasslands may be part of a terrestrial carbon (C) sink.[1-3] The terrestrial or "missing" sink, which includes cropland and forests, may play a role in offsetting CO_2 emissions generated from fossil fuel use and global changes in land management (e.g., deforestation) through C sequestration in soils, vegetation, and residues.[1,2,4] Fan et al.[5] estimated the northern boundary of the terrestrial sink to be approximately 51° N lat.

The potential sink size of grasslands may be significant because they cover a large area.[4] World grasslands represent 32% of global vegetation on an area basis.[1] Estimated area of U.S. pasture and rangeland is 51 and 161 million ha, respectively, compared to cropland at 155 million ha. Area of public grazing land in the U.S. is 124 million ha.[4] Canadian pastureland was estimated at 10 million ha, of which 9 million ha is located in western Canada.[6-9] Entz et al.[10] estimated an area of 44 million ha of range in the Northern Great Plains region of Manitoba, Saskatchewan, Alberta, North Dakota, South Dakota, and Montana. There are approximately 6.5 million ha of native rangeland mostly south of 52° N lat in the semi-arid region of the Prairie Provinces.[11]

Rangelands have relatively low potential C sequestration rates per hectare, but cover large areas of the North American continent and, thus, could act as a large potential C sink. Improved pastures have larger potential rates of C sequestration as they are located in humid and subhumid regions and receive greater management inputs; most are managed below production potential,[4,12] but are also subject to greater removal of biomass-C as conserved ruminant feed.

Most pastureland in the U.S. is located east of 98° W long.,[12] where annual precipitation balances or exceeds evapotranspiration.[13] In Canada, the majority of pasture is located on the black and gray wooded soils in the subhumid Aspen Parkland and Boreal Transition zone of Western Canada[8] and in all regions of eastern Canada.[6-9] Alberta contains the largest area of pasture of the Prairie Provinces at 2.2 million ha as well as 6.6 million ha of rangeland consisting of naturalized and native species.[14]

Micrometeorological studies carried out by the USDA-Agricultural Research Service Rangeland Carbon Dioxide Project[15] documented ecosystem CO_2 balance for various grasslands.[16-23] These studies provided ecological insights into fundamental processes that affect C sequestration. All CO_2 uptake occurs as a result of photosynthesis during the growing season. Uptake is countered by respiration, resulting in CO_2 emission from the crop canopy, root, and soil microbial degradation of litter and crop organic residues.[24,25] Bremer et al.[25] estimated that CO_2 respired from a tall-grass prairie ecosystem in Kansas was three to four times that accumulated in new biomass during the season; a fraction of the CO_2 respired by the system is re-assimilated.

Most of this research, involving aspects of grassland CO_2 flux, was conducted during the growing season in the Great Plains region. While several short-term studies indicate that grassland ecosystems act as small sinks, there are only a few studies that evaluated ecosystems annually. Climate, season, species, phenology, and management (e.g., grazing, fire, etc.) influence both CO_2 uptake and ecosystem respiration.[16,22,24-26] In particular, drought[20,24] and the dormant season[19,20] are periods of net CO_2 loss. Thus, in areas with short growing seasons, intermittent drought, and long dormant periods such as the Aspen Parkland region,[27] C sequestration may be limited.

Conversion of cropland to grassland and intensification of grassland management are key methods to enhance rate of C sequestration in soils.[4,12] Studies about the ecosystem CO_2 exchange process are needed to assess the Aspen Parkland region for potential as a C sink. Our objective was to evaluate net ecosystem CO_2 exchange, within season and annually, during the establishment or seedling year and during the first production year of a forage stand in transition from a cereal to forage sequence to be used for pasture in a mixed farm crop rotation.

8.2 METHODS

Net CO_2 ecosystem exchange measurements were made on a black chernozemic sandy loam soil in transition from cereal to perennial pasture, at Lacombe, Alberta, Canada (52° 26′ N: 113° 45′ W). The site had been in a cereal-forage rotation over the past 20 years. Farming practices prevalent in the region were used — silage and hay were removed in the seedling and first production years and grazing of the forage species mixture occurred in September or October. For the seedling year the field was sown (May 15, 2002) to barley (*Hordeum vulgare* L.), a nurse crop, and under sown to a mixture of meadow bromegrass (*Bromus riparius* Rehm.) and alfalfa (*Medicaigo sativa* L.). The nurse crop was removed as silage on August 1, 2002. In 2002, the seedling forage stand was allowed to regrow until fall when it was grazed severely (4.2 Animal Unit Months, AUM, over 19 days) between September 27 and October 15. For the first production year, in 2003, the field was harvested as hay on July 16, allowed to regrow, and was grazed lightly (0.63 AUM over 14 days) between September 2 and September 15. Hereafter reference to heavy and light grazing are stocking rates of 4.2 and 0.63 AUM ha^{-1}, respectively, over 19 and 14 days during the seedling and first production years, respectively. Each year 100 kg ha^{-1} N was applied in the spring.

8.2.1 Net Ecosystem CO_2 Exchange

A tower with Bowen ratio/energy balance (BREB) instrumentation (Model 023/CO2 Bowen ratio system, Campbell Scientific, Inc., Logan, UT) was placed on a 2% northwest-facing slope after seeding in spring 2002. Borders of the field provided a minimum 200-m fetch from the tower. Bowen ratios were calculated as described previously.[16,18,19,28] Measurements began on May 26, 2002. When the BREB method was not valid for calculating turbulent diffusivity, because of differences in the sign of the sensible heat flux and the temperature/humidity gradient, it was calculated using wind speed, atmospheric stability, and canopy height.[16] This occurred 12% of the time. Daily net CO_2 flux measurements between May 15 (seeding date) and May 26 were estimated from the average daily flux for the first week of measurements. Carbon dioxide and water vapor concentration gradients were measured at 1.4 and 2.4 m above the soil surface with an infrared gas analyzer (Model 6262, LICOR, Inc., Lincoln, NE). Methods were described previously by Frank and Dugas[18] and Frank et al.[19]

The seedling year includes year-round CO_2 exchange data from May 15, 2002 to May 14, 2003 and the first production year includes data from May 15, 2003 to May 14, 2004.

8.2.2 SOIL CO_2 FLUX

Soil respiration measurements were made with a vented closed system using a LICOR 6200 portable infrared gas analyzer fitted with a soil respiration chamber (LICOR 6000-09) with a volume of 1000.3 cm^3 and a diameter of 10.3 cm allowing an area of exposed soil of 71.5 cm^2. Rings were inserted 25 mm into bare soil in three positions around the BREB tower and were kept free of vegetation. Flux measurements were made by placing the chamber over the rings for periods of 3 to 5 min. Measurements were made at approximately 2-week intervals during the growing season, spring and fall from 1100 to 1300 hours, but not in winter.

BREB measurements are presented as a daily total of net CO_2 flux (g m^2/d^{-1}). Soil respiration measurements are presented in the same units, but are averaged over at least three subsamples. Net CO_2 uptake into the ecosystem is indicated by a positive sign and net efflux by a negative sign.

Linear regression was used to relate midday soil respiration (independent variable) to BREB nighttime CO_2 flux (dependent variable). A significant linear regression coefficient ($P \leq 0.05$) was indicative of a relationship between the fluxes of different origins. The coefficient of determination (R^2) was used to quantify the extent to which the variation in BREB nighttime CO_2 flux was explained by soil respiration.

8.3 RESULTS AND DISCUSSION

8.3.1 CLIMATE

The seedling year season could be described as a dry spring and summer with a moist fall (Figure 8.1); April to October precipitation was 277 mm and the long-term average for this period was 363 mm. The first production year season had a wet spring with a dry summer; April to October precipitation was 262 mm. Temperatures during the June to August period were generally above average. Mean winter temperatures (October to March) were above average for the seedling year and average to below average for the first production year. Spring time air temperatures were average.

8.3.2 BREB CO_2 FLUX

Periods of initial growth, regrowth, grazing, dormancy, and transition between dormancy and growth could be identified from daily net CO_2 exchange in both seedling and first production years (Figure 8.1). From May 15 to September 30 (entire growth period), net uptake occurred on only 65 and 61% of the 140 days for seedling and first production years, respectively (Figure 8.1A and B). Daily flux values were highly variable due to effects of variation in radiation, temperature, and precipitation on plant and soil processes. Occasional spikes of soil CO_2 efflux often occurred immediately after rainfall events that flush CO_2 from soil pores.[24,29]

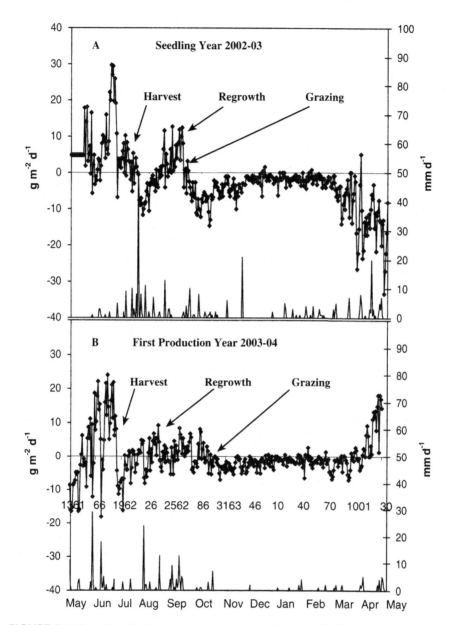

FIGURE 8.1 Net daily CO_2 flux and precipitation for (A) May 15, 2002 to May 14, 2003 (seedling year) and (B) May 15, 2003 to May 14, 2004 (first production year) for meadow bromegrass–alfalfa.

8.3.2.1 Initial Growth Period

For both seedling and first production year, maximum daily BREB CO_2 flux coincided with periods of initial growth in June and July (Figure 8.1A and B). From May 15 until Aug. 1, 2002 seedling year average daily net flux was 6.7 g m² d⁻¹

and from May 15 until July 16, 2003 the average flux in the first production year was 1.7 g m^2 d^{-1}.

8.3.2.2 Regrowth

Harvest was followed immediately by a period of net CO_2 loss that lasted 25 days in the seedling year and 14 days in the first production year. McGinn and King[30] observed 12 days of ecosystem respiration after cutting alfalfa during midsummer in Ontario. Mean daily CO_2 flux during these periods was −4.9 and −5.5 g m^2 d^{-1} during seedling and first production year, respectively.

Once a plant canopy was reestablished, net CO_2 uptake into the ecosystem occurred well into September of both years. In the seedling year, August and September rainfall resulted in regrowth of forage, but the perennial stand had to establish a canopy after the barley was removed in spite of dry early summer conditions. Regrowth was slow in the first production year due to drought, but generally net CO_2 uptake occurred (Figure 8.1B). Average net CO_2 flux after harvest until September 30 was −0.75 and 0.41 g m^2 d^{-1} during seedling and first production years, respectively. Kim et al.[24] observed net CO_2 release at a rate of −3 g m^2 d^{-1} from a tall-grass prairie ecosystem in Kansas during drought conditions.

8.3.2.3 Grazing

A season-end grazing removed most of the live vegetative material from the seedling year stand during early October. A very light grazing in September of the first production year left residual biomass for light interception. During hard grazing (September 27 to October 30) in the seedling year an average net CO_2 release of −7.9 g m^2 d^{-1} was observed. During the light grazing (August 27 to September 15) in first production year an average net CO_2 loss of −0.8 g m^2 d^{-1} was observed. Fall residual dry matter was less than 50 and approximately 300 g m^{-2} in seedling and first production years, respectively.

8.3.2.4 Dormant Period

Almost no growth occurred during the dormant period, October 1 to March 31, although 6 and 17 days of net uptake occurred in the seedling and first production years, respectively, mostly in October. Over the October 1 to March 31 period, average daily CO_2 release was −3.6 and −1.6 g m^2 d^{-1} during seedling and first production years, respectively. Relatively large losses from the system occurred during October, November, and March of the seedling year. Over the entire dormant period the range in CO_2 flux was 1.5 to −14.7 g m^2 d^{-1} in the seedling year (Figure 8.1A) and 7.2 to −8.2 g m^2 d^{-1} in the first production year (Figure 8.1B).

8.3.2.5 Spring

The daily net CO_2 loss from the ecosystem in March of the seedling year (Figure 8.1A) lasting until late May 2003 (Figure 8.1B) was large. From April 1 until May 15 there were only 4 days of net uptake. This is partly due to the lack of leaf area

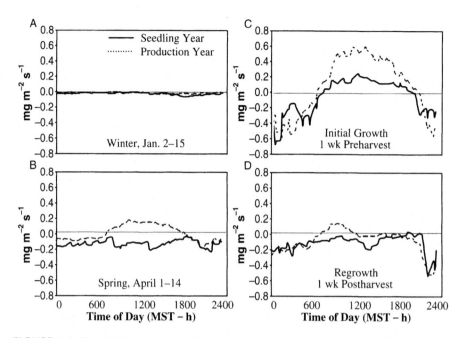

FIGURE 8.2 The 20-minute average CO_2 flux for 24-h periods of seedling and first production year. (A) Winter, averaged over 14 days, (B) spring, averaged over 14 days, (C) initial growth, averaged over 7 days prior to harvest, and (D) regrowth, averaged over 7 days after harvest.

as a result of heavy grazing the previous fall. Thus, net CO_2 loss occurred from respiration required to generate new leaf material just as it did after cutting. Pasture dry matter yields on April 18, May 1, and May 15 were 0.0, 100, and 500 kg ha^{-1}, respectively, in 2003 (seedling year). By contrast there was net uptake during 80% of the days between April 1 and May 14, 2004 for the first production year (Figure 8.1B). Dry matter yield on May 15, 2004 was 1600 kg ha^{-1}. The CO_2 flux averaged -12.7 and 3.3 g m^2 d^{-1} from April 1 to May 14 during the seedling year (2003) and first production year (2004), respectively.

8.3.2.6 Diurnal CO_2 Flux

Diurnal CO_2 fluxes were representative of the contrasting periods within years (Figure 8.2). The 20 min average CO_2 fluxes averaged over a 14-day period for winter (Figure 8.2A) showed small CO_2 flux loss during day and night in early January for both seedling and first production years. During spring (Figure 8.2B), CO_2 fluxes were negative during day and night for the seedling year, reflecting high soil respiration rates (Table 8.1) and lack of a crop canopy, following intense grazing during the fall of 2002. By contrast, spring diurnal CO_2 fluxes during the first production year resembled those of the initial growth period (Figure 8.2C), although the peak values were smaller. For the examples given (Figure 8.2B and C), average daytime CO_2 flux for the first production year-spring was approximately 39% of the

TABLE 8.1
Midday Soil CO_2 Flux Measurements Taken during Selected Periods from 2002 to 2004 from Seedling Year and First Production Year Stands of Meadow Bromegrass–Alfalfa Mixtures at Lacombe, Alberta

Period	Sample No., n	Mean, g m² d⁻¹	Range
		2002	
June 6 to Sept. 30	27	–11.8	–5.5 to –19.8
Oct. 1 to Dec. 4	18	–3.7	–0.1 to –7.9
		2003	
April 1 to May 15	9	–13.6	–6.5 to –25.9
May 16 to Sept. 30	30	–8.9	–4.6 to –21.0
Oct. 1 to May 15	6	–4.6	–2.6 to –6.7
		2004	
April 1 to May 15	12	–6.5	–3.8 to –8.5

average of both years' initial growth daytime flux; spring nighttime flux (Figure 8.2B) was 53% of the nighttime CO_2 flux for the initial growth period (Figure 8.2C). Diurnal CO_2 flux for the initial growing period (averaged over 7 days prior to harvest) were comparable to those from other studies (e.g., Frank et al.[19] and Sims and Bradford[20]) for periods of active growth. Diurnal CO_2 flux for the regrowth period (averaged over a 7-day period immediately after harvest) showed negative daytime and nighttime flux for the seedling year (Figure 8.2C). Regrowth daytime CO_2 flux for the first production year averaged near zero. The low daytime CO_2 flux for both years reflected a small leaf area during regrowth. However, average nighttime regrowth CO_2 flux for both years was approximately 82% of those observed during initial growth, explaining the tendency for relatively large net losses from the ecosystem during regrowth periods of both years. A decline in afternoon CO_2 flux rates for seedling and first production year initial growth (Figure 8.2C) and first production year regrowth (Figure 8.2D) was likely due to tissue water deficit, resulting in stomatal closure, as suggested in a similar example by Sims and Bradford.[20] The growing seasons used in this study were drier than normal.

8.3.3 SOIL RESPIRATION

Our measurements of respiration are the sum of root and soil microbial respiration.[25] During the night, the BREB system measures aboveground (including canopy and litter) and belowground respiration. As expected, nighttime BREB CO_2 flux was highest during periods of maximum plant growth. Averaged over initial growth periods BREB nighttime respiration averaged –4.96 and –8.58 CO_2 flux g m² d⁻¹ for seedling (May 15 to August 1) and first production years (May 15 to July 16), respectively, compared to –2.5 and –2.0 g m² d⁻¹, respectively, during dormant periods.

TABLE 8.2
Linear Regression Relationship between Midday Soil (independent variable) and BREB Nighttime CO_2 Flux for Selected Periods in a Meadow Bromegrass–Alfalfa Stand during Seedling Year and First Production Year Combined at Lacombe, Alberta

Relationship	P Sig.[a]	R^2	RMSE[b]	n
	Entire Growing Season (May 15 to Sept. 30)			
$-4.60 + 0.22x$	0.09	0.11	3.3	26
	Initial Growth Period[c]			
$-5.99 + 0.10x$	0.67	0.02	4.0	13
	Re-growth Period[c]			
$-2.19 + 0.60x$	0.01	0.46	2.3	8
	Entire Dormant Period (Oct. 1 to March 31)			
$-1.45 + 0.86x$	0.01	0.56	2.2	9
	Spring Period (April 1 to May 15)			
$-2.64 + 0.18x$	0.01	0.39	1.7	8

[a] Probability of a significant regression.
[b] RMSE is root of the mean square error of the linear regression.
[c] Initial growth period for 2002–2003 is May 15 to Aug. 1 and for 2003–2004 is May 15 to July 16; regrowth period for 2002–2003 is Aug. 2 to Sept. 30 and 2003–2004 is 17 July 17 to Sept. 30.

Mean soil CO_2 flux varied depending upon on time of year and stage of crop development (Table 8.1). Soil CO_2 flux during the growing season (May 16 to September 30) was approximately three times greater than the dormant season (November 1 to March 31), but growing season fluxes were more variable. Spring 2003 (April 1 to September 15) soil respiration losses were highest, supporting net BREB losses for the seedling year during April and early May (Figure 8.1A). Losses during spring 2004 (first production year) were about half the CO_2 flux values of spring 2003. Similar results were shown by De Jong et al.[31] in southern Saskatchewan, who concluded that high soil respiration rates following a drought were due to wetting and drying cycles, which stimulated soil microflora to accelerate respiration rate. Dormant season daily mean soil CO_2 flux were in agreement with Kim et al.[24] Frank and Dugas[18] determined daily average soil flux to be -1.7 g CO_2 m^{-2} from measurements made throughout fall and winter in North Dakota.

Linear regression analyses between soil respiration CO_2 flux rates and BREB night time fluxes indicated that soil CO_2 fluxes did not predict BREB nighttime fluxes well (Table 8.2). Frank et al.[22] found that a single flux rate taken at midday overestimated the average of five sequential measurements taken at 3-hour intervals by 9%. However, they[22] determined that single, midday soil CO_2 flux rates were most representative of daily soil flux CO_2 compared to soil flux rates taken at other times of the day.

During the growing season, soil respiration did not explain variation in ecosystem BREB respiration well, as indicated by low R^2 values and regression equations with low probability of significance (Table 8.2). A large proportion of ecosystem respiration is derived from canopy dark respiration when dry matter yield is close to maximum.[32] However, during the regrowth phase, which included periods of net loss after harvest (Figure 8.1A and B) and drought, soil respiration explained 46% of the variation in BREB nighttime CO_2 flux. During the dormant period (essentially October and November measurements) soil CO_2 flux explained 56% of the BREB nighttime CO_2 flux. During the spring period there was a transition from dormancy to growth; both plant and soil microorganism metabolism should be high, but the crop leaf area is relatively low compared to mid-June. Over the spring period soil respiration explained 39% of the variability in BREB nighttime respiration (Table 8.2). Soil CO_2 flux should underestimate CO_2 flux from ecosystem respiration as it does not account for litter decomposition and dark respiration of live vegetative material.

8.3.4 ECOSYSTEM SINK OR SOURCE

During the seedling year there was an average annual net daily CO_2 flux of -2.0 g m^{-2} d^{-1} for a net loss from the ecosystem of 730 g CO_2 m^{-2} yr^{-1}. During first production year the average annual net daily CO_2 flux was -0.01 g m^{-2} d^{-1} for a net annual loss of 3.65 g CO_2 m^{-2} yr^{-1}. Year to year variation in annual CO_2 flux is likely to be high. Long-term annual ecosystem equilibrium should be bounded by a standard deviation of CO_2 flux. In most cases enough years of data have not been collected to determine what this statistical boundary might be. Studies such as the current one are often not replicated or numbers of replicates are very small. Statistical rigor improves with the number of years involved in the study. This may not be economically possible, so it is important to be aware that small annual net losses might be indicative of an equilibrium state. Dugas et al.[16] estimated that annual fluxes of 183 to 293 g CO_2 m^{-2} yr^{-1} at a Temple, TX site as in "approximate equilibrium." Sims and Singh[33] as cited by Frank et al.[19] estimated that the CO_2 budget of native grasslands should be near equilibrium.

Losses in the current study indicate that on the basis of annual BREB CO_2 flux data that the ecosystem acted as a CO_2 source during the seedling year and that it was close to equilibrium during the first production year. However, approximately 530 g m^{-2} yr^{-1} of dry matter was removed as silage and during grazing in the seedling year and approximately 450 g m^{-2} yr^{-1} was removed in the first production year, equating to an additional 805 and 684 g CO_2 m^{-2} of C loss from the ecosystem besides that determined by BREB CO_2 flux measurements. Thus, in both years the forage stand acted as a CO_2 source. The removal of C from the ecosystem as harvested dry matter does not complete the C accounting because return of materials as manure after feeding or as feces during grazing may bring a fraction of the original crop-C back to the ecosystem.

8.4 CONCLUSION

More years of annual measurement of CO_2 flux are required to determine precise patterns for CO_2-dynamics on cropland pasture in mixed cropping systems. This study indicates that seedling year establishment of forage crops in the Aspen Parkland region is vulnerable to CO_2 loss from the crop rotation ecosystem. Even in a dry season first production year productivity was sufficient to achieve annual CO_2-equilibrium. However, removal of hay and silage resulted in losses of CO_2 from the ecosystem that were as large as or much larger than from natural processes. Overgrazing predisposed the ecosystem to large CO_2 loss as well.

In a region with a short growing season the number of days when net CO_2 uptake occurs is inherently lower than in the dormant period. Periods of drought and lag phases of growth after cutting and hard grazing are periods when soil respiratory losses may be larger than net plant uptake on a daily basis. These periods reduce the number of growing days further, seriously eroding potential ecosystem C sequestration. To balance ecosystem CO_2 exchange in this short-season temperate pasture system every effort must be made to extend the duration of green vegetative cover. Productivity must be large enough to sequester as much C as is harvested for agricultural purposes.

ACKNOWLEDGMENT

Financial support from Campbell Scientific (Canada) Corp. is appreciated.

REFERENCES

1. Adams, J.M. et al., Increases in terrestrial carbon storage from the last glacial maximum to the present, *Nature*, 348, 711, 1990.
2. Sundquist, E.T., The global carbon dioxide budget, *Science*, 259, 934, 1993.
3. Houghton, R.A., Davidson, E.A., and Woodwell, G.M., Missing sinks, feedbacks, and understanding the role of terrestrial ecosystems in the global carbon source, *Global Biogeochem. Cycles*, 12, 25, 1998.
4. Follet, R.F., Kimble, J.M., and Lal., R., The potential for U.S. grazing lands to sequester soil carbon, in *The Potential of U.S. Grazing Lands to Sequester Carbon and Mitigate the Greenhouse Effect*, R.F. Follet et al., Eds., Lewis Publishers, Boca Raton, FL, 2001, 401.
5. Fan, S. et al., *Science*, 282, 442, 1998.
6. Clark, E.A., Buchanan-Smith, J.G., and Weise, F.S., Intensively managed pastures in the Great Lakes Basin. A future oriented review, *Can. J. Anim. Sci.*, 73, 725, 1993.
7. Papadopoulos, Y.A., Kuneilius, H.T., and Fredeen, A.H., Factors influencing pasture productivity in Atlantic Canada, *Can. J. Anim. Sci.*, 73, 699, 1993.
8. McCartney, D.H., History of grazing research in the Aspen Parkland, *Can. J. Anim. Sci.*, 73, 749, 1993.
9. Petit, H.V., Pasture management in animal production in Quebec: a review, *Can. J. Anim. Sci.*, 73, 724, 1993.
10. Entz, M.H. et al., Potential of forages to diversify Canadian and American northern great plain cropping systems, *Agron. J.*, 94, 240, 2002.
11. Willms, W.D. and Jefferson, P.G., Production characteristics of the mixed prairie. Constraints and potential, *Can. J. Anim. Sci.*, 73, 665, 1993.

12. Schnabel, R.R. et al., The effects of pasture management practices. In *The Potential of U.S. Grazing Lands to Sequester Carbon and Mitigate the Greenhouse Effect*, R.F. Follet et al., Eds. Lewis Publishers, Boca Raton, FL, 2001, 291.

13. Bailey, R.G., *Ecoregions. The Ecosystem Geography of Oceans and Continents*, Springer, New York, 1998.

14. MacAlpine, N.D. et al., Resources for Beef Industry Expansion in Alberta, Alberta Agriculture Food and Rural Development, Edmonton, Alberta, 1997.

15. Svejcar, T., Mayeux, H., and Angell, R., The rangeland carbon dioxide flux project, *Rangelands*, 19, 16, 1997.

16. Dugas, W.A., Huer, M.L., and Mayeux, H.S., Carbon dioxide fluxes over bermudagrass, native prairie and sorghum, *Agric. For. Meteorol.*, 93, 121, 1999.

17. Angell, R.F. et al., Bowen ratio and closed chamber carbon dioxide flux measurements over sagebrush steppe vegetation, *Agric. For. Meteorol.*, 108, 153, 2001.

18. Frank, A.B. and Dugas, W.A., Carbon dioxide fluxes over a northern, semi arid mixed grass prairie, *Agric. For. Meteorol.*, 108, 317, 2001.

19. Frank, A.B. et al., Carbon dioxide fluxes over three Great Plains grasslands. In *The Potential of U.S. Grazing Lands to Sequester Carbon and Mitigate the Greenhouse Effect*, R.F. Follet et al., Eds. Lewis Publishers, Boca Raton, FL, 2001, 167.

20. Sims, P.L. and Bradford, J.L., Carbon dioxide fluxes in a southern plains prairie, *Agric. For. Meteorol.*, 109, 117, 2001.

21. Frank, A.B., Carbon dioxide fluxes over prairie and seeded pasture in the Northern Great Plains, *Environ. Pollut.*, 116, 397, 2002.

22. Frank, A.B., Liebig, M.A., and Hanson, J.D., Soil carbon dioxide fluxes in northern semi arid grasslands, *Soil Biol. Biochem.*, 34, 1235, 2002.

23. Johnson, D.A. et al., *J. Range Manage.*, 56, 517, 2003.

24. Kim, J., Verma, S.B., and Clement, R.J., Carbon dioxide budget in a temperate grassland ecosystem, *J. Geophys. Res.*, 97, 6057, 1992.

25. Bremer, D.J. et al., Responses of soil respiration to clipping and grazing in tall grass prairie, *J. Environ. Qual.*, 27, 1539, 1998.

26. Haferkamp, M.R. and MacNiel, M.D., Grazing effects on carbon dynamics in the northern mixed-grass prairie. *Environ. Manage.*, 33, S462, 2004.

27. Padbury, G. et al., Agroecosystems and land resources of the Northern Great Plains, *Agron. J.*, 94, 251, 2002.

28. Dugas, W.A., Micrometeorological and chamber measurements of CO_2 flux from bare soil, *Agric. For. Meteorol.*, 67, 115, 1993.

29. Akinremi, O.O., McGinn, S.M., and McLean, H.D.J., Effects of soil temperature and moisture on soil respiration in barley and fallow plots, *Can. J. Soil Sci.*, 79, 5, 1999.

30. McGinn, S.M. and King, K.M., Simultaneous measurements of heat, water vapour and CO_2 fluxes above alfalfa and maize, *Agric. For. Meteorol.*, 49, 331, 1990.

31. De Jong, E., Schapperd, H.J.V., and MacDonald, R.B., Carbon dioxide evolution and cultivated soil as affected by management practices and climate, *Can. J. Soil Sci.*, 54, 299, 1974.

32. Pattey, E. et al., Measuring nighttime CO_2 flux over terrestrial ecosystems using eddy covariance and nocturnal boundary layer methods, *Agric. For. Meteorol.*, 113, 145, 2002.

33. Sims, P.L. and Singh, J.S., The structure and function of ten western North American grasslands IV. Compartmental transfers and energy flow within the ecosystem, *J. Ecol.*, 66, 983, 1978.

9 Forests in the Global Carbon Cycle: Implications of Climate Change

M.J. Apps, P. Bernier, and J.S. Bhatti

CONTENTS

9.1 INTRODUCTION

As a consequence of human activity, Earth's climate has changed during the last 100 years and will change significantly for centuries to come.[1] The predicted changes for the next 50 to 100 years and beyond are both larger and faster than previously thought,[2,3] and also more certain.[4] Recent assessments indicate that, in the absence of purposeful mitigation interventions, it is likely that changes in the global mean temperature over the next 100 years will be at the high end of, or even exceed the IPCC 2001 predictions of +1.4 to 5.8°C above 1990 temperatures[4] — itself a decade of record-breaking temperature.[5]

175

The change is not expected to be a simple linear increase in temperature or other climatic variable: abrupt and likely unpredictable changes similar to those seen in the geological record must be anticipated in the future. The impacts that have already been reported through the 20th century can be expected to intensify over the 21st, disrupting natural ecosystems and the services society has come to depend on, at all spatial scales from local to regional and global.

Moreover, the change has not been — and will not be in the future — distributed evenly over the Earth; climate change is greatest at mid- to high latitudes and over the continental landmasses found in North America, Europe, and Asia where large carbon pools are currently found in forest ecosystems. In these regions, local biogeochemical processes will likely experience profound changes in prolonged growing season, intensified incidence of drought and fire, systematic changes in annual snow accumulations, and an overall mobilization of large pools of ecosystem C, from forested uplands to forested wetlands.[6]

Climate change is arguably the most important environmental issue of the 21st century. It will have significant implications for resource management strategies. Are forests and forestry part of the problem or part of the solution?[6] This chapter examines the contribution of northern forest ecosystems, especially the contribution of their management to the global carbon cycle.

9.2 CLIMATE CHANGE AND THE GLOBAL CARBON CYCLE

Throughout at least the last four glacial cycles, spanning nearly 1.5 million years prior to the 20th century, the atmospheric concentration of CO_2 only varied between ~180 ppmv during glaciations, when the global temperature was 8 to 9°C colder than today, and ~280 ppmv during the interglacial periods when the temperature was similar to present values (Figure 9.1A). This narrow range of variation in atmospheric CO_2 is remarkable given that its concentration is determined by a highly dynamic biogeochemical cycle. Every year, approximately 16% of the CO_2 in the atmosphere (approximately 760 Gt C) is taken up through photosynthesis by vegetation, and an almost identical amount is released by the respiration of vegetation and heterotrophs feeding on that vegetation. A similar exchange of ~90 Gt C yr[1] takes place at the ocean surface where phytoplankton provide the photosynthetic engine driving the exchange.[7]

This generally tight domain of stability between variations in CO_2 and global temperature (Figure 9.1B) suggests that the global carbon cycle has been controlled by powerful biological feedback processes that have maintained the climate in a habitable range. The biosphere appears to play a central role in regulating Earth's climate, a suggestion strongly reinforced by the physics of the greenhouse gas feedbacks. The biosphere–climate system coupling includes other factors, such as surface reflectance properties (albedo), that have effects both regional and global in extent (see, e.g., Reference 8), but here our focus is restricted to the global carbon cycle.

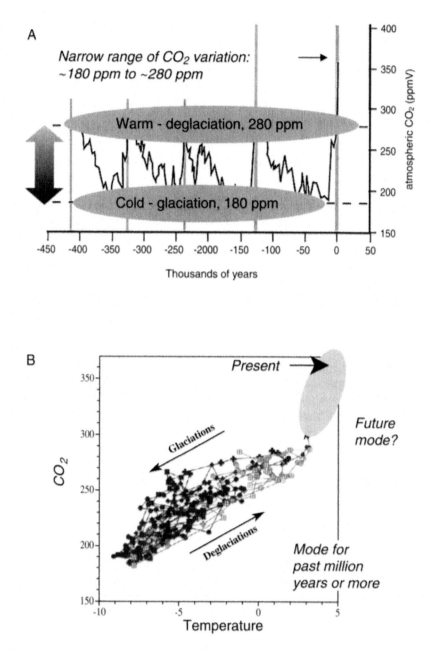

FIGURE 9.1 (A) Variation in atmospheric CO_2 from analysis of ice cores over four glacial cycles during the last 420,000 years. Present levels (>360 ppmv) are indicated by the arrow. (B) The stability domain of atmospheric CO_2 and global temperature over the last four glacial cycles, showing recent departures and possible shift to a new domain of unknown stability. (Adapted from Falkowski et al.[7])

In contrast to the long-term record, the atmospheric CO_2 concentration today is ~370 ppmv — nearly 100 ppmv higher than at any time in at least the past 1.5 million years — as a result of human perturbations to the global carbon cycle. The concentration is also rising at a rate that is at least 10, and perhaps as much as 100, times faster than ever before observed.[7] Clearly, the biosphere's ability to regulate the global carbon cycle — and hence the climate system — has been exceeded by human-induced carbon emissions.

9.3 HUMAN PERTURBATIONS TO THE GLOBAL CARBON CYCLE

Human perturbations to the carbon cycle have been both direct and indirect (Figure 9.2). On land, human activities have modified vegetation patterns and functioning in global proportions, while changes to freshwater inputs and pollutant eutrophication of the oceans have altered their ecology as well. In other words, humans have changed the very nature of the biospheric systems that are responsible for biospheric exchange of CO_2. In addition, and more significantly, human use of fossil fuels has introduced additional, new carbon into the active* global carbon cycle through the combustion of fossil fuels. Deforestation — removal of forest vegetation and replacement by other surface cover — has had a twofold impact on the carbon cycle: the loss of photosynthetic capacity in forest vegetation, and the release of the large carbon stocks that had accumulated in these forest ecosystems over long periods. Indirect human impacts on the carbon cycle include changes in other major global biogeochemical cycles (especially nitrogen),[9] alteration of the atmospheric composition through the additions of pollutants as well as CO_2, and changes in the biodiversity of landscapes and species — all of which are believed to significantly influence the functioning of the biosphere.

9.4 FOREST SOURCES AND SINKS AT THE STAND AND LANDSCAPE SCALE

A forest ecosystem is a sink (source) when it effects a net removal (release) of atmospheric CO_2. The sink results when the uptake through photosynthesis results in an increase in the sum of the carbon stocks retained in the forest vegetation itself and in the stocks of organic carbon in other material derived from the forest. The most important of these derived reservoirs are the detritus and soil organic matter pools. The net carbon balance of the ecosystem may be calculated as the net change over time in total ecosystem carbon stocks (dC_{ecosys}/dt, where C_{ecosys} is the sum of carbon stocks in vegetation, forest floor and soil). Ignoring for the moment any export of organic carbon from the ecosystem, the net carbon balance is identical to the net ecosystem productivity (NEP):

---

* "Active" is used here to distinguish the carbon pools and processes that dominate the exchange that occurs on time scales of order of years to decades from those that are important on geological time scales, such as the accumulation of organic carbon in fossil fuel deposits.

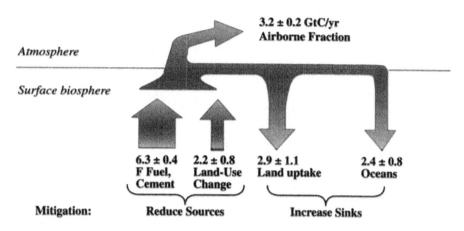

FIGURE 9.2 Human-induced perturbations (Gt C yr⁻¹) to the global carbon cycle during the 1990s. The arrow widths are proportional to the fluxes. Land uptake is inferred as the residual required to balance the other fluxes with the observed accumulation (airborne fraction) in the atmosphere. (Data from Houghton.[10])

$$\text{Net Carbon Balance} = dC_{ecosys}/dt \qquad (9.1)$$

$$dC_{ecosys}/dt = \text{NEP} = \text{GPP} - R \qquad (9.2)$$

where GPP (gross primary production) is the rate of CO_2 uptake by foliage through photosynthesis and $R = R_a + R_h$ is the total ecosystem respiration flux comprising autotrophic (plant) respiration R_a and heterotrophic respiration R_h (decomposition) of the accumulated detritus and soil pools.

The term net biome production (NBP) is sometimes used to account for exported carbon and its subsequent decomposition outside the ecosystem:[11]

$$\text{NBP} = \text{NEP} - R_{exp} \qquad (9.3)$$

where R_{exp} is the flux of carbon transferred out of the ecosystem. Forest products form an important part of the offsite carbon pools in that the timing and manner of their decomposition is (in principle at least) under human control.

Figure 9.3 shows the conceptual pools and transfers of carbon involved in forest ecosystems and the forest sector. To provide a comprehensive system, the ecosystem compartments (vegetation and detritus and soil pools), the exported pools that are located offsite (including forest products and the waste created during their manufacture and abandonment in landfills), and the influence of the forest sector on fossil fuel use are all included.

The net accumulation of carbon in the ecosystem (or the larger system shown in the figure) is thus a summation over time of the difference between a large ingoing CO_2 flux (GPP) and a nearly equal outgoing flux (R). Different processes, whose rates differ over time and space and vary both with environmental conditions and the state of the ecosystem, control the two fluxes. The processes involved include

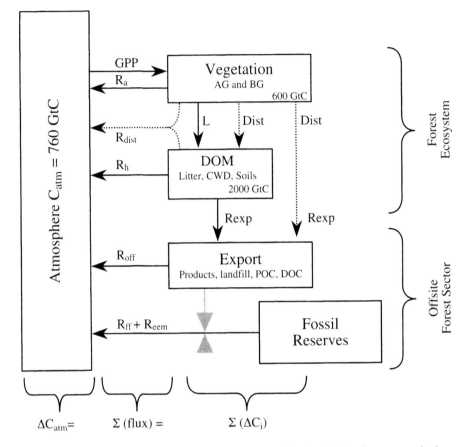

FIGURE 9.3 Carbon fluxes (arrows) and pools (boxes) involved in the forest sector budget. Smoothly varying fluxes include GPP = gross primary production, R_a = autotrophic respiration, R_h = heterotrophic respiration, R_{off} = offsite respiration, L = litter fall (above- and belowground AG and BG) and leaching from DOM (dead organic matter) on the forest floor and in soils. Pulsed fluxes (dotted lines) are associated with disturbances. R_{exp}, the carbon flux that is exported to offsite carbon pools, has both a smooth component (leaching) and a pulsed component (from disturbances). Fluxes from offsite carbon pools (products, landfills, POC = particulate organic carbon, DOC = dissolved organic carbon in water or air) are lumped into one flux R_{off}. The influence of bioenergy and use of forest products on fossil fuel use is shown as a control valve on fluxes from fossil fuel use (R_{ff}) and cement production (R_{cem}) production.

both those regulating the internal redistribution of organic carbon within the ecosystem, such as allocation of photosynthate within the plants and breakdown of fresh litter into less decomposable forms of soil organic matter, and disturbances (such as windthrow, insect predation, harvest, or fire).

Disturbances are discrete events that are particularly interesting because they generate large pulses of internal transfers of carbon between pools within the ecosystem or out of it (e.g., harvest). They therefore bequeath a legacy of increased decomposition emissions in the future. In addition, disturbances such as fire may also generate large, immediate CO_2 releases to the atmosphere. The complex set of processes — operating independently over a range of timescales — gives rise to rich variation in NEP (and NBP) in both time and space.

The net carbon balance in a forest ecosystem (NEP) can be estimated by summing all the changes in ecosystem carbon stocks (the "stock inventory" method), direct measurement of the net exchange of CO_2 with the atmosphere (using, for example, eddy covariance techniques), or a combination of these methods. Provided all stocks and fluxes are accounted for, the approaches must give identical answers (a result of the principle of conservation of mass), as has been shown by careful experiments at the Harvard forest and several other locations.[12]

The net carbon balance of a stand of trees or patch of forest varies with the prevailing conditions that affect both the rates of CO_2 uptake and release (Figure 9.4A). It also depends very strongly on the past history of the stand or site. For example, the net carbon balance (NBP) of a clear-cut stand is initially highly negative (when the harvest carbon is removed from the site — an export flux not directly captured by net ecosystem exchange flux measurements) and remains so for several years while the releases of CO_2 from decomposition of slash and soil carbon exceeds the CO_2 uptake of regrowing vegetation. Eventually the uptake through regrowth exceeds decomposition efflux, at which time above- and belowground detrital production starts to rebuild the depleted stocks on the forest floor and in the soil. NEP then rises steeply to a maximum rate that typically occurs around or shortly after canopy closure. As the ecosystem continues to age and more organic carbon accumulates in the vegetation, forest floor, and soils, the respiration efflux from these reservoirs also increases. Rates of photosynthetic input tend to level off as the stand approaches maturity, and net primary productivity may even decline when stand-breakup occurs in overmature stands.[13,14] Thus in older stands, the net carbon balance (NEP) tends toward zero (or even becomes negative) as decomposition of the soil and detritus layer approaches that of the photosynthetic inputs. In some ecosystems, such a decrease in NEP may take a very long time after the last carbon-removing disturbance.[11]

At the landscape (or biome) scale, a forest comprises many stands of trees (individual ecosystems) in various stages of development (Figure 9.4A), and the net carbon balance at this scale is the integration across all such ecosystems in the landscape. Here, for illustrative purposes, only even-aged forests such as are found in disturbance-dominated natural forests or in clear-cut plantations are considered: the principles apply, however, to all forests. For forests dominated by even-aged stands, the stand age-class distribution can be used to facilitate the summing across ecosystems in different stages of development. For a forest comprising only one ecosystem type, the total ecosystem carbon in the landscape is

$$C_{landscape} = \sum_{i=1}^{N} C_i A_i \tag{9.4}$$

and its change over time is

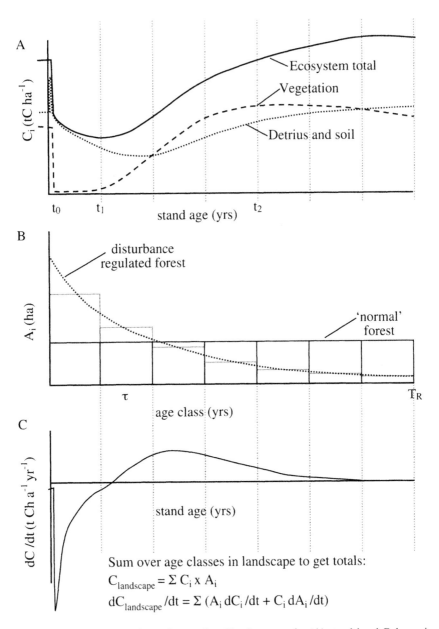

FIGURE 9.4 Carbon dynamics at the stand and landscape scale: (A) stand-level C dynamics after disturbance at t_0. The stand is a source until t_1, but does not recover C lost at an immediately after the disturbance until t_2; (B) stable age-class distributions for "normal forest" (rotation T_R) and random disturbance-regulated forest (return interval t); (C) stand-level accumulation rate. For landscape pools, sum product of a * b over all age classes; similarly sum product of c * b for changes in pools in unchanging conditions.

$$dC_{landscape}/dt = \sum_{i=1}^{N}(A_i \, dC_i/dt + C_i dA_i/dt) \qquad (9.5)$$

where A_i is the area (ha) of forest in age class i, and C_i is the carbon concentration (Kg C ha^{-1}) of this age class. For a more general heterogeneous forest, the total landscape carbon involves additional summations over all the distinguishable ecosystem types (each characterized by a different carbon accumulation curve). Moreover, the actual carbon accumulation curve (Figure 9.4A and C) changes with disturbance type and intensity as each may leave different amounts of litter and hence different legacies of decomposition pulses; the actual site history has a direct effect. This generally involves additional summations over disturbance types and inevitably requires historical information about past disturbance regimes.[15]

Changes in the net carbon accumulation at the landscape scale (Equation 9.5) thus has two components:

1. Changes in productivity of the individual ecosystem growth and respiration responses to environmental variations (functional response, alterations to curves in Figure 9.4A and C)
2. Changes in the age-class distribution associated with landscape variation in mortality and recruitment (structural response, alterations to curve in Figure 9.4B)

Over long enough times, succession alteration to the distribution of vegetation types will also take place, providing further structural and functional responses and changes in NEP.

At any given time, the age-class distribution is a direct result of the cumulative effects of mortality and recruitment to that point in time, and for the even-aged forests discussed here, is a direct reflection of the history of past disturbances. Under a steady disturbance rate (such as a constant fire return interval, or a fixed harvest rotation), the balance between mortality and recruitment leads to a stable age-class distribution that can maintain its shape over time. An example of such distributions is the managed "normal forest"[*][16] associated with sustainable harvesting and regeneration of stands in a plantation, in which each age-class occupies an equal area up to the rotation age T_R (Figure 9.4B). Another example is the (approximately) exponential age-class distribution (also shown in Figure 9.4B) that is associated with randomly occurring disturbances, applied with equal probability to all ages, and having a constant mean return rate and variance. Such distributions are often found (but not always) with naturally occurring disturbances such as wild fire, windstorms, and some insect outbreaks.[17,18]

Sources and sinks at the landscape scale are created when the disturbance rate changes. If the disturbance rate increases, the age-class distribution shifts to the

---

[*] The term "normal" is used here in a technical sense (see MacLaren[16]) and not as the common adjective to imply "usual" or "average."

left (younger), and the total carbon retained in the ecosystems in the landscape decreases. The landscape becomes a net source of CO_2 to the atmosphere while its age-class distribution adapts to the new disturbance regime. (If some of the lost carbon is transported out of the ecosystem landscape to decompose in offsite reservoirs, such as the case of forest products, the landscape source is reduced by that amount — in essence this component of the source is exported.) Similarly, if disturbances are suppressed, the ages shift to the right, the forest ages and carbon stocks increase with a net removal of CO_2 from the atmosphere. Taking changes in disturbance regimes into account is clearly important in predicting the future carbon budgets of forested regions.

9.5 LAND-BASED CARBON SINK AND ITS FUTURE

Until recently, the net land-based carbon sink required to balance the perturbed global carbon budget (Figure 9.2) was thought to be fully explained by changes in ecosystem functioning. Enhanced forest uptake rates (increased GPP) associated with elevated atmospheric CO_2, increased nutrient inputs from pollution, and a positive response to global temperature increases were used to close the global budget. However, although physiological mechanisms and normal climate variations may explain some of the short-term changes (seasonal to inter-annual) in forest ecosystem uptake (GPP), their ability to cause longer-term net uptake and retention (GPP-R) has been questioned by a number of authors (e.g., References 19 and 20).

It is now recognized that changes in the structure of ecosystems, especially the age-class structure of forests, are at least as important as the functional changes. For example, changes in land-use practices, such as abandonment of marginal agricultural lands to forest and the rehabilitation of previously degraded or deforested lands has been shown to be largely responsible for the putative North American sink,[21] and a much larger contributor than any of the proposed physiological mechanisms such as CO_2 fertilization.[22]

Change in the climate regime may also affect current carbon pools of forests, although the direction and magnitude of these changes is still uncertain and difficult to predict. Over periods of years to decades, the stimulation of GPP through longer growing seasons should result in increased vegetation biomass, an effect that may already be apparent in the global atmospheric CO_2 record.[23] However, although GPP may increase with increased temperature, so may the heterotrophic decomposition rate — approximately doubling for each 10°C increase in soil temperature. Given the very large size of the C stocks in forest litter and soil pools, this gives rise to concern that increased heterotrophic respiration may generate a positive feedback mechanism to climate change by releasing additional quantities of CO_2 in the atmosphere. Recent work, however, suggests that in some ecosystems, increased heterotrophic respiration may be largely offset by increased detrital production by trees, leaving detrital and soil carbon content relatively unchanged as long as the forest composition remains unaltered.[24]

At longer timescales (decades to centuries and longer), changes in the vegetation itself take place through successional processes as the ecosystems adapt to changing conditions. These longer-term changes may lead to either greater carbon stocks, as in more productive forest ecosystems, or smaller stocks, as in a transition to a grassland ecosystem. Comparison of relative pool sizes for boreal, temperate, and tropical forests suggests a general shift of dominance from belowground to aboveground stocks as temperatures warm. Over the intermediate term, the expansion of forests into existing nonforest regions, such as the northward expansion of the boreal forest, may provide some additional uptake. This expansion, however, is a slow and uncertain process (e.g., Reference 25) and over the short term will likely be more than offset by possible dieback of forests at the other end of their range. Such dieback and transition to grasslands in southern boreal forests in south central Canada have been suggested by several authors,[26–28] and can happen extremely quickly if driven by more frequent or more intense large-scale disturbances such as fire.[29] One of the major causes of uncertainty is the unprecedented rate of current climatic changes that are taking place over timescales that are out of synchrony with the dominant processes of some ecosystems and beyond their adaptive capacity.[30]

In addition, for each of the stimulation mechanisms there typically exists limiting factors that eventually counteract it over time.[19] Elevated levels of ambient CO_2 increase the photosynthetic efficiency of foliage, but as the concentration increases, this stimulation decreases and saturates at atmospheric concentrations that may be reached in the next 50 to 100 years.[31] To date, *in situ* fumigation of stands with elevated CO_2 for periods for 3 years has yielded consistently high GPP values, but the future of this effect remains uncertain.[32] Although initial response to increased N inputs associated with atmospheric pollution is growth enhancement, at higher loadings (already reached in some areas), the effects of acidification may lead to net decreases.[33,34] Moreover, there is good evidence that the response of forest ecosystems to either CO_2 or N fertilization will be short-lived when other required resources, such as water or other nutrients, become limited. Results from stand-level fumigation studies also show that tropospheric ozone may counteract the growth enhancement offered by increases in CO_2.[35] Thus, while many ecosystems studied to date indicate an initial positive response in NEP to these manipulations, they also show an acclimation over time to these stimuli — usually interpreted as a combination of subtle changes in the ecosystem structure and the onset of another limiting factor.[3]

Finally, there are concerns that climate change will bring about changes in the disturbance regimes (rate, intensity, and form). Although fire[36] is the best known of these disturbances, changes in insect dynamics, drought stress, ozone and ultraviolet damage, and damage from hurricane or severe storms may be more important in some regions.[37–40] The impact of changes in disturbance regime over the last few decades in Canada's forests — suggestive of, but not definitively shown to be due to climate change — appears responsible for a shift of these forests from a significant sink to a small source of atmospheric CO_2.[15]

9.6 MITIGATION OPPORTUNITIES

There are two fundamental mitigation interventions:

1. Reduce emission sources, or
2. Increase sinks

Land management, and especially forestry and forest management, can contribute to both of these opportunities. Interventions that maintain healthy ecosystems can also maintain, or even increase, land-based carbon stocks. Using forest goods and services can simultaneously help to reduce anthropogenic emissions of CO_2 typically generated by alternative supplies of these goods and services. These two opportunities are not mutually exclusive, and will be briefly described in very broad terms.

9.6.1 FOREST MANAGEMENT TO INCREASE OR MAINTAIN TERRESTRIAL ECOSYSTEM CARBON

The ultimate aim of mitigation strategies, such as put in place by the Kyoto Protocol, is "the stabilization of greenhouse gas concentrations in the atmosphere at a level that would prevent dangerous anthropogenic interference with the climate system."[41] Mitigation strategies that promote the preservation and maintenance of healthy ecosystem functioning may therefore be as valuable as land-management strategies that aim to enhance the net uptake, and decrease the releases of CO_2 in terrestrial ecosystems, the so-called terrestrial sinks (and sources) of the Kyoto Protocol.

It is beyond the scope of this chapter to attempt a detailed review of the different schemes for ecosystem carbon management that have been proposed, or their economic, ecological, and social impacts. The IPCC has provided in-depth analyses in two major reports released in 2000[42] and 2001,[43] and good practice guidelines for managing terrestrial forest ecosystems in the context of carbon sequestration.[44] The various forest ecosystem management activities that have been proposed[43,45] can be grouped into three broad approaches:

1. Strategies to maintain and preserve existing forests
2. Strategies to increase the area of land under forest
3. Strategies to increase the carbon stock density on the forested land (C ha^{-1})

Much of the focus on carbon sequestration in forests ecosystems has been on enhancement of aboveground biomass as a natural extension of timber production forestry. Recently, a shift to more comprehensive ecosystem management appears to be taking place, together with renewed opportunities and interest in rehabilitating degraded lands, mitigating the effects of deforestation, and managing for natural values (such as wildlife or water quality), not merely for timber. The success of different approaches in any given region depends on prevailing social, economic, and historical conditions. In some regions such as in the central part of Canada, slowing, halting, and mitigating deforestation associated with infrastructure may provide the most efficient strategy, while in other regions, such as central British Columbia, more traditional timber production approaches combined with protection from disturbance may be more attractive.

Protection against disturbance is not, of itself, an efficient or long-term mitigation measure. This is especially true of wildfire where large expenditures simply protect carbon are analogous to paying high rent: the carbon is retained only as long as the protection continues and is lost when the next fire comes along. However, if the protected area is subsequently harvested, transferring carbon to long-lived forest products, and successfully regenerated, a potentially significant carbon gain can be realized — both within the offsite and within the forest ecosystem pools (Figure 9.3). In the forest ecosystem, the combination of protection and harvesting can be visualized as an increase in the effective rotation length (see Figure 9.4), and as demonstrated by Kurz et al.[46] the transition from a natural disturbance regime to a managed one (including protection) can have positive carbon benefits.

Increased carbon stocks can also be accomplished through techniques that reduce the time for stand establishment (such as site preparation, planting, and weed control), increasing resources (especially nutrients) required for growth, or through the selection of species that are more productive for a particular area. Decreasing the losses can be accomplished through modification of harvesting practices such as low-impact harvesting (reduce damage to residual trees and soil structure), increased efficiency (reduced logging residue), and managing residues to leave carbon on site.[45] Despite the interest in all of these techniques, fundamental scientific questions remain about how the ecosystems will respond to a rapidly changing climate, including the allocation of photosynthate between above- and belowground compartments, regeneration success, growth vs. respiration responses — all of which have a direct influence on the carbon benefit a given technique will achieve.

Nutrient fertilization has long been used to enhance stand productivity and can result in increased C stocks in trees and soils,[47] but its success is dependent on the site conditions, and is therefore potentially susceptible to rapid climate changes. For example, on more fertile sites the effect of fertilization is reduced as other factors begin to limit growth.[48] For planting, species selection and stocking are important considerations and, depending on the management objective, planting fast-growing species such as hybrid poplar can yield high carbon accumulation rates in early years.[49] For long-term sequestration, however, planting species adapted to the local climate may be more effective.[49] But what will be the local climate as the trees approach maturity?

In all such interventions aimed at increasing forest ecosystem, or offsite, carbon stocks, it is necessary to account for fossil fuel consumed in plantation establishment, maintenance, and harvesting (see Figure 9.3). Thus while short rotation plantations can provide an excellent opportunity to displace fossil fuel and, at the same time provide (on average) a significant carbon stock in the plantation, significant direct and indirect (fertilizer production) expenditures of fossil fuel are usually required to realize these gains. An overview of forest-relevant, C sink-source interventions is provided in Table 9.1.

9.6.2 MANAGING PRODUCTS AND SERVICES DERIVED FROM FORESTS FOR C BENEFITS

Products extracted from managed forest ecosystems play multiple roles in the global carbon cycle:

TABLE 9.1
Classes of Management Activities, Cost and Benefits

Intervention	Cost	Comments	Short Term (<25 yrs)	Longer Term
		Maintain and Preserve Existing Forests		
Preserve primary forests[a]	Future opportunity costs	No new sink added / Sink already accounted for		+
Halt/slow deforestation	Eliminate causes	Big avoided emissions	++++	+++ / -Va
Halt logging[a]	Forgone economic activity	Loss of forest services	+	-- Va
		Increase Forest Area		
Afforestation and reforestation[a,b]	Loss of land for other purposes	One-time C gain	++	-Va,c
Establish and manage reserves	Future opportunity costs?	One-time C gain	+	-Va,c
Multiple use (e.g., agroforestry, sheterbelts)		Cross sectoral benefits	+	+
Restoration of degraded lands	Feasibility?	Reason for degradation ameliorated?	++	-Va
Urban forestry		Energy costs/benefits	+	-Vc
		Increase Carbon Density (C ha⁻¹)		
Longer rotation length		Reduced short-term yield of products?	++	-Va,c
Enhance tree productivity				+Va, -Vc
Control stand density (thinning)	Implementation cost	Timber benefit but total biomass and DOM C?	-?	-Vc
Enhance nutrient availability	Implementation cost Feasibility	Energy costs	++	Va
Control water table	Implementation cost. Increased CO_2, reduced CH_4	Soil respiration, tree growth	+	-Vc

		Response to climate? Cost; diversity impacts		
Selected species and genotypes			+	$--$Vc
Protect from natural disturbance, reduce risk	Implementation cost Feasibility	Requires ongoing maintenance	+++	$--$Va,c
Reduced impact logging	Implementation cost	Cost, extent	?	+Va
Reduce regeneration delay	Implementation cost	Small one-time gain	+	+Va
Manage onsite logging residues	Implementation cost	Forgone use in products?	+	+Va,c

Note: Number of +'s ($-$'s) indicates expected magnitude of C benefit (decrement). Vc indicates expected vulnerable to climate change, and Va indicates potentially vulnerable to changes in human activity.

[a] Kauppi et al.,[43] Fig. 4.8.
[b] Includes savannah thickening as a special case.

1. They act as an offsite, manageable carbon reservoir.
2. They can be burned to provide a renewable source of energy (direct substitution).
3. They substitute for competing materials having a larger atmospheric CO_2 footprint (indirect substitution).

Both direct and indirect substitution can add significantly to the mitigation potential of forest products.

9.6.3 FOREST PRODUCTS AS A MANAGEABLE CARBON POOL

From a global perspective, the export of organic carbon from the forest ecosystem where the CO_2 is initially withdrawn from the atmosphere by photosynthesis to a different location where it subsequently decomposes and releases the CO_2 back to the atmosphere, results in a spatial displacement of the source component (at the site of the decomposing product) relative to a comparable sink component (in the forest ecosystem). The net effect on atmospheric concentration is negligible unless the rate of decomposition in the geographically displaced product pools is different from that in the forest ecosystem from which it was removed. This separation of apparent source (forest product) and sink (forest ecosystem) has interesting political implications that have, to date, led to an impasse in attempts to incorporate forest product carbon management in the Kyoto Protocol (who gets the credit — the exporting country in whose forest the uptake of CO_2 took place, or the receiving country where the forest product reservoir management occurs?).

Despite these political difficulties, the carbon contained in forest products makes a small, and manageable, contribution to the global carbon balance. The geographical displacement of forest ecosystem uptake (sink) from the forest product decomposition (source) may also be required for reconciliation of observed geographical distributions of atmospheric CO_2 concentrations with atmospheric transport of CO_2 from known emission sources and sinks.[50,51]

As a carbon reservoir, the size of exported product pools is the cumulative difference between harvest inputs and depletions through decomposition and combustion that release CO_2 back to the atmosphere. Estimating the size of this pool and its change over time is complicated by at least three factors: the difficulty of tracking the flows of forest products through the multitude of uses society has found for wood products; accounting for the changes over time in the reuse and recycling of woody materials (including pulp and paper); and the wide geographical dispersal of the products through trade (increasingly international). These factors make it difficult to compile inventories of products with widely different half-lives, to estimate the rates of product recycling between different half-lives, and to determine the rates of decomposition and combustion (releasing CO_2 back to the atmosphere) at each stage in the product life cycle, each of which depends on the nature of the product, its use, level of protection, and the local environment in which it is used and discarded.

Despite these inherent uncertainties, estimates of the forest product pools have been made at the global scale, where the pool is thought to be relatively small —

between 5 and 10 Gt C.[43] Despite its small size, the IPCC concluded that the potential for an increased contribution to mitigating human perturbations to the global carbon cycle are not insignificant: wood products "already contribute somewhat to climate mitigation, but if infrastructures and incentives can be developed, wood and agricultural products may become vital elements of a sustainable economy: they are among the few renewable resources available on a large scale."[43]

An increasing products pool releases proportionally more CO_2. For a steady rate of harvest inputs, the forest products pool in any given region eventually tends to reach a plateau, at which point the accumulated forest products' releases of CO_2 become equal to the harvest inputs derived from the forest uptake of CO_2. This may be the reason analyses of forest product contributions to the national carbon balance for countries with a long history of forestry, such as in Fennoscandia[52] and the U.S.,[53] tend to be relatively small. Where the forest product pools are young (and not yet saturated), or where the harvest rate is increasing, the increases in the forest product pool may still be significant. In Canada, for example, both of these factors may be responsible for the increases of 23.5 Tg C yr⁻¹ during the late 1980s — of the same magnitude and nearly offsetting the net decreases (due to increased natural disturbances) in the carbon stocks of Canada forests for the same period.[54]

Although the decision has not yet been made on whether, or how, forest product carbon pools will be accounted for under the Kyoto Protocol, there is little doubt that their wise management can offer some degree of mitigation of the increases in atmospheric CO_2. Some general observations may help to guide management decisions.

- Once harvest inputs cease, product pools can only act as a source of atmospheric CO_2 as they decompose or are incinerated as waste. On the other hand, the CO_2 sink generated in the regrowing forest was "created" by the very act of harvesting and over time exactly balances the source term, provided there was no degradation or improvement of site productivity. The source (decomposing or burned product) and the sink (regrowing forest) are inherently linked; they are autocorrelated with a time delay.
- Forest products and the regrowing forest also constitute a spatially displaced source–sink pair with the emission and uptake occurring at different geographical locations. This spatial displacement presents political challenges as the source and sink may therefore be accredited or debited to different parties. It also arises in atmospheric inversion studies, which must deal with the spatially separated CO_2 emissions–receptor pair.[50,51]
- Retention of carbon in forest products is functionally similar to retention in ecosystem detrital pools: if the half-life of carbon in the products is greater than the natural half-life in the ecosystem, there is a net gain in retained carbon in the forest managed for timber supply relative to the natural ecosystem having the same age distribution.[55]
- The rate of loss from product pools is, in principle, under the control of society through decisions made on the duration of use of the products and their recycling fate. This includes also their final use as a source of energy (see below).

- As a rule of thumb, using long-lived forest carbon stocks to generate short-lived forest products has a disproportionately positive impact on CO_2 emissions, relative to preserving the forest ecosystem stocks,[56] but this conclusion does not hold if the end use in the product chain is bioenergy that substitutes for fossil fuel.

9.6.4 USE OF FOREST BIOMASS FOR BIOENERGY

In addition to their modest role as a manageable carbon reservoir, forest-derived organic materials can also serve to reduce anthropogenic emissions in two important ways: by supplying essential products and services that otherwise entail greater fossil fuel CO_2 emissions, and by directly supplying energy services (bioenergy). Figure 9.3 shows this emission reduction role as a control on the fossil fuel emissions.

Forest biomass is one of the oldest harnessed sources of energy for human activities, providing both domestic heating and cooking functions, and as an industrial source of energy (see Reference 57 and references therein). Globally, bioenergy at present supplies about 14% of the primary energy needs.[42] Where sustainably produced bioenergy replaces, or avoids, the combustion of fossil fuel, it has a lasting influence on the global carbon cycle, as explained below. The extent to which sustainably produced forest products supply essential services that otherwise would result in higher emissions from fossil fuel use, in their manufacture, or their operation, or their maintenance, makes a similar contribution. Moreover, the avoidance of emission sources (Figure 9.2) can be additional to the role of forest products as a managed carbon reservoir discussed above. Both the manufacturing residues generated during their production, and the forest products themselves after their serviceable life, can be used to feed bioenergy supply systems.[57-59]

The trend of increasing replacement of traditional wood-based construction products by cement, metals such as steel and aluminum, and plastics has an adverse impact on the global carbon cycle by increasing the combustion of fossil fuel for their production. For example, the CO_2 emissions associated with electrical transmission line towers is estimated at ~ 10 t C km^{-1} when manufactured from tubular steel and ~ 4.3 t C km^{-1} from concrete, in contrast to the ~ 1 t C km^{-1} estimated for roundwood poles.[60] Similar ratios are found for other materials such as aluminum and plastic, which require expenditures of energy in their production,[60] but which are increasingly becoming substitutes for traditional wood products.

Halting the increase in use of metal and plastic products in replacement of wood products or increasing the substitution of these energy-intensive products by wood benefits the carbon budget in multiple ways. The first is the energy expenditure avoided, which is the net difference in CO_2 emissions required to generate the product from the raw materials. The second is the accrual of carbon in the forest products pool. The third, with a longer-lasting impact, is the use of discarded forest products for the production of bioenergy.

The importance of the contributions of forest products to emission reductions lies in the relative permanence of the CO_2 influence on the global carbon cycle. The combustion of fossil fuels and forest biomass for energy both release comparable amounts of CO_2 to the atmosphere for the similar amounts of energy,[57] and both fuel sources ultimately derive from the same source: the conversion of solar energy to chemical bonds in organic carbon compounds through the process of photosynthesis. Fossil reserves, however, were accumulated over millennia, with natural inputs to and emissions from these deeply buried reservoirs occurring only slowly on geological timescales. Until recently, the fossil reserves have played a negligible part in the active global carbon cycle. Human withdrawal of fossil fuel from these relatively inert reservoirs has effectively added new carbon to the active global carbon cycle at a rate that has increased dramatically over the last 100 years. In contrast, the burning of (modern) biomass simply returns to the atmosphere the CO_2 that was accumulated from the atmosphere in recent times, adding no new carbon to the active global carbon cycle. Provided the forest ecosystems providing the feedstock are managed sustainably,* there is no direct global change in the atmospheric CO_2 concentration from the combusting modern biomass for energy, although there may be additional emissions associated with the infrastructure for bioenergy systems.

Bioenergy derived from forest ecosystems takes many forms, ranging from dedicated bioenergy plantations to co-generation of heat and electricity as a by-product of product manufacture, and the capture and combustion of methane from landfills. The net impact on the global carbon cycle varies with the efficiency of these production systems and the extent to which the expenditure of fossil fuel is required in their production, distribution, and use.[57] In addition, the economic feasibility depends strongly on the availability of land for bioenergy purposes, with costs rising steeply if other production uses are displaced.[57,61]

Including both forest and agricultural systems, global bioenergy production in the year 2050 could be between 95 and 280 EJ (1 EJ is 10^{18} joules = 2.28 10^{15} KWhrs).[62] This would supply 5 to 25% of the projected energy needs under some future development scenarios,[63] and potentially avoid fossil fuel emissions of 1.4 to 4.2 Gt C yr^{-1} in 2050.[64] The maximum potential of bioenergy could be as much as five times greater,[62] but this would require significant infrastructure development. Similar projections of ~4 Gt C yr^{-1} avoided fossil fuel emissions and carbon sequestration by about 2040 were estimated in computer simulations of an ambitious global program of sustainable development of community-scale short-rotation bioenergy plantations estimated by Read 1999, as reported by Sampson et al.[65]

9.7 CONCLUSIONS: THE GLOBAL FOREST SECTOR AND THE GLOBAL CARBON CYCLE

We asked initially if forests were part of the problem or part of the solution, and have tried to show that they are part of both. Forests and their management are not the largest source of the problem, nor can they be its sole solution. However, our past and present use of the forest land base, especially through deforestation, has had and continues to make a double contribution to the increase in atmospheric CO_2 through the reduction in the planet's photosynthetic capacity, and through the elimination or dramatic reduction of the carbon stocks associated with the former forests.

---

* "Sustainable" in this context means that the net forest ecosystem uptake of CO_2 (NEP) is at least as great as the net CO_2 emissions from the combustion of the exported biomass.

This is an important part of the problem. Reduction of the rate of deforestation will have an immediate and lasting impact on CO_2 emissions and on atmospheric CO_2 concentrations, in addition to other associated environmental benefits.

Forest responses to changes in the global environment, including Earth's climate, may also contribute to both the problem and its solution.

Although terrestrial ecosystems appear to currently accommodate nearly 60% of the direct anthropogenic perturbation inputs of CO_2 to the atmosphere, the natural physiological mechanisms that are thought to be responsible for this increased uptake are not likely to function as effectively in the future. Thus, in the absence of purposeful mitigation, the land-based CO_2 sink will likely decrease and could even become a source over the coming century,[66] leading to even greater climate changes.

Sustainable development in forestry has an important role to play in reversing these trends. This role is not restricted to the maintenance or enhancement of carbon stocks in forest ecosystems, but also can help to alleviate the underlying causes of deforestation by providing economic benefits. Although there are many activities that can be undertaken in the management of forest ecosystems to this end, their specific costs will vary. As climate change proceeds, more expensive activities will become necessary for additional mitigation.

The sustainable use of forest products, including the production of energy supplies that displace the use of fossil fuels, may make a significant contribution to mitigate climate change in the longer term because such use avoids the entry of new carbon into the active part of that cycle, while supplying essential goods and services to society. These avoided emissions accrue both from the use of forest biomass to supply energy (either directly or as a last stage in the life cycle of forest products) and from the use of forest-derived products as substitutes for materials that require large expenditure of energy (typically from the combustion of fossil fuels).

Management activities that enhance or protect carbon stocks in forest ecosystems include reducing the regeneration delay through seeding and planting, enhancing forest productivity, changing the harvest rotation length, the judicious use of forest products, and forest protection through control and suppression of disturbance by fire, pests, and disease. At the same time, the flow of material goods and services from a thriving forest products sector not only reduces the dependence on more energy-intensive products, such as cement, but also provides economic benefits that can help pay for such forest-enhancing activities. The sustainable use of forests thereby offers a potential win–win situation: maintenance of carbon stocks in healthy forest ecosystems, the cost of which can be offset by the continuous stream of forest products, which themselves help avoid the direct input of new carbon into the atmosphere. Good forestry can be part of the solution.

Protection of forest carbon stocks from intensifying and recurring disturbance events solely as a mitigation strategy is likely neither efficient nor effective as a long-term measure. This is especially true of wildfires where increasingly large financial expenditures will be needed to protect vulnerable forests. The situation is analogous to paying high rent: C pools are only retained as long as the protection continues, and are lost when the next fire comes along. On the other hand, if protection is coupled with sustainable forest utilization, transferring C to long-lived forest products, then potentially significant gains in terrestrial C retention can be realized, both on- and offsite.

Although it is straightforward to quantify the direct anthropogenic inputs of CO_2 to the atmosphere, a quantitative understanding of the rates of atmospheric increase remains a challenge, precisely because of the strong feedbacks exerted by terrestrial and ocean ecosystems to the changes. Understanding the biospheric feedback — the response of the world's biota to the perturbations — is needed:

- To gauge the magnitude of future impacts
- To identify realistic mitigation opportunities that can help reduce or avoid further adverse perturbation of the C cycle–climate system
- To design, implement, and monitor appropriate mitigation activities
- To design and implement adaptation strategies that can help society to cope with those changes that are unavoidable

The quantification of C-related costs and benefits from sustainable forest management, and of the impacts of climate change on forest C sinks and sources also remains a challenge, requiring:

- Improved methods and data for assessing vulnerable C pools and the processes affecting them
- New and improved models for predicting the fate of these pools in a changing environment
- New tools and data to monitor and verify the predictions over large scales

Over the past decade, there has been a dramatic improvement in the science of interactions between climate and forests. Particular advances have been made in understanding landscape-level carbon dynamics through the implementation of large-scale manipulative experiments and advanced monitoring programs, and in the development of practical forest-oriented remote-sensing technologies. Further advances are critically important to understand the dynamics and impacts of human activities on changing carbon uptake by terrestrial ecosystems. Part of this challenge is the identification of mitigation opportunities that can help reduce or avoid further adverse perturbation of the carbon cycle–climate system. A significant component of this is the development of predictive tools that incorporate human decision making and social behavior as an integral part of the analytical process. This task has recently been initiated by the Global Carbon Project of the Earth Systems Partnership.[67]

ACKNOWLEDGMENTS

This chapter is based in part on an earlier presentation by M.J.A. at the XII World Forestry Congress. The authors thankfully acknowledge the thoughtful comments by two anonymous reviewers.

REFERENCES

1. Zwiers, F.W. 2002. The 20-year forecast. *Nature* 416:690–691.
2. IPCC. 1990. *Climate Change 1990: IPCC Scientific Assessment.* Cambridge University Press, New York.
3. IPCC. 2001. *Climate Change 2001: The Scientific Basis. Contribution of Working Group I to the Third Assessment Report of the Intergovernmental Panel on Climate Change.* Cambridge University Press, Cambridge.
4. Reilly, J., P.H. Stone, C.E. Forest, M.D. Webster, G.C. Jacoby, and R.G. Prinn. 2001. Uncertainty and climate change assessments. *Science* 293:430–433.
5. WMO. 2002. WMO Statement on the status of the global climate in 2001. WMO-No. 940, World Meteorological Organisation, Geneva.
6. Apps, M.J. and W.A. Kurz. 1991. The carbon budget of Canadian forests in a changing climate: can forestry be part of the solution? In *Proceedings 3rd International Symposium on Cold Region Development,* "ISCORD '91," Edmonton, June 16–20, 1991. Alberta Research Council, 48 pp.
7. Falkowski, P., R.J. Scholes, E. Boyle, J. Canadell, D. Canfield, J. Elser, N. Gruber, K. Hibbard, P. Hogbeg, S. Linder, A.F. MacKenzie, B. Moore III, T.F. Pedersen, Y. Rosenthal, S. Seitzinger, V. Smetacek, and W. Steffen. 2000. The global carbon cycle: a test of our knowledge of Earth as a system. *Science* 290:291–296.
8. Marland, G., R. Pielke, Sr., M. Apps, R. Avissar, R.A. Betts, K.J. Davis, P.C. Frumhoff, S.T. Jackson, L.A. Joyce, P. Kauppi, J. Katzenberger, K.G. MacDicken, R.P. Neilson, J.O. Niles, D.S. Niyogi, R.J. Norby, N. Pena, N. Sampson, and Y. Xue. 2003. The climatic impacts of land surface change and carbon management, and the implications for climate-change mitigation policy. *Climate Policy* 3:149–157.
9. Melillo, J.R., C.B. Field, and B. Moldan. 2003. Element interactions and the cycles of life: An overview. In *Interactions of the Major Biogeochemical Cycles: Global Change and Human Impacts,* J.M. Metillo, C.B. Field, and B. Moldan, Eds. Island Press, Washington, D.C., 1–14.
10. Houghton, R.A. 2003. Revised estimates of the annual net flux of carbon to the atmosphere from changes in land use and land management 1850–2000. *Tellus* 55B(2):378–390.
11. Schulze, E., D.C. Wirth, and M. Heimann. 2000. Managing forests after Kyoto. *Science* 289:2058–2059.
12. Barford, C.C., S.C. Wofsy, et al. 2001. Factors controlling long- and short-term sequestration of atmospheric CO_2 in a mid-latitude forest. *Science* 294:1688–1691.
13. Gholz, H.L. and R.F. Fisher. 1982. Organic matter production and distribution in slash pine (*Pinus elliottii*) plantations. *Ecology* 63:1827–1839.
14. Gower, S.T., O.N. Krankina, R.J. Olson, M.J. Apps, S. Linder, and C. Wang. 2000. Net primary production and carbon allocation patterns of boreal forest ecosystems. *Ecol. Appl.* 11:1395–1411.
15. Kurz, W.A. and M.J. Apps. 1999. A 70-year retrospective analysis of carbon fluxes in the Canadian forest sector. *Ecol. Appl.* 9:526–547.
16. MacLaren, J.P. 1996. Plantation forestry — its role as a carbon sink: conclusions from calculations based on New Zealand's planted forest estate. Pages 257–270 In *Forest Ecosystems, Forest Management and the Global Carbon Cycle,* M.J. Apps and D.T. Price, Eds. Springer-Verlag, Berlin, 257–270.
17. Van Wagner, C.E. 1978. Age-class distribution and the forest fire cycle. *Can. J. For. Res.* 8:220–227.

18. Harrington, J. 1982. A Statistical Study of Area Burned by Wildfire in Canada 1953–1980. Information Report PI-X-16, Petawawa National Forestry Institute, Canadian Forestry Service, Department of Environment.

19. Canadell, J.G., H.A. Mooney, D.D. Baldocchi, J.A. Berry, J.R. Ehleringer, C.B. Field, S.T. Gower, D.Y. Hollinger, J.E. Hunt, R.B. Jackson, S.W. Running, G.R. Shaver, W. Steffen, S.E. Trumbore, R. Valentini, and B.Y. Bond. 2000. Carbon metabolism of the terrestrial biosphere: a multi-technique approach for improved understanding. *Ecosystems* 3:115–130.

20. Steffen, W. and P. Tyson. 2001. Global Change and the Earth System: A Planet under Pressure. The Global Environmental Programmes. IGBP Science No. 4, IGBP (International Geosphere-Biosphere Programme), Stockholm.

21. Pacala, S.W., G.C. Hurtt, D. Baker, P. Peylin, R.A. Houghton, R.A. Birdsey, L. Heath, E.T. Sundquist, R.F. Stallard, P. Ciais, P. Moorcroft, J.P. Caspersen, E. Shevliakova, B. Moore, G. Kohlmaier, E. Holland, M. Gloor, M.E. Harmon, S.-M. Fan, J.L. Sarmiento, C.L. Goodale, D. Schimel, and C.B. Field. 2001. Consistent land- and atmosphere-based U.S. carbon sink estimates. *Science* 292:2316–2320.

22. Caspersen, J.P., S.W. Pacala, J.C. Jenkins, G.C. Hurtt, P.C. Moorcroft, and R.A. Birdsey. 2000. Contributions of land-use history to carbon accumulation in U.S. forests. *Science* 290:1148–1151.

23. Keeling, C.D., J.F.S. Chin, and T.P. Whorf. Increased activity on northern vegetation inferred from atmospheric CO_2 measurements. *Nature* 382:146–149.

24. Lavigne, M.B., R. Boutin, R.J. Foster, G. Goodine, P.Y. Bernier, and G. Robitaille. 2003. Soil respiration responses to temperature are controlled more by roots than by decomposition in balsam fir ecosystems. *Can. J. For. Res.* 33:1744–1753.

25. Lavoie, C. and S. Payette. 1996. The long-term stability of the boreal forest limit in subarctic Quebec. *Ecology* 77:1226–1233.

26. Zoltai, S.C., T. Singh, and M.J. Apps. 1991. Aspen in a changing climate. In *Aspen Management for the 21st Century,* S. Navratil and P.B. Chapman, Eds. Forestry Canada, Northwest Region, Northern Forestry Centre and Poplar Council of Canada, Edmonton, Alberta, 143–152.

27. Hogg, E.H. and P.A Hurdle. 1995. The aspen parkland in western Canada: A dry-climate analogue for the future boreal forest? *Water Air Soil Pollut.* 82:391–400.

28. Hogg, E.H. 2001. Modelling aspen responses to climatic warming and insect defoliation in western Canada. In *Sustaining Aspen in Western Landscapes: Symposium Proceedings,* Shepperd, W.D., Binkley, D., Bartos, D.L., Stohlgren, T.J., and Eskew, L.G., Comps. 13–15 June 2000, Grand Junction, Colorado. U.S. Department of Agriculture, Forest Service, Rocky Mountain Research Station, Fort Collins, Colorado, Proceedings RMRS-P-18, 325–338.

29. Kurz, W.A., M.J. Apps, B.J. Stocks, and W.J.A. Volney. 1995. Global climatic change: disturbance regimes and biospheric feedbacks of temperate and boreal forests. In *Biotic Feedbacks in the Global Climatic System: Will the Warming Feed the Warming?* G.M. Woodwell and F.T. Mackenzie, Eds. Oxford University Press, 119–133.

30. Holling, C. S. 2001. Understanding the complexity of economic, ecological, and social systems. *Ecosystems* 4:390–405.

31. Luo, Y., L.W. White, J.G. Canadell, E.H. DeLucia, D.S. Ellsworth, A. Finzi, J. Lichter, and W.E. Schlesinger. 2003. Sustainability of terrestrial carbon sequestration: a case study in Duke Forest with inversion approach. *Global Biogeochem. Cycles* 17:1021–1034.

32. DeLucia, E.H., K. George, and J.G. Hamilton. 2002. Radiation-use efficiency of a forest exposed to elevated concentrations of atmospheric carbon dioxide. *Tree Physiol.* 22:1003–1010.

33. Oren, R., D.S. Ellsworth, K.H. Johnsen, N. Phillips, B.E. Ewers, C. Maier, K.V.R. Schafer, H. McCarthy, G. Hendrey, S.G. McNulty, and G.G. Katul. 2001. Soil fertility limits carbon sequestration by forest ecosystem in a CO_2-enriched atmosphere. *Nature* 411:469–472.

34. Schindler, D.W. 1998. A dim future for boreal waters and landscapes. *BioScience* 48:157–164.

35. Isebrands, J.G., Dickson, R.E., Rebbeck, J., and Karnosky, D.F. 2000. Interacting effects of multiple stresses on growth and physiological processes in northern forests. In *Responses of Northern U.S. Forests to Environmental Change*, R.E. Mickler, R.A. Birdsey, and J. Hom, Eds. Ecological Studies 139, Springer-Verlag, Berlin, 149–180.

36. Flannigan, M.D., I. Campbell, M. Wotton, C. Carcaillet, P. Richard, and Y. Bergeron. 2001. Future fire in Canada's boreal forest: paleoecology results and general circulation model — regional climate model simulations. *Can. J. For. Res.* 31:854–864.

37. Fleming, R.A. and W.J.A. Volney. 1995. Effects of climate change on insect defoliator population processes in Canada's boreal forest. *Water Air Soil Pollut.* 82:445–454.

38. Dale, V.H., L.A. Joyce, S. McNulty, R.P. Neilson, M. Ayres, M.D. Flannigan, P.J. Hanson, L. Irland, A.E. Lugo, C. Peterson, D. Simberloff, F. Swanson, B.J. Stocks, and B.M. Wotton. 2001. Climate Change and forest disturbances. *BioScience* 51:723–734.

39. McNulty, S. 2002. Hurricane impacts on US forest carbon sequestration. *Environ. Pollut.* 116:S17–S24.

40. Percy, K.E., C.S. Awmack, R.L. Lindroth, M.E. Kubiske, B.J. Kopper, J.G. Isebrands, K.S. Pregitzer, G.R. Hendry, R.E. Dickson, D.R. Zak, E. Okansen, J. Sober, R. Harrington, and D.F. Karnosky. 2002. Altered performance of forest pests under atmospheres enriched by CO_2 and O_3. *Nature* 420:403–407.

41. UNFCCC. 1997. Kyoto Protocol to the United Nations Framework Convention on Climate Change. In FCCC/CP/1997/7/Add.1.http://WWW.UNFCCC.DE.

42. Watson, R., I. Noble, B. Bolin, M.J. Apps et al. 2000. Summary for Policy Makers, *IPCC Special Report on Land Use, Land-Use Change and Forestry.* Cambridge University Press, New York.

43. Kauppi, P.E., R.A. Sedjo, M.J. Apps, C.C. Cerri, T. Fujimori, H. Janzen, O.N. Krankina, W. Makundi, G. Marland, O. Masera, G.J. Nabuurs, W. Razali, and N.H. Ravindranath. 2001. Technical and economic potential of options to enhance, maintain and manage biological carbon reservoirs and geo-engineering. In *Climate Change 2001: Mitigation. Contribution of Working Group III to the Third Assessment Report of the Intergovernmental Panel on Climate Change,* B. Metz, O. Davidson, R. Swart, and J. Pan, Eds. Cambridge University Press, Cambridge, 310–343.

44. Penman, J., M. Gytarsky, T. Hiraishi, T. Krug, D. Kruger, R. Pipatti, L. Buendia, K. Miwa, T. Ngara, K. Tanabe, and F. Wagner, Eds. 2003. *Good Practice Guidance for Land Use, Land-Use Change and Forestry.* IPCC National Greenhouse Gas Inventories Program, Institute for Global Environmental Strategies, Kanagawa, Japan.

45. Binkley, C.S., M.J. Apps, R.K. Dixon, P. Kauppi, and L.-O. Nilsson. 1998. Sequestering carbon in natural forests. *Crit. Rev. Environ. Sci. Technol.* 27:S23–45.

46. Kurz, W.A., S.J. Beukema, and M.J. Apps. 1998. Carbon budget implications of the transition from natural to managed disturbance regimes in forest landscapes. *Mitigation Adaptation Strat. Global Change* 2:405–421.

47. Nohrstedt, H.O. 2001. Response of coniferous forest ecosystems on mineral soils to nutrient additions: a review of Swedish experiences. *Scand. J. For. Res.* 16:555–573.
48. Saarsalmi, A. and E. Malkonen. E.2001. Forest fertilization research in Finland: a literature review. *Scand. J. For. Res.* 16:514–535.
49. Schroeder, W. and J. Kort. 2001. Temperate agroforestry: adaptive and mitigative roles in a changing physical and socio-economic climate. In *Proceedings of the Seventh Biennial Conference on Agroforestry in North America*, Saskatchewan, Canada, August 12–15, 2001, 350 pp.
50. Fan, S., M. Gloor, J. Mahlman, S. Pacala, J.L. Sarmiento, T. Takahashi, and P. Tans. 1999. The North American Sink. *Science* 483:1815a.
51. Bousquet, P., P. Peylin, P. Ciais, C.L. Querre, P. Friedlingstein, and P. Tans. 2000. Regional changes in carbon dioxide fluxes of land and oceans since 1980. *Science* 290:1342–1346.
52. Karjalainen, T., S. Kellomaki, and A. Pussinen. 1994. Role of wood-based products in absorbing atmospheric carbon. *Silva Fenn.* 28:67–80.
53. Harmon, M.E., J.M. Harmon, W.K. Ferrell, and D. Brooks. 1996. Modeling carbon stores in Oregon and Washington forest products: 1900–1992. *Climatic Change* 33:521–550.
54. Apps, M.J., W.A. Kurz, S.J. Beukema, and J.S. Bhatti. 1999. Carbon budget of the Canadian forest product sector. *Environ. Sci. Policy* 2:25–41.
55. Hendrickson, O.Q. 1990. How does forestry influence atmospheric carbon? *For. Chron.* 66:469–472.
56. Dewar, R.C. 1990. Analytical model of carbon storage in the trees, soils and wood products of managed forests. *Tree Physiology* 6:417–428.
57. Schlamadinger, B., M. Apps, F. Bohlin, L. Gustavsson, G. Jungmeier, G. Marland, K. Pingoud, and I. Savolainen. 1997. Towards a standard methodology for greenhouse gas balances of bioenergy systems in comparison with fossil energy systems. *Biomass Bioenerg.* 13:359–375.
58. Pingoud, K., A. Lehitila, and I. Savlolainen. 1999. Bioenergy and the forest industry in Finland after the adoption of the Kyoto Protocol. *Environ. Sci. Policy* 2:153–163.
59. Gustavsson, L., T. Karajalainen, G. Marland, I. Savolainen, B. Schlamadinger, and M.J. Apps. 2000. Project-based greenhouse gas accounting: Guiding principles with focus on baselines. *Energ. Policy* 28:935–946.
60. Richter, K. 1998. Life cycle assessment of wood products. In *Carbon Dioxide Mitigation in Forestry and Wood Industry*, G. Kohlmaier, K. Weber, and R. Houghton, Eds. Springer Verlag, Berlin, 219–248.
61. Schlamadinger, B. and G. Marland. 1996. Full fuel cycle carbon balances of bioenergy and forestry options. *Energ. Conserv. Manage.* 37(6–8):813–818.
62. Hall, D.O. and J.I. Scrase. 1998. "Will biomass be the environmentally friendly fuel of the future?" *Biomass Bioenerg.* 15(4/5):357–367.
63. Nakicenovic, N., J. Alcamo, G. Davis, B. de Vries, J. Fenhann, S. Gaffin, K. Gregory, A. Grübler, T.Y. Jung, T. Kram, E.L. La Rovere, L. Michaelis, S. Mori, T. Morita, W. Pepper, H. Pitcher, L. Price, K. Riahi, A. Roehrl, H.-H. Rogner, A. Sankovski, M. Schlesinger, P. Shukla, S. Smith, R. Swart, S. van Rooijen, N. Victor, and Z. Dadi. 2000. Special Report on Emissions Scenarios. Intergovernmental Panel on Climate Change, Cambridge.
64. Matthews, R. and K. Robertson. 2001. Answers to ten frequently asked questions about bioenergy, carbon sinks and their role in global climate change. Graz, Austria, IEA Bioenergy Task 38, 8 pp.

65. Sampson, R.N., R.J. Scholes, C.C. Cerri, L. Erda, D.O. Hall, M. Handa, P. Hill, M. Howden, H. Janzen, J.M. Kimble, R. Lal, G. Marland, K. Minami, K. Paustian, P. Read, P.A. Sanchez, C. Scoppa, B. Solberg, M.A. Trossero, S.E. Trumbore, O. van Cleemput, A.P. Whitmore, and D. Xu. 2000. Additional human-induced activities, Article 3.4. In *Special Report on Land Use, Land-Use Change and Forestry,* R.T. Watson, I.R. Noble, B. Bolin, N.H. Ravindranath, D.J. Verardo, and D.J. Dokken, Eds. Cambridge University Press, Cambridge, 183–281.

66. Cox, P.M., R.A. Betts, C.D. Jones, S.A. Spall, and I.J. Totterdell. 2000. Acceleration of global warming due to carbon-cycle feedbacks in a coupled climate model. *Nature* 409:184–187.

67. GCP Steering Committee, 2003. Global Carbon Project: The Science Framework and Implementation. Canberra, Earth Systems Partnership (IGBP, IHDP, WCRP, Diversitas): Report 1, Canberra, 69 pp.

10 Peatlands: Canada's Past and Future Carbon Legacy

D.H. Vitt

CONTENTS

10.1 INTRODUCTION

Peatland ecosystems are characterized by the accumulation of organic matter in soil and, if following Joosten and Clark's[1] definition of having at least 30 cm of peat with a minimum organic content of 30%, then peatlands cover over 4 million km^2 — about 3% of the Earth's land surface. Nearly 70% of this peatland area lies in the boreal regions of Canada and Russia. Canada alone contains just over 1,200,000 km^2 of peat. Peatlands are significant in that they provide a wide diversity of ecosystem services, not the least of which is the accumulation of large stores of carbon. Joosten and Clark[1] estimated that since 1800, 10 to 20% of the world's peatlands have been lost, but it has been the view of many that Canada's peatlands remain in pristine condition, undisturbed by human activities.[2] However, as we will see, this is certainly not the case.

Globally, wetlands (especially peatlands) represent a large carbon stock, with estimates varying from 200 to 860 Pg (= Petagrams) of carbon (see for example

Gorham,[2] Bohn,[3,4] Sjörs,[5] Post et al.,[6] Houghton et al.,[7] Armentano and Menges,[8] and Markov et al.[9]). Generally, carbon-rich peatland soils are thought to represent about one third of the world's soil carbon, yet cover only about 3% of the land surface. Release of this store of carbon into the atmosphere would increase atmospheric CO_2 concentrations by more than 50%. Canada's peat inventory has been estimated to contain up to 170 Pg of carbon[2] and is approximately 38% of the carbon stock in northern peatlands. The western boreal forest region of Alberta, Saskatchewan, and Manitoba contain 365,157 km^2 of peatlands and along with British Columbia these four provinces have about 40% of Canada's peatland area, while eastern Canada (Ontario eastward) contains about 37%, and northern Canada (the three territories) contains approximately 23%.[10] In terms of carbon, western and northern Canada store at least 83 Pg, whereas eastern Canada has a minimum of 52 Pg. In Alberta, peatlands may contain as much as 70% of the province's soil carbon (13.5 Pg C in peatlands; 2.3 Pg C in lakes, 2.7 Pg C in forests, and 0.8 Pg C in grasslands (data from J. Bhatti[11]). In general, the western Canadian peatlands have sequestered about 48 Pg of carbon during the past 10,000 years, with about half of this accumulated in the last 4000 years.[12]

Peat accumulates on the landscape when annual net primary production exceeds the sum of annual decomposition and the loss of carbon that is dissolved in the pore water and exported from peatlands. The initiation, development, and succession of ecosystems that sequester carbon, as well as the rate of peat accumulation in boreal peatlands, are dependent on regional allogenic factors such as climate, substrate chemistry, and landscape and hydrological position. These regional driving factors in turn determine a suite of local factors that influence the form and function of individual peatland sites (Figure 10.1). These local factors include water chemistry,

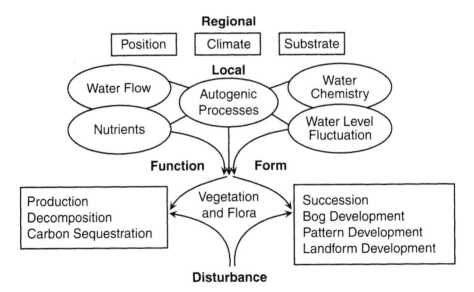

FIGURE 10.1 Diagrammatic representation of the interactions between regional, local, and ecological factors that control function and form of peatlands.

water level fluctuation, water flow rates, and nutrient inputs. Peatland form is determined by this interacting suite of local and regional factors through the development of ombrotrophy wherein the peatland receives all water and nutrients from the atmosphere, evolution of internal landforms and landscape pattern, and the direction of succession. Additionally, once established, peatlands have strong autogenic controls (acidification, eutrophication; Vitt[13]) that also help regulate form and function.[14]

The functioning of peatland ecosystems centers on the process of peat accumulation. Yu et al.[15] provided conceptual diagrams of carbon cycling in peatlands. Peat accumulation is dependent on the rate of input of organic matter into the anaerobic peat column (the catotelm) and on the rate of the slow decomposition of this material over time.[16] Climate is the most important regional factor, mainly through its regulation of local water regimes. Among climatic variables, Halsey et al.[17] demonstrated for wetlands of Manitoba that temperature and aridity are the most critical limiting factors at the landscape scale for peatland ecosystems.

Climate affects carbon sequestration by limiting photosynthesis and aerobic (acrotelm) decomposition rates, thus influencing the amount and quality of the organic material reaching the catotelm. Climate also affects carbon stocks within the catotelm by limiting anaerobic processes (methenogenesis, sulfate reduction, and N_2O production), as well as controlling the position of the acrotelm–catotelm boundary. Thus, climatic change can affect current peat accumulation as well as persistence of the peat column itself.

10.2 LIMITATIONS ON CARBON SEQUESTRATION IN BOREAL PEATLANDS

Four factors contribute to limiting carbon sequestration in pristine boreal peatlands: (1) The formation of permafrost in boreal peatlands reduces the input of carbon to the peatland. (2) Ground layer biomass contributes high-quality organic matter that is resistant to decay to the peat column and, along with vascular plant roots and litter from aboveground vascular plant biomass, compose the carbon inputs to peat-forming ecosystems. These inputs are limited by net annual primary production (NPP). (3) Rates of aerobic respiration (occurring in the acrotelm) limit peat accumulation. Rates of aerobic decomposition may be determined by substrate quality and by temperature. (4) The amount of time the decomposing plant material spends in aerobic conditions. In summary, cold, dry climatic conditions favor permafrost aggradation; warm, dry (arid) conditions limit ground-layer NPP and increase acrotelm depth, while warm, wet conditions increase microbial respiration. Peat accumulation decreases with aridity and increases under cool, moist conditions (Figure 10.2). Corollaries to these relationships provide us with four mechanistic statements:

1. Increases in precipitation increase the ground layer production, and these, coupled with a rise in water table, decrease residence time in the acrotelm lowering initial catotelmic bulk densities: Carbon sequestration increases.
2. Decreases in precipitation decrease ground layer production and are coupled to a lowering of the water table. These factors increase the residence

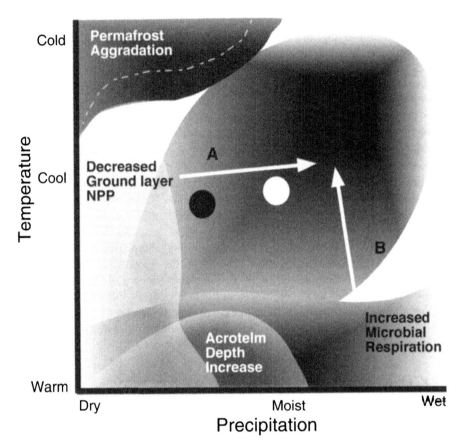

FIGURE 10.2 Factors limiting carbon sequestration in boreal peatlands plotted over climatic space represented by temperature (inverse on *y* axis) and precipitation (*x* axis). Shading of central ellipse indicates increased rates of carbon accumulation. A = Direction of increase in ground-layer NPP. B = Direction of bulk density decrease. White circle = Estimated peat accumulation during Holocene wet period.[43] Black circle = Estimated peat accumulation during Holocene dry period.[43] Details of long-term peatland dynamics are available in Yu et al.[15,50] Dotted line is degrading permafrost.

    time in the acrotelm, thus increasing the initial catotelmic bulk densities: Carbon sequestration decreases.

3. Increases in temperatures increase acrotelmic respiration, hence increasing initial catotelmic bulk densities: Carbon sequestration decreases.

4. Decreases in temperatures decrease acrotelmic respiration, hence decreasing initial catotelmic bulk densities: Carbon sequestration increases.

10.3 THE ECOLOGY OF BOREAL PEAT ACCUMULATION

Accumulation of peat in the boreal region appears to occur under three somewhat different ecological regimes.

10.3.1 Bog Accumulation

Bogs are ombrogenous, receiving their nutrients and water supply solely from the atmosphere as precipitation. Bogs that occur in the boreal region are generally treed and possess a continuous ground layer of *Sphagnum* (peat mosses). These peat mosses develop an extensive, undulating microrelief of hummocks and hollows. Hummocks attain heights of nearly 1 m above the water surface. Thus the aerobic zone of decomposing peat (the acrotelm) is well developed and organic matter spends a relatively large amount of time in this zone; however, rates of decomposition are reduced here largely due to factors inherent in the *Sphagnum* species themselves.[18–20] Furthermore, catotelmic bulk densities are relatively low due to the fibrous nature of the hummock-occurring *Sphagnum* species and low number of graminoid roots.

10.3.2 Poor Fen Accumulation

Fens are geogenous, receiving waters and nutrients that have been in contact with the surrounding uplands as well as from precipitation. Poor fens have ground layers dominated by species of *Sphagnum* and are acidic ecosystems. The *Sphagnum* species of poor fens occur in carpets and lawns forming extensive flat areas relatively close to the water's surface. Thus, the acrotelm is poorly developed and the residence time of organic material in the aerobic zone is low, with the organic matter reaching the catotelm rather quickly. Catotelmic bulk densities are higher than in bogs, but due to the fibrous nature of *Sphagnum* and the low root biomass, are less than those of the rich fens.

10.3.3 Rich Fen Accumulation

Geogenous rich fens have ground layers dominated by true mosses (generally referred to as "brown mosses"). These plants, like the sphagna of poor fens, form lawns and carpets, water tables are high, and acrotelms are poorly developed. Ground-layer canopies of rich fens differ from those of poor fens and bogs by the difference between true moss and peat moss plant architectures. The rich fen acrotelm has denser canopies because it is dominated by true mosses. As a result of the high water table in rich fens, this relatively dense (carbon-rich) ground layer spends little time in the acrotelm, and reaches the catotelm as high-quality peat with high bulk densities. Additionally, rich fens have higher abundances of graminoids, and sedge roots also contribute to the high bulk densities.

All three peatland types effectively sequester carbon, each in a somewhat different manner. Fundamental differences in vegetation, hydrology, and chemistry between these three peatland types[13,21] lead to three generalizations about how climate change can affect the ecology of these peatland systems.

1. Bogs require a local positive climatic water balance. The large *Sphagnum* hummocks and well-developed acrotelms must be maintained through precipitation input. Nutrient supplies for these ombrogenous peatlands are dependent on atmospheric influxes.

2. Fens require a constant groundwater source, and the acrotelm–catotelm boundary (so critical for peat accumulation in fens) must be maintained at a relatively stable elevation. Lowering of water tables or changes in annual water table fluctuations alter the boundary conditions.
3. Changes in upland and surrounding nutrient fluxes strongly affect fens, whereas changes in atmospheric nutrient inputs strongly affect bogs.

In conclusion, peat accumulation and the sequestration of carbon from the atmosphere (as CO_2) to solid organic matter (as CHO) is determined and controlled by a series of interacting processes. I argue that four of these processes are of most importance (Figure 10.2) and that all of these are climatically controlled. Two are more affected by temperature, while the other two are more affected by precipitation. How these four factors interrelate and how they are individually affected by climate change is still poorly understood and needs to be a priority research goal.

10.4 NORTHERN PEATLANDS: SINKS OR SOURCES OF CARBON?

Although I believe that it is generally acknowledged that northern peatlands are a present-day sink for atmospheric CO_2,[2] several complicating factors exist that may severely limit their role in maintaining this large carbon sink.

Local temporal and spatial variation in carbon sequestration is high. Annual changes from net carbon sinks to net carbon sources have been demonstrated for an oligotrophic pine fen in Finland by Alm et al.,[22] a Minnesota peatland by Shurpali et al.,[23] a Manitoba rich fen by Suyker et al.[24] and Lafleur et al.,[25] and a temperate poor fen by Carroll and Crill.[26] Likewise, spatially local microhabitats may be either net sinks or sources.[27]

The concept of the Canadian peatlands and the boreal forest being of a pristine nature is doubtful. Long-term carbon accumulation rates have been estimated at between 19.4 g m^2 yr^{-1} for western Canada[12] and 28.1 g m^2 yr^{-1} for northern peatlands in general.[2] These rates, however, are based on apparent long-term accumulation in pristine peatlands (and as well may not be representative of current net rates), and do not take into consideration peat lost from the direct and indirect effects of fire and other natural disturbances. If peat losses due to the effects of the historical fire regime are added back into the 19.4 g m^2 yr^{-1} estimates of peat accumulation the actual rate of peat accumulation is 24.5.[28] In the only cumulative effects study of which I am aware, Turetsky et al.[28] estimated that 13% of western Canada's peatlands are disturbed. She estimated that of the 8940 Gg C yr^{-1} of carbon that should be sequestered annually under a no-disturbance regime, 48 Gg C yr^{-1} are lost to oil sands mining, 80 Gg C yr^{-1} to flooding from hydro-electric projects, 135 Gg C yr^{-1} to peat extraction activities, 4704 Gg C yr^{-1} from the direct effects of fire (carbon released from the fire itself), and 1578 Gg C yr^{-1} are lost to the indirect effects of fire (due to decreased sequestration of carbon from vegetation recovery, plus decomposition during recovery). On the positive side, melting of boreal permafrost yields a return of +100 Gg C yr^{-1} (see Turetsky et al.[29] for explanation) and undisturbed peatlands sequester 7781 Gg yr^{-1} of carbon. Overall, disturbance and development across western Canada has reduced the annual carbon sequestration to +1319 Gg C yr^{-1} — only 14% of the long-term carbon sink rate. Increases in any of the anthropogenic disturbances or in the future fire regime, or a decrease of only 17% in the carbon sequestered in undisturbed peatlands because of drought and or temperature increases, will move western Canadian peatlands from a sink to a source of CO_2.

Additionally, peatland types differ in the forms of gaseous carbon release and have different global warming potentials. Anaerobic respiration releases include methane (among other gases). Methane is a greenhouse gas with different absorptive properties and different atmospheric lifetimes from those of CO_2. Boreal wetlands release an estimated 34 Tg of CH_4 annually.[30,31] Joosten and Clark[1] provided global warming potentials (GWP) for northern pristine bogs and fens that were calculated for different time horizons into the future. Their data indicate that currently pristine fens remove 250 kg C ha^{-1} yr^{-1} (as CO_2) and release 297 kg C ha^{-1} yr^{-1} (as CH_4), while bogs currently remove 310 kg C ha^{-1} yr^{-1} and release 53 kg C ha^{-1} yr^{-1}. So, even though a carbon sink is indicated by 560 kg C ha^{-1} yr^{-1} being sequestered and only 350 kg C ha^{-1} yr^{-1} released, differences in atmospheric properties of CO_2 and CH_4 produce a positive GWP when calculated per hectare for bogs and fens over the next 20- and 100-year intervals, but a negative GWP at 500 years due to different atmospheric residence times of the gases involved.

10.5 POTENTIAL CLIMATIC EFFECTS ON PEATLAND FORM AND VEGETATION

Gignac et al.[32] established response surfaces for a number of the indicator and keystone species of western Canadian peatlands for climate (using an aridity index), pH, and height above the water surface table. Of the 31 indicator species that were examined, all but 8 are climatically limited in western Canada. Using these response surfaces, Gignac and Vitt[33] developed peatland indicator bryophyte communities and constructed two peatland communities for contemporary climate; one at Athabasca, Alberta and one at Wandering River, Alberta. These communities encompassed a range of peatland types from bogs to rich fens. Then, by using the Canadian CCC General Circulation model 2 × CO_2 scenario that predicted an increase of 4°C for these southern boreal sites, an increase in the growing season of 19 days, and no increase in precipitation, two future climate peatland bryophyte communities were constructed. The resulting indicator communities for these two locations show the complete absence of all peatland species at Athabasca and a reduction of cover at Wandering River from 14 species with 77% cover to 5 species with less than 1% cover. Essentially, peatland communities would cease to exist at both of these southern boreal locations.

Nicholson and Gignac[34] and Nicholson et al.[35,36] examined the current and future occurrences of fens and bogs in the Mackenzie River Basin. They constructed three-dimensional response surfaces for 21 indicator species spanning the rich fen, poor

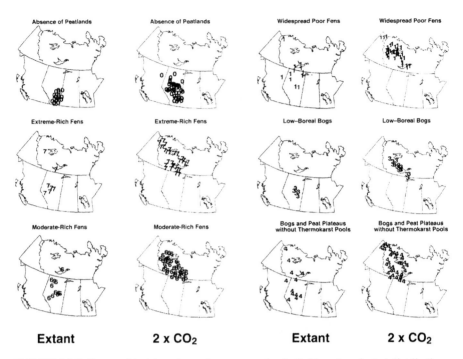

FIGURE 10.3 Geographical locations of extant peatlands (left) and projected distributions (right) by peatland type of sites in the Mackenzie River Basin as a result of global warming. Climatic data that were used by the model to generate the projected distribution of peatlands were obtained from the Geophysics Fluid Dynamics Laboratory (GFDL) Model for $2 \times$ present CO_2 concentrations. (From Nicholson, B.J. et al., in Mackenzie Basin Impact Study (MBIS), Final Report, Cohen, S.J., Ed., Environment Canada, Downsview, Ontario, 1997, 295. With permission.)

fen, boreal bog, and peat plateau (ombrotrophic sites with extensive permafrost) gradient. Under $2 \times CO_2$ climatic scenarios (using both the GFDL and CCC General Circulation Models), peatland ecosystems of all types were displaced northward approximately 780 km (Figure 10.3). The southern limit of peat-forming ecosystems was predicted to be at about 60° N latitude. Bryophyte species are especially sensitive indicators of water level changes, and Nicholson et al.[36] utilized these sensitivities to predict projected changes in depth to the water table relative to the peat surface. Predictions of changes ranged from –7 dm in northeastern Alberta, to –5 dm in north central Alberta, decreasing to a –3 to –1 dm change north of 60° N latitude (Figure 10.4). The use of plant indicators to predict water table position is a site-specific modeled response that has much more ecosystem relevance than predictions made from landscape-scale hydrology. Present-day vegetation response for drawdown is clearly evident in fens of the Athabasca area, which is situated north of central Alberta. Furthermore, the latitudinal position of the parkland–boreal forest boundary may react to increasing temperature through a parallel northward migration.

FIGURE 10.4 Projected minimum mean changes in depth of water table relative to peat surface (dm) for peatlands in the Mackenzie Basin based on climatic data obtained from the Canadian Climate Centre (CCC) Model for $1 \times$ and $2 \times$ present CO_2 concentrations. (From Nicholson, B.J. et al., in Mackenzie Basin Impact Study (MBIS), Final Report, Cohen, S.J., Ed., Environment Canada, Downsview, Ontario, 1997, 295. With permission.)

10.6 PERMAFROST MELTING IN THE BOREAL FOREST

In 1994, Vitt et al.[37] described a series of landforms associated with permafrost features (frost mounds) found in boreal peatlands. When these frost mounds melt, they collapse and form melt features termed *internal lawns*. Boreal permafrost melt is in disequilibrium with present-day climate,[38] owing to the insulative features of peat and of living *Sphagnum*, as well as the local microclimatic variation due to tree and shrub cover. Over the last millennium, permafrost distribution in the boreal forest has fluctuated in a sensitive zone that is 672,000 km^2 in extent across western continental Canada. During the Little Ice Age, about 28,800 km^2 of permafrost were present.[39] With the climate warming over the past 100 to 150 years, 9% (or 2627 km^2) of this permafrost has degraded. Additionally, 22% (5813 km^2) is currently in disequilibrium and actively degrading. Only 69% of boreal permafrost exists today in an equilibrium undegraded state.

Collapse of a frost mound is followed by extremely rapid recolonization of the resulting internal lawn by sedges and species of *Sphagnum* that form wet carpets and lawns.[37,40,41] Over the subsequent 100 to 200 years, vegetation of the internal lawn gradually increases in height and the system regenerates to the hummocky microrelief of a continental bog. For at least the first 100 years, carbon sequestration in internal lawns is greater than that of both nonpermafrost boreal bogs and permafrost mounds,[29] and the melting of permafrost results in an increase in the storage of organic matter. Turetsky et al.[29] reported organic matter accumulation in internal lawns (formed when permafrost melts) are 1.6 times higher than the close-by frost mounds and boreal bogs. Organic matter accumulation in boreal western Canada (where at least 90% of the permafrost has melted) has increased by 5% (or 2×10^{-11} g yr^{-1}) when compared to Little Ice Age amounts.[12]

10.7 GLOBAL CLIMATE CHANGE VS. CUMULATIVE DISTURBANCE

Across the boreal and subarctic regions of the world large amounts of carbon are sequestered in different places in our natural ecosystems. Carbon can be sequestered in lakes and buried in lake sediments where it is effectively permanently removed from the global carbon cycle. In Alberta, Campbell et al.[42] estimated that about 2.3 Pg of carbon are stored in this long-term sink.

Forests and croplands, on the other hand, sequester new carbon that is released and recirculated to the atmosphere within a short-term time range of decades to a few hundred years and have current soil carbon stocks estimated at 3.5 Pg C.[11] These two ecosystem types have carbon stored largely in living biomass and in the uppermost relatively shallow soil profile. These systems are generally intensively managed with harvest and planting cycles planned and implemented following tight management schedules.

Northern peatlands contain one third of the world's soil carbon. In Alberta, they contain about 13.5 Pg of carbon, while in continental western Canada they store 48 Pg of carbon of which only 0.1 Pg is found in living vascular plant aboveground biomass.[12] Of this large carbon stock, 50% was developed in the last 4000 years. Vitt et al.[12] estimated that in the last 1000 years, the western Canadian carbon store increased by 7.1 Pg or 14.8%. Both rates of peat accumulation and peatland initiation apparently are highly sensitive to natural Holocene climatic changes. In western Canada, carbon sequestration has been highly sensitive to millennial wet climate cycles.[43,44] These periods of increased moisture, rapid organic matter accumulation, and increased peatland initiation in western Canada appear to be related to warm periods in the North Atlantic,[45] as well as to global atmospheric CO_2 concentrations in the past.[43] Peat accumulation rates at one rich fen in western Alberta varied from means of about 183 g m^2 yr^{-1} during wet periods to a low of 7 g m^2 yr^{-1} during dry periods (with the long-term time-weighted mean of 31.3).[43] These data suggest that even minor climatic fluctuations in the past have had strong affects on peatland function and they may alter the rates of peat accumulation considerably. Furthermore, northern peatlands appear to be strongly coupled to natural global climatic changes.

From these data it is apparent that boreal peatlands have been strongly affected by climate change in the past; however, it is also important to realize that the cumulative effects of disturbance may actually have more of an impact on these carbon-rich ecosystems. Given current disturbance in the western boreal forest,[28] reduction in carbon sequestration rates by only 17% will convert these northern peatlands to a net CO_2 source to the atmosphere. This reduction in carbon sequestration is closer to reality then one might expect. Currently, it appears that much of the area in boreal western Canada is too dry for new peatland initiation, and current peatlands are largely relicts of a once wetter (and perhaps cooler) climatic regime.[46,47] Increased atmospheric CO_2 concentrations are predicted to increase temperatures and perhaps decrease precipitation, or at least increase drought across the boreal zone, where peatlands are abundant. This global climate change will potentially reduce carbon sequestration through a series of progressively more severe mechanisms.

- Net primary production of the ground layer will decrease, thus new high quality (= highly recalcitrant) carbon input to the ecosystem will be reduced.
- Belowground net primary production (vascular plant roots) will increase in some peatland types, thus new low-quality (= easily decomposed) carbon input to the ecosystem will be increased and may increase methane production.
- Rates of microbial respiration will be increased; thus existing carbon will be released to the atmosphere at an increased rate. This assumes that decreased moisture availability in the acrotelm will not be so severe as to decrease microbial respiration.
- Initially, acrotelms will increase in depth; thus the residence time of young peat in an aerobic atmosphere will increase the amount of CO_2 released to the atmosphere, but will also serve to oxidize the methane produced in the catotelm.
- With lowered water tables, hummock-growing plant species will die in place and recolonize the previous hollow/lawn — the result will be enhanced decomposition (and loss of) the uppermost peat column until equilibrium with the new water table is established.
- Continued aridity will limit keystone bog species. Peatlands will only continue to serve as a carbon sink in discharge sites served with stable groundwater flows; however, these peatlands will be at risk or reduced due to increased decomposition.
- It remains unclear whether new peatland initiation and development north of 60° N latitude will be as extensive as the loss of peatlands to the south and whether the peatland carbon stock will remain intact.
- Altered disturbance regimes, especially increases in fire frequency, may lead to catastrophic carbon losses from peatlands, especially bogs.

10.8 MITIGATION

It is my opinion that even under current climate conditions of the western Canadian boreal region, most boreal Canadian peatlands may not be able to continue to sequester carbon. What their ultimate fate will be is currently unknown and should be a high research priority. The worst-case scenario is clearly shown in predictions by Nicholson et al.[35,36] The best-case scenario may be a "resetting" of the peat surface at some distance below the present surface, with recolonization and continued carbon sequestration. What is not known is how much, and in what gaseous form, the uppermost peat column will be lost.

Clearly, mitigation for these loosely managed boreal peatlands is difficult. However, several priorities are suggested here:

- Develop a long-range plan of corridors and reserves that includes predicted future occurrences of peatlands. Since our future peatlands may only exist in a fully functional condition north of 60° N latitude, we should begin now to incorporate a reserve system that examines these northern sites.
- Restoration of wetlands after oil sands extraction may only be possible by examining wetlands that currently exist under our future predicted climatic regime. Examination of how these wetlands have initiated and continue to exist may provide valuable insights into our wetland environments. For example, a key indicator species of rich fens is *Meesia triquetra*. Examination of herbarium specimens and distribution maps[48] of the occurrence of *M. triquetra* in southern Saskatchewan and the midwestern states may be useful in developing landscapes for rich fen development under future climatic regimes.
- Maintain our peatlands in as pristine condition as possible. Use of peatlands for agriculture increases GWP (global warming potential) of fens and bogs substantially. Whereas the GWP of pristine bogs is negative and of fens is only slightly positive (less than 100 kg CO_2-C equiv ha^{-1} yr^{-1}), when peatlands are drained for pasture or tilled the GWP increases to 4000 to 5000 kg CO_2-C equiv. ha^{-1} yr^{-1} for the former and more than 10,000 for the latter for fens.[49]
- Disturbance in peatlands has two effects: direct effects of the disturbance itself (peat removal by the peat harvesting industry, clearing areas for oil exploration vehicles) and indirect effects (the effect of returning to the pre-disturbance condition). Mitigation for indirect effects can be as follows:
  Do not remove the actively growing top few centimeters of the ground layer when grading access lines.
  Keep the time from the end of peat harvesting activity to revegetation as short as possible. In western Canada, develop a clear management plan for restoration of cut over bogs back to fens.
  Avoid nutrient inputs to peatlands during construction activities; these include keeping to a minimum the introduction of mineral soil to peatland areas.
  Adequate buffer zones should be maintained around peatland complexes. Higher water tables from increased upland runoff after forest harvest or wildfire increase nutrients and decrease acrotelms resulting in complete

successional turnover of keystone species and this may be as devastating for peatlands as lowered water tables due to climate change. Buffer zones should be designed relative to peatland size, runoff amount, and watershed extent in order to protect small, sensitive peatlands as well as larger, less sensitive peatland complexes.

Road construction engineering should endeavor to understand peatland hydrology in order to avoid changes in water levels.

ACKNOWLEDGMENTS

Many of the ideas and data presented here were developed and collected during periods of funding from The Natural Sciences and Engineering Research Council of Canada and from The National Science Foundation (U.S.), for which I am grateful. In particular, I thank Ilka Bauer, Jagtar Bhatti, Suzanne Bayley, Kevin Devito, Dennis Gignac, Linda Halsey, Barbara Nicholson, Merritt Turetsky, R. Kelman Wieder, and Zicheng Yu for providing data and stimulating discussions, for offering many ideas that I have liberally used, and for friendship over the years. Portions of the text were extracted from joint manuscripts of R. Kelman Wieder and myself. Sandi Vitt prepared the graphics, for which I am grateful.

REFERENCES

1. Joosten, H. and Clark, D., *Wise Use of Mires and Peatlands — Background and Principles Including a Framework for Decision-Making*, International Mire Conservation Group and International Peat Society, Saarijärven Offset Oy, Saarijärvi, Finland, 2002.
2. Gorham, E., Northern peatlands: role in the carbon cycle and probable responses to climatic warming, *Ecol. Appl.*, 1, 182, 1991.
3. Bohn, H.L., Estimate of organic carbon in world soils, *Soil Sci. Soc. Am. J.*, 40, 468, 1976.
4. Bohn, H.L., Organic carbon in world soils, *Soil Sci. Soc. Am. J.*, 46, 1118, 1982.
5. Sjörs, H., An arrangement of changes along gradients, with examples from successions in boreal peatlands, *Vegetatio*, 43, 1, 1980.
6. Post, W.M., Emanuel, W.R., Zinke, P.J., and Stanberger, A.G., Soil carbon pools and world life zones, *Nature*, 298, 156, 1982.
7. Houghton, J.T., Filho, L.G.M., Bruce, J., Lee, H., Callander, B.A., Haites, E., Harris, N., and Maskell, K., Radiative forcing of climate change, in *Climate Change 1994*, Houghton, J.T., Filho, L.G.M., Bruce, J., Lee, H., Callander, B.A., and Haites, E., Eds., Cambridge University Press, Cambridge, 1995, 231 pp.
8. Armentano, T.V. and Menges, E.S., Patterns of change in the carbon balance of organic soil wetlands of the temperate zone, *J. Ecol.*, 74, 755, 1986.
9. Markov, V.D., Olunin, A.S., Ospennikova, L.A., Skobeeva, E.I., and Khoroshev, P.I., *World Peat Resources*, Moscow 'Nedra,' 1988, 383 pp. [in Russian].

10. National Wetlands Working Group. *Wetlands of Canada*. Ecological Land Classification Series, No. 24. Sustainable Development Branch, Environment Canada, Ottawa, Ontario, and Polyscience Publications, Inc., Montreal, Quebec, 1988, 452 pp.

11. Bhatti, J.S., personal communication, 2004.

12. Vitt, D.H., Halsey, L.A., Bauer, I.E., and Campbell, C., Spatial and temporal trends of carbon sequestration in peatlands of continental western Canada through the Holocene, *Can. J. Earth Sci.*, 37, 683, 2000.

13. Vitt, D.H., Peatlands: Ecosystems dominated by bryophytes, in *Bryophyte Biology*, Shaw, A.J. and Goffinet, B., Eds., Cambridge University Press, Cambridge, 2000, 312.

14. Bauer, I.E. and Vitt, D.H., Autogenic succession and its importance for the peatlands of Canada's western boreal forests, in *Peatlands. Proceedings of the Peatland Conference 2002 in Hannover, Germany*, Bauerochse, A. and Hassmann, H., Eds., Leidorf, Rahden/Westf., 2003, 170.

15. Yu, Z., Campbell, I.D., Vitt, D.H., and Apps, M.J., Modelling long-term peatland dynamics. I. Concepts, review, and proposed design, *Ecol. Modelling*, 145, 197, 2001.

16. Clymo, R.S., The limits to peat bog growth, *Philos. Trans. R. Soc. Lond. B*, 303, 605, 1984.

17. Halsey, L.A., Vitt, D.H., and Zoltai, S.C., Climatic and physiographic controls on wetland type and distribution in Manitoba, Canada, *Wetlands*, 17, 243, 1997.

18. Johnson, L.C. and Damman, A.W.H., Species-controlled *Sphagnum* decay on a South Swedish raised bog, *Oikos*, 61, 234, 1991.

19. Johnson, L.C. and Damman, A.W.H., Decay and its regulation in *Sphagnum* peatlands, *Adv. Bryol.*, 5, 249, 1993.

20. Turetsky, M.R., The role of bryophytes in carbon and nitrogen cycling, *Bryologist*, 106, 395, 2003.

21. Vitt, D.H., An overview of factors that influence the development of Canadian peatlands, *Mem. Entomol. Soc. Can.*, 169, 7, 1994.

22. Alm, J., Talanov, A., Saarnio, S. Silvola, J., Ilkkonen, E., Aaltonen, H., Nykänen, H., and Martikainen, P.J., Reconstruction of carbon balance for microsites in a boreal oligotrophic pine fen, Finland, *Oecologia*, 110, 423, 1997.

23. Shurpali, N.J., Verma, J.K., and Arkebauer, T.J., Carbon dioxide exchange in a peatland ecosystem, *J. Geophys. Res.*, 100, 14319, 1995.

24. Suyker, A.E., Verma, S.B., and Arkebauer, T.J., Season-long measurement of carbon dioxide exchange in a boreal fen, *J. Geophys. Res.*, 102, 29021, 1997.

25. Lafleur, P.M., McCaughey, J.H., Jelinsky, D.E., Joiner D., and Bartlett, P., Seasonal trends in energy, water and carbon dioxide fluxes from a northern boreal wetland, *J. Geophys. Res.*, 102, 29009, 1997.

26. Carroll, P.J. and Crill, P.M., Carbon balance of a temperate poor fen, *Global Biogeochem. Cycles*, 11, 349, 1997.

27. Waddington, J.M. and Roulet, N.T., Atmosphere-wetland carbon exchange CO_2 and CH_4 exchange on the developmental topography of a peatland, *Global Biogeochem. Cycles*, 10, 233, 1996.

28. Turetsky, M.R., Wieder, R.K., Halsey, L.A., and Vitt, D.H., Current disturbance and the diminishing peatland carbon sink, *Geophys. Res. Lett.*, 10.1029/2001GL014000, 2002.

29. Turetsky, M.R., Wieder, R.K., Williams, C.J., and Vitt, D.H., Organic matter accumulation, peat chemistry, and permafrost melting in peatlands of boreal Alberta, *Ecoscience*, 7, 279, 2000.

30. Milich, L., The role of methane in global warming: where might mitigation strategies be focused? *Global Environ. Change,* 9, 179, 1999.
31. Scholes, M.C., Matrai, P.A., Smith, K.A., Andreae, M.C., and Guenther, A., Biosphere-atmosphere interactions, in *The Changing Atmosphere.* Available online at http://medias.obs-mip.fr/igac/html/book/chap2/chap2.html, 2000.
32. Gignac, L.D., Vitt, D.H., Zoltai, S.C., and Bayley, S.E., Bryophyte response surfaces along climatic, chemical, and physical gradients in peatlands of western Canada, *Nova Hedw.,* 53, 27, 1991.
33. Gignac, L.D. and Vitt, D.H., Responses of northern peatlands to climate change: effects on bryophytes, *J. Hattori Bot. Lab.,* 75, 119, 1993.
34. Nicholson, B.J. and Gignac, L.D., Ecotope dimensions of peatland bryophyte indicator species along environmental and climatic gradient in the Mackenzie River Basin, *Bryologist,* 98, 437, 1995.
35. Nicholson, B.J., Gignac, L.D., and Bayley, S.E., Peatland distribution along a north-south transect in the Mackenzie River Basin in relation to climatic and environmental gradients, *Vegetatio,* 126, 119, 1996.
36. Nicholson, B.J., Gignac, L.D., Bayley, S.E., and Vitt, D.H., Vegetation response to global warming: Interactions between boreal forest, wetlands, and regional hydrology, in Mackenzie Basin Impact Study (MBIS), Final Report, Cohen, S.J., Ed., Environment Canada, Downsview, Ontario, 1997, 295.
37. Vitt, D.H., Halsey, L.A., and Zoltai, S.C., The bog landforms of continental western Canada, relative to climate and permafrost patterns, *Arctic Alpine Res.,* 26, 1, 1994.
38. Halsey, L.A., Vitt, D.H., and Zoltai, S.C., Disequilibrium response of permafrost in boreal continental western Canada to climatic change, *Climatic Change,* 30, 57, 1995.
39. Vitt, D.H., Halsey, L.A., and Zoltai, S.C., The changing landscape of Canada's western boreal forest: the dynamics of permafrost, *Can. J. For. Res.,* 30, 283, 2000.
40. Camill, P. and Clark, J.S., Climate change disequilibrium of boreal permafrost peatlands caused by local processes, *Am. Nat.,* 1551, 202, 1998.
41. Beilman, D.W., Vitt, D.H., and Halsey, L.A., Localized permafrost peatlands in western Canada: definition, distributions, and degradation, *Arctic Antarctic Alpine Res.,* 33, 70, 2001.
42. Campbell, I.D., Campbell, C., Vitt, D.H., Kelker, D., Laird, L.D., Trew, D., Kotak, B. LeClair, D., and Bayley, S., A first estimate of organic carbon storage in Holocene lake sediments in Alberta, Canada, *J. Paleolimnol.,* 24, 395, 2000.
43. Yu, Z., Campbell, I.D., Campbell, C., Vitt, D.H., Bond, G.C., and Apps, M.J., Carbon sequestration in western Canadian peat highly sensitive to Holocene wet-dry climate cycles at millennial timescales, *Holocene,* 13, 801, 2003.
44. Campbell, I.D., Campbell, C., Yu, Z., Vitt, D.H., and Apps, M.J., Millennial-scale rhythms in peatlands in the western interior of Canada and in the Global Carbon Cycle, *Quaternary. Res.,* 54, 155, 2000.
45. Bond, G., Showers, W., Cheseby, M., Lotti, R., Almasi, P., de Menocal, P., Priore, P., Cullen, H., Hajdas, I., and Bonani, G., A pervasive millennial-scale cycle in North Atlantic Holocene and glacial climates, *Science,* 278, 11257, 1999.
46. Devito, K.J., Creed, I.F., and Fraser, C., Controls on runoff from a partially harvested aspen forested headwater catchment, boreal plain, *Can. Hydrol. Process,* 19, 3, 2005.
47. Winter, T.C. and Woo, M.-K., Hydrology of lakes and wetlands, in *Surface Water Hydrology,* Wolman, M.G. and Riggs, H.C., Eds., Geological Society of America, Boulder, CO, 1990, 159.
48. Montagnes, R.J.S., The habitat and distribution of *Meesia triquetra* in North America and Greenland, *Bryologist,* 93, 349, 1990.

49. Höper, H., Nitrogen and carbon mineralization rates in German agriculturally used fenlands, in *Soil Ecological Processes in Wetlands of Germany*, Broll, G., Merbach, W., and Pfeiffer, E.M., Eds., Springer, Berlin. 2000.

50. Yu, Z., Turetsky, M.R., Campbell, I.D., and Vitt, D.H., Modelling long-term peatland dynamics. II. Processes and rates as inferred from litter and peat-core data, *Ecol. Modelling*, 145, 159, 2001.

11 Linking Biomass Energy to Biosphere Greenhouse Gas Management

D.B. Layzell and J. Stephen

CONTENTS

11.1 INTRODUCTION

Plants have been in the business of managing greenhouse gases (GHGs) and solar energy for hundreds of millions of years. Therefore, in a world where there are concerns about climate change and energy supply, it is not unreasonable to look to biological systems and the biosphere both for solutions to these challenges and for a better understanding of the processes regulating atmospheric levels of greenhouse gases and the Earth's climate.

This is especially true for Canada, a nation with a vast biosphere and a relatively small population. In Canada, the annual flux of carbon into and out of biological systems is more than ten times the nation's emissions of CO_2 from fossil fuel energy use. The biosphere is also a large natural source and sink for the other greenhouse gases, nitrous oxide (N_2O) and methane (CH_4).

This chapter provides a brief overview of the role that the biosphere could play in helping Canada address the challenges of climate change and energy supply. It also explores the potential for using biomass to meet Canada's energy demands, and the relevance of doing so on biosphere management in a climatically different future.

11.2 BIOSPHERE SOLUTIONS

Biosphere solutions to the challenges of climate change and energy supply can be classified into four options for action.

11.2.1 REDUCE BIOSPHERE GHG EMISSIONS

Agricultural systems in Canada are responsible for 8 to 10% of the nation's annual GHG emissions. Approximately half of these agricultural emissions are associated with N_2O production from cropping systems, whereas the remainder involve N_2O and CH_4 production associated with animal production and manure management.[1]

Through improved management practices for cropping systems, animal production, or manure management,[2] it is possible to reduce these emissions. Such "beneficial management practices" may also reduce input costs, water pollution, soil degradation, or farm odors, so there can be benefits that are additional to GHGs and climate change.

New technologies may also play a role in reducing GHG emissions. Precision farming[2] could ensure more efficient use of fertilizers, new fertilizer formulations could reduce the nitrification and denitrification processes that lead to N_2O emissions, and new crops could be selected or engineered for improved N use, which may also reduce N_2O emissions. Similarly, improved feed formulations or antimethanogenic feed additives can reduce CH_4 production and manure production in ruminants,[3] and a range of improved manure-handling technologies will reduce GHG emissions.[4]

11.2.2 SEQUESTER ATMOSPHERIC CO_2

Through photosynthesis, plants use the energy of the sun to remove CO_2 from the atmosphere and sequester it into energy-rich biomass. Initially, this biomass is living, but when it dies, it can enhance soil carbon stocks before it eventually returns to the atmosphere through either respiration or fire. Increases in carbon stocks, be it the living biomass in trees and perennial crops or the organic carbon in soils, could be used to remove CO_2 from the atmosphere.

Improved land management strategies and a range of technologies are well known to build or preserve biosphere carbon stocks and could play a major, albeit temporary role in offsetting fossil fuel emissions. Examples in the forest sector include afforestation, fire and pest control in forests, use of faster-growing tree genotypes, selective harvest, rapid replanting after harvest and pre-commercial thinning. In agriculture, reduced tillage, crop rotation, incorporation of char, the cultivation of perennial biomass crops, or the development of new crops that build soil carbon pools have all been proposed as strategies for biosphere sequestration of atmospheric CO_2.

11.2.3 COMPLEMENT FOSSIL ENERGY STREAMS

Biomass can also be used to provide an energy resource that can complement our existing fossil energy streams, thereby reducing net GHG emissions so long as the biosphere carbon stocks are not depleted. This option is considered in more detail below.

11.2.4 ADAPT OUR BIOSPHERE TO A CHANGING ATMOSPHERE AND CLIMATE

If Canada is to manage its biosphere to reduce GHG emissions, sequester more carbon into biomass, and use a portion of that biomass to provide an energy resource, we will need to do this in a future in which the atmospheric CO_2 is higher and the climate is changed from what it has been in the past. Such a future will have both positive and negative effects on the health and vitality of the biosphere, and in the case of the negative impact, human intervention in the form of new management strategies and technologies will be needed to help the biosphere adapt.

Examples include replanting after forest harvest or natural disturbance with genotypes or species that are better suited to future climate and the stresses associated with climate change, the establishment of north–south corridors to facilitate the movement of plant and animal species, or the selection or engineering of crops so they can take better advantage of a high-CO_2 future.

11.3 THE BIOENERGY CHALLENGE

Canada's primary energy consumption in 2000 was about 12.6 exajoules (EJ), with approximately 8.24 EJ yr^{-1} coming from the combustion of fossil fuels.[5,6] In Canada, the CO_2 emissions from these fossil fuels are about 150 Mt C yr^{-1}.

Because biomass has lower energy content than fossil fuels,[6] providing the same energy output requires more of the feedstock. This is especially the case with wet biomass being processed thermochemically because energy must be expended to dry the biomass before processing. In addition, the carbon content of most biomass is typically lower than that of fossil fuels, so to provide a given amount of energy about 1.5 to 2 times the weight of biomass is required compared to that of coal.

Assuming that biomass has 50% water, the effective energy content would be about 30 GJ per tonne of biomass carbon.[6] Therefore, to provide the 8.24 EJ of energy that Canada now derives from fossil fuels would require about 275 Mt yr^{-1} biomass carbon. The aboveground carbon content of Canada's forests has been estimated at 15,800 Mt C,[7] equivalent to approximately 57 years of our current energy demand from fossil fuels.

Perhaps a more relevant reference point is that currently biomass has been estimated to provide about 6% of Canada's energy needs,[6] a sink for about 16 Mt C yr^{-1}, primarily through home heating and power generation. The total annual agriculture and forest harvest in Canada is about 143 Mt biomass C,[7] a value equivalent to 50% of the biomass needed to meet the nation's current fossil fuel energy demand. Not all of this harvested biomass makes its way into products; recent estimates[7] suggest that up to 50 Mt C yr^{-1} is residual, some of which could be used as a bioenergy resource. Economic and environmental concerns are key factors in determining the proportion that is accessible, and to date no consensus exists whether this proportion is as low as 20% (10 Mt C yr^{-1}) or as high as 70% (35 Mt C yr^{-1}).

Even considering this resource, for Canada to meet its entire fossil fuel demand would require a threefold increase in the size of the forestry and agricultural harvest, with two thirds of the total harvest directed to bioenergy. It is difficult to imagine an increase in harvest of this magnitude, especially if it were to be done with minimal environmental impact. Nevertheless, as discussed in the next section, there is capacity for a significant increase in the forestry and agricultural harvest in Canada, and when this is coupled with waste or residual carbon coming from municipalities, agriculture, and forestry, it is clear that biomass energy could complement our fossil energy streams, while helping the nation meet its climate change commitments. Of course, to do this, biomass resources in the range of many tens of megatonnes of biomass C must be identified and directed toward bioenergy.

11.4 SUSTAINABLE SOURCES OF BIOMASS

If Canada is to expand its use of biomass energy, it will be very important to ensure that the effort is sustainable, not only in economic and social terms, but in environmental terms. For example, maintaining or enhancing biodiversity and minimizing environmental damage from pesticides and fertilizers are key issues. Also, to address concerns about climate change, it will be imperative that biosphere carbon stocks are maintained in the face of an increased harvest and removal of biomass from ecosystems. Because biomass has a lower energy content than coal or other fossil fuels and because biomass C stocks are more likely than fossil fuels to provide habitat for other organisms, the use of biomass energy in a way that depletes C stocks would be less sustainable, and have a greater climate change impact, than the use of any fossil fuel resource.

New and improved biomass processing technologies will also be important, as it will be necessary to avoid the particulate emissions and other environmental

TABLE 11.1

A Summary of the Magnitude of Possible Sources of Biomass That Could Be Used as a Sustainable Bioenergy Resource

	Mt C yr^{-1}
Municipalities	
Municipal solid waste, landfill gas, industrial waste	5 to 15
Agriculture	
Waste (animal and crop), residues	2 to 10
Fast-growing species on unused lands	10 to 20
Forestry	
Unused harvest and mill residues	10 to 30
Enhanced forest management	20 to 40
Harvest remaining carbon stock after severe disturbance	5 to 30
Total	52 to 145

impacts associated with simple combustion and other traditional biomass energy technologies.

So what are the sustainable sources of biomass and how much carbon could be obtained from each? At the present time, this information can only be estimated (Table 11.1), and very little work has been done to explore the costs and feasibility of accessing these biomass energy streams for use in power, transportation fuels, or heavy industry. Consequently, the analysis presented below explores only the biomass energy resources, not the feasibility of accessing these carbon streams, economically or otherwise.

11.4.1 Municipalities

Municipal solid wastes (MSW) that are currently discarded (rather than recycled) have been estimated to contain about 6 Mt C yr^{-1}. Together with industrial waste, biosolids, and landfill gases, a waste carbon stream of up to 15 Mt C could be made available.[7]

11.4.2 Agriculture

Of the 55 Mt C that are harvested from agricultural lands every year, about 33 Mt C are in the primary products for which the crops were grown.[7] There are other uses for the crop residues, including the maintenance of soil carbon stocks, but when the surplus is combined with animal waste, a bio-based carbon stream of up to 10 Mt C yr^{-1} or more should be available.

In addition, Canada has more than 7 million hectares of unused agricultural land, a significant portion of which could be brought into production for biomass crops. Such crops are typically fast-growing perennials, and can produce lignocellulosic carbon at two to four times the average carbon accumulation rate of Canada's current food crops ($\sim$1.7 tC ha^{-1} yr^{-1})[7]. Assuming 7 million hectares producing lignocellulosic biomass at a conservative 3 tC ha^{-1} yr^{-1}, there would be an annual biomass carbon stream of more than 20 Mt C.

11.4.3 FORESTRY

Each year Canada's forest industry harvests about 88 Mt C on 1 million of the 245 million hectares of timber productive forest. Approximately 48 Mt C are removed from the site, and some of the remaining 40 Mt C are needed to maintain soil carbon stocks, other nutrients, and biodiversity.[7] Unfortunately, there is no consensus regarding how much of this could be removed sustainably and at an acceptable cost. In Table 11.1, we have provided an estimated range of 10 to 30 Mt C yr^{-1}.

Enhanced forest management could be used to build biomass carbon stocks to provide a biomass energy resource. Pre-commercial thinning, replanting after harvest, improved pest or fire control, or the judicious use of fertilizers could all contribute to the health and vitality of forests, making it possible to increase the harvest of biomass on a landscape or regional scale with little or no negative impact on net carbon stocks or biodiversity. Assuming current forest management practices are sustainable from the perspective of carbon stocks and energy input, improvements in management practices could increase forest growth and yield by 25 to 80%.[8,9] This could provide a bioenergy resource of 20 to 40 Mt C yr^{-1} (Table 11.1).

Large natural disturbances, such as those caused by fire, insect, wind, or floods, could also provide a source of forest biomass to support a bioenergy future for Canada. Given that an actively growing Canadian forest accumulates carbon at a rate of 1 to 2 tC yr^{-1},[10] the fact that Canada has 240 M ha of timber productive forest, and that disturbance plays a major role in determining the net carbon balance of our forest ecosystems, a conservative estimate of biomass carbon that could flow to bioenergy from disturbed sites is 5 to 30 Mt C yr^{-1} (Table 11.1). A case study exploring options for the possible use of this biomass from natural disturbance follows.

11.5 CASE STUDY: ACCESSING BIOMASS FROM DISTURBED FOREST SITES

In developing strategies for mitigating climate change through biosphere greenhouse gas management, it is also important to recognize and plan effective responses to the predicted effects of a future world with high atmospheric CO_2 and a different climate. The International Panel on Climate Change[11] has predicted significant climate impacts on North American forests over the next 50 years, including greater frequency and impact of "catastrophic events (e.g. fire, insect outbreaks, pathogens, storms) that have marked effects on ecosystem structure ... leading to changes in composition as forests regenerate under altered conditions."

In recent decades, 2.8 million hectares of forested lands have been burned every year in Canada, representing approximately 27 Mt C.[12] Large fires of greater than 1000 ha in size account for only 1.4% of forest fires in Canada, but represent 93.1%

of the total area burned.[8] These statistics account for the fact that there is a large year-to-year variability in losses to fires, ranging from 3 Mt C to over 115 Mt C.[13]

Fire activity, including land area and total biomass burned, has been increasing in Canada over the past three decades. On average, increased fire activity is the result of decreased moisture conditions, brought about by increased temperatures and reductions in rain/snow fall.[14] An earlier start date to the fire season and an increase in the area under high to extreme fire risk have already been seen as temperatures have risen in recent years.[15] Based on general circulation models (set to two times pre-industrial atmospheric CO_2), the trend is expected to continue.[15]

Insect infestations are the single largest cause of losses in Canada's forests, accounting for 1.4 to 2 times that lost to fire. Since climate change over the next 50 years has been predicted to accelerate insect development, expand current ranges (particularly northward), and raise overwinter survival rates,[16] losses due to insect infestation may increase. For example, stresses put on trees by climate, such as unusually high temperatures or drought conditions, make them more susceptible to attack. An earlier spring enhances the likelihood of synchronous development between insects and buds of their host, leading to increased survival rates.[16] Insect infestations cause changes in ecosystem carbon and nutrient cycling and increase rates of decomposition, due to tree mortality. In addition, they significantly increase the risk of wildfires.[10]

Approximately 6.3 million hectares of forest (at a conservative 25 t C ha^{-1}, this would be equivalent to 160 Mt C) are currently infected by insects, and tree mortality rates on these sites can reach 85%.[16,17] Insect infestation is a leading cause of pathogen infection, and fungi play the most important role in enhancing death rates.

Storms and extreme weather events, such as the 1998 ice storm in eastern Ontario and Quebec,[10] are also likely to increase in occurrence and intensity[12] and may be a third source of adverse climate impacts on Canada's forests. In future decades, there may be significant forest death due to temperature and water stress linked directly to climate change.[18]

There are several options for the excess timber that will be generated as a result of climate change and the associated impacts of temperature, fire, infestation, and disease.

11.5.1 OPTION 1. HARVEST BIOMASS FOR FIBER MARKETS

This is currently done with some of the dead biomass, especially that left from insect infestation. However, if the harvest is not carried out reasonably soon after disturbance, there is a decline in the suitability of the timber for higher-value wood products. This fact, coupled with the sheer magnitude of supply from heavily disturbed ecosystems, means that the biomass supply from disturbance is likely to swamp either the processing capacity in the region of the disturbance or the regional and global markets for the forest products, especially in a climate-different future.

On the basis of climate models, higher-latitude timber producers are expected to be the hardest hit by climate change because of a large predicted increase in the availability of timber, but a relatively low predicted growth in demand for wood products.[19] An increase in harvest scope to include more damaged and "at-risk" trees would only widen this difference between supply and demand.

11.5.2 Option 2: Leave the Biomass to Decompose

This strategy is currently being used in many situations involving major disturbances. On the positive side, the nutrients are being left in the field to be recycled, and the carbon will remain for perhaps a few years or decades where it will continue to act as a carbon sink. However, the biomass carbon will eventually decompose or be burned, and in the latter case, the fire will take with it much of the biomass carbon that has regrown in the meantime. Even though fire-affected sites provide an opportunity for forest managers to replant them with tree species that are better adapted to the climate-different future in which they must grow, it seems likely that there would be an overall decline in carbon stocks on affected lands over time, and it could be many decades or centuries before this trend would be reversed.

11.5.3 Option 3: Harvest for Bioenergy

Once a tree is dead, it will no longer gain carbon, but it begins to decompose or it will burn in a forest fire. Either way, the carbon and the energy it contains will be released to the atmosphere. By harvesting the biomass and processing it to extract its energy content, the biomass could replace fossil fuels in the energy streams, thereby providing energy for human needs and reducing the associated greenhouse gas emissions. This strategy would reduce forest fire risk, remove diseased or infected trees that might reinfect others, and make it possible for forest managers to better manage the affected area to ensure rapid forest regrowth, thereby rebuilding carbon stocks. Recent studies have calculated that, in the long run, managed bioenergy systems will have a much greater impact on reducing greenhouse gas emissions than forest management practices focused solely on building carbon stocks.[20]

11.6 CASE STUDY: IMPACT OF VARIOUS FEEDSTOCK-TO-PRODUCT THREADS

Compared to other developed nations, Canada has very large biomass reserves, and has the capacity to produce more biomass that could feed into human energy systems. Indeed, Table 11.1 identifies potential sources of 52 to 145 Mt biomass carbon. If this magnitude of biomass were to be used for energy production in Canada, bioenergy would make a significant and lasting contribution to Canada's energy needs and the challenges of climate change.

To illustrate this point, we present below three case studies of feedstock-to-product threads. For each, we explore what could be achieved with 10 Mt biomass carbon in terms of Canada's energy demands and GHG benefits.

11.6.1 Biopower

Biomass for power, or biopower, is a technology that already exists and is being implemented, especially by forestry companies using their mill wastes. Canada produces 576 TWhr of electricity each year (Table 11.2, Item 1a), only 15 of which is from biomass, whereas 19% comes from coal combustion that leads to significant

TABLE 11.2

Comparison of the Contribution that 10 Mt Biomass Carbon Could Make to Canada's Energy Demand for Electrical Power and Liquid Fuels, and the Calculated Net Impact on Greenhouse Gas Emissions

Item	Units	Value	%	Notes [Ref.]
1. Biopower				
a. Current Cdn Power Consumption	TW hr	576	100%	From Cdn Electrical Association (www.canelect.ca)
Resource Potential of 10 Mt C				
b. 10 Mt C yr^{-1} at 16% conversion efficiency	TW hr 10Mt C^{-1}	13.3	2.3%	Assumes 100% energy conversion would yield 8.33 MWhr tC^{-1} (30 GJ t C^{-1} and 3.6 GJ $MWhr^{-1}$) [22]
c. 10 Mt C yr^{-1} at 40% conversion efficiency	TW hr 10Mt C^{-1}	33.3	5.8%	Assumes 100% energy conversion would yield 8.33 MWhr tC^{-1} (30 GJ t C^{-1} and 3.6 GJ $MWhr^{-1}$) [22]
d. GHG emissions from coal power generation	g CO_2e KW hr^{-1}	1042		[23]
GHG Mitigation Potential Assuming Coal Displacement				
e. 10 Mt C yr^{-1} at 16% conversion efficiency	Mt CO_2e yr^{-1}	13.9		Calculated as (1d) × (1b) 1000^{-1}
f. 10 Mt C yr^{-1} at 40% conversion efficiency	Mt CO_2e yr^{-1}	34.7		Calculated as (1d) × (1c) 1000^{-1}
2. Bioethanol				
a. Current Canadian gasoline use	GL yr^{-1}	40.3	100%	[24]
Resource Potential of 10 Mt C				
b. Starch crop: 10 Mt C yr^{-1} at 844 l tC^{-1}	GL yr^{-1}	8.4	21%	Assumes 380 l ethanol t^{-1} biomass at 45% C [25]
c. Lignocellulose crop: 10 Mt C yr^{-1} at 620 l tC^{-1}	GL yr^{-1}	6.2	15%	Assumes 310 l ethanol t^{-1} biomass at 50% C [25]
d Canadian GHG emissions from gasoline	Mt CO_2e yr^{-1}	111		Assumes [Item 2a × 2.75 kg CO_2 l^{-1} [1]

TABLE 11.2 (continued)

Comparison of the Contribution that 10 Mt Biomass Carbon Could Make to Canada's Energy Demand for Electrical Power and Liquid Fuels, and the Calculated Net Impact on Greenhouse Gas Emissions

Item	Units	Value	%	Notes [Ref.]
GHG Mitigation Potential Assuming Gasoline Displacement (per km traveled)				
e. Starch crop: 10 Mt C yr⁻¹ at 844 l tC⁻¹	Mt CO$_2$e yr⁻¹	8.1		Calculated as (2d) * (2b%) × Life cycle factor [F(LCA)], where F(LCA) = 35% [22]
f. Lignocellulose crop: 10 Mt C yr⁻¹ at 620 l tC⁻¹	Mt CO$_2$e yr⁻¹	13.6		Calculated as (2d) * (2c%) × Life cycle factor [F(LCA)], where F(LCA) = 80% [26]
3. Biodiesel				
a. Current Canadian petroleum diesel use	GL yr⁻¹	26.6	100%	[24]
Resource Potential of 10 Mt C				
b. Oil seed crop: 10 Mt C yr⁻¹ at 813 l tC⁻¹	GL yr⁻¹	8.1	31%	Assumes 528 l t⁻¹ seed at 65% C [27]
c. Lignocellulose crop: 10 Mt C yr⁻¹ at 300 l tC⁻¹	GL yr⁻¹	3.0	11%	Assumes 150 l t⁻¹ biomass at 50% C [28]
d. Canadian GHG emissions from diesel	Mt CO$_2$e yr⁻¹	73		Assumes Item 2a × 2.75 kg CO$_2$ l⁻¹ [1]
GHG Mitigation Potential Assuming Diesel Displacement (per km traveled)				
e. Oil seed crop: 10 Mt C yr⁻¹ at 813 l tC⁻¹	Mt CO$_2$e yr⁻¹	11.2		Calculated as (3d) * (3b%) × Life cycle factor [F(LCA)], where F(LCA) = 50% [25]
f. Lignocellulose crop: 10 Mt C yr⁻¹ at 300 l tC⁻¹	Mt CO$_2$e yr⁻¹	8.3		Calculated as (3d) * (3c%) × Life cycle factor [F(LCA)], where F(LCA) = 100% [29]

greenhouse gas emissions. Increased biomass use for power generation could involve either co-firing biomass with coal or the construction of power facilities entirely dependent on biomass as a fuel.

If 10 Mt of biomass C were to be used for electrical power and the conversion technology was low (16%) efficiency (e.g., basic combustion technology), approximately 13.3 TWhr of power would be generated, accounting for about 2.3% of Canada's electric power consumption (Table 11.2, Item 1b). By increasing conversion efficiency to 40% (typically of a large biomass gasification system[21]), 10 Mt of biomass carbon would generate 33.3 TWhr and provide about 5.8% of Canada's electric power consumption.

If coal were to be displaced by co-firing with biomass or through the use of dedicated biopower facilities (Table 11.2, Item d), the greenhouse gas benefit associated with the low- or high-efficiency use of 10 MT biomass C would be 14 or 35 Mt CO_2e yr^{-1}, respectively. The upper of these two numbers approximates the maximum greenhouse gas benefit (37 Mt CO_2e yr^{-1}) that could be gained as a temporary carbon credit by storing the carbon and building biosphere carbon stocks rather than diverting it for bioenergy. One advantage of bioenergy over the building of carbon stocks is that the greenhouse gas emission reductions from bioenergy are permanent, whereas biosphere carbon stocks are not.

11.6.2 BIOETHANOL

Bioethanol can be made by fermentation from sugars derived from either starch or lignocellulosic biomass, and then be added to gasoline as a transportation fuel. A 90:10 mixture of gasoline to ethanol performs well in automobiles and, because of its higher oxygen content, reduces the emissions of both greenhouse gases and various pollutants. However, with improvements in engine design, it is possible to achieve much higher ethanol levels in gasoline, even up to 85% or more.

Given current conversion efficiencies of 380 l tonne^{-1} starch (Table 11.2, Item 2b) or 310 l tonne^{-1} lingo-cellulosic biomass (Table 11.2, Item 2c), 10 Mt biomass carbon could provide 6.2 to 8.4% of the 40.3 billion liters of gasoline Canadians currently use per year.

The greenhouse gas benefit associated with this process would be about 8.1 Mt CO_2e yr^{-1} for starch-based bioethanol production and 13.6 Mt CO_2e yr^{-1} for bioethanol from lignocellulosic feedstocks. The greater GHG benefit associated with the lignocellulosic feedstocks reflects the additional greenhouse gas costs associated with managing and fertilizing the starch crops such as soybean or wheat.

11.6.3 BIODIESEL

Lipid-based biodiesel is another established fuel, but with no significant Canadian production at the present time. It is produced via transesterification of oils and lipids, from sources such as soybean, canola, and animal tallow. Biodiesel has advantages as an additive (now approved for 5%, but higher mixtures are possible) in petroleum diesel, as it has greater lubricity than regular diesel, and when it burns there are less particulate emissions.[22] With 10 Mt of carbon as plant oil, it would be possible to provide about 31% (Table 11.2, Item 3b) of the current petroleum diesel use in Canada (26.6 GL yr^{-1}, Table 11.2, Item 3a). However, such a target is unreasonably high as the current canola production is only 6 Mt seed per year, and there is limited capacity to increase production to the scale required to take advantage of a significant proportion of the diesel market.

For biomass-derived diesel to make a more significant contribution to Canada diesel fuel demand, it will be necessary to convert lignocellulosic feedstocks to a hydrocarbon diesel fuel through a process such as gasification and Fischer-Tropsch synthesis.[24] This technology is still developing, but assuming a conservative 150 l diesel per tonne biomass, 10 Mt biomass could provide about 11% of Canada diesel fuel market (Table 11.2, Item 3c).

The greenhouse gas benefits of processing 10 Mt biomass carbon through to biodiesel have been estimated to fall between 8.3 and 11.2 Mt CO_2e yr^{-1} using conversion efficiencies that have been achieved to date (Table 11.2, Items 3e and 3f). Presumably, with research and development driven by commercialization, these efficiencies will improve.

11.6.4 CONCLUSION

From these case studies, it can be seen that 10 Mt biomass carbon could make a significant contribution to a variety of energy streams, and integrate well with the current fossil fuel economy. However, the overall greenhouse gas benefit associated with the use of a given quantity of biomass will depend on the process used and the fossil fuel source from which the energy supply is diverted. Among the case studies explored here, the greatest greenhouse gas benefit would come from biomass use to displace coal in power generation (Table 11.2, Items 1e and 1f), whereas the least benefit was associated with the production of ethanol from a starch crop (Table 11.2, Item 2e). Such insights should be considered in developing policies for climate change mitigation.

11.7 SOCIOECONOMICS OF BIOMASS ENERGY

The above analyses have not considered the socioeconomic factors affecting the access to and processing of biomass feedstocks into energy products. This must be done, as the socioeconomic issues will be among the most important in determining the ultimate role that biomass can and will play in addressing Canada's energy and climate change needs. Although such an analysis is beyond the scope of this chapter, it is worth noting that when the analysis is carried out, it needs to consider the following factors:

- **Other options for greenhouse gas management.** Movement to an economy more dependent on bioenergy is not without its costs, but in a world that is truly concerned about greenhouse gas emissions, these costs may prove to be less than other options for mitigating climate change, especially for Canada.

- **Societal costs and benefits.** If carried out at a scale to make a significant impact on Canada's energy needs and greenhouse gas emissions, movement toward bioenergy could be a major stimulus for the rural economy and native peoples. It will also generate much larger domestic markets for agricultural and forest production, thereby giving these sectors some protection from the vagaries of international markets that are affected by subsidies and trade barriers imposed by other nations. On the other hand, it could lead to the greater industrialization of the rural landscape and have a negative impact on some small rural communities.
- **Environmental costs and benefits**. Using and managing our ecosystems for a sustainable bioenergy supply could provide the resources, technologies, and incentives needed to clean up waste streams or to increase productivity of some forests while leaving others as a preserve for biodiversity and carbon stocks. It could also develop greatly improved opportunities for helping ecosystems adapt to our changing atmosphere and climate. On the other hand, poor implementation of bioenergy solutions could lead to overexploitation of ecosystems and the loss of biodiversity and biosphere carbon stocks.
- **A level playing field for clean energy options.** Most energy analysts predict that energy prices will continue to rise in coming years as the world's ability to provide energy is outstripped by the demand. As we near the end of the era of cheap oil, other energy options will be considered to fill demand. Many of the climate-friendly energy options are very expensive, and some require large amounts of public money for building, insuring, or decommissioning energy facilities. In considering biomass as an alternative, it will be important to ensure that the playing field is level as the cases are made for and against various alternatives.
- **Sociopolitical factors.** Not all climate-friendly energy alternatives have the widespread support of society. Issues of health and safety, environmental impact, waste disposal, terrorist threat, nuclear disarmament, and public perception will all be important factors and must be explored.

11.8 CONCLUSIONS

As global climate change threatens to fundamentally change our biosphere, Canada needs to improve the way it manages its natural resources to mitigate greenhouse gas emissions and adapt to a climate-different future. With its extensive boreal forests, Canada has both opportunities and liabilities associated with its vast biosphere.

For example, while overall forest growth is anticipated to increase, the climate-induced changes that are predicted to occur will dramatically alter ecosystem structure and the biomass carbon stocks within these ecosystems. It is essential that Canada's forests are managed to maintain or increase carbon stocks. They should also be managed to optimize the potential use of biomass as a source of energy to complement our existing fossil fuel streams.

There are many potential sources of biomass for bioenergy conversion, but the stocks that are most attractive for bioenergy from an ecosystem management and biodiversity perspective are the residual or dead carbon stocks, including those associated with harvest, pre-commercial thinning, fire, insect infestation, or climatic change itself. Outside of the forestry sector, large biomass resources for bioenergy can also be found, including agricultural residues, new biomass crops, municipal solid wastes, animal wastes, and municipal biosolids. If cost-effective technologies could be developed to access and process these carbon stocks in a sustainable way, there would be significant and long-lasting positive impacts on Canada's greenhouse gas emissions and its supply of clean energy.

ACKNOWLEDGMENT

This work was made possible with the support of Queen's University and the BIOCAP Canada Foundation.

REFERENCES

1. Olsen, K., Wellisch, M., Boileau, P., Blain, D., Ha, C., Henderson, L., Liang, C., McCarthy, J., and McKibbon, S. (2003) Canada's Greenhouse Gas Inventory 1990–2001. Environment Canada.
2. U.S. Climate Change Technology Program. (2003) Technology Option for the Near and Long Term: A Compendium of Technology Profiles and Ongoing Research and Development at Participating Federal Agencies. http://www.climatetechnology.gov/library/2003/tech-options/tech-options.pdf.
3. Mathison, G., Okine, E., McAllister, T., Dong, Y., Galbraith, J., and Dmytruk, O. (1998) Reducing methane emissions from ruminant animals. *J. Appl. Anim. Res.* 14: 1–28.
4. Hao, X., Chang, C., and Larney, F. (2004) Carbon, nitrogen balances and greenhouse gas emission during cattle feedlot manure composting. *J. Environ. Qual.* 33: 37–44.
5. Natural Resources Canada (1996) Canada's Energy Outlook, 1996–2020, Ottawa, ON (http://www.nrcan.gc.ca/es/ceo/toc-96E.html).
6. Klass, D.L (1998) *Biomass for Renewable Energy, Fuels, and Chemicals.* Academic Press, Toronto.
7. Wood, S.M. and Layzell, D.B. (2003) A Canadian Biomass Inventory: Feedstocks for a Bio-based Economy. Prepared for Industry Canada: Contract 5006125. BIOCAP Canada Foundation. Available online at http://www.biocap.ca/images/pdfs/BIOCAP_Biomass_Inventory.pdf.
8. Chen, W., Chen, J., and Cihlar, J. (2000) An integrated terrestrial ecosystem carbon-budget model based on changes in disturbance, climate, and atmospheric chemistry. *Ecol. Modelling* 135: 55–79.
9. ArborVitae Environmental Services Ltd & Woodrising Consulting. (1999) Estimating the Carbon Sequestration Benefits of Reforestation in Eastern Canada. Forest Sector Table and Sinks Table, Canadian National Climate Change Process.

10. Cohen, S. and Miller, K. (2001) North America, in *Climate Change 2001: Impacts, Adaptation, and Vulnerability. Contribution of Working Group II to the Third Assessment Report of the Intergovernmental Panel on Climate Change*, McCarthy, J., Canziani, O., Leary, N., Dokken, D., and White, K., Eds. Cambridge University Press, Cambridge, 735–800.

11. Stocks, B., Fosberg, M., Lynham, T., Mearns, L., Wotton, B., Yang, Q., Jin, J., Lawrence, K, Hartley, G., Mason, J., and McKenney, D. (1998) Climate change and forest fire potential in Russian and Canadian boreal forests. *Climatic Change* 38: 1–13.

12. Climate Change Impacts and Adaptation Directorate (2002) Climate Change Impacts and Adaptation: A Canadian Perspective. Natural Resources Canada, Ottawa, ON (ISBN: 0-662-33123-0).

13. Amiro, B., Todd, J., Wotton, B., Logan, K., Flannigan, M., Stocks, B., Mason, J., Martell, D., and Hirsch, K. (2001) Direct carbon emissions from Canadian forest fires, 1959–1999. *Can. J. For. Res.* 31: 512–525.

14. Wotton, B., Martell, D., and Logan, K. (2003) Climate change and people-caused forest fire occurrence in Ontario. *Climatic Change* 60: 275–295.

15. Stocks, B., Fosberg, M., Lynham, T., Mearns, L., Wotton, B., Yang, Q., Jin, J., Lawrence, K, Hartley, G., Mason, J., and McKenney, D. (1998) Climate change and forest fire potential in Russian and Canadian boreal forests. *Climatic Change* 38: 1–13.

16. Volney, W. and Fleming, R. (2000) Climate change and impacts of boreal forest insects. *Agric. Ecosyst. Environ.* 82: 283–294.

17. Natural Resources Canada (2002) The State of Canada's Forests 2001–2002.

18. Albritton, D. and Filho, L. (2001) Technical summary in climate change 2001: the scientific basis, in *Contribution of Working Group I to the Third Assessment of the Intergovernmental Panel on Climate Change*, Houghton, J., Ding, Y., Griggs, D., Noguer, M., van der Linden, P., Dai, X., Maskell, K., and Johnson, C., Eds. Cambridge University Press, Cambridge, 21–84.

19. Sohngen, B., Mendelsohn, R., and Sedjo, R. (2001) A global model of climate change impacts on timber markets. *J. Agric. Resour. Econ.* 26: 326–343.

20. Baral, A. and Guha, G. (2004) Trees for carbon sequestration or fossil fuel substitution: the issue of cost vs. carbon benefit. *Biomass Bioenerg.* 27: 41–55.

21. Bridgwater, A., Toft, A., and Brammer, J. (2002) A techno-economic comparison of power production by biomass fast pyrolysis with gasification and combustion. *Renewable Sustainable Energ. Rev.* 6: 181–248.

22. Oak Ridge National Laboratory (2004) Bioenergy Conversion Factors. Available online at http://bioenergy.ornl.gov/papers/misc/energy_conv.html.

23. Mann, M. and Spath, P. (2002) A Comparison of the Environmental Consequences of Power from Biomass, Coal, and Natural Gas. National Renewable Energy Laboratory, Golden, CO.

24. International Energy Agency (IEA), Energy Statistics-Oil-Canada. Accessed July 2004. Available online at http://www.iea.org/Textbase/stats/oiloecd.asp?oecd=Canada&SubmitB=Submit.

25. Levelton Engineering Ltd, R-2000-2, with (S&T)2 Consultants Inc. Assessment of Net Emissions of Greenhouse Gases from Ethanol-Blended Gasolines in Canada: Lignocellulosic Feedstock, Richmond, B.C., 2000.

26. Fulton, L., Howes, T., and Hardy, J. (2004) Biofuels for Transport: An International Perspective. International Energy Agency (IEA), Paris, France (ISBN: 92-64-01512-4).

27. Canola Council of Canada, Markets and Statistics. Accessed July 2004. Available online at http://www.canola-council.org/.
28. Tijmensen, M. (2000) *The Production of Fischer Tropsch Liquids and Power through Biomass Gasification.* Universiteit Utrecht Science Technology Society, Utrecht, the Netherlands.
29. Boerrigter, H., den Uil, H., and Calis, H.P. (2003) Green diesel from biomass by Fischer-Tropsch synthesis: new insights in gas cleaning and process design. Available online at http://www.ecn.nl/docs/library/report/2003/rx03047.pdf.

12 Ruminant Contributions to Methane and Global Warming — A New Zealand Perspective

G.C. Waghorn and S.L. Woodward

CONTENTS

12.1 INTRODUCTION

An overview of the implications, research, and policies concerning greenhouse gas (GHG) emissions from New Zealand agriculture is presented. Most emphasis is given to methane from ruminants and to opportunities for mitigation in forage-based feeding systems. The opportunities for practical reductions in both methane and nitrous oxide emissions are indicated.

The underlying principles affecting levels of methane emissions from ruminants are examined and compared with values obtained from sheep and cattle fed fresh forages. Opportunities for mitigation are presented as short-, medium-, and long-term strategies. Topics include the bases for animal variance, effects of management and diet, as well as potential mitigation through rumen additives.

The risks associated with mitigating a single GHG in isolation from others are demonstrated using a model of CO_2 and CH_4 emissions from contrasting dairy systems and the importance of maintaining economic viability in addition to environmental improvement is central to all considerations. The information presented here is based primarily on New Zealand experience. Our mixture of sheep, dairy and beef cattle, and deer is farmed outdoors all year on pastures varying in topography, fertility, and quality with diverse climatic conditions. New Zealand has a substantial challenge to determine agricultural GHG inventory and to mitigate emissions.

12.2 RELEVANCE OF GREENHOUSE GASES FOR NEW ZEALAND PRODUCERS

Methane accounts for 38% of New Zealand greenhouse gas emissions (based on Tier II estimations), which is a higher percentage than emissions in Australia (24%), Canada (13%), the U.S. (9%), and most industrialized countries, which emit only 5 to 10% of GHG as methane.[1] Nitrous oxide (N_2O) accounts for 17% (largely Tier I estimates) and CO_2 44% of our national GHG inventory (Table 12.1). Total annual emissions are 72.4 million tonnes of CO_2 equivalents, or about 18 tonnes per human.[2] Countries with higher emissions (tonnes head^{-1} of population) include Australia (25.1), the U.S. (23.6), and Canada (22.6).

In New Zealand 88% of CH_4 emissions are associated with animal agriculture, of which 98% is from digestion, primarily in the rumen. A single source of CH_4 provides an excellent focus for both measurement and mitigation, especially as energy losses account for about 10% of metabolizable energy (ME) intake of ruminants grazing grass-dominant pasture. Mitigation should be investigated on the basis of improved performance and efficiency of feed utilization as well as GHG inventory. Examples include halving CH_4 production to provide sufficient energy for an additional 400 kg milk cow^{-1} lactation^{-1} (average annual milk production from pasture fed cows is 3700 kg cow^{-1}). Alternatively, if total emissions could be collected from an adult cow over 1 year, the energy would fuel a midsize car for 1000 km!

The New Zealand government had intended to raise a ruminant tax (dubbed the "fart tax" by farmers and media) to generate research revenue. Planned taxation (per annum) was about US$0.50 per cow and US$0.08 per sheep, but this was abandoned

TABLE 12.1

Annual (2001) New Zealand Greenhouse Gas Emissions (as CO_2 equivalents)[2]

	Total CO_2 Equivalents (tonnes $\times 10^6$)	% of Total
New Zealand		
Carbon dioxide	32.43	44.6
Methane	27.06	37.5
Nitrous oxides	12.58	17.4
PFC, HFC, SF_6	0.31	0.4
Agriculture	35.85	51.0
Energy	30.93	39.0
Industrial	3.18	5.0
Waste	2.31	5.0
Agricultural Emissions		**% of CH_4 or N_2O**
Methane		
From digestion	23.12	84.5
From manure	0.55	2.0
Nitrous oxides[a]		
From animal production	7.12	56.6
Indirect from agricultural soils	3.13	24.9
Direct from agricultural soils	1.81	14.4

Abbreviations: PFC, perfluorocarbons; HFC, hydrofluorocarbons; SF_6, sulfur hexafluoride

[a]Nitrous oxide emissions apply to all agriculture, with some direct and indirect emissions attributable to animal agriculture.

in the face of farmer protest and current annual investment (NZ$4.7m) supports about 32 full-time equivalent researchers. Approximately 55% of funds are directed toward inventory, 20% to fundamental, and 25% to abatement research. A comprehensive report on Abatement of Agricultural non-CO_2 GHG Emissions in New Zealand[3] summarizes all current research and identifies research priorities.

There is good and increasing collaboration between Australian and New Zealand researchers with annual conferences and reports receiving direct government support. This collaboration is essential, given the relatively small investment in GHG research in both countries. Although Australia is not a signatory to the Kyoto Protocol, there is a strong commitment by federal and state governments to GHG reduction.

Promotion of benefits from lower GHG emissions in terms of productivity and environmental sustainability are receiving guarded support from farmers and the public. The concept of energy wastage provides an appropriate avenue for lobbying farmers and agricultural professionals to secure their support for funding. New Zealand farmers are sensitive to their role as guardians of their land and to the need to maintain or improve their environment. Successful mitigation (abatement) will require a mixture of consultation, education, and awareness as well as research if it is to be successful in the longer term. Ironically, the threat of an emission ("fart") tax has contributed awareness, although it was of little benefit for research funding.

12.3 NEW ZEALAND GHG INVENTORY

12.3.1 METHANE

New Zealand agricultural production is not subsidized and follows market demands, with significant reductions in sheep numbers over the past 20 years and concomitant increases in dairy cattle and deer. The census data (undertaken every 5 years) are crucial to the Tier II method for estimating CH_4 production, from livestock numbers, feed requirements, and estimated feed intakes. This Tier II inventory calculation is based on monthly measurements of animal requirements and feed dry matter (DM) intakes.[2,4]

Briefly, the ruminant population is defined in terms of dairy cattle, beef cattle, sheep, and deer (numbers of goats, horses, and swine are very low; Table 12.2). Each group is subdivided into categories based on farming systems, with monthly adjustment of numbers to account for births, deaths, and transfer between age groups. Productivity and performance data required to estimate feed intakes include average live weights of all categories, milk yields and composition from dairy cows, growth rates of all categories, and wool production from ewes and lambs. The ME content of diets consumed is measured and the DM intake determined from ME requirements for each population, using CSIRO algorithms.[5]

TABLE 12.2
Animal Numbers (3-year average), CH_4 Emission Rates, and Total Annual Emissions for New Zealand in 2001

Species	Numbers ($\times 10^6$)[a]	CH_4 /Head (kg)	Total CH_4 Emissions[b] tonne $\times 10^3$	%
Sheep	41.36	10.6	438.7	40.0
Dairy	4.98	74.7	372.5	33.8
Beef	4.54	56.0	254.0	23.0
Deer	1.55	20.9	32.7	3.0
Goats	0.17	8.9	1.5	0.1
Swine	0.35	1.5	0.5	0.0
Horses	0.08	18.0	1.4	0.1

Note: Data are calculated from census data, monthly feed requirements, estimated intakes, and methane emissions unit^{-1} intake[2].

[a] Adult equivalents.
[b] Excludes contribution from manure.

These data form the basis of the Tier II inventory, with current emissions (g CH_4 kg^{-1} DM intake) of 21.6 for adult dairy cattle, 20.9 for adult sheep, and 16.8 for sheep aged less than 1 year grazing pasture — 6.5, 6.3, and 5.1% of the gross energy (GE) intakes. The accuracy of methane emissions is given as ±50%, with a coefficient of variation of 23%.[2] The census data are accurate but concerns remain over the accuracy of predicted DM intakes and CH_4 emission $unit^{-1}$ DM intake (DMI).

Manure CH_4 emissions are low and are based on calculation of total animal manure production. Annual emissions from manure are calculated to be about 0.9 kg for cattle, 0.18 kg for sheep, and 0.37 kg for deer.[2]

12.3.2 NITROUS OXIDES

Pastoral agriculture is the source of most N_2O in New Zealand. Emission estimates have been revised[6] on the basis of the Inter Governmental Panel on Climate Change[7] and use default values of 0.0125 kg N_2O-N kg^{-1} N for N_2O from all origins (Tier I). Emissions are derived primarily from N in animal excreta (about 53% of total) and nitrogenous fertilizers (10%) as well as other direct and indirect (leaching, runoff, volatilization) emissions. Current research suggests N_2O-N losses kg^{-1} N of 0.007 and 0.003 are appropriate for dung and urine, respectively,[8] which is substantially lower than values used in calculations of inventory. All sheep, deer, beef, and most dairy cattle waste is deposited on pasture.

12.4 DEFINING MITIGATION

Methane emissions can be expressed in several ways:

- Gross emissions, which have significant meaning for inventory but little indication of the animals' performance or physiological status. Low emissions may be due to low performance, and vice versa.
- Expressions as a function of feed intake, for example, DMI or digestible DMI. This expression enables comparisons between feeds, but high intakes by animals consuming good-quality diets (with low CH_4 kg^{-1} DMI) may result in high gross emissions.
- Methane per unit of production. This appears to be a useful expression of "GHG efficiency," especially from a systems perspective because total emissions can be judged on the basis of performance. This is a good procedure providing emissions are totaled over a cycle of events, e.g., growth of a lamb from conception to slaughter, or annual milk production from dairy cows. This procedure is easily abused, for example, when expressing CH_4 $unit^{-1}$ milk production, because values will be low in early lactation when maintenance is a small proportion of energy intake (and the cow has lost weight) but high in late lactation as milk yield declines and the cow (and fetus) is gaining weight.
- Methane mitigation should be expressed in association with other GHG and economical scenarios. For example, feeding grains with forages will

lower CH_4 yields kg^{-1} DMI and CH_4 kg^{-1} milk production but large CO_2 emissions are associated with soil organic matter losses (from cultivation), use of fuel, fertilizers, harvesting, drying, and transport of grain. Furthermore, costly mitigation must not disadvantage producers in a competitive world economy.

Table 12.3 lists options for methane mitigation, with an indication of applicability, risk, and a timescale for commercial availability. Most consideration will be given to forages and feeding, constituent nutrients, animal management, variations among individuals, and the importance of a whole system analysis. These options can be applied in the short term with a high level of acceptability.

12.5 METHANE MITIGATION

Opportunities for methane mitigation[3,9–16] include short-, medium-, and long-term strategies (Table 12.3). Mitigation must also be economical, sustainable, and relatively inexpensive; persistent and high levels of methane production should not be viewed as an inevitable consequence of ruminant digestion. It can be reduced by 90% through daily administration of halogenated methane analogues[13] with minor effects on performance.[17] However, total elimination of methane production during digestion is unlikely to be sustainable, acceptable, or economical. Although halogenated methane compounds are potentially carcinogenic, less toxic alternatives for methanogen inhibition may become available and achieve consumer acceptance for registration and industry use.

Successful mitigation strategies can either lower production of the hydrogen substrate used for methane synthesis or increase available sinks for hydrogen disposal. Rumen bacterial degradation of fiber to acetate will inevitably release hydrogen ions and sinks must be available to prevent microbial inhibition.

Dairy cattle and feedlot animals provide excellent opportunities for mitigation because daily administration of methane suppressors, mitigators, or hydrogen "sinks/users" (acetogens) is practical and potentially cost-effective in animals producing high-value commodities. However, the majority of ruminants are raised under extensive grazing and mitigation can only involve occasional intervention, hence the attraction of vaccination against methanogens[18] or protozoa.

Animal management techniques to improve productivity may offer benefits to producers as well as lower methane emissions per unit of product (e.g., milk or live weight gain) but options will depend on government policies. For example, one solution is inclusion of grains and concentrates in ruminant diets to boost production; however, a full system appraisal of grain production, considering fertilizer, cultivation, fuel and other energy inputs, and consequent emissions of CH_4, N_2O, and especially CO_2 shows very high net GHG emissions per unit of ruminant production, compared to production from ruminants grazing pasture.[19] Any consideration of methane abatement should consider other GHG costs, economics, and environmental consequences of change.

TABLE 12.3
Options for Reducing Methane Emissions, in Total or per Feed Intake or per Unit Product from Ruminants Fed Forages

Technique	Application	Limitations	Consequences[a]	Potential Uptake
Short-Term Options				
Maintain forage quality	Medium-high fertility grazing	No limitations; require skilled management	Improved animal performance, must limit excess fertilizer use	High
Feed legumes/herbs, high-quality grasses	All situations depending on species	Costs of establishment and maintenance lower yields could lower profitability	Improved animal performance but more agronomic care needed	Moderate
Incorporate condensed tannin into diet	Widespread, especially with lotuses, sainfoin	Lower yield and persistence except lotus in low fertility	Very good animal performance, 13–17% reduction in methane and lower N_2O emissions	Moderate
Specific lipids	Currently limited to dairy unless expressed in forage plants	Cost-effectiveness	May affect product flavor	High with incentive
Balance rations to meet animal needs	Systems involving supplementary feeding	Requires nutritional knowledge and advice	Improved performance from high producers. Could lessen N_2O emissions by lowering N intake	Moderate
Select high-producing animals	Normal practice	High producers require good feeding and management	Lower stock numbers, increased profitability	High
Optimal farm management	Widespread but requires good skills	Depends on commodity prices; need consultant advice	Potential for high profitability	Moderate
Medium-Term Options				
Selection of low methane producing animals	Widespread if trait is heritable	None known but low CH_4 producers may only apply to some diets	Unlikely to have detrimental consequences	High with incentive[b]

TABLE 12.3 (continued)
Options for Reducing Methane Emissions, in Total or Per Feed Intake or Per Unit Product from Ruminants Fed Forages

Technique	Application	Limitations	Consequences[a]	Potential Uptake
Use of ionophores	Widespread if viable	Current data show inconsistent responses, variable persistence with forage diets	If viable, an added benefit is protection from bloat and possible improved feed conversion	Low to medium
Probiotics	Dairy, unless available as slow release	Minimal evidence of efficacy *in vivo*	Unknown	Unknown
Halogenated compounds	Could be widespread if in slow-release form	Need approval and verification of persistence	Consumer avoidance of products	High with incentive
Acetogens	Dairy cows	Require daily administration	Responses not defined; excess acetate will not benefit high-producing ruminants fed forage	Low unless incentive
Defaunation	Moderate, depending on diet	Current technology risky, a vaccine would help.	Beneficial for animals fed poor forage	Moderate if safe
High-efficiency animals	Widespread	Require selection of animals with efficient nutrient utilization	Selections may be feed specific	Moderate
Long-Term Options				
Vaccines — methanogens	Widespread	Good opportunities hampered by lack of funding	Potential for improved animal performance	High
Vaccines — protozoal	Moderate	Probably minimal	OK when poor feed is available	Moderate
Specific methanogen inhibitors (HMG-S-CoA[c] and Phage)	Widespread	Depends on specific inhibition of methanogens	Improved performance if intakes maintained	High with incentive

[a] Consequences refer to the animal or environment; a net reduction in CH_4 kg^{-1} feed or product is implied.

[b] If performance is not enhanced an incentive may be required to use these materials.

[c] HMG-CoA, hydroxymethyl glutaryl-S-CoA.

12.6 RELATIONSHIP BETWEEN DIET COMPOSITION AND METHANOGENESIS

The analysis of methane data by Blaxter and Clapperton[9] has served as reference for effects of intake, digestibility, feed type, and animal species on CH_4 emissions unit[-1] feed intake. These data appear to be based on dried feeds but relationships between methane (as a percentage of GE) and digestible energy (DE) content or level of intake were not consistent across dietary types. For example, there was no relationship between feed quality (DE content) and energy loss to methane for concentrate–roughage mixtures fed at maintenance, despite a significant correlation for dried roughages. These details appear to have been overlooked by some researchers.

A more recent analysis[20] failed to demonstrate any relationship ($r^2 = 0.052$) between observed GE loss to CH_4 (range 2.5 to 11.5%) and DE of the diet (range 50 to 87% of GE). These authors also showed a very poor relationship ($r^2 = 0.23$) between the Blaxter and Clapperton[9] predictions of CH_4 losses from beef cattle fed a diverse range of diets and actual values.

An alternative equation derived from trials with dairy cows fed mixed rations,[21] based on intakes of hemicellulose, cellulose, and nonfiber carbohydrate (NFC), enabled 67% of the variance in predicted methane production to be explained:

$$CH_4 \text{ (MJ day}^{-1}) = 3.406 + 0.510 \text{ (NFC)} + 1.736 \text{ (hemicellulose)} + 2.648 \text{ (cellulose)}$$

where NFC (DM less fiber, crude protein (CP), ash, and lipid), hemicellulose, and cellulose are daily intakes (kg). The prediction was improved by using digestible NFC, hemicellulose, and cellulose intakes, explaining 74% of the variance, but measurements of digestibility are not always available.

These authors[21] concluded that methane production by adult cattle at maintenance could be predicted from dry matter or total digestible carbohydrate intake, but accurate prediction at higher intakes, typical of lactating cows, requires the type of dietary carbohydrate to be determined. The intercept of equations based on fiber and digestible fiber did not pass through zero, which emphasizes the empirical nature of the relationship and precludes expression on the basis of GE intake.

Complex equations developed for lactating dairy cows[22] did not improve predictions over those based on carbohydrate fractions,[21] and Wilkerson et al.[23] concluded that estimates based on cellulose, hemicellulose, and NFC provided the highest correlation with actual methane emissions, and had the lowest errors. Use of either intakes or digestible intakes of carbohydrate fractions provided similar levels of accuracy for predicting energy loss to CH_4.

Prediction of emissions from animals fed contrasting diets are complicated by differences among individuals (e.g., References 24 through 26; Figure 12.1). There is also some evidence that increasing the proportion of concentrates in a diet will increase the variation between individuals.[9,14,27]

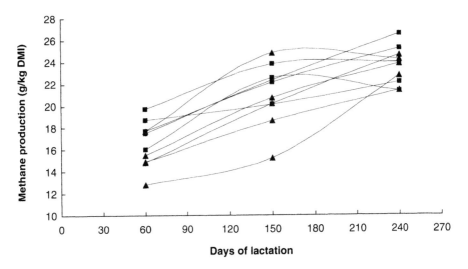

FIGURE 12.1 Methane production (g kg^{-1} dry matter intake) from five cows with a New Zealand Friesian genotype (■) and five with a North American/Dutch genotype (▲) genotype grazing pasture and measured at 60, 150, and 240 days of lactation. (Waghorn, Unpublished data.)

12.7 METHANE EMISSIONS FROM RUMINANTS FED FRESH FORAGES

12.7.1 NEW ZEALAND MEASUREMENTS

New Zealand research has focused on measurement of methane emissions from sheep and cattle fed fresh forage diets (usually perennial ryegrass-dominant pasture) throughout the season and with animals differing in age and physiological status. Four data sets have been analyzed using multiple regression to define relationships among treatment means (CH$_4$ kg^{-1} DMI) and linear combinations of dietary components — soluble sugars, NFC, CP, ash, lipid, condensed tannin (CT), neutral detergent fiber (NDF), acid detergent fiber (ADF), hemicellulose (H), and cellulose (C). Analyses have been undertaken for sheep fed ryegrass-based pasture (15 data sets), sheep fed legumes and herbs alone or in mixtures (12 data sets), lactating Friesian cows fed pasture (12 data sets), and lactating Friesian cows fed a range of diets including pasture ($n = 22$).

Perennial ryegrass feeding with sheep included *ad libitum* grazing[24,25,28,29] and indoor feeding[30,31] with forage quality ranging from immature to mature (CP 29 to 11%, NDF, 36 to 51%). Methane emissions ranged from 13 to 26 g kg^{-1} DMI (Table 12.4; 3.8 to 7.6% of GE). Correlation coefficients (r^2) between CH$_4$ kg^{-1} DMI and NFC, NDF, and ADF concentrations were 0.47, 0.28, and 0.58, respectively. Multiple regression using the criteria developed by Moe and Tyrrell[21] for cattle showed only 51% of the variance in methane yield was explained by NFC, hemicellulose (H), and cellulose (C) concentrations in the DM:

TABLE 12.4

Composition, Digestibility, and Methane Production from Sheep Fed a Range of Legumes and Herbs[30]

Forage	DM Composition (%)				DM Digestibility	Methane (g kg^{-1} DMI)
	CP	NFC[a]	Hemi-cellulose	Cellulose	(%)	
Lucerne	24.0	30.4	2.9	18.1	71.3	20.6
Sulla	17.5	41.5	0.0	10.3	72.8	17.5
Sulla/Lucerne[b]	25.9	36.3	0.6	12.5	71.1	19.0
Chicory	12.3	58.8	0.0	4.0	79.3	16.2
Red clover	24.4	28.6	10.0	15.4	75.6	17.7
Sulla[c]	19.7	37.9	2.8	11.0	63.2	17.5
Chicory/sulla[b]	15.5	46.8	12.0	0	71.1	16.9
Chicory/red clover[b]	19.5	42.7	2.1	8.1	76.5	19.7
White clover	26.9	31.2	6.3	11.5	78.8	12.9
Lotus	26.4	23.6	8.4	12.4	70.0	11.5
Lotus + PEG[d]	26.4	23.6	8.4	12.4	76.9	13.8

[a] NFC, non-fiber carbohydrate.

[b] Mixtures are 50:50, DM basis.

[c] Mean of two trials each including sulla, one year apart.

[d] PEG, polyethylene glycol, preferentially binds to and inactivates tannin.

$$CH_4 \text{ (g kg}^{-1}\text{ DMI)} = 0.468 \text{ NFC} - 0.075 \text{ H} + 0.737 \text{ C} \quad r^2 = 0.51$$

A similar analysis was undertaken for legumes and herbs fed to sheep held indoors as single components or mixtures (Table 12.4). These forages usually yielded lower CH_4 emissions than ryegrass-dominant pastures ranging from 12.0 g kg^{-1} DMI for white clover to 20.6 g kg^{-1} DMI for alfalfa. There were no significant correlations between methane production and feed components and the equation incorporating NFC, NDF, and ADF accounted for 18% of the variance between diets (NS).

Analyses of methane production from cows were also compared with diet composition. A total of 12 data sets were based on perennial ryegrass given as a sole diet, either grazing or cut and fed indoors, and a further 6 data sets included mixtures of perennial ryegrass pasture fed with maize or pasture silage or fresh white clover. Two trials involved *Lotus corniculatus* and sulla (*Hedysarum coronarium*) fed as sole diets. Analyses of either the 12 ryegrass data sets or the 22 data sets including pasture, pasture with legumes, or silage did not demonstrate any significant relationships between CH_4 emission kg^{-1} DMI for any component or combination of components in the diets.

In summary, legumes and herbs usually resulted in lower CH_4 emissions from rumen fermentation than ryegrass pastures, but the chemical composition of the feed eaten, including the concentration of condensed tannin, did not explain variations in CH_4 production. Chemical composition explained about 50% of the variance in emissions from sheep fed perennial ryegrass-dominant pasture but did not explain the variance in methane production from cow trials, even though indoor measurements enable an accurate determination of feed eaten. These data suggest a poor understanding of methanogenesis in sheep and cattle fed fresh forages, exacerbated in some (but not all) situations by difficulty in determining intakes. Research needs to revisit the physiology of digestion to better explain the formation of methane during digestion.

12.7.2 PASTURE METHANE MEASUREMENTS OUTSIDE NEW ZEALAND

Although the focus on fresh forages has been with New Zealand measurements, data are available from Australia, the U.K., Canada, the U.S., and elsewhere. Data from cattle research do little to clarify the confusion associated with our analyses. For example, Boadi et al.[32] reported CH_4 yields of 15.5 and 27.3 g CH_4 kg^{-1} DMI (4.7 and 8.4% of GE) from steers grazing alfalfa/brome grass pastures containing 50 and 54% NDF and 19.2 and 17.9% of CP the DM, respectively. Boadi and Wittenberg[33] reported CH_4 emissions of 6.0, 7.1, and 6.9% of gross energy intake (GEI) from beef and dairy heifers fed *ad libitum* legume and grass hays containing 41.8, 58.1, and 68.8% NDF in the DM, respectively. Methanogenesis was not related to feed quality. These values are higher than those reported by McCaughey et al.[34] for steers grazing alfalfa/meadow-brome grass pastures (4.1 to 5.2% of GEI) with widely differing composition (31 to 64% NDF) but similar to a later trial with grazing cattle.[35] This range of values and the apparently minimal relationship to fiber and other components of forage highlight the need to better understand processes affecting methanogenesis in ruminants grazing pasture.

12.8 CONDENSED TANNINS AND METHANOGENESIS

Waghorn et al.[30] reported a 16% depression in CH_4 emissions kg^{-1} DMI (from 13.8 to 11.5 g kg^{-1} DMI) due to the presence of CT in a diet of *Lotus pedunculatus* fed to sheep housed indoors. The sheep were fed at about 1.4 × maintenance to ensure minimum selection of plant components (leaf vs. stem) and given a twice-daily oral administration of polyethylene glycol (PEG), which preferentially binds to and inactivates CT. The PEG does not affect other aspects of digestion, so daily dosing effectively creates a CT-free lotus, and enables evaluation of CT per se. More recently, Woodward et al.[36] carried out a similar trial with cows fed *Lotus corniculatus*, containing a lower concentration of CT in the DM (2.62 g 100g^{-1}) compared to 5.3% in the *L. pedunculatus* fed to sheep. This trial comprised four treatments, ryegrass/white clover without and with PEG, and *L. corniculatus* without and with PEG. Methane was 24.2, 24.7, 19.9, and 22.9 g kg^{-1} DMI for the respective treatments (Table 12.5). The CT in lotus reduced methane kg^{-1} DMI by 13% ($p < 0.05$) and the cows fed lotus produced 32% less methane kg^{-1} milksolids (fat + protein) compared to those fed good-quality ryegrass.

The difference in GE loss to CH_4 for lotus vs. ryegrass (Table 12.4) enables a calculation of energy potentially available for milk production. For cows consuming 15 kg pasture DM day^{-1}, there would be 64 g less CH_4 day^{-1} from the lotus diet, which if absorbed as VFA, could contribute 0.6 kg milk or 48 g milksolids day^{-1}.

TABLE 12.5
Effect of Diets Containing Condensed Tannins on Milk and Methane Production by Holstein-Friesian Cows in Late Lactation[36,37]

	Ryegrass		Lotus corniculatus		SED
		+ PEG[a]		+ PEG[a]	
Trial 1					
DM intake (kg cow^{-1} d^{-1})	14.9	14.9	17.4	17.1	0.46
Milk (kg cow^{-1} d^{-1})	18.5	19.0	24.4	22.1	0.70
Milk protein (%)	3.59	3.56	3.63	3.61	0.05
Methane					
Total (g cow^{-1} d^{-1})	360	368	343	392	12.40
g kg^{-1} DMI[b]	24.2	24.7	19.9	22.9	0.78
g kg^{-1} milksolids[c]	250	244	171	216	10.6
% of GEI[d]	7.50	7.66	5.98	6.89	
		—	**Sulla**	—	
Trial 2					
DM intake (kg cow^{-1} d^{-1})	10.7	—	13.1	—	0.6
Milk (kg cow^{-1} d^{-1})	8.4	—	11.2	—	0.35
Milk protein (%)	3.76	—	4.05	—	0.06
Methane		—		—	
Total	260	—	254	—	24.7
g kg^{-1} DM 1	24.6	—	19.5	—	1.6
g kg^{-1} milksolids	327		243		24.7
% of GEI	7.2		6.1		0.4

[a] PEG, polyethylene glycol to remove effects of condensed tannins.
[b] DMI, dry matter intake.
[c] Milk solids is fat + protein.
[d] GEI, gross energy intake.

The lower CH_4 losses attributed to CT are supported by lower CH_4 production unit^{-1} feed intake from cows fed sulla containing 2.7% CT in the DM vs. ryegrass pasture.[37] Emissions were 19.5 vs. 24.6 g CH_4 kg^{-1} DMI for the respective feeds (6.1 vs. 7.2% of GEI). Puchala et al.[38] have also reported low CH_4 emissions from goats fed Serecia lespedeza (*Lespedeza cuneata*) containing 6% CT in the DM, compared to grass dominant forage (6 vs. 14.1 g kg^{-1} DMI for the respective diets).

Mechanisms for CT inhibition of methanogenesis are largely hypothetical. Animal trials have shown that the CT in temperate legumes containing CT protect dietary protein from rumen degradation and can increase absorption of essential amino acids from the intestine, to give very good animal performance.[39,40] CT inhibit microbial activity *in vitro*[41] and *in vivo*[42,43] but proportions of VFA are unchanged, so there will be a similar yield of hydrogen with or without CT. Mechanisms by which polyphenolics affect a reduction in methanogenesis are speculative.

12.9 ANIMAL VARIATION IN METHANOGENESIS

Within groups of sheep or cattle fed fresh forages, about 10% have very high and 10% low methane emissions (per kg DMI) and the difference between the two groups is about 40%. For example, Pinares-Patino et al.[25] showed mean methane production from four highest and four lowest producing sheep (selected from a random group of 20 animals) over a 4-month period was 3.75 vs. 5.15% of GEI. Earlier reports[24] found 86% of variation in methane production by sheep consuming 900 to 1700 g DM day^{-1} was due to animal variation and only 14% was attributable to diet. Ulyatt et al.[44] summarized data from six trials involving either sheep or cattle fed forages and showed that 71 to 95% of variation between days was attributable to animals even though intakes and composition of each diet were relatively constant.

The impact of genotype was highlighted in a trial involving New Zealand Friesian (NZHF) and Overseas Holstein (OSHF) cows fed either pasture or total mixed rations (TMR; Table 12.6). The OSHF genotypes produced significantly less CH_4 kg^{-1} DMI when fed both TMR and pasture diets at both 60 and 150 days of lactation.[26] Genotype differences had disappeared by day 240. Individual cow data, summarized in Figure 12.1, demonstrate a persistent high or low methanogenesis for some but not all cows fed pasture. A similar variation between individuals was evident for TMR diets fed to cows.

Animal differences in methane yield kg^{-1} DMI provide an ideal opportunity for selection of low methane producers, providing the trait is heritable. Pinares-Patino

TABLE 12.6
Effect of Cow Genotype (overseas Holstein, OSHF vs. New Zealand Friesian, NZHF) on Methane Production When Grazing Pasture (five cows per treatment)[26]

	Days of Lactation					
	60		150		240	
	Mean	sd	Mean	sd	Mean	sd
DM intake (kg d^{-1})						
NZHF	17.2	1.70	17.0	0.80	15.0	2.44
OSHF	17.7	1.58	17.6	3.15	16.3	1.75
Milk production (kg d^{-1})						
NZHF	26.5	1.69	19.5	1.78	14.7	1.74
OSHF	27.9	3.72	20.0	3.69	16.1	4.46
CH_4 production (g d^{-1})						
NZHF	308	19.7	376	20.4	353	33.5
OSHF	267	33.2	345	59.8	379	33.5
CH_4 g kg^{-1} DMI						
NZHF	18.0	1.41	22.2	1.32	23.8	2.15
OSHF	15.1	1.76	19.9	3.48	23.4	1.30
CH_4 g kg^{-1} milk						
NZHF	11.7	1.01	19.4	1.88	24.3	3.62
OSHF	9.7	1.38	17.4	1.36	24.9	6.44

et al.[25] showed sheep with high CH_4 yields had larger rumen volumes, a slower particulate outflow rate, higher fiber digestibility, and longer retention times than sheep with low CH_4 kg^{-1} DMI. Methane yield was best predicted as a function of particulate fractional outflow rate, organic matter intake (g kg $LW^{-0.75}$) and molar proportion of butyrate ($r^2 = 0.88$). Smuts et al.[45] suggested that rumen retention time was a heritable characteristic in sheep.

Differences between animals may be affected by salivation, feed communition (or eating rate), as well as rumen pool size, turnover, and outflow. Animal effects on rumen microflora have been demonstrated by widely differing *in sacco* degradation rates and contrasting populations of cellulolytic bacteria.[46,47] Variation in susceptibility to bloat appears affected by salivary proteins and bloat prone cattle produce bloat prone offspring.[48,49] This capacity to affect their microflora offers potential for development of antimethanogen or antiprotozoal vaccines.

12.10 MANAGEMENT TO MITIGATE METHANE IN GRAZING ANIMALS

Effective management to mitigate methane could be viewed in terms of animal productivity vs. animal methane emissions. Expression could be on an annual basis to avoid short-term bias; for example, cows grazing ryegrass pastures produced 11.7, 19.4, and 24.3 g CH_4 kg^{-1} milk at day 60, 150, and 240 of lactation.[26] The difference in emissions was largely due to a live-weight loss contributing energy to milk synthesis in early lactation and use of dietary energy to restore live weight in late lactation. A similar scenario applies to sheep, with very high CH_4 emissions associated with wool growth (typically 10 to 12 g day^{-1}) in adult animals, but a lesser emission cost associated with growing lambs and reproduction.

Mitigation can be achieved by minimizing maintenance costs as a proportion of feed intake and maximizing the productive worth of livestock. High intakes of high-producing animals dilute their maintenance cost and lower the methane emissions per unit of production. This will be best achieved by offering high-quality diets to animals of high genetic merit and imposing good livestock and pasture management practices.

These effects are illustrated[3] for 30-kg lambs growing at 100, 200, and 300 g day^{-1} with methane emissions of 166, 115, and 98 g kg^{-1} live-weight gain, respectively. Comparative values for 450-kg grazing dairy cows producing 12, 20, or 24 kg milk day^{-1} were 17.2, 13.6, and 12.7 g CH_4 kg^{-1} milk. The methane emissions associated with production increased from 49 to 61 and 66% for the respective treatments.

Animal performance can be improved by selection for a high metabolic efficiency or by using rumen modifiers to alter products of digestion. Any factor able to improve feed conversion efficiency will lower CH_4 emissions $unit^{-1}$ production. However, farmers need to achieve a balance between increasing efficiency of feed utilization and the efficiency of pasture utilization.

12.11   FEED ADDITIVES

There is extensive literature concerning the impact of feed additives on methanogenesis (e.g., References 12, 15, and 50 through 52), so a brief summary of viable
options is presented here. Feed additives may be hydrogen sinks, influence the rumen
microflora to lower hydrogen production, or influence the methanogenic archaea
directly. Antibiotics, bacteriocins, and probiotics seem to have short-term
effectiveness[15] and all need to be evaluated *in vivo*. Consistent responses are essential
for commercial application. Products must be acceptable to consumers and increased
use of antibiotics is likely to be restricted by legislation.

12.11.1   Oils

Oils offer a practical approach to reducing methane in situations where animals can
be given daily feed supplements, but excess oil is detrimental to fiber digestion and
production. Oils may act as hydrogen sinks but medium-chain-length oils appear to
act directly on methanogens and reduce numbers of ciliate protozoa.[53] These
researchers reported a methane suppression of 10 to 26% with a variety of oils given
to sheep, although these values were about half of their effect *in vitro*.

A 27% reduction in methane emission kg^{-1} DM intake has been demonstrated
at this laboratory from lactating cows fed pasture and receiving a daily dose of 500
ml of sunflower/fish oil mixture (Woodward et al., unpublished). In contrast, Johnson
et al.[54] found no response to diets containing 2.3, 4.0, and 5.6% fat (cottonseed and
canola) fed over an entire lactation.

12.11.2   Ionophors

Ionophors (e.g., monensin) improve the net feed efficiency of cattle fed total mixed
rations by increasing the proportion of propionate:acetate from rumen fermentation so
that daily gain is maintained but with 5 to 6% lower feed consumption.[55] However,
responses to monensin by cows fed forage diets are usually low, often variable, and
sometimes there are no performance gains in either feed utilization or milk production.[56]

Monensin is available in a slow release (100 day) formulation and is used to
reduce the risk of bloat in cattle and can lower methane emissions. Clark et al.[57]
reported emissions of 158 and 179 g CH_4 day^{-1} from cows fed ryegrass-based pasture
with and without monensin treatment. Intakes were not affected by monensin and
there was a significant reduction in methane kg^{-1} milk solids (milk fat + protein)
for monensin (375 g kg^{-1}) vs. control (420 g kg^{-1}; $p = 0.05$) cows. In that study the
monensin treatment continued to lower methane emission after 60 days, but persistence of methane suppression by ionophors is variable[14,58,59] and often not sustained.[13]

12.11.3   Removing the Protozoa (Defaunation)

Hegarty[60] reviewed the impact of total or partial defaunation to improve ruminant
performance and lower methane emissions. Improved performance has been associated
with increased microbial flow to the intestine (protozoa consume bacteria) and
increased proportions of propionate (protozoa produce acetate, butyrate as well as

hydrogen gas). There is also a close (symbiotic) association between protozoa and methanogens, and defaunation is likely to lower methane emissions by 10 to 30%.

Defaunation is somewhat risky, and is frequently incomplete, with a return of protozoa within weeks or months even if defaunated animals are kept separate from faunated livestock. However, even partial defaunation is likely to reduce CH_4 and benefit animal performance, especially when grazing diets with a medium-low protein content. Australian research is investigating an antiprotozoal vaccine[61] that would have wide applicability and minimal toxicity for ruminants.

12.12 TARGETING METHANOGENS

Halogenated methane analogues can be very potent methane inhibitors, including chloralhydrate, chloroform, bromochloromethane, and bromoethanesulfonic acid. CSIRO (Australia) has patented an antimethanogen comprising bromochloromethane in a cyclodextrin matrix. In a trial with steers, Tomkins and Hunter[17] showed dose rates of 0, 0.15, 0.3, and 0.6 g 100 kg^{-1} live weight reduced methane from 3.9 to 1.0, 0.6, and 0.3% GEI. Dry matter intakes for the respective treatments were 6.2, 7.4, 5.6, and 5.5 kg. In a separate trial average daily gain (1.5 kg day^{-1}) was unaffected by a twice daily dose of 0.3 g kg^{-1} DMI for 85 days of treatment.

In a review of data from sheep and cattle trials involving administration of halogenated methane compounds,[62] intake reduction (0 to 13%) was minor and did not always occur. In most *in vivo* studies feed conversion efficiency was increased by 0 to 11% and live-weight gain tended to be 5% lower, due to reduced intakes. They concluded that a partial inhibition of methanogenesis could have beneficial effects on animal production, especially if acetogens could utilize the hydrogen arising from fermentation.

The unique membrane lipids of methanogens and other *Archaea* contain glycerol linked to long-chain isoprenoid alcohols. A key precursor of isoprenoid synthesis is mevalonate formed by reduction of hydroxymethyl glutaryl–S-CoA (HMG-CoA). The HMG-CoA reductase enzyme (which enables the formation of mevalonate) is a target of drugs used to lower cholesterol in humans, and these compounds (lovastatin; mevinolin) are potentially able to inhibit growth of methenogenic archaea in the rumen.[63] Other bacteria do not contain HMG-CoA and should be unaffected by these inhibitors. These authors demonstrated an *in vitro* inhibition of *Methanobrevibacter* strains using HMG-CoA inhibitors without affecting a range of rumen cellulolytic and other bacteria. The concentration of inhibitor is equivalent to about 400 mg 100 kg^{-1} rumen content, but unlike halogenated methane analogues, lovastatin is prescribed to humans (i.e., safe) and it is relatively inexpensive.

Other specific targets for methanogens include phage and vaccines.

12.12.1 VACCINE

A vaccine developed from a three-methanogen mixture produced a 7.7% reduction (kg^{-1} DM) in methane emissions from sheep ($P = 0.051$) despite that only one antigen was effective against the methanogenic species in the sheep. The vaccine[18] was much more effective than the seven-methanogen mix tested previously and was able to increase saliva and plasma antibody titers by four- to ninefold over the seven-methanogen mixture. Successful elevation of antibody titers in saliva and a significant reduction in methane emissions offer real potential for a widespread application to ruminants in all environments. At present, vaccines do not have sufficient efficacy for commercial use and funding has recently been curtailed.

Opportunities through rumen additives, defaunation, and specific compounds targeting methanogens provide several routes for reducing methane production. However, these agents have not addressed the inevitable production of hydrogen from fermentation of fiber. Ruminants are able to utilize fiber because of their microflora and hydrogen production is an unavoidable consequence. Excess hydrogen accumulation will inhibit microbial growth, but acetogens offer an opportunity for production of acetate as well as removing accumulated hydrogen. Acetogens are present in moderate concentrations in the digestive tract of horses, llamas, and buffalo (10^4 to 10^5 ml^{-1}) but values for sheep and cattle have been very low.[64] Acetogens require a higher partial pressure of hydrogen to become active[11,60] and could become important hydrogen users in the event of methanogen suppression.

12.13 AGRONOMY AND COMPLEMENTARY FEEDS

The commonly held view is that fibrous, low-quality pastures yield a higher proportion of CH_4 GEI^{-1} than good-quality, low-fiber forages. This is probably true but other factors have contributed a great deal of variability, so that Johnson and Johnson[20] found no relationship between CH_4 and digestible energy (both %GEI) for cattle ($r^2 = 0.05$) and results presented in this paper were unable to account for between-trial variations in CH_4 production on the basis of diet composition.

These findings present a serious challenge to researchers attempting to create an inventory or account for effects of diet on energy losses to methane. Good relationships between methane production and animal or forage factors (for example, rumen and outflow rates) have been obtained within trials.[25] Pinares-Patino et al.[65] also demonstrated consistent methane emissions unit^{-1} digestible NDF intake (53 g kg^{-1}) from Charolais cattle grazing timothy (*Phleum pratense*) grass of widely different quality (4 to 31% CP in the DM).

At this laboratory we have investigated the effect of substituting ryegrass with increasing amounts (0 to 60%) of white clover, for dairy cows. White clover diets resulted in very low CH_4 emissions when fed to sheep (12.5 g kg^{-1} DMI; Table 12.4) but in this trial a 60% substitution reduced emissions by only 16% (21.7 to 18.1 g kg^{-1} DMI; $P = 0.004$). In contrast, substitution of pasture with maize silage (to 38% of DMI) increased methane emissions by 16% (16.3 to 19.0 g kg^{-1} DMI; $P = 0.14$). Significant reductions in NDF content and increases in starch for the respective diets had minor effects on net methane production. Diet is able to affect methane production but greatest benefits may be from lowering CH_4 unit^{-1} product when high-quality diets are fed.

12.14 NITROUS OXIDE EMISSIONS AND ABATEMENT

Nearly all N_2O emissions arise from agricultural soils in New Zealand[66] and 85% of these are grazed by livestock. Emissions of N_2O arise from both reduction of soil nitrates (denitrification) and also from oxidation of ammonium to nitrite and nitrate. The extent and type of processes are determined mainly by mineral N availability and aeration (or water logging) of soils.[8]

The processes are as follows:

$$\text{Denitrification}: \quad \underset{\text{nitrate}}{NO_3^-} \;\to\; \underset{\text{nitrate}}{NO_2^-} \;\to\; \underset{\text{nitric oxide}}{NO} \;\to\; \underset{\text{nitric oxide}}{N_2O} \;\to\; \underset{\text{nitrogen}}{N_2}$$

$$\text{Nitrification}: \quad NH_4^+ \;\to\; NH_2OH \;\to\; [HNO] \;\to\; \underset{\downarrow}{NO_2^-} \;\to\; NO_3^-$$
$$N_2O$$

In general, the proportion of soil N released as N_2O vs. N_2 increases as nitrate concentration increases especially in saturated, anaerobic soil conditions. Mitigation is achieved by either reducing soil N availability (less inputs as fertilizer, urine, dung), limiting water saturation by provision of drainage, and especially by minimizing treading damage (pugging) in wet conditions. Application of lime to raise soil pH can also lessen N_2O emissions. A brief overview of emissions, mitigation options, and the extent to which emissions may be reduced[67] is presented with emphasis for grazing animals in Table 12.7.

12.14.1 MITIGATION OPTIONS

Improved N fertilizer management can be achieved by application on the basis of requirement. Soil testing and skilled management will enable the correct amount of N fertilizer to be applied to best meet plant requirements and minimize wastage. Controlled release fertilizers and those containing nitrification inhibitors (e.g.,

TABLE 12.7
Nitrous Oxide Mitigation Options and Potential Reductions for Dairy Farms in New Zealand[67,70]

Mitigating Option	Approximate Decrease in N_2O (%)
Improve performance, lower numbers	4–5
Alter diet to reduce N contents or enable better N capture by rumen bacteria and production	7–14
Improve cow winter management to protect pastures	6–7
Improve spread of excreta from sheds and feed pads	4–5
Liming to raise soil pH	4–5
Improve fertilizer management	6–8
Improved drainage and lessen compaction	5–10

dicyandiamide[68]) can lessen losses and improve plant N utilization, especially with strategic placement beneath the surface, or on the basis of need using global positioning systems technology. These technologies will lower fertilizer use and improve profitability as well as reduce environmental pollution from N runoff to streams and waterways, volatilization, and N_2O emissions.

In confined animal systems, manure (feces and urine) management has important consequences for GHG emissions, but New Zealand management is through fertilizer and dietary manipulation because animals graze outdoors year round. Apart from reducing stock numbers, viable options for limiting N_2O emissions include increasing productivity per animal, management to lessen pasture and soil damage, and lowering dietary (and therefore waste) nitrogen concentrations. Plants containing condensed tannins alter digestion and repartition N from urine (with high N_2O emissions) toward feces.

Nutritional management offers good opportunities to mitigate N_2O, especially in dairy farming where pasture supplementation (e.g., with maize silage) is becoming standard practice. Animal management to minimize pasture damage is also becoming an attractive option for farmers and this affects N_2O emissions. All of these changes to traditional farming practice are driven by acceptable prices for farm commodities and a general desire for both cost-effective agriculture and environmental sustainability by most farmers.

Cropping and irrigation play a small but increasingly significant role in New Zealand agriculture, and good water management to avoid deficits and excess will minimize N_2O losses as well as make best use of irrigation water. Cultivars either requiring less N or making better use of applied N will minimize losses to N_2O. However, many forage species are dependent on high levels of fertility (N, P, S, K) to achieve the performance claimed by breeders and marketers.

Nitrogen fixing forages (such as white clover) used to be the principal source of N for grasses in New Zealand pasture, but the advent of relatively inexpensive urea has contributed to increased N_2O losses. Clovers remain an important component of pastures, but urea application in early spring provides a rapid and early grass growth to meet needs of dairy cows and of lambing ewes in some regions. The combination of N fixation and urea fertilizer has resulted in high concentrations of dietary N (often in excess of 4% of dietary DM) and a large amount is voided in urine and feces.

12.14.2 Animal Management and Feeding

A major aspect of N_2O research concerns measurement of N_2O emissions from a range of soil types, water contents, and from dung and urine to better predict emission. Saggar et al.[69] emphasized the impact of uneven deposition of excreted N, with low emissions in dry periods and high values in winter. Dairy grazed pastures yielded about five times as much N_2O as those grazed by sheep. Saggar et al.[69] consider the IPCC[7] default methodology underpredicts urinary losses from dairy pastures and overpredicts losses from sheep urine. N_2O emissions have been defined for a range of soil types[8] and in association with drainage using sheep and cow urine.[70] They have demonstrated a range from 0.3 to 2.5% of N loss to N_2O for cow urine with lower values for sheep. Values for dung are half (or less) of those for urine.

These variations emphasize the difficulty in attaining an accurate, predictable, and defensible inventory of N_2O from grazing animals. However, dietary manipulation does offer a viable option to lessen N_2O. Typical spring diets for all ruminants contain 22 to 29% crude protein. This far exceeds optimal or desirable concentrations for ruminant nutrition, but extensive degradation of protein by rumen microflora causes a high loss of protein to ammonia, which is absorbed and excreted in the urine. Methods for lowering the protein (N) content of the diet, without inducing N limitations for performance, include use of forages containing lower N concentrations (e.g., maize silage) to be fed with pasture, selecting forage species with a slower rate of protein degradation (e.g., containing condensed tannins), or feeding forages with a higher proportion of nonstructural carbohydrates (high-sugar grasses). The caveat to all of these options is that the species must be competitive once sown, highly productive, disease resistant, and persistent.

From a nitrogen viewpoint, maize grown for silage offers good advantages as a stock feed. Clark et al.[67] calculated that 1000 kg N fertilizer applied to a maize crop would produce about 100 tonnes of DM. This represents a much more efficient N capture compared to the response from pasture to urea N application. When pasture is highly productive, the marginal response to N application is low and losses to leaching, volatilization, and from animal waste is high. Maize silage can complement spring pasture for cows and reduce the amount of N deposited in urine by about 30% compared to a pasture diet. Use of N for maize silage production will reduce both fertilizer and urinary N inputs and outputs compared to an all-grass system. Cow performance will be maintained, but a whole systems analysis would indicate high CO_2 emissions associated with maize production.

Table 12.7 summarizes the impact of nutrition and other forms of intervention on N_2O emissions for dairy cattle. Impacts on sheep and beef industries are likely to be less, because farms are usually less fertile and have a hilly terrain, so fertilizer inputs will be lower and there is less likelihood of saturated soils having treading damage.

12.15 WHOLE-FARM SYSTEMS

Concern about individual vs. all GHG emissions resulted in a partial life cycle analysis of emissions from a conventional New Zealand pasture-based (with silage supplements) dairy farm and one in which total mixed rations were fed.[19] This analysis assumed typical dairy herd sizes (250 cows) but only estimated CH_4 and CO_2 emissions over an entire lactation. Inputs to the TMR[71] were based as far as possible on grains, silages, and forages grown in New Zealand with appropriate use of herbicides, fuel, cultivation, and fertilizer. Protein supplements (fishmeal, soy and cottonseed meals) were imported and calculations made for production costs with adjustments where crops yielded multiple products (e.g., cotton fiber as well as meal) to achieve a fair distribution of GHG costs across products.[19] Inputs to the pastoral system included costs of renovation (every 15 years) with a maize silage crop grown on cultivated pasture prior to re-establishment.

Milk production and cow data[71] were from the same animals used previously to measure methane emissions from both pasture (Table 12.6) and TMR.[26] Inputs and emissions for this model are summarized in Table 12.8.

Principal findings were a higher intake of cows fed TMR compared to pasture, a doubling of milk production, and 58% increase in CH_4 emissions from cows fed the TMR ration. When expressed in terms of milk production, TMR yielded significantly less methane (19.5 g kg^{-1} milk) than pasture (24.6 g kg^{-1}) suggesting benefits for the grain-based ration. However, this is a shortsighted appraisal because pastoral grazing is based on *in situ* harvesting by cows with minimal inputs to energy or carbon losses to cultivation.

When CO_2 emissions from soils, machinery, fuel for cultivation, harvesting, transport, processing, and drying are accounted for (Table 12.8), relative emissions are altered considerably. Carbon loss from soils was 3 to 4 tonnes ha^{-1} per annum.[72,73] Summation of total carbon and methane emissions as CO_2 equivalents suggested losses of 0.84 kg kg^{-1} milk from conventional pastoral dairying compared to 1.51 kg kg^{-1} milk for TMR systems.

These data illustrate the dangers of a narrow focus for GHG calculations. While it could be argued that CO_2 emission does not apply to agricultural inventory, CO_2 is a significant greenhouse gas and any change of land use (e.g., from pastoral to cultivated systems) will incur emissions costs/taxes. The data of van der Nagel et al.[19] provide a basis for modeling whole-farm systems to include nitrous oxide emissions in addition to carbon dioxide and methane. Recent experimental findings enabling more accurate accounting of N_2O emissions from dung and urine patches under a range of environmental conditions and soil types[6,8,70] will improve inventory. It is important to base modeling and systems predictions on actual data with minimal assumptions and speculation, because small changes in agricultural procedures can have major impacts on overall greenhouse gas emissions.

TABLE 12.8
Comparison of Pastoral-Based Dairying and Total Mixed Ration (TMR) Systems for Feed DM Intake, Milk Production, and Methane and Carbon Dioxide Emissions[19]

	Pasture	TMR
Feed DM intake (kg cow^{-1}p.a.)	4560	6050
Milk yield (kg cow^{-1}p.a.)	3650	7300
Methane (kg cow^{-1}p.a.)	90	142
Methane/milk (g kg^{-1})	24.6	19.5
CO_2 equivalent emissions from herds (tonnes p.a.)		
From soils	186	1784
Machinery, fuel, fertilizer etc	91	198
Methane	495	783
Total	772	2765
CO_2 equivalent milk^{-1}(kg kg^{-1})	0.84	1.51

12.16 SUMMARY AND CONCLUSIONS

New Zealand GHG emissions include a high percentage of methane (37%) mainly derived from ruminant animals. Methane inventory calculations are based on animal census, physiological status, feed intakes, and methane production per kilogram dry matter intake. The New Zealand farming community supports environmental sustainability and recognizes nitrogen and methane pollution, in part because of publicity surrounding an attempt to levy livestock farmers to fund GHG research. Mitigation can be expressed in terms of total emissions, a proportion of gross energy intake, or on the basis of production. Principal opportunities for short-term methane mitigation include improved feed quality, animal performance, and pasture management. Long-term strategies include selection of low methane producers, vaccination, and use of slow release, nontoxic methanogen inhibitors. Analysis of experimental data from sheep and cattle fed fresh forage diets showed a poor prediction of methane emissions on the basis of diet composition; an improved understanding of rumen digestive physiology should complement mitigation strategies. Nitrous oxide emissions are dependent on nitrogen inputs from urine, feces, and fertilizer and are exacerbated by soil moisture content. Strategic placement of appropriate fertilizers and matching ruminant requirements to feed composition will lessen nitrous oxide losses. Practical solutions for GHG mitigation require an integrated assessment of all GHG and costs of implementation must not penalize producers.

Mitigation can be measured in absolute terms or in terms of production, and one GHG should not be lowered at the expense of others. Mitigation must not add to costs of production. These constraints, and those associated with food safety, limit opportunities for major reductions in methane emissions from ruminants in the short term, but there are good options for mitigation in the longer term.

Selection of highly productive animals will minimize the proportion of methane associated with maintenance and good-quality balanced diets, including the use of legumes, will lessen methane costs per unit production. Future options include selecting animals with low emissions, prudent fertilizer application, and development of chemical inhibitors and vaccines in the longer term. Mitigation should apply to animals under both intensive and extensive farming.

Central to successful methane mitigation will be an improved understanding of digestive physiology, including contributions of animal, feed, and microbial components to methanogenesis. Producers should apply multiple technologies to mitigate GHG emissions and these may compliment future developments to target rumen methanogenesis.

ACKNOWLEDGMENTS

The authors thank Barbara Dow for statistical analyses of methane production from forages fed to sheep and cattle.

REFERENCES

1. United Nations Framework Convention on Climate Change. Available online at http://ghg.unfccc.int/default1.htf.
2. New Zealand Climate Change Office, National Inventory report. Greenhouse gas inventory 1990–2001, Wellington, New Zealand, 2003, 174 pp.
3. O'Hara, P., Freney, J., and Ulyatt, M., Abatement of Agricultural Non-Carbon Dioxide Greenhouse Gas Emissions. A Study of Research Requirements. Ministry of Agriculture and Forestry, Wellington, New Zealand, 2003, 170 pp. Available online at www.maf.govt.nz/publications.
4. Clark, H., Brookes, I., and Walcroft, A. Enteric methane emissions from New Zealand ruminants 1990–2001. Calculated using an IPCC Tier 2 approach. Unpublished report prepared for the New Zealand Ministry of Agriculture and Forestry, 2003.
5. CSIRO, *Feeding Standards for Australian Livestock: Ruminants*, CSIRO, East Melbourne, 1990, 266 pp..
6. Kelliher, F.M. et al., A revised nitrous oxide emissions inventory for New Zealand 1990–2001. A report prepared for the Ministry of Agriculture and Forestry, Wellington, New Zealand, 2003.
7. IPCC, *Climate Change 1995. Impacts, Adaptations and Mitigation of Climate Change, Scientific-Technical Analyses*, Watson, R.T., Zinyowera, M.C., and Moss, R.H., Eds. Cambridge University Press, Cambridge, 1996.
8. de Klein, C.A.M. et al., Estimating a nitrous oxide emission factor for animal urine from some New Zealand pastoral soils, *Aust. J. Soil. Res.*, 41, 381, 2003.
9. Blaxter, K.L. and Clapperton, J.L., Prediction of the amount of methane produced by ruminants, *Br. J. Nutr.*, 19, 511, 1965.
10. Moss, A.R. *Methane Global Warming and Production by Animals*, Chalcombe, Canterbury, Kent, England, 1993.
11. Joblin, K.N., Options for reducing methane emissions from ruminants in New Zealand and Australia, in *Greenhouse: Coping with Climate Change,* Bouma, W.J., Pearman, G.I., and Manning, M.R., Eds. CSIRO Publishing, Collingwood, Australia, 1996, 437.
12. Van Nevel, C.J. and Demeyer, D.I., Control of rumen methanogenesis, *Environ. Monit. Assess.*, 42, 73, 1996.
13. Johnson, D.E., Ward, G.M., and Ramsey, S.J., Livestock methane: current emissions and mitigation potential, in *Nutrient Management of Food Animals to Enhance and Protect the Environment.* Kornegay, E.T., Ed., CRC Press, Boca Raton, FL, 1996, 219.
14. Johnson, D.E., Johnson, K.A., Ward, G.M., and Branine, M.E. Ruminants and other animals, in *Atmosphere Methane. Its Role in the Global Environment.* Khalil, M.A.K., Ed. Springer-Verlag, Berlin, 2000.
15. Moss, A.R., Jouany, J.-P., and Newbold, J., Methane production by ruminants: its contribution to global warming, *Ann. Zootech.*, 49, 231, 2000.
16. Hegarty, R.S., Greenhouse gas emissions from the Australian livestock sector. Australian Greenhouse Office. Canberra, 2001, 32 pp. Available online at http://www.greenhouse.gov.au.
17. Tomkins, N.W. and Hunter, R.A., Methane mitigation in beef cattle using a patented anti-methanogen, in *Proceedings of the Second Joint Australia and New Zealand Forum on Non-CO$_2$ Greenhouse Emission from Agriculture,* Eckard, R., Ed. CRC for Greenhouse Accounting, Canberra, Australia, 2003, F3.
18. Wright, A.D.G. et al., Reducing methane emissions in sheep by immunization against rumen methanogens, *Vaccine*, 22 (29–30), 3976, 2004.

19. Van der Nagel, L.S., Waghorn, G.C., and Forgie, V.E., Methane and carbon emissions from conventional pasture and grain-based total mixed rations for dairying, *Proc. N.Z. Soc. Anim. Prod.*, 6, 128, 2003.

20. Johnson, K.A. and Johnson, D.E., Methane emissions from cattle, *J. Anim. Sci.*, 73, 2483, 1995.

21. Moe, P.W. and Tyrrell, H.F., Methane production in dairy cows, *J. Dairy Sci.*, 62, 1583, 1979.

22. Holter, J.B. and Young, A.J., Methane prediction in dry and lactating Holstein cows, *J. Dairy Sci.*, 75, 2165, 1992.

23. Wilkerson, V.A., Casper, D.P., and Mertens, D.R., The prediction of methane production of Holstein cows by several equations, *J. Dairy Sci.*, 78, 2402, 1995.

24. Lassey, K.R. et al., Methane emissions measured directly from grazing livestock in New Zealand, *Atmos. Environ.*, 31, 2905, 1997.

25. Pinares-Patino, C.S. et al., Persistence of differences between sheep in methane emission under generous grazing conditions, *J. Agric. Sci.*, 140, 227, 2003.

26. Robertson, L.R. and Waghorn, G.C., Dairy industry perspectives on methane emissions and production from cattle fed pasture or total mixed rations in New Zealand, *Proc. N.Z. Soc. Anim. Prod.*, 62, 213, 2002.

27. Martin, C. et al., Influence of cereal supplementation on methane production by sheep measured by the SF_6 tracer method, in Challenges for microbial digestive ecology at the beginning of the third millennium, *Reprod. Nutr. Dev.*, 40 (2), 211, 2000.

28. Ulyatt, M.J. et al., Seasonal variation in methane emission from dairy cows and breeding ewes grazing ryegrass/white clover pasture in New Zealand, *N.Z. J. Agric. Res.*, 45, 217, 2002.

29. Pinares-Patino, C.S. et al., Methane emission by alpaca and sheep fed on lucerne hay or grazed on pastures of perennial ryegrass/white clovers or birdsfoot trefoil, *J. Agric. Sci.*, 140, 215, 2003.

30. Waghorn, G.C., Tavendale, M.H., and Woodfield, D.R., Methanogenesis from forages fed to sheep, *Proc. N.Z. Grass Assoc.*, 64, 167, 2002.

31. Molano, G., Renard, T., and Clark, H., The effect of level of feeding and forage quality on methane emissions by wether lambs, in *Proceedings of the Trace Gas Workshop*. NIWA Technical report 125, Wellington, New Zealand, 2004, 86.

32. Boadi, D.A., Wittenberg, K.M., and McCaughey, W.P., Effects of grain supplementation on methane production of grazing steers using the sulphur hexaflouride (SF_6) tracer gas technique, *Can. J. Anim. Sci.*, 82, 151, 2002.

33. Boadi, D.A. and Wittenberg, K.M., Methane production from dairy and beef heifers fed forages differing in nutrient density using the sulphur hexafluoride (SF_6) tracer gas technique, *Can. J. Anim. Sci.*, 82, 201, 2002.

34. McCaughey, W.P., Wittenberg, K., and Corrigan, D., Methane production by steers on pasture, *Can. J. Anim. Sci.*, 77, 519, 1997.

35. McCaughey, W.P., Wittenberg, K., and Corrigan, D., Impact of pasture type on methane production by lactating beef cows, *Can. J. Anim. Sci.*, 79, 221, 1999.

36. Woodward, S.L., Waghorn, G.C., and Laboyrie, P.G., Condensed tannins in birdsfoot trefoil (*Lotus cornicalatus*) reduce methane emissions from dairy cows, *Proc. N.Z. Soc. Anim. Prod.*, 64, 160, 2004.

37. Woodward, S.L. et al., Does feeding sulla (*Hedysarum coronarium*) reduce methane emissions from dairy cows? *Proc. N.Z. Soc. Anim. Prod.*, 62, 227, 2002.

38. Puchala, R. et al., The effect of a condensed tannin-containing forage on methane emission by goats, *J. Anim. Sci.*, 81(Suppl. 2), 2003.

39. Waghorn, G.C. et al., Forages with condensed tannins — their management and nutritive value for ruminants, *Proc. N.Z. Grass. Assoc.*, 60, 89, 1998.
40. Waghorn, G.C., Reed, J.D., and Ndlovu, L.R., Condensed tannins and herbivore nutrition, in *Proceedings of the XVIII International Grasslands Congress*, Buchanan-Smith, J.G., Bailey, L.D., and McCaughey, P., Eds. Association Management Centre, Calgary, Alberta, Canada, 1999, vol. III, 153.
41. Jones, G.A. et al., Effects of Sainfoin (*Onobrychus viciifolia* Scop.) condensed tannins on growth and proteolysis by four strains of ruminal bacteria, *Appl. Environ. Microbiol.*, 60, 1374, 1994.
42. Waghorn, G.C. et al., Effect of condensed tannins in *Lotus pedunculatus* on its nutritive value for sheep. 2. Nitrogenous aspects, *J. Agric. Sci.* (Cambridge), 123, 109, 1994.
43. McNabb, W.C. et al., The effect of condensed tannins in *Lotus pedunculatus* on the solubilisation and degradation of ribulose-1,5-*bis* phosphate carboxylase (EC 4.1.1.39; Rubisco) protein in the rumen and the sites of Rubisco digestion, *Br. J. Nutr.*, 76, 535, 1996.
44. Ulyatt, M.J. et al., Accuracy of the SF_6 tracer technology and alternatives for field measurements, *Aust. J. Agric. Res.*, 50, 1329, 1999.
45. Smuts, M., Meissner, H.H., and Cronje, P.B., Retention time of digestion in the rumen: its repeatability and relationship with wool production in merino rams, *J. Anim. Sci.*, 73, 206, 1995.
46. Waghorn, G.C. and Caradus, J.R., Screening white clover cultivars for improved nutritive value — development of a method, *Proc. N.Z. Grass Assoc.*, 56, 49, 1994.
47. Weimer, P.J. et al., Effect of diet on populations of three species of rumen cellulolytic bacteria in lactating dairy cows, *J. Dairy Sci.*, 82, 122, 1999.
48. Reid, C.S.W. et al., Physiological and genetical aspects of pasture (legume) bloat, in *Digestion and Metabolism in the Ruminant. Proceedings of the IV International Symposium on Ruminant Physiology*, McDonald, I.W. and Warner, A.C.I., Eds. University of New England Publishing Unit, Armidale, Australia, 1975, 524.
49. Morris, C.A., Cullen, N.G., and Geertsema, H.G., Genetic studies of bloat susceptibility in cattle, *Proc. N.Z. Soc. Anim. Prod.*, 57, 19, 1997.
50. McAllister, T.A. et al., Dietary, environmental and microbial aspects of methane production in ruminants, *Can. J. Anim. Sci.*, 76, 231, 1996.
51. Mathison, G.W. et al., Reducing methane emissions from ruminant animals, *J. Appl. Anim. Res.*, 14, 1, 1998.
52. Hegarty, R.S., Strategies for mitigating methane emissions from livestock — Australian options and opportunities, in *Greenhouse Gases and Animal Agriculture*, Takahashi, J. and Young, B.A., Eds. Elsevier Science, Amsterdam, 2002, 61.
53. Machmuller, A., Ossowski, D.A., and Kreuzer, M., Comparative evaluation of the effects of coconut oil, oil seeds and crystalline fat on methane release, digestion and energy balance in lambs, *Anim. Feed Sci. Tech.*, 85, 41, 2000.
54. Johnson, K.A. et al., The effect of oil seeds in diets of lactating cows on milk production and methane emissions, *J. Dairy Sci.*, 85, 1509, 2002.
55. Goodrich, R.D. et al., The influence of monensin on the performance of cattle, *J. Anim. Sci.*, 58, 1484, 1984.
56. Lean, I.J. and Wade, L., Effects of monensin on metabolism, production and health of dairy cattle, presented at a symposium on the usefulness of ionophores in lactating dairy cattle, Ontario Veterinary College, Guelph, June 25–26, 1997, 50.

57. Clark, D.A. et al., The effect of monensin on methane emissions from identical twin dairy cows fed pasture, presented at the 2nd Joint Australia and New Zealand Forum on Non-CO_2 Greenhouse Emissions from Agriculture, Melbourne, Australia, October 20–21, 2003, F7.

58. Fellner, V. et al., The effect of Rumensin® on milk fatty acid profiles and methane production in lactating dairy cows, presented at a symposium on the usefulness of ionophores in lactating dairy cattle, Ontario Veterinary College, Guelph, June 25–26, 1997, 59.

59. Johnson, D.E. et al., Persistence of methane suppression by propionate enhancers in cattle diets, in *Energy Metabolism of Farm Animals, Proceedings of the 13th Symposium.* Mojacar, Spain, Aquilera, J.J., Ed. EAPP Publication No. 76, 1994, 339.

60. Hegarty, R.S., Reducing rumen methane emissions through elimination of rumen protozoa, *Aust. J. Agric. Res.,* 50, 1321, 1999.

61. CSIRO, http://CSIRO.au/index/protozoa.

62. McSweeney, G.S. and McCrabb, G.J., Inhibition of rumen methanogenesis and its effects on feed intake, digestion and animal production, in *Greenhouse Gases and Animal Agriculture,* Takahashi, J. and Young, B.A., Eds. Elsevier Science, Amsterdam, 2002, 129.

63. Miller, T.L. and Wolin, M.J., Inhibition of growth of methane-producing bacteria of the rumen forestomach by hydroxymethyl glutaryl-SCoA reductase inhibitors, *J. Dairy Sci.,* 84, 1445, 2001.

64. Morvan, B. et al., Quantitative determination of H_2- utilizing acetogenic and sulphate reducing bacteria and methanogenic archaea from the digestive tract of different mammals, *Curr. Microbiol.,* 32, 129, 1996.

65. Pinares-Patino, C.S., Baumont, R., and Martin, C., Methane emissions by Charolais cows grazing a monospecific pasture of timothy at four stages of maturity, *Can. J. Anim. Sci.,* 83, 769, 2003.

66. MFE, New Zealand Greenhouse Gas Inventory 1990–1998, Ministry for the Environment, Wellington, New Zealand, 2000.

67. Clark, H., de Klein, C.A.M., and Newton, P., Potential management practices and technologies to reduce nitrous oxide, methane and carbon dioxide emissions from New Zealand agriculture. A report prepared for the Ministry of Agriculture and Forestry, September 2001. Available online at http://www.maf.govt.nz/mafnet/rural-NZ/sustainable-resource-use/climate/green-house.

68. Di, H.J. and Cameron, K.C., The use of a nitrification inhibitor, dicyandiamide (DCD), to decrease nitrate leaching and nitrous oxide emissions in a simulated grazing and irrigated grassland, *Soil Use Manage.,* 18, 395, 2002.

69. Saggar, S. et al., Nitrous oxide emissions from New Zealand dairy and sheep-grazed pastures, presented at the 2nd Joint Australia and New Zealand Forum on Non-CO_2 Greenhouse Emissions from Agriculture, Melbourne, Australia, October 20–21, 2003, E7.

70. de Klein, C.A.M. et al., Evaluation of two potential mitigation options for reducing nitrous oxide emissions, presented at the 2nd Joint Australia and New Zealand Forum on Non-CO_2 Greenhouse Emissions from Agriculture, Melbourne, Australia, October 20–21, 2003, E3.

71. Kolver, E. et al., Total mixed rations versus pasture diets: evidence for a genotype x diet interaction in dairy cow performance, *Proc. N.Z. Soc. Anim. Prod.,* 62, 246, 2002.

72. Crush, J.R., Waghorn, G.C., and Rolston, M.P., Greenhouse gas emissions from pasture and arable crops grown on a Kairanga soil in the Manawatu, North Island, New Zealand, *N.Z. J. Agric. Res.*, 35, 253, 1992.

73. McLaren, R.G. and Cameron, K.C., Soil science, in *Sustainable Production and Environmental Protection*, Oxford University Press, Auckland, New Zealand, 1996, chap. 10.

13 Strategies for Reducing Enteric Methane Emissions in Forage-Based Beef Production Systems

K.H. Ominski and K.M. Wittenberg

CONTENTS

13.1 INTRODUCTION

The Canadian agricultural landscape includes some 4,804,496 ha of tame or seeded pasture and 15,391,072 ha of natural land for pasture.[1] A significant portion of this forage is used by the Canadian beef cattle industry as a source of feed for cows, bulls, and growing/young stock. Microbial breakdown of forage and other feedstuffs in the rumen, also known as enteric fermentation, results in the production of methane. Approximately 87% of enteric methane originates in the reticulo-rumen while the remainder is produced in the hindgut. A significant portion of the latter, approximately 89%, is absorbed and expired through the lungs, with the remainder being excreted through the anus,[2,3] Losses in gross energy intake associated with methane production range from 2 to 3% of gross energy intake (GEI) when animals are fed high-grain diets[4] to 11.3% of GEI when consuming low-quality forage.[5]

In Canada, enteric fermentation, as calculated using Intergovernmental Panel on Climate Change (IPCC) Tier I values, contributes approximately 19,000 kt CO_2 equivalents year^{-1} — approximately 32% of total agricultural emissions.[6] An understanding of the microbial processes responsible for the production of enteric methane production, coupled with the identification of management strategies leading to reduced methane emissions and improved animal performance, will help facilitate the efforts of the Canadian Government to achieve a 6% reduction in greenhouse gas emissions by 2008–2012, as outlined in the Kyoto Protocol.[7]

Several comprehensive reviews have examined enteric methane production by methanogenic bacteria.[8,9] As a consequence, this chapter does not examine this area in detail but instead summarizes the applied research that has been conducted in Canada, in an attempt to identify potential mitigation strategies for forage-based beef cattle production systems.

13.2 ENTERIC FERMENTATION

Methane production in the rumen occurs as a consequence of the presence of a group of microorganisms called methanogens that reside in the reticulo-rumen and large intestine of ruminant livestock. These organisms play an important role in converting organic matter to methane. As described in a detailed review by McAllister et al.,[8] proteins, starch, and plant cell-wall polymers consumed by the animal are hydrolyzed to amino acids and simple sugars by the bacteria, protozoa, and fungi that reside in the rumen. Primary and secondary digestive microorganisms further ferment the amino acids and sugars into volatile fatty acids, hydrogen, carbon dioxide, and other end products. Methanogens then reduce carbon dioxide to methane, preventing the accumulation of hydrogen. Excessive quantities of hydrogen ions or protons, when allowed to accumulate in the rumen environment, result in a decline in pH, and subsequent inhibition of many organisms that are essential for fiber digestion.

13.3 MECHANISMS BY WHICH METHANE PRODUCTION MAY BE REDUCED

Several mechanisms influence the availability of hydrogen in the rumen and subsequent production of enteric methane emissions by cattle. Processes that yield propionate act as net proton-using reactions while those that yield acetate result in a net increase in protons.[10] That is, the proportion of volatile fatty acids, specifically acetate:propionate, produced as a consequence of microbial fermentation in the rumen has a significant influence on methane production. This ratio may, for example, be influenced by the type of carbohydrate consumed by the animal. Cereal-based diets that are high in starch favor propionate production and consequently tend to produce less methane per unit of feed consumed than forage-based diets.[8]

In addition to the type of carbohydrate in the diet, other dietary factors influence the acetate:propionate profile in the rumen, including residence time in the rumen. Okine et al.[11] have demonstrated a 29% reduction in methane production when weights were added to the rumen to stimulate contraction of the rumen wall in order to decrease residence time of the feed in the digestive tract.

Digestibility of dietary energy may also influence enteric methane production. Boadi and Wittenberg[12] have demonstrated that a reduction in forage *in vitro* organic matter digestibility (IVOMD) from 61.5 to 38.5% tended ($P = 0.14$) to lead to an increase in GEI lost as methane from 6.0 ± 0.38 to $6.9 \pm 0.98\%$ when animals were fed *ad libitum*. When intake was restricted to 2% of bodyweight, an increase in methane production, expressed as a percent of GEI, was no longer evident. As changes in diet digestibility and residence time in the rumen are associated with intake, it is not unexpected that level of intake also influences enteric methane emissions. Johnson and Johnson[4] concluded that feeding highly available carbohydrates at limited intakes results in high fractional methane losses while feeding highly available carbohydrates at high intakes leads to low fractional methane losses. Further, Blaxter[13] has described an increase in total production of methane with increases in intake, from maintenance to twice maintenance. However, when expressed as amount of energy lost per unit of feed, a reduction is realized.

Another mechanism by which methane production may be reduced during the rumen fermentation process is through the provision of alternative hydrogen acceptors or sinks. Compounds such as unsaturated fatty acids provide alternative hydrogen acceptors, consuming hydrogen in limited quantities, during biohydration.[14] Dicarboxylic acids (such as fumaric and malic acids), which are intermediates in the propionic acid pathway, may also serve as alternative electron sinks for H_2, as described in a recent review by Boadi et al.[15] Bayaru et al.[16] observed a 23% reduction in methane production, and increased propionic acid formation with no effect on DM digestibility when fumaric acid was added to whole crop sorghum silage fed to Holstein steers.

13.4 MANAGEMENT STRATEGIES LEADING TO A REDUCTION IN ENTERIC METHANE EMISSIONS

There are several management strategies that may be employed in the Canadian beef cattle industry to reduce enteric methane emissions via the mechanisms described above. These management strategies may be categorized as follows: forage utilization, feed additives, and improved production efficiencies. Each is addressed in the subsequent sections.

13.4.1 FORAGE UTILIZATION

13.4.1.1 Quality

Boadi and Wittenberg[12] have demonstrated that forage quality has a significant impact on enteric methane emissions. Cattle given hay of high (61.5% IVOMD), medium (50.7% IVOMD), and low (38.5% IVOMD) quality differed ($P < 0.01$) in dry matter intake, as animals consumed 9.7 ± 0.23, 8.9 ± 0.23, and 6.3 ± 0.23 kg d^{-1}, respectively. Further, differences existed in enteric methane emissions ($P < 0.01$), as 47.8 ± 4.02, 63.7 ± 4.02, and 83.2 ± 4.02 CH_4 L kg^{-1} digestible organic matter intake was produced from cattle consuming the high-, medium-, and low-quality forages, respectively.

These same authors subsequently demonstrated this same phenomenon on pasture.[17] Steers grazing during the early period of the grazing season had 44 and 29% less energy lost as methane ($P < 0.01$) compared to steers grazing during the mid and late grazing periods, respectively. Further, steers experienced a 54% reduction ($P < 0.01$) in enteric emissions upon entry vs. exit of a paddock. Efficiency of forage fermentation was linked to biomass availability and quality of pasture.

The impact of pasture forage quality and availability on enteric methane emissions from cattle in grass-based production systems has been studied by Ominski et al.[5] Emissions were influenced by pasture dry matter availability and quality, in that emissions were highest (11% of GEI) when pasture quality and availability were low. Emissions were lower when pasture quality was high and availability was low (6.9% of GEI) or when quality was low and availability was high (7.1 to 9.4% of GEI). Unfortunately, neither pasture ever attained a status of high forage quality and high pasture availability. It can be concluded that enteric emissions are highest when the animal is presented with poor-quality forage and has limited opportunity to select higher-quality forage as a consequence of reduced dry matter availability.

The impact of pasture quality on enteric emissions has recently been examined by Pinares-Patiño et al.[18] In this study, beef cows were grazed on a monospecific pasture of timothy at four stages of maturity: early vegetative, heading, flowering, and senescence. Although the crude protein and NDF values were 31.4 and 52.6; 13.2 and 59.8; 7.8 and 68.4; and 4.4 and 75.4 at vegetative, heading, flowering, and senescent stages, respectively; organic matter intake (kg) and methane emissions (g d $^{-1}$) were lower only at heading (11.3 ± 1.4; 273.3 ± 28.7) but not at vegetative (9.1 ± 0.7; 204.4 ± 28.1), flowering (10.1 ± 1.5; 232.2 ± 25.4), or senescent (10.1 ± 1.3; 228.4 ± 32.9) stages. Further, methane emissions when expressed as a percent of gross energy intake did not differ among treatments. Although the trial was designed to decrease species selection, it did not limit selection of plant parts. Therefore, the lack of response associated with maturity that was observed by the authors may be attributed to animal selection during grazing. Although the area of pasture allocated daily was calculated using required herbage area (set as twice intake capacity), herbage mass, and the number of cows, post-grazing sward surface height ranged from 11.3 ± 1.4 to 51.2 ± 5.0 cm. Further, as the animals were strip grazed, they had access to the pasture grazed in the previous 12 hours. Under these conditions, animals could have selected the more vegetative plants or more digestible plant parts. Thus, it is paramount that potential mitigation strategies be assessed under conditions that parallel those observed in the Canadian production environment.

13.4.1.2 Species

McCaughey et al.[19] have demonstrated that the species present in a pasture may significantly influence enteric methane emissions. Pasture types examined were alfalfa–grass mix (78% alfalfa and 22% meadow bromegrass) or 100% meadow bromegrass. Although cows grazing the alfalfa–grass pastures had significantly greater dry matter intake (11.4 ± 0.4 vs. 9.7 ± 0.4 kg DM d^{-1}), lower methane production was observed (373.8 ± 10.1 vs. 411.0 ± 10.1 L CH$_4$ d^{-1}) compared to their counterparts grazing grass-only pastures. As a consequence, cows grazing the alfalfa-grass pastures lost less energy ($P = 0.001$) through eructation of CH$_4$ ($7.1 \pm 0.4\%$ vs. $9.5 \pm 0.4\%$ of GEI). This reduction in CH$_4$ emissions may be attributed to a reduction in the proportion of structural carbohydrates. The NDF of the alfalfa-based pasture ($58.4 \pm 0.8\%$) was lower than the NDF of the grass-based pasture ($73.1 \pm 0.8\%$). Inclusion of legume-based forages in the diet is associated with higher digestibility and faster rate of passage,[20] resulting in a shift toward high propionate in the rumen and reduced methane production. The improved feed utilization observed in the cow as a consequence of the reduced enteric emission proved to be beneficial to the calves in this study as calf growth rate was 11% higher on the alfalfa–grass pasture compared to the grass-only pasture.

The need to examine animal performance concomitant with enteric emissions has been illustrated by Olson,[21] who examined methane production of animals grazing native pasture compared to those grazing several alternative forage species ("Nordan" crested wheatgrass, "Hycrest" crested wheatgrass, "Vinall" Russian wild-rye, and "Syn-A" Russian wildrye) during a 30-day grazing season in the fall and spring. Enteric emissions from cattle grazing these species in the fall were the same for all pasture species even though intake was greater ($P = 0.01$) and bodyweight loss was less ($P = 0.05$) for the wildrye varieties compared to the wheatgrass varieties.[22] This same author observed that daily enteric methane emissions increased by approximately 70% ($P < 0.01$) for lactating cows grazing the same pastures in the spring because animal intake, expressed as % bodyweight, was significantly higher. Although pasture species did not affect grazing cow methane emissions, bodyweight gains of animals consuming crested wheatgrass varieties were higher than those of animals consuming Russian wildrye varieties. This work clearly demonstrates the opportunity to use management strategies that match pasture forage with animal requirements as a means of optimizing performance with no increase in enteric emissions.

Although studies examining the mitigation potential of other species in Canada have not been published to date, such studies have been conducted in New Zealand using ram lambs.[23] Forages examined included fresh ryegrass/white clover pasture, lucerne, sulla, chicory, red clover, and lotus, as well as sulla/lucerne, chicory/sulla, and chicory/red clover mixes. All forages, which were cut on a daily basis, were of good quality with crude protein concentrations ranging from 12.3% (chicory) to 26.4% (lotus), fiber ranging from 12.7% (chicory) to 44.4% (pasture), and DM digestibility in excess of 65%. A twofold range in methane emissions was observed from 11.5 g kg^{-1} DMI with lotus to 25.7 g kg^{-1} DMI with rye grass/white clover pasture. These studies not only demonstrated that species selection may play an important role in reducing enteric methane emissions, but more specifically, that condensed tannins present in several forage species are associated with reduced methanogenesis and thus may be an effective technique to lower methane production. Other benefits associated with feeding condensed tannins in ruminant diets include reduced incidence of bloat and intestinal worm populations.[24] Further, as described by Jones et al.,[25] ingestion of tannin-containing forages leads to the formation of insoluble tannin–protein complexes as a consequence of the ability of tannins to bind to plant complexes at pH ranges of 3.5 to 7.0. As these plant complexes dissociate at pH values below 3.0, as is characteristic of the postruminal environment, they serve to protect protein from microbial degradation in the rumen, thereby increasing the proportion of plant amino acids absorbed postruminally. To date, no data have been published to establish the mitigation potential of tannin-containing legumes such as sainfoin or bird's-foot trefoil in the Canadian production environment.

13.4.1.3 Pasture Management

Several studies have been conducted in Canada that examine the impact of pasture management on enteric methane emissions.[17,26] McCaughey et al.[26] examined the impact of two grazing systems (continuous vs. rotational) and two stocking densities (low, 1.1 steer ha^{-1} or high, 2.2 steer ha^{-1}) on animal intake and methane production. In this alfalfa–grass-based production system, neither intake nor methane production (expressed as emissions per unit gain or as a percent of gross energy intake) were affected by the management strategies described above.

Boadi et al.[17] examined the use of supplemental grain as a means of reducing enteric methane emissions in an alfalfa–grass pasture environment. Steers were fed 2, 4, and 4 kg of rolled barley on a daily basis during the early, mid and late periods of the grazing season, respectively. Although supplementation reduced forage dry matter intake by an average of 11% ($P < 0.03$) and increased total organic matter intake by 14% ($P < 0.001$), daily enteric emissions (L d^{-1}) were similar in the supplemented and control steers. Further, there was no significant difference between the two treatments in terms of energy lost as methane (6.4 and 6.7% of GEI for supplemented and control steers, respectively). These data suggest that the benefits of grain supplementation in terms of mitigation potential on good-quality pasture are limited, as pasture quality had a greater impact on methane production than did grain supplementation.

13.4.1.4 Forage Preservation and Processing

To the authors' knowledge, no work has been conducted in Canada to date to establish the impact of ensiling on enteric methane emissions. A decrease in methane production as a consequence of ensiling has been reported.[9] This finding has not been reported elsewhere. In fact, in New Zealand, Woodward et al.[27] observed some of the highest methane losses reported in the literature associated with feeding ryegrass silage (10.8% of gross energy) and lotus silage (8.6% of gross energy). Thus, additional research is needed to assess the methane mitigating potential of silages, as well as processing or chemical treatment of forages in the Canadian production environment.

13.4.2 FEED ADDITIVES

Ionophores are frequently utilized in beef cattle production systems to improve animal performance, as well as to reduce the incidence of bloat and prevent outbreaks of coccidiosis. Improvements in feed efficiency of 5 to 6% have been attributed to a shift in the fermentation pathway from acetate to propionate.[28] Although ionophore supplementation may reduce methane emissions by 20 to 25%, work conducted at the University of Colorado has demonstrated that an adaptive response occurs in both forage and grain diets, resulting in a return to baseline methane levels in approximately 2 weeks.[29]

The use of ionophores to reduce enteric methane emissions on pasture has been examined by McCaughey et al.[26] As described above, steers in this trial were grazed under two management regimens — continuous or rotational. Half of the animals on each pasture were given a monensin controlled-release capsule (CRC) delivering 270 mg d^{-1}. Neither voluntary intake nor methane production was affected by the presence of monensin.

The mitigation potential of other feed additives, including the addition of salts to alter the dietary cation-anion balance, has been explored.[30] In this trial, salts were added to the rumen (via cannula) to achieve a dietary cation-anion balance of 10, 30, 50, and 70 mEq 100 g^{-1} DM. Methane emissions, expressed as g hd^{-1} d^{-1} ($P < 0.009$) and as a percent of gross energy intake ($P < 0.02$), were lower for cows receiving the diet containing 70 mEq 100 g^{-1} DM compared to those receiving the 10 and 30 mEq 100 g^{-1} DM diet. Diet dry matter intake and rumen fermentation characteristics were not affected by the change in dietary cation-anion balance. Research conducted by Müller-Özkan[31] has demonstrated that the addition of calcium ions *in vitro* had a tendency to reduce activity of methane-producing bacteria, as evidenced by a decrease in methane production.

The addition of fat supplementation in high-energy finishing diets has been examined in several commercial production systems in Canada. Mathison[32] demonstrated that daily enteric methane emissions could be reduced by 33% when canola oil was added to an 85% concentrate feedlot diet. Further, a Manitoba study[33] demonstrated a 30% reduction in daily enteric methane emissions when comparing a typical feedlot diet (88.5% concentrate) with a ration of equal energy density containing a 44:42:14 ratio of concentrate, silage, and whole sunflower seed. These studies demonstrate that fat supplementation may effectively reduce enteric emissions for finishing cattle. To the authors' knowledge, the mitigation potential of supplemental fat in low-quality forage diets has not been addressed. Although this strategy of reducing enteric methane emissions may not seem viable from an economic perspective at the present time, future trading of carbon offsets may prove otherwise. In addition to economic viability, other factors that require exploration include practical techniques for inclusion of fat in forage-based diets and appropriate levels of supplementation to ensure that fiber degradation is not compromised.

13.4.3 IMPROVED PRODUCTION EFFICIENCIES

A summary of enteric methane emissions from numerous classes of cattle in the U.S. led Johnson and Johnson[4] to conclude that methane losses (% GEI) in commercial situations do not deviate considerably from 6%. As a consequence, these authors have suggested that the best strategy for mitigation is to decrease methane loss per unit of product. Using this approach, strategies to decrease methane emissions in the cattle industry should include effective management of feed resources other than forage, such as water quality, mineral supplementation, and ration balancing.

The impact of water quality on cattle performance has been demonstrated in a series of trials conducted by Willms et al.[34] Calves from cows having access to water from a natural water source delivered to a trough (clean water) gained 9% more than calves from cows that had direct access to a pond. Yearling heifers having access to clean water gained 23 and 20% more weight than their counterparts drinking directly from a pond, or from a pond pumped to a trough, respectively. Thus, adopting management strategies that serve to improve water quality and subsequent animal performance should serve to reduce methane emissions per unit of product.

Adequate mineral supplementation is another avenue by which the cattle industry may realize a net reduction in enteric methane emissions. Minerals, which serve numerous functions in the body — structural, physiological, catalytic, and regulatory — are necessary if optimal growth, health, and productivity of the animal are to be achieved.[35] An extensive study of pasture quality in the Eastern Interlake Region of Manitoba from 1996–1998[36] provided evidence that many of the surveyed pastures were deficient in several trace minerals including copper, manganese, and zinc. It is anticipated that lack of mineral supplementation or inadequate intake of supplemental mineral on these pastures would result in less-than-optimal performance and thus increased emissions per unit of product.

Other than effective management of feeding programs, there are several other management strategies that would serve to improve animal productivity. These include animal selection for improved production, management for improved reproduction, and use of growth promoting agents.

Genetic selection of animals that consume less feed or produce less methane per unit of feed is another management strategy that may be employed to reduce enteric methane emissions. Trials conducted at the University of Manitoba have shown that as much as 27% of the variation in methane production from cattle on all-forage diets was associated with animal-to-animal variation.[12] Considerable variation among grazing animals has also been reported by Pinares-Patiño et al.,[18] Lassey et al.,[37] and Ulyatt et al.,[38] where animal-to-animal variation accounted for 70 and 85% of the variation in daily methane production. Two traits that are actively being investigated as a means of identifying genetically superior animals are net feed efficiency and mean retention time of digesta in the rumen.[39]

13.5 A SYSTEMS-BASED APPROACH TO MANAGEMENT

In addition to the above management strategies related to management of the forage and livestock components of beef production systems in Canada, it is important to consider opportunities for greenhouse gas mitigation as they relate to other facets of forage-based beef production including the potential for carbon sequestration associated with perennial cropping systems, wetlands, and shelterbelts. A recent agreement among Agrosuper, the world's eighth-largest pork producer, Japan's Tokyo Electric Power Company, and Canada's TransAlta Corp[40] is a good example of the partnerships that are being formed between large corporations and the livestock sector in an attempt to meet the targeted reductions set out in the Kyoto Protocol.

As a consequence of such agreements, a paradigm shift may be required in forage-based beef production. In the future, revenue from cow-calf operations may include traditional commodities such as forage and weaned/backgrounded calves, but may also include revenue generated from carbon offsets. Accordingly, management decisions should be made on the basis of net return to all facets of the production system.

A systems-based approach to greenhouse gas mitigation is currently being explored in western Canada through several multidisciplinary research projects.[41,42] The objective of these projects is to bring together scientists from a variety of disciplines, as well as leading conservation and producer groups, such that net greenhouse gas production may be assessed in a production system, and further that mitigation strategies that are economically and environmentally sustainable may be identified and disseminated to interested parties (producers, government officials, etc.).

This strategy is currently being employed by a team of animal, plant, soil, and food scientists from the University of Manitoba who have teamed up with several livestock commodity groups, as well as members of the provincial and federal governments, to explore the net greenhouse gas emissions, as well as nutrient and pathogen movement associated with the application of liquid hog manure in grassland pasture systems.

13.6 SUMMARY AND CONCLUSIONS

Research conducted to date has demonstrated that a reduction in enteric methane emissions from cattle in forage-based production systems is possible in commercial production systems. These strategies include feeding management strategies such as inclusion of legumes in forage mixes and feeding highly digestible forages. Adoption of strategies that serve to improve production efficiency including feed analyses and ration balancing, pregnancy testing, and provision of good quality water will not only serve to reduce enteric methane emissions but will also prove to be economically beneficial. The mitigation potential of several commercially accepted practices, including the use of ionophores, probiotics, and silage-based feeding systems, requires further evaluation, as does the use of novel production practices such as the inclusion of fat in forage-based diets. In addition, long-term research support is required to examine the potential for selection of low methane-emitting animals in forage-based production environments.

REFERENCES

1. Statistics Canada, Table 5.1, Land use, by province, Census Agricultural Region (CAR) and Census Division (CD), Catalogue No. 95F0301XIE, 2001.
2. Murray, R.M., Bryant, A.M., and Leng, R.A., Rates of production of methane in the rumen and large intestine of sheep, *Br. J. Nutr.*, 36, 1, 1976.
3. Torrent, J. and Johnson, D.E., Methane production in the large intestine of sheep, in *Energy Metabolism of Farm Animals,* Aquilera, J.F., Ed., EAAP Publication No. 76, CSIC Publishing Service, Spain, 1994.
4. Johnson, K.A. and Johnson, D.E., Methane emissions from cattle, *J. Anim. Sci.*, 73, 2483, 1995.
5. Ominski, K.H., Wittenberg, K.M., and Boadi, D., Examination of economically and environmentally sustainable management practices in forage-based beef production systems, presented at the CCFIA Final Workshop, Winnipeg, January 2004.
6. Environment Canada, Canada's Greenhouse Gas Inventory 1990–2001, Greenhouse Gas Division, Environment Canada, Ottawa, Canada, 2003.
7. English Conference of the Parties, Third Session Kyoto, 1-10, Kyoto Protocol to the United Nations Framework Convention on Climate Change. Available online at www.cnn.com/specials/1997/global.warming/stories/treaty/, December, 1997.
8. McAllister, T.A., Okine, E.K., Mathison, G.W., and Cheng, K.-J., Dietary, environmental and microbiological aspects of methane production in ruminants, *Can. J. Anim. Sci.,* 76, 231, 1996.
9. Moss, A.R., Jouany, J.-P., and Newbold, J., Methane production by ruminants: its contribution to global warming, *Ann. Zootech.,* 49, 231, 2000.
10. Hegarty, R.S., Mechanisms for competitively reducing ruminal methanogenesis, *Aust. J. Agric. Res.,* 50, 1299, 1999.
11. Okine, E.K., Mathison, G.W., and Hardin, R.T., Effects of changes in frequency of reticular contractions on fluid and particulate passage rates in cattle, *Can. J. Anim. Sci.,* 67, 3388, 1989.
12. Boadi, D. and Wittenberg, K.M., Methane production from dairy and beef heifers fed forages differing in nutrient density using the sulphur hexafluoride (SF_6) tracer gas technique, *Can. J. Anim. Sci.,* 82, 201, 2002.
13. Blaxter, K.L., *The Energy Metabolism of Ruminants,* Hutchinson, London, U.K., 1967, 332 pp.
14. Czerkawski, J.W., Fate of metabolic hydrogen in the rumen, *Proc. Nutr. Soc.,* 31, 141, 1972.
15. Boadi, D., Benchaar, C., Chiquette, J., and Massé, D., Mitigation strategies to reduce enteric methane emissions from dairy cows: update review, *Can. J. Anim. Sci.,* 84, 319, 2004.

16. Bayaru, E., Kanda, S., Kamada, T., Itabashi, H., Andoh, S., Nishida, T., Ishida, M., Itoh, T., Nagara, K., and Isobe, Y., Effect of fumaric acid on methane production, rumen fermentation and digestibility of cattle fed roughage alone, *Anim. Sci. J.*, 72, 139, 2001.
17. Boadi, D., Wittenberg, K.M., and McCaughey, W.P., Effects of grain supplementation on methane production of grazing steers using the sulphur hexafluoride (SF_6) tracer gas technique, *Can. J. Anim. Sci.*, 82, 151, 2002.
18. Pinares-Patiño, C.S., Baumont, R., and Martin, C., Methane emissions by Charolais cows grazing a monospecific pasture of timothy at four stages of maturity, *Can. J. Anim. Sci.*, 83, 769, 2003.
19. McCaughey, W.P., Wittenberg, K., and Corrigan, D., Impact of pasture type on methane production by lactating beef cows, *Can. J. Anim. Sci.*, 79, 221, 1999.
20. Minson, D.J. and Wilson, J.R., Prediction of intakes as an element of forage quality, in *Forage Quality, Evaluation and Utilization*, Fahey, G.C., Ed., ASA, Madison, WI, 1994.
21. Olson, K., Reducing methane emissions from beef cow herds in range-based management systems, in Ruminant Livestock Efficiency Program Conference Proceedings, EPA, USDA, Knoxville, TN, 1997.
22. Olson, K., personal communication, 2004.
23. Waghorn, G.C., Tavendale, M.H., and Woodfield, D.R., Methanogenesis from forages fed to sheep, *Proc. N.Z. Grassland Assoc.*, 64, 167, 2002.
24. Waghorn, G.C., Douglas, G.B., Niezen, J.H., McNabb, W.C., and Foote, A.G., Forages with condensed tannins — their nutritive value for ruminants, *Proc. N.Z. Grassland Assoc.*, 60, 89, 1998.
25. Jones, G.A., McAllister, T.A., Muir, A.D., and Cheng, K.-J., Effects of sainfoin (*Onobrychis viciifolia* Scop.) condensed tannins on growth and proteolysis by four strains of ruminal bacteria, *Appl. Environ. Microbiol.*, 60, 1374, 1994.
26. McCaughey, W.P., Wittenberg, K.M., and Corrigan, D., Methane production by steers on pasture, *Can. J. Anim. Sci.*, 77, 519, 1997.
27. Woodward, S.L., Waghorn, G.C., Ulyatt, M.J., and Lassey, K.R., Early indications that feeding *Lotus* will reduce methane emissions from ruminants, *Proc. N.Z. Soc. Anim. Prod.*, 61, 23, 2001.
28. Goodrich, R.D., Garrett, J.E., Gast, D.R., Kirick, M.A., Larson, D.A., and Meiske, J.C., Influence of monensin of the performance of cattle, *J. Anim. Sci.*, 58, 1484, 1984.
29. Johnson, D.E., Ward, G.M., and Ramsey, J.J., Livestock methane: current emissions and mitigation potential, in *Nutrient Management of Food Animals*, CRC Press, Boca Raton, FL, 1996, chap. 15.
30. Johnson, K.A., Westberg, H.H., Lamb, B.K., and Kincaid, R.L., Quantifying methane emissions from ruminant livestock and examination of methane reduction strategies, in Ruminant Livestock Efficiency Program Conference Proceedings, EPA, USDA, Knoxville, TN, 1997.
31. Müller-Özkan, E., Investigation of the Effects of Anion Rich Diets on the Carbohydrate Metabolism in the Bovine Rumen (*in vitro*), Ph.D. thesis, School of Veterinary Medicine, Hannover, 2002.
32. Mathison, G., Effect of canola oil on methane production in steers, *Can. J. Anim. Sci.*, 77, 545, 1997 (abstr.).
33. Boadi, D., Wittenburg, K.M., Scott, S.L., Burton, D., Buckley, K., Small, J.A., and Ominski, K.H., Effect of low and high forage diets on enteric and manure pack greenhouse gas emissions from a feedlot, *Can. J. Anim. Sci.*, 84, 445, 2004.

34. Willms, W.D., Kenzie, O.R., McAllister, T.A., Colwell, D., Veira, D., Wilmshurst, J.F., Entz, T., and Olson, M.E., Effects of water quality on cattle performance, *J. Range Manage.,* 55, 452, 2002.

35. Underwood, E.J. and Suttle, N.F., The *Mineral Nutrition of Livestock,* 3rd ed., CABI Publishing/CAB International, Midlothian, U.K.

36. Wittenberg, K., Making sense of minerals, in *Proc. Manitoba Grazing School,* Portage la Prairie, Manitoba, Canada, December 9–10, 1997, 56.

37. Lassey, K.R., Ulyatt, M.J., Martin, R.J., Walker, C.F., and Shelton, I.D., Methane emissions measured directly from grazing livestock in New Zealand, *Atmos. Environ.,* 31, 2905, 1997.

38. Ulyatt, M.J., Baker, S.K., McCrabb, G.J., and Lassey, K.R., Accuracy of SF_6 tracer technology and alternatives for field measurement, *Aust. J. Agric. Res.,* 50, 1329, 1999.

39. Hagarty, R.S., Strategies for mitigating methane emissions from livestock — Australian options and opportunities, in *Greenhouse Gases and Animal Agriculture,* Takahashi, J. and Young, B.A., Eds., Elsevier Science, Amsterdam, 2002.

40. Delaney, R., Chile's Agrosuper sells credits from pig waste to utilities. Available online at http://quote.bloomberg.com/apps/news?pid=71000001&refer=news_index &sid=aw7XyyBL9xXU, 2004.

41. McDougal, R., The carbon connection, *Conservator,* 25(2), 33, 2004.

42. Ominski, K.H., Tenuta, M., Flaten, D.N., Holley, R., Wittenberg, K.M., Entz, M., and Krause, D., Best management practices to improve environmental sustainability and productivity of grassland systems using hog manure, in *Living with Livestock Conference Proceedings,* Winnipeg, Manitoba, Canada, October 5–7, 2004, 42.

14 Mitigating Environmental Pollution from Swine Production

A.L. Sutton, B.T. Richert, and B.C. Joern

CONTENTS

14.1 INTRODUCTION

Nutrients, pathogens, and organic sources reaching our nations waters can adversely affect the tropic status and potential uses of the water body. If nutrients from manures and other sources are applied at excessive rates to cropland, increased accumulations of the nutrients in the soil can result in significant losses to bodies of water. Gaseous emissions of volatile compounds from manure are also a threat to our atmosphere. This chapter provides an overview of issues related to excess nitrogen and phosphorus in the environment, and agronomic, dietary, and managerial practices that may be used to reduce nutrient and gas emission impacts on the environment and sustain environmental stewardship.

14.2 ENVIRONMENTAL IMPACTS

Nitrogen (N), phosphorus (P), and other nutrients are essential elements for normal growth, development, and reproduction of both plants and animals. However, excessive nutrient levels, especially N and P, applied to cropland can potentially impair surface water and groundwater quality. It is well established that P is the limiting nutrient for phytoplankton production in lakes.[1-3] Although fewer data exist for streams and rivers, research indicates P also is a key element controlling productivity in these systems.[4,5] High P levels in surface waters accelerate the eutrophication process and often result in the excessive production of phytoplankton such as algae and cyanobacteria. The respiration of these organisms leads to decreased oxygen levels in bottom waters and, under certain circumstances (at night under calm, warm conditions), in surface waters.[6] These decreased oxygen levels can lead to fish kills and significantly reduce aquatic organism diversity.

Similarly N, especially in the ammonium form, can stress aquatic life at a very low concentration and is toxic to fish at excessive levels. The enrichment of N in water will enhance the biological degradation of organic matter resulting in algal growth and oxygen reduction in the waters. Excessive NO_3^- levels in drinking water can cause methemoglobinemia in young infants[7] and, at excessive concentrations, even in livestock. Ammonia emissions and gases created from digestion of manure slurry in pit systems of confinement facilities can lead to nasal and lung irritation in workers caring for livestock in these facilities. Zhang et al.[8] also reported that the air quality of confinement swine housing can have significant effects on respiration, as well as cause an increase in white blood cell count of humans subjected to typical confinement conditions. Pig manure contains a variety of organic compounds, complex to simple in nature, inorganic compounds, including considerable amounts of N, P, Ca, K, Zn, Cu, Cl, Mn, Mg, S, and Se, and indigenous microorganisms. Fecal N arises from undigested dietary protein, intestinal secretions (mucin, enzymes, etc.), sloughed intestinal cells, and intestinal bacteria. Urinary N, largely in the form of urea, arises from the breakdown of absorbed dietary amino acids that are in excess of the amounts needed for lean tissue protein synthesis and maintenance functions, and from the normal turnover of body tissue proteins. Most P excreted by pigs is in the feces. Fecal P arises from the dietary P that is undigested (mainly phytate) and/or unabsorbed, and from endogenous P secretions. Normally, only small amounts of urinary P are excreted unless the diet is grossly excessive in P. Other mineral concentrations in excreta depend on their absorption, retention, and release after metabolism in the animal. Currently N and P are the major nutrients of primary environmental concern. However, because of performance enhancement, higher levels of Zn and Cu may be fed to pigs. Thus, limiting Zn and Cu excretion also may become an important feeding practice to minimize their potential as environmental pollutants.

Odors and gaseous compounds emitted from swine operations are a major deterrent for the growth on the industry because of neighbor complaints, potential health concerns, and deposition of particulates (acid rain) on the ecosystem. Odorous

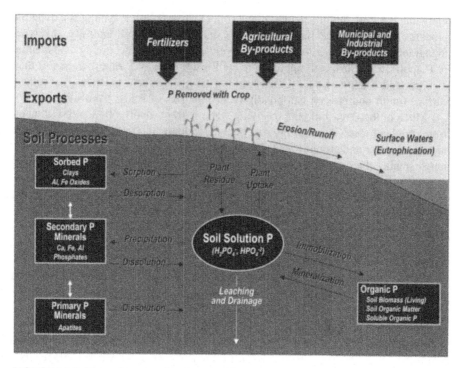

FIGURE 14.1 The soil–water phosphorus cycle.

and gaseous compounds are emitted from manure immediately after excretion due to microbial metabolism in the digestive tract of the animal. Further decomposition occurs in storage, resulting in significant gaseous emissions and odors that have an impact on air quality. These include nitrogenous and sulfur compounds, volatile organic compounds, greenhouse gases (CH_4, CO_2, NO_x), and particulates.

14.3 AGRONOMIC CONSIDERATIONS

14.3.1 Phosphorus

The soil-water P cycle is illustrated in Figure 14.1. Both organic and inorganic P are present in soil, but only inorganic P is the form taken up by plants. Soil P dynamics are largely influenced by soil pH, clay content and mineralogy, amorphous iron and aluminum, and organic matter. Inorganic P is the predominant P form in both manures and commercial fertilizers. Depending on soil pH and mineralogy, inorganic P can be sorbed on the surface of clays and amorphous iron and aluminum compounds or precipitated as mineral salts until utilized by plants. Organic forms of P from crop residues, soil organic matter, and manures can be mineralized by soil microorganisms and become available for plant uptake. Conversely, inorganic P can be immobilized to organic P forms not available for plant uptake. In addition, some organic P forms excreted in manure may displace sorbed inorganic and increase inorganic P runoff and/or leaching in the soil. Obviously, soil P cycling is a dynamic process. The extent of P runoff from soils depends on rainfall intensity, soil type, topography, soil moisture content, crop cover, and the form, rate, timing, and method of P application. Surface P applications will result in more P runoff from soil than incorporated P applications.[9] Conservation best management practices that reduce surface runoff and erosion can greatly reduce the risk of P loss from soils.

Much of the P reaching the receiving water is from runoff, often with sediment, from cropland receiving high rates of manure or inorganic fertilizers. While P loss to surface water and groundwater via P leaching through the soil profile is generally much smaller than runoff P losses, excessive P applications to soils over time will move P to lower portions of the soil profile, and this P can discharge into tile drains, ditches, and eventually streams (Figure 14.1). Significant tile discharges of P also can occur via macropore transport of manure to tile lines after land application, especially during the dry season when cracks form in the topsoil. Additionally, sandy soils with rapid drainage and low anion exchange sites generally have greater P leaching potential than heavier textured clay-type soils.

Swine manure N, P, and potassium (K) composition is not properly balanced for plant uptake by typical crops grown in production agriculture. The relative ratio of N, P_2O_5, and K_2O in manure from pigs fed commercial diets after storage in an under-floor liquid pit is approximately 1:1:1. When based on fertilizer recommendations for N and crop removal rates for P_2O_5 and K_2O, corn grain production requires roughly a 3:1:1 ratio, and if corn is grown for silage, then approximately a 2:1:2 ratio of N, P_2O_5, and K_2O is required. Therefore, if under-floor liquid pit manure is applied to meet the N requirement of corn grain pro-duction, manure P application will be approximately three times crop P removal under an ideal manure application scenario. Uncovered earthen pits and lagoons will typically lose more N than under-floor pits, and if agitated prior to manure application, will have manure N:P_2O_5 ratios less than 1:1. Nitrogen losses for applied manure that is not injected or immediately incorporated can be up to 30%[10] within 4 days of application. Additional N losses occur as the time between manure application and crop utilization increases. In addition, excessive P levels in animal diets increase animal manure P excretion, and land application of this manure to soil can increase potential P losses from fields to surface water and groundwater resources. Ideally, if the ratio of N, P, and K in manure could be altered by nutritional means to more closely meet specific crop nutrient requirements, it would alleviate a significant problem currently facing many pork producers uti-lizing manure as a crop nutrient resource.

Current regulations are forcing pork producers to apply manure at agronomic rates based on the most limiting nutrient, which in most cases is P. However, there is the potential that producers can "bank" P for short periods of time if there is sufficient land available to rotate the fields for manure application in subsequent years. A common practice may be to apply the manure to meet the N requirement of the crop, but apply the manure to the field only every 3 years. A rotation for manure application to crop fields must be established for manure applications to meet crop P needs for the crop rotation grown on specific fields.

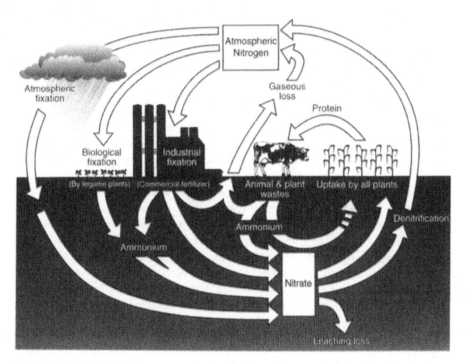

FIGURE 14.2 The soil–water–atmospheric nitrogen cycle.

14.3.2 NITROGEN

The soil–water–atmosphere N cycle,[11] presented in Figure 14.2, is only a part of the overall N cycle. Most soil N is sequestered in soil organic matter and only about 1% of soil N is available to plants as nitrate (NO_3^-) or exchangeable ammonium (NH_4^+) at any one time. Soil organic matter decomposition, manures, and commercial fertilizers are the primary inputs to the soil N cycle. Organic nitrogen present in organic matter, manures, and other organic N sources must be mineralized to ammonium (NH_4^+) before it can be taken up by plants, held in an exchangeable form on soil cation exchange sites, or fixed by various clay minerals. If mineralization takes place at the soil surface, ammonia volatilization can be a significant loss pathway. The ammonium fraction of manure also can be lost via ammonia volatilization if manure is left on the surface, especially under warm, windy conditions or if the soil pH is greater than 7.0. During the normal crop growing season, solution and exchangeable NH_4^+ is converted to NO_3^- fairly rapidly in the soil environment. Nitrate N may be taken up by plants, leached below the root zone, or lost to the atmosphere as NO_x or N_2 gas via denitrification. Both nitric oxide and nitrous oxide gases contribute to greenhouse warming while nitric oxide also plays a role in the production of tropospheric ozone and is known to be the main component of acid rain.[12] With the current interest in greenhouse gas emissions, gaseous N losses will likely be more closely scrutinized and potentially subject to regulation in the future.

Current agricultural practices in the Mississippi River Basin contribute approximately 2.25 to 3.6 kg of nitrogen per agricultural hectare to the Mississippi River each year. Similar loss of nutrients is occurring in the livestock dense areas of Europe. Vitousek et al.[12] report the rate of nitrogen deposition in the Netherlands is the highest in the world at a rate averaging 40 to 90 $kg \cdot ha^{-1} \cdot yr^{-1}$. Timing and method of manure application can significantly affect potential N loss to the environment. In the midwestern U.S., much of the manure is applied in the fall and early winter when crops are either not present or not actively growing. In general, the greater the length of time between manure application and crop uptake, the greater the risk of N loss. For fall applications of manure, cover crops can take up some N that may otherwise leach or be denitrified during the winter and early spring prior to planting grain crops. Timing of manure application is also important from the aspect of commercial fertilizer application as is reported by Torstensson and Aronsson.[13]

A comparison of N leaching from manure or commercial fertilizer applied to ground covered with or without a catch crop was conducted in Sweden. Catch crops are fall planted crops, such as perennial ryegrass or winter rye used in this study, which serve as sources for nutrient uptake during manure application while the commodity crop is not being grown. The catch crop is then tilled back into the soil and the nutrients captured in the catch crop may be recycled back into the nutrient cycle for the next growing season. The authors report that when either a single or double application of manure was applied to ground without a catch crop there was a 15 and 34% increase in average N leaching, respectively, compared to commercial fertilizer application. It was observed that while catch crops reduced N leaching from commercial fertilizer application 60%, when a double application of manure was applied to a catch crop there was only a 35% reduction in leaching due to greater applications of mineral N in the spring with manured treatments compared to fertilized treatments.

As is expected, ammonia release is subject to temperature as well as the above-mentioned time of manure application and other factors. In a heavily concentrated swine and poultry production area in North Carolina, ammonia emission was directly correlated with air temperature and it was reported that as much as 50% of the total amount of ammonia lost from swine effluent lagoons in a year is lost during summer months.[14] Robarge et al.[14] suggest that because the partial pressure of ammonia increases with an increase in temperature and this leads to increased ammonium ions in the aqueous phase, the increased temperature during the summer months would cause greater deposition of ammonium ions in rain water and potentially greater deposition in alternative ecosystems.

Nitrification inhibitors can aid in retaining fertilizer and manure N in the soil and minimize nitrate leaching by inhibiting the microbial conversion of ammonium N to nitrate N. Early research has shown that use of commercial nitrification inhibitors will reduce nitrate leaching from injected swine slurry manure applications when applied in the fall and summer seasons. Varel[15] showed that phosphoryl diamide and triamide compounds can be added to manure slurries and inhibit urease activity resulting in minimal volatile N losses. Immediate injection of manure to cropland results in <5% volatile N losses compared to 20 to 30% volatile N losses with surface application in a 48-h period after application.[16]

TABLE 14.1
Theoretical Model of Effects of Dietary P Level and Phytase Supplementation, 91-kg Pig

Dietary P, %	Intake	P, g d^{-1} Retained	Excreted	Change from Industry Avg, %
.70	21.0	4.8	16.2	+57
.60	18.0	4.8	13.2	+32
.50	15.0	4.7	10.3	0
.40 (NRC, 1988)	12.0	4.5	7.5	−27
.30	9.0	2.5	6.5	−37
.30 + phytase	9.0	4.5	4.5	−56

Source: Adapted from Cromwell, G.L. and Coffey, R.D., *Proc. Pork Acad.*, 1995, 43.

14.4 FEED FORMULATION

14.4.1 PHOSPHORUS

Many feed ingredients in swine diets are high in phytate, and certain small grains (wheat, rye, triticale, and barley) contain endogenous phytase that can release the phytic P. This creates a wide variation in the bioavailability of P in feed ingredients. For example, the P in corn is only 14% available while the P in wheat is 50% available.[17] The P in dehulled soybean meal is more available than the P in cottonseed meal (23 vs. 1%), but neither source of P is as highly available as the P in meat and bone meal (90%), fishmeal (93%), or dicalcium phosphate (100%). Due to this great variation in the availability of P in feed ingredients coupled with a lack of precise information on the requirements of P for pigs, nutritionists have great difficulty in estimating the available P levels in the diet. Consequently, additional supplemental P is added to the diet, oftentimes in excess for a safety margin and excess P is excreted in manure. Reducing the safety margin alone would potentially decrease P input by 8 to 10% in the diet and excretion by 20 to 30% (Table 14.1).[18]

Supplementing the diet with the enzyme, phytase, is an effective means of increasing the breakdown of phytate P in the digestive tract and reducing the P excretion in the feces. Using phytase allows a lower P diet to be fed because a portion of the unavailable phytate P in the grain and soybean meal is made available by the phytase enzyme to help meet the pig's P needs. Table 14.1 shows the theoretical model for using dietary P levels and phytase supplementation on P excretion. Numerous studies have indicated that the inclusion of phytase increased the availability of P in a corn–soybean meal diet by threefold, from 15% up to 45%.[19,20] Phosphorus excretion was reduced from 31 to 62% when the diets for growing through finishing pigs were changed with a lower inorganic P level and addition of phytase or wheat bran (10 to 20% of the diet) compared to typical corn–soybean meal diets.[21] The availability and utilization of amino acids and other trace minerals have been shown to increase in pig studies with phytase supplementation resulting in lower excretion of elements such as N, Zn, Cu, Mn, and Ca.[22] Radcliffe et al.[23] showed an increase in P and Ca digestibility with the addition of phytase in low P and low Ca diet. Qian et al.[24] showed that maintaining a relative narrow Ca:P ratio (1.2:1 vs. 2:1) is critical with low P diets and when phytase is used. In their study, performance and P and Ca digestibility were reduced with the wider ratio.

Smith et al.[25] and Baxter et al.[26] showed that use of phytase and/or LPA corn will change the form of P excreted with an increased percentage of water-soluble or soluble reactive P (SRP) in the manure. In the Smith et al.[25] study, use of phytase in the diet reduced SRP by 22%. There has been an environmental concern about increased SRP in poultry manure and potentially in manure from pigs fed phytase, especially if surface applied to cropland or grassland, since it has been shown to increase runoff potential. However, if incorporated in the soil, this impact was not a concern.[9]

If P is the limiting nutrient for land application, a 50% reduction in excreted P by pigs would mean that pork producers would need 50% less land for manure application and minimize any potential impact on water quality. Obviously, this will have a major impact if environmental regulations are being proposed to regulate swine waste application on a P basis. While the impact of reducing dietary P below NRC requirements, utilizing exogenous phytase and more available P sources seems like a partial solution, its impact on whole-body P including lean tissue mass and bone health as well as on other essential minerals still needs to be investigated. The available P requirements and mineral composition of today's genetic lines of pigs has not been determined and must be researched to produce greater reductions of P excretion from diet manipulation in the future.

14.4.2 NITROGEN

In the review by Kerr,[27] the impact of amino acid supplementation with low crude protein (CP) diets to reduce N excretion ranged from 3.2 to 62% depending on the size of the pig, level of dietary CP reduction, and initial CP level in the control diet. The average reduction in N excretion per unit of dietary CP reduction was 8.4%. Table 14.2 shows the theoretical model for the impact of reducing dietary protein and supplementing with amino acids in a 91-kg pig. Sutton et al.[28] showed that reducing the CP level in corn–soybean meal growing–finishing diets by 3% (from 13 to 10% CP) and supplementing the diet with lysine, tryptophan, threonine, and methionine reduced ammonium and total N each in freshly excreted manure and stored manure by 28 and 43%, respectively (Table 14.3 and Table 14.4). Hobbs et al.[29] showed that reducing the CP in practical diets from 21 to 14% CP plus synthetic amino acids in growing diets and from 19 to 13% CP plus synthetic amino acids in finishing diets reduced N excretion by 40% and also reduced concentrations of a majority of odorants in the slurry. In a practical feeding study, Kay and Lee[30] used the same diets and showed 41% total reduction in slurry N output. Reducing the intact CP content of the diet (generally soybean meal) and replacing it with crystalline lysine and corn will reduce N input to the diet by 13.2%. In studies at Missouri[31,32] the protein content of diets for early finishing barrows (50 to 80 kg) was reduced four percentage units (15.34 vs. 11.43%) with the addition of lysine, threonine, tryptophan, and methionine with

TABLE 14.2
Theoretical Model of the Effects of Reducing Dietary Protein and Supplementing with Amino Acids on N Excretion by 91-kg Finishing Pig[a]

Diet Concentration	14% CP	12 % CP + Lysine	10% CP + Lysine + Threonine + Tryptophan + Methionine
N Balance			
N intake, g d[-1]	67	58	50
N digestested and absorbed, g d[-1]	60	51	43
N excreted in feces, g d[-1]	7	7	7
N retained, g d[-1]	26	26	26
N excreted in urine, g d[-1]	34	25	17
N excreted, total, g d[-1]	41	32	24
Reduction in N excretion, %	—	22	41
Diet costs,[b] $ kg[-1]	0.142	0.138	0.151
Change in dietary costs,[b] $ kg[-1]	0	–$0.004	+$0.009

[a] Assumes an intake of 3.0 kg d[-1], a growth rate of 900g d[-1].

[b] Delivered prices used as of 6/1/04: Corn, $0.09 kg[-1]; SBM (48%), $0.302 kg[-1]; Choice White Grease, $0.364 kg[-1]; Dical. Phos., $0.346 kg[-1]; Limestone, $0.064 kg[-1]; Salt, $0.161 kg[-1]; Swine Vitamin Premix, $1.050 kg[-1]; Swine Trace Mineral Premix, $0.706 kg[-1]; Se Premix, $0.273 kg[-1]; Tylan 40, $0.273 kg[-1]; Lysine-HCl, $3.226 kg[-1]; DL-Methionine, $3.043 kg[-1]; Threonine, $3.391 kg[-1]; Tryptophan, $35 kg[-1].

Source: Adapted from Cromwell, G.L. and Coffey, R.D., *Proc. Pork Academy,* 1995, 39.

no differences in any performance criteria. Pigs fed the control diet and those fed the low-protein diet had similar carcass protein and fat, and N retention. However, N excretion of pigs fed the low-protein diets was 38% lower (31.5 vs. 51.2 g d[-1]). Results from late-finishing pigs (85 to 120 kg) demonstrated that an all corn diet supplemented with lysine, threonine, tryptophan, methionine, isoleucine, and valine gave similar pig performance with similar carcass protein and fat, and N retention.[33,34] Nitrogen excretion was reduced 48% with the low-protein amino acid supplemented diet. However, due to the cost of isoleucine and valine, addition of soybean meal to meet these amino acids and the addition of lysine, threonine, tryptophan, and methionine would be more cost-effective and result in a 30 to 40% reduction of N excretion without affecting pig performance.

Kendall et al.[35] used a reduced CP (12.2% CP) corn–soy diet with synthetic lysine, methionine, tryptophan, and threonine fed to 27 kg pigs for 9 weeks and compared to pigs fed a high CP corn–soy diet (16.7% CP). Slurry manure contents

TABLE 14.3
Effect of Diet on pH and Nitrogen
Components in Fresh Manure[a]

Diet (%CP)	pH	DM	NH$_3$-N	TKN[b]
		%	%DM	%DM
Deficient (10)	7.80[a]	17.3[a,b]	3.47[b]	7.40[b]
Suppl. (10 + AA)	7.33[b]	18.4[a]	2.61[c]	5.90[c]
Standard (13)	7.84[a]	16.0[b]	3.61[b]	8.16[b]
Excess (18)	8.13[a]	12.9[c]	4.35[a]	10.13[a]

[a] Different letter superscripts within a column are significant ($P < 0.05$).
[b] TKN = total Kjeldahl nitrogen.

Source: Sutton, A.L., et al., *J. Anim. Sci.*, 77, 430, 1999.

TABLE 14.4
Effect of Diet on pH and Nitrogen
Components in Stored Manure[a]

Diet (%CP)	pH	DM	NH$_3$-N	TKN[b]
		%	Mg l^{-1}	Mg l^{-1}
Deficient (10)	7.58[a]	5.40[b]	4375[c]	5631[c]
Suppl. (10 + AA)	6.94[b]	6.47[a]	2986[d]	4026[d]
Standard (13)	7.80[a]	5.64[b]	5239[b]	7012[b]
Excess (18)	7.97[a]	5.75[b]	6789[a]	8912[a]

[a] Different letter superscripts within a column are significant ($P < 0.05$).
[b] TKN = total Kjeldahl nitrogen.

Source: Sutton, A.L., et al., *J. Anim. Sci.*, 77, 430, 1999.

had a lower pH (0.4 units), lower total N (40%), and lower ammonium N (20%) from pigs fed the reduced CP diet compared to the slurry manure from pigs fed the high CP diet.

14.4.3 OTHER MINERALS

Copper sulfate addition to the diet (125 to 250 ppm) has been shown to improve feed efficiency 5 to 10% and to reduce odors[36] but will significantly affect copper excretion.[37] Adding copper sulfate at 125 or 250 ppm to the diet will increase Cu dietary input by 7.8 and 16.7 times a control diet (15 ppm of Cu), respectively. Use of lower levels of organic forms of Cu that provide similar growth promotion benefits increases the Cu excretion levels only 2.1 times the control. Use of chelated minerals or organic forms can reduce the excretion of a variety of minerals by 15 to nearly 50%. Researchers at Michigan State University and North Carolina State University[38] have reported that reduced dietary Cu, Zn, Mn, and Fe concentrations fed throughout the life cycle of pigs for three parities did not depress growth or alter feed efficiency and enzymatic activities indicative of health parameters.

14.5 FEED MANAGEMENT

Diet ingredient sources, forms, and levels greatly influence nutrient availability and excretion levels. Understanding the bioavailability of nutrients from feed sources is critical for formulating diets that will meet the productive needs of the animal without excesses. Any management procedure that improves the overall efficiency of feed utilization in a swine herd will generally reduce the total amount of manure produced, and should reduce nutrient excretion. Controlling feed wastage improves herd feed conversion and reduces nutrient losses. Use of wet–dry feeding systems will reduce manure volume by 30 to 50%; however, nutrient contents in the manure can increase by about 30 to 50%. Maintaining pigs under comfortable environmental conditions with proper ventilation, temperature, humidity, space, and general well-being will improve feed utilization and reduce nutrient excretions. Raising genetically lean pigs, using growth promoters such as antibiotics, β-agonists, growth hormones, controlling diseases and parasites, and using good management practices are further examples of how one can improve feed conversion efficiency and reduce nutrient excretion by 10 to 15%.

Fine grinding and pelleting are also effective in improving feed utilization and decreasing dry matter (DM), N, P, and other mineral excretion. By reducing the particle size, the surface area of the grain particles is increased, allowing for greater interaction with digestive enzymes. Hancock et al.,[39] based on a summary of eight pelleting trials for swine, reported that pelleting resulted in an average 6% improvement in average daily gain and a 7% improvement in feed efficiency. Wondra et al.[40] reported a 23% decrease in DM excretion and a 22% decrease in N excretion when finishing diets were pelleted. Henry and Dourmad[41] reported for growing–finishing pigs that for each 0.1 percentage unit decrease in feed-to-gain ratio there was a 3% decrease in N output.

Dividing the growth period into more phases with less spread in weight between groups allows producers to more closely meet the pig's protein and other nutrient requirements. Also, since gilts require more protein than barrows, penning barrows separate from gilts will allow lower protein levels to be fed to barrows without compromising leanness and performance efficiency in gilts.[42] Henry and Dourmad[43] reported that N excretion could be reduced approximately 15% when feeding of 14% CP diet was initiated at 60 kg bodyweight, rather than the continuous feeding of 16% CP grower diet to market weight. A 14.7% reduction in urinary N excretion was reported when a multiphase feeding program was compared to a two-phase feeding program.[44] Ammonia emission also was reduced 16.8%.

14.5.1 BY-PRODUCT FEEDS AND ADDITIVES

By-product feeds can serve as a source of nutrients in pig diets. Often, by-product feeds, such as distiller's grains, corn gluten meal, wheat middlings, etc., are included in the diet if they are readily available and economically justified, especially when there is a shortage or increase in prices of conventional feed sources (i.e., corn and soybean meal). Also, due to the processing methods employed, nutrients in the by-products become more biologically available and can potentially reduce nutrient excretion if the by-product nutrients can be balanced in the diet.

Distillers' dried grains with solubles (DDGS), which contain from 0.62 to 0.87% P, have a higher concentration of available P than corn, other cereal grains, and cereal co-products. For DDGS, available P averages 77% compared to a range of 12 to 30% for corn, other cereal grains, and cereal co-products. Studies by Spiehs et al.[45] showed that when formulating diets on a total P basis, the percentage of P retained tends to increase when 10 and 20% DDGS is added to grower pig diets compared to a control corn–soybean meal diet. Similar results were observed when feeding finisher pig diets containing up to 30% DDGS. These results suggest that the P in DDGS is more available than in the corn–soybean meal diet. Spiehs et al.[45] suggested that adding up to 20% DDGS to grower and finisher diets will reduce supplemental inorganic P needs in the diet and, consequently, could reduce P levels in the manure assuming that the diet is formulated on an available P basis.

Milling processes to degerm and dehull corn have created new interest as feed ingredient sources with reduced levels of phytic P. The removal of fiber and germ from corn has recently been reported to result in a 56% reduction in dry matter excretion and 39% reduction in nitrogen excretion in short-term digestion trials.[46] Indigestible P was reduced by 15% with degermed, dehulled corn.[47] Lee et al.[48] formulated a low excretion diet by reducing the CP, adding synthetic amino acids and 5% soybean oil, but omitting vitamins and mineral premixes. Processing of the low excretion diet included grinding to 600 μm, stream conditioning, expander processing, and pelleting. Phytase was spray-applied post-pelleting. Dry matter excretion was reduced by 35%, N intake and excretion were decreased by 22 and 39%, and P intake and excretion were reduced by 27 and 51%, respectively, for pigs fed the low excretion diet compare to a standard diet. Lysine and threonine digest-ibility and energy parameters were also improved by the low excretion diet. Smiricky et al.[49] used feed steeping techniques and feed degrading enzymes (0.3% α-galac-tosidase, 0.1% cellulose, 0.2% hemicellulase, 0.1% pectinase/arabinase, and 0.05% xylanase) to improve the digestibility of the diet for growing pigs. They noticed an improved ileal and total tract digestibility for DM, N, and amino acids with the feed degrading enzymes, steeping, and reduced feed particle size. Future processing techniques may result in feed products that will allow the swine industry to precision-feed pigs with higher digestible nutrient levels and lower nutrient excretion; however, economic issues may limit the implementation of these technologies.

Ractopamine (Paylean®) is a feed additive for swine that has the potential to further increase the rate and efficiency of lean tissue growth. DeCamp et al.[50] and Hankins et al.[51] fed pigs a 16.1% CP-ractopamine (RAC) diet (20 mg kg^{-1}) that excreted 14.9% less total N compared to the 13.8% CP control diet, with the majority of the N reduction accounted for by reduced urinary N excretion. Slurry pH was reduced 0.5 units and NH_4-N was reduced 8 to 21% from pigs fed ractopamine. In an attempt to maximize N utilization and minimize N excretion, a 13.8% CP + RAC and synthetic amino acid diet was also fed that reduced N excretion by 35.7% compared to the 13.8% CP control diet.

14.6   GENETIC MODIFICATIONS

The development of genetically modified grains can potentially enhance the utilization of nutrients and reduce nutrient outputs. An example of a genetically modified grain is the low phytic acid (LPA) corn that has a lower concentration of phytic P, which is not available to the pig. Thus, the LPA corn provides more available P for the pig and consequently less P excreted. Recent work with LPA corn hybrids[26,52,53] has shown that P excretion can be reduced 20 to nearly 50% depending upon the combination of P reducing techniques used. In research at Purdue,[54] use of LPA corn to replace normal corn increased the utilization of P in pigs by 30%. With phytase alone, P utilization was increased by 27%. When both LPA corn and phytase were fed together in the diet, P utilization was increased by 53%. Phosphorus excretion was reduced 21% with phytase alone with normal corn, was reduced 23% with LPA corn only, but was reduced by 41% when both LPA corn and phytase were combined together. LPA soybean meal was fed to growing pigs and reductions in P excretion were 17% as compared to normal soybean meal. The combination of LPA corn, LPA soybeans, and phytase fed to growing pigs enhanced P digestibility by 78% and reduced P excretion by 43% compared to pigs fed normal corn and soybeans without phytase.[55] The combination of LPA corn and phytase can nearly eliminate all supplemental dietary P in pig grow-finish diets. To date, the genetically enhanced corn has not been commercially produced so it is not available as a current viable solution for the reduction of P excretion from commercial pork production.

Researchers at the University of Guelph[56] have developed a genetically modified pig that uses plant P more efficiently. These transgenic pigs have the capability of secreting significant quantities of phytase from the salivary glands, which can react in the stomach of the pigs and release phytic P from the cereal grain diet. Compared to non-transgenic pigs fed a normal diet, results have shown a 75% reduction in P excretion from nursery pigs and from 56 to 67% reduction in P excretion from finishing pigs with no supplemental P in the diet. In addition, mineral availability in the diet was enhanced with the transgenic pig. Currently, the transgenic pig is not commercially available.

14.7   ODOR REDUCTION

Odorous volatile organic compounds (VOC), short-chain volatile fatty acids (VFA), and other volatile carbon-, nitrogen-, and sulfur-containing compounds in feces from microbial fermentation in the gastrointestinal tract (GIT) of the pig can be emitted immediately after excretion. Furthermore, the release of ammonia (NH_3) from the urine due to the enzymatic conversion of urea can occur within a short time after excretion. Several chemical, biological, and physical technologies have been developed for the control of odors and gaseous emissions from swine operations. These technologies are not discussed in this chapter. Recently, the effect of diet composition on excretion products related to odorous compounds was studied. One approach is to provide the pig, as closely as possible, with essential available nutrients based on its genetic potential and stage of growth, so that nutrient excretion is minimal and a lower potential for creating compounds responsible for odor production. Another concept is to manipulate the bacteria in the GIT of the pig by inhibiting certain microbial groups or altering the fermentation of existing bacteria, thus, controlling odorous end products. Finally, changing diet composition may change the physical characteristics of urine and feces that would control odor production.

14.7.1 NITROGEN MANIPULATION

Several recent studies have shown that with a 1% reduction in the CP of a pig diet and supplementation with the limiting synthetic amino acids will reduce ammonia emission into the air by 10 to 12%.[57] In addition, reduced concentrations of a majority of odorants in pig slurry and the air samples were observed. In practical feeding studies with CP levels reduced from 19 to 13% showed a 47 to 59% reduction in NH_3 emissions from building air.[30] This reduction in dietary N also reduced manure odors by 40 to 86% and decreased p-cresol by 43%.[29] In group feeding studies, Kendall et al.[35] showed that reducing the CP (4.5%) and supplementing the diets with synthetic amino acids can effectively reduce ammonia and hydrogen sulfide by 40% each, and odor emissions by 30% from confinement buildings. Growing–finishing gilts were fed different levels of sulfur-containing amino acids and sulfur mineral sources to determine odors and nutrients from fresh manure and manure stored in anaerobic systems.[54,58] The pH, ammonia N, and total N in fresh manure were lower for pigs fed low CP-amino acid supplemented diet compared to the standard commercial diet. In addition, hydrogen sulfide emission was less (by up to 48%) with the low crude protein and low sulfur mineral diet.

14.7.2 ADDING FERMENTABLE CARBOHYDRATES

Another possibility to reduce emission of NH_3 would be to alter the ratio of N excretion in urine and feces by addition of fermentable carbohydrates. By reducing the N excretion in urine as urea, and shifting the N excretion in feces in the form of bacterial protein, NH_3 volatilization can be reduced. Complex carbohydrates such as β-glucans, and other non-starch polysaccharides (NSP) can influence endogenous N excretion at the terminal ileum and microbial fermentation in the large colon resulting in increased bacterial protein production.[59,60] Mroz et al.[59] showed that manure from pigs fed cellulose significantly reduced NH_3 emissions from the manure compared to manure from pigs fed cornstarch, hemicellulose, and pectin. This was attributed to the partitioning of the N excretion more into feces, compared to urine and an increase in bacterial protein in feces.

Canh et al.[57] obtained similar results with the inclusion of 30% dried sugar beet pulp that reduced NH_3 emission from pig slurry by 47% compared to a control (Table 14.5). In another study, Canh et al.[61] showed that, as the levels of NSP (coconut expeller, soybean hulls, dried sugar beet pulp) were increased in the diet (from 15 to 49%), greater reductions in levels of NH_3 emission were possible (from 6.4 to

TABLE 14.5
Composition of Feces, Urine, and Slurry from Pigs Fed Different Diets and Ammonia Emissions from Slurry

	Diet				
Composition and Source (4)[a]	Grain (4)[a]	By-Products (4)	Tapioca (4)	Sugar Beet Pulp (4)	p^b SEM
NH$_4$-N, g kg^{-1}					
Feces	0.67	0.76	0.66	0.66	NS0.06
Urine	0.22[c]	0.13[d]	0.31[e]	0.40[f]	***0.03
Total N, g kg^{-1}					
Feces	7.99[c]	9.32[d]	9.18[d]	8.59[c]	*0.47
Urine	6.61[c]	5.30[d]	6.63[c]	4.90[d]	***0.30
Urea, mmol kg^{-1}					
Urine	195.2[c]	157.0[d]	196.1[c]	122.9[e]	**11.0
pH					
Feces	6.84[c]	6.85[c]	6.95[c]	6.47[d]	*0.10
Urine	7.48[c]	8.19[d]	7.03[c.e]	6.77[e]	***0.19
Slurry	7.64[c]	7.80[c]	7.11[d]	6.67[e]	***0.13
NH$_3$ emission, mmol d^{-1}					
Slurry	32.69[c]	30.10[c]	21.59[c]	14.03[d] ***3.18	
N lost[g]	23.69[c]	23.76[c]	21.59[c]	14.03[d]	***1.3

[a] Number of observations in parentheses.
[b] Probability of a significant treatment effect. * $P < 0.05$; ** $P < 0.01$; *** $P < 0.001$; NS = not significant.
[c,d,e,f] Means within a row lacking a common superscript letter differ.
[g] N lost after 7 d, calculated as percentage of daily N excretion.
Source: Adapted from Canh.[57,60,61]

35.8%). Soybean hulls had the greatest effects on reduced NH$_3$ emission (from 16.9 to 35.8%) but also increased VFA in feces. Hawe et al.[62] reported that increased dietary fiber from sugar beet pulp (400 mg g^{-1}) increased the daily elimination of skatole and indole, and indole concentration of feces. When lactose (25 m g^{-1}) was added to the diet, it did not affect indole concentrations but did reduce skatole concentrations and daily excretion. Kendall et al.[54] reduced the dietary CP (by 3.25%), added 5% soybean hulls, and used a nonsulfur trace mineral premix and LPA corn in low nutrient excretion (LNE) diets for growing pigs that resulted in a 49.8% reduction in exhaust air ammonia, 43.3% lower hydrogen sulfide, 38.6% lower odor detection threshold for building air. The stored manure from pigs fed the LNE diet had 27.0% less total-N, 29.5% lower ammonium-N, and 51.7% less excreted P on a DM basis. DeCamp et al.[63] showed that the addition of 10% soy hulls with 3.4% fat to practical corn–soybean meal diets reduced aerial ammonia emission by 20% in room air, 32% of hydrogen sulfide, and 11% lower odor detection threshold. Nitrogen accumulation in manure was increased 21%, pH of the manure was decreased, and volatile fatty acid concentrations were increased by 32% in manure from pigs fed diets with soy hulls inclusion.

14.7.3 MICROBIAL MANIPULATION

Attempts have been made to isolate and identify the microbial populations in the digestive systems of pigs. An excellent review by Mackie et al.[64] presented the role and impact of microbial metabolism on odor generation from livestock. They stated that the bacterial genera involved with deamination of amino acids were *Bacteroides, Prevotella, Selenomonas, Butyrivibrio, Lachnospira, Eubactreium, Fusobacterium, Clostridium, Peptostreptococcus,* and *Acidaminococcus.* The production of indoles and phenols was primarily from microbial metabolism of amino acids. In another review, Yokoyama and Carlson[65] reported that several *Clostridia* sp., *Escherichia coli,* and *Bacteroides thetaiotaomicron* can be involved with indole and skatole production. Ward et al.[66] isolated an obligate anaerobe of the *Lactobacillus* sp. that decaboxylated *p*-hydroxyphenylacetic acid to 4-methylphenol (*p*-cresol) in swine feces. Compounds such as oligosaccharides (fructooligosaccharides, mannoligosac-charides, sucrose thermal oligosaccharide caramel, inulin, arabinogalactan, galac-tan), dairy by-products (lactulose, lactitol, lactose, whey), and organic acids (propi-onic, fumaric, citric) have been added to manipulate the microflora populations but with variable results. Fructooligosaccharides have been shown to alter VFA patterns in the lower GIT (reduce proportion of acetate and increase the proportion of propionate), reduce total aerobes, predominantly coliforms, increase bifidobacteria,[67] and reduce odorous compounds from swine manure.[68]

14.7.4 PHYSICAL CHARACTERISTICS

Gaseous emissions from slurries are affected by environmental conditions such as temperature, oxygen content, humidity, and air exchange rate, as well as pH, buff-ering capacity, and DM content of the slurry. Canh et al.[60] showed that increasing the NSP content and decreasing the electrolyte balance (dEB) of the diet reduced the pH of pig slurry. Inclusion of 30% sugar beet pulp (with 31.2% NSP) reduced the pH of slurry by 0.44 to 1.13 pH units lower than a by-product diet (with 18.2% NSP), grain-based diet (with 13.8% NSP), and a tapioca-based diet (with 13.5% NSP). The decreased dietary electrolyte balance (expressed as mEq Na + K − Cl) in the diet reduced the pH of urine and subsequent slurry.

Geisting and Easter[69] summarized studies incorporating acids (citric, hydrochlo-ric, propionic, fumaric, and sulfuric) at dietary levels of 1 to 4% showing variable results in pH effects on digesta and growth effects on swine. Risley et al.[70] also showed that addition of fumaric or citric acids (1.5%) had very little effect on pH, volatile fatty acids, or chlorine concentrations of intestinal contents of swine. Rad-cliffe et al.[23] showed that inclusion of 1.5 and 3.0% citric acid reduced stomach pH, improved gains and feed efficiency, and increased Ca digestibility, but not P digest-ibility. In another trial, 2.0% citric acid had no effect on growth parameters or diet nutrient availability. Canh et al.[71] and Mroz et al.[72] showed that dietary calcium salts and electrolyte balance significantly influenced urinary pH and subsequent pH and

NH_3 emission from pig slurry. Mroz et al.[73] showed that increasing the levels of calcium benzoate (2, 4, 8 g kg^{-1} feed) in the diet of sows significantly reduced pH of urine from 7.7 to 5.5 and reduced NH_3 emissions up to 53%. In nursery diets, Colina et al.[74] observed a significant reduction in ammonia emissions in the rooms housing the nursery pigs. However, feed intakes were lowered and performance was reduced when 1.96% calcium chloride was added to the diet. Van Kempen et al.[75,76] showed the benefits of using adipic acid in pig diets to reduce pH and ammonia emissions but it did not improve lysine utilization in the pig.

14.8 SUMMARY

Nutrients, pathogens, and organic sources reaching our nation's waters can adversely affect the tropic status and potential uses of the water body. If nutrients from manures and other sources are applied at excessive rates to cropland, increased accumulations of the nutrients in the soil can result in significant losses to bodies of water. Gaseous emissions of volatile compounds from manure are also a threat to our atmosphere. Swine producers who reduce the potential of polluting the environment with excess nutrients and pathogens can implement a number of agronomic, nutritional, and managerial practices. The potential risk of nutrient loss from manured fields, including volatilization, can be mitigated by balancing and reducing the nutrient loading rate, use of proper methods and timing of manure application, controlling the chemical form of the nutrients in the manure, and understanding the soil-water conditions of the application site. Avoiding excessive dietary protein, using high-quality protein sources, and feeding low-protein, amino acid–supplemented diets are practices that will reduce the N in excreta. Avoiding excessive dietary P, balancing diets on an available P basis, and use of phytase as a dietary supplement will reduce the P in manure. Use of reduced or organic forms of Cu, Zn, Fe, and Mg will reduce excretion of these nutrients in manure. Feeding management technologies that enhance feed efficiencies and reduce nutrient excretion include feeding for phase, sex, and genetic ability of the animal. Reducing the intact protein levels in diets and balancing with synthetic amino acids, use of low levels of specific NSP (soybean hulls, sugar beet pulp), and maintaining the proper acid–base balance and buffering in the diet can significantly reduce odorous compounds. Greater nutrient reductions may be possible through the development of specialty feed ingredients that will be used for specific animal diets and potentially specialized animals that can be more efficient in nutrient utilization of our current diets. Diets may be modified to create nutrient-balanced manure for sustainable crop production. In some situations, nutrient management and animal performance can be compatible; however, in some situations optimal whole-system performance may not coincide with maximum production, daily weight gain, and feed efficiency. Future eco-nutrition research needs to be conducted with a multidisciplinary approach in a wider context including consideration of available land, animal health and welfare, quality of animal products produced, and nutrient balance in the whole-farm system.

REFERENCES

1. Vollenweider, R.A., Water Management Research, OECD Paris, DAS/CS1/68.27, 1968.
2. Schindler, D.W., Eutrophication and recovery in experimental lakes: Implications for lake management, *Science*, 184, 897, 1974.
3. Schindler, D.W., Evolution of phosphorus limitation in lakes, *Science*, 195, 260, 1977.
4. Stockner, J.G. and Shortreed, K.R.S., Enhancement of autotrophic production by nutrient addition in a coastal rainforest stream on Vancouver Island, *J. Fish Res. Board Can.*, 35, 28, 1978.
5. Peterson, B.J., Hobbie, A.E., Hershey, A.E., Lock, M.A., Ford, T.E., Vestal, J.R., McKinly, V.L., Hullar, M.A.J., Miller, M.C., Ventullo, R.M., and Volk, G.S., Transformation of a tundra river from heterotrophy to autotrophy by addition of phosphorus, *Science*, 229, 1383, 1985.
6. Correll, D.L., The role of phosphorus in the eutrophication of receiving waters: a review, *J. Environ. Qual.*, 27, 261, 1998.
7. Johnson, C.J. and Kross, B.C., Continuing importance of nitrate contamination of groundwater and wells in rural areas, *Am. J. Indust. Med.,* 18, 449, 1990.
8. Zhang, Y., Tanaka, A., Dosman, J.A., Senthilselvan, A., Barber, E.M., Kirychuck, S.P., Holifeld, L.E., and Hurst, T.S., Acute respiratory responses of human subjects to air quality in a swine building, *J. Agric. Eng. Res.,* 70, 367, 1998.
9. Nussbaum-Wagler, D., Phosphorus Runoff from Tilled Soils under Simulated Rainfall, M.S. thesis, Purdue University, West Lafayette, IN, 2003.
10. MWPS, Livestock Waste Facilities Handbook, Midwest Plan Service, Iowa State University, Ames, 1993.
11. Foth, H.D. and Ellis, B.G., *Soil Fertility*, 2nd ed., CRC Press, Boca Raton, FL, 1997.
12. Vitousek, P.M., Aber, J.D.R., Howarth, W., Likens, G.E., Matson, P.A., Schindler, D.W., Schlesinger, W.H., and Tilman, D.G., Human alteration of the global nitrogen cycle: sources and consequences, *Econ. Appl.*, 7(3), 737, 1997.
13. Torstensson, G. and Aronsson, H., Nitrogen leaching and crop availability in manured catch crop systems in Sweden, *Nutr. Cycling Agroecosyst.*, 56, 139, 2000.
14. Robarge, W.P., McCulloch, R.B., and Cure, W., Atmospheric concentrations of ammonia and ammonium aerosols in eastern North Carolina, in *Proc. 2nd Int. Conf. on Air Pollution from Agric. Operations*, ASAE, St. Joseph, MI, 2000, 10.
15. Varel, V.H., Effect of urease inhibitors on reducing ammonia emissions in cattle waste, in *Int. Conf. on Air Pollution from Agric. Operations*, Midwest Plan Service, Iowa State University, Ames, 1996, 459.
16. Hoff, J.D., Nelson, D.W., and Sutton, A.L., Ammonia volatilization from swine manure applied to cropland, *J. Environ. Qual.*, 10, 90, 1981.
17. NRC, Nutrient Requirements of Swine, 10th ed., National Academy Press, Washington, D.C., 1998.
18. Cromwell, G.L. and Coffery, R.D., Nutritional technologies to reduce the nutrient content of swine manure, in *Proc. Pork Academy*, National Pork Producers Council, Des Moines, IA, 1994, 27.
19. Cromwell, G.L., Coffey, R.D., Monegue, H.J., and Randolph, J.H., Efficacy of low-activity, microbial phytase in improving the bioavailability of phosphorus in corn-soybean meal diets for pigs, *J. Anim. Sci.*, 73, 449, 1995.
20. Kemme, P., Jongbloed, A.W., Mroz, Z., and Beynen, A.C., The efficacy of *Aspergillus niger* phytase in rendering phytate phosphorus available for absorption in pigs is influenced by pig physiological status, *J. Anim. Sci.*, 75, 2129, 1997.

21. Han, Y.M., Yang, F., Zhou, A.G., Miller, E.R., Ku, P.K., Hogberg, M.G., and Lei, X.G., Supplemental phytases of microbial and cereal sources improve dietary phytate phosphorus utilization by pigs from weaning through finishing, *J. Anim. Sci.*, 75, 1017, 1997.

22. Adeola, O.L., Lawrence, B.V., Sutton, A.L., and Cline, T.R., Phytase-induced changes in mineral utilization in zinc-supplemented diets for pigs, *J. Anim. Sci.*, 73, 3384, 1995.

23. Radcliffe, J.S., Zhang, Z., and Kornegay, E.T., The effects of microbial phytase, citric acid, and their interaction in a corn-soybean meal diet for weanling pigs, *J. Anim. Sci.*, 76, 1880, 1998.

24. Qian, H., Kornegay, E.T., and Conner, D.E., Jr., Adverse effects of wide calcium:phosphorus ratios on supplemental phytase efficacy for weanling pigs fed two dietary phosphorus levels, *J. Anim. Sci.*, 74, 1288, 1996.

25. Smith, D.R., Moore, P.A., Jr., Maxwell, C.V., and Daniel, T.C., Effects of dietary phytase and aluminum chloride manure amendments on phosphorus in swine manure, *J. Anim. Sci.*, 79, 252, 2001.

26. Baxter, C.A., Joern, B.C., and Adeola, O.L., Dietary P management to reduce soil P loading form pig manure, in *Proc. NC Extension-Industry Soil Fertility Conf.*, 14, 104, 1998.

27. Kerr, B.J., Nutritional strategies for waste reduction-management: nitrogen, in *Proc. New Horizons in Animal Nutrition and Health*, Raleigh, NC, 1995.

28. Sutton, A.L., Kephart, K.B., Verstegen, M.W.A., Canh, T.T., and Hobbs, P.J., Potential for reduction of odorous compounds in swine manure through diet modification, *J. Anim. Sci.*, 77, 430, 1999.

29. Hobbs, P.J., Pain, B.F., Kay, R.M., and Lee, P.A., Reduction of odorous compounds in fresh pig slurry by dietary control of crude protein, *J. Sci. Food Agric.*, 71, 508, 1996.

30. Kay, R.M. and Lee, P.A., Ammonia emission from pig buildings and characteristics of slurry produced by pigs offered low crude protein diets in *Proc. Symp. Ammonia and Odour Control from Anim. Prod. Facilities*, Vinkeloord, the Netherlands, 1997, 253.

31. Liu, H., Allee, G.L., Touchette, K.J., Frank, J.W., and Spencer, J.D., Effect of reducing protein and adding amino acids on performance, carcass characteristics, and nitrogen excretion, and the valine requirement of early finishing barrows, *J. Anim. Sci.*, 78(Suppl. 2), 66, 2000.

32. Allee, G.L., Liu, H., Spencer, J.D., Touchette, K.J., and Frank, J.W., Effect of reducing dietary protein level and adding amino acids on performance and nitrogen excretion of early-finishing barrows, in *Proc. Am. Assoc. of Swine Veterinarians*, 2001, 527.

33. Liu, H., Allee, G.L., Berg, E.P., Touchette, K.J., Spencer, J.D., and Frank, J.W., Amino acid fortified corn diets for late-finishing barrows, *J. Anim. Sci.*, 78(Suppl. 2), 66, 2000.

34. Liu, H., Yi, G.F., Spencer, J.D., Frank, J.W., and Allee, G.L., Amino acid fortified all-corn diets for late-finishing gilts, *J. Anim. Sci.*, 79(Suppl. 2), 63, 2001.

35. Kendall, D.C., Lemenager, K.M., Richert, B.T., Sutton, A.L., Frank, J.W., and Belstra, B.A., Effects of intact protein diets versus reduced crude protein diets supplemented with synthetic amino acids on pig performance and ammonia levels in swine buildings, *J. Anim. Sci.*, 76(Suppl. 1), 173, 1998.

36. Armstrong, T.A., Williams, C.M., Spears, J.W., and Shiffman, S.S., High dietary copper improves odor characteristics of swine waste, *J. Anim. Sci.*, 78, 859, 2000.

37. Prince, T.J., Sutton, A.L., von Bernuth, R.D., and Verstegen, M.W.A., Application of nutrition knowledge for developing eco-nutrition feeding programs on commercial swine farms, in *Proc. Am. Soc. Anim. Sci.* Available online at http://www.asas.org/jas/symposia/proceedings/0931.pdf, 1999.

38. Hill, G.M., Cromwell, G.L., Crenshaw, T. D., Dove, C.R., Ewan, R.C., Knabe, D.A., Lewis, A.J., Libal, G.W., Mahan, D.C., Shurson, G.C., Southern, L.L., and Veum, T.L., Growth promotion effects and plasma changes from feeding high dietary concentrations of zinc and copper to weanling pigs (regional study), *J. Anim. Sci.*, 78, 1010, 2000.

39. Hancock, J.D., Wondra, K.J., Traylor, S.L., and Mavromichalis, I., Feed processing and diet modifications affect growth performance and economics of swine production, in *Proc. Carolina Swine Nutr. Conf.*, 1996, 90.

40. Wondra, K.J., Hancock, J.D., Behnke, K.C., Hines, R.H., and Stark, C.R., Effects of particle size and pelleting on growth performance, nutrient digestibility and stomach morphology in finishing pigs, *J. Anim. Sci.*, 73, 757, 1995.

41. Henry, Y. and Dourmad, J.Y., Protein nutrition and N pollution, *Feed Mix*, May 25, 1992.

42. Cromwell, G.L., Cline, T.R., Crenshaw, J.D. Crenshaw, T.D., Ewan, R.C., Hamilton, C.R, Lewis, A.J., Mahan, D.C., Miller, E.R., Pettigrew, J.E., Tribble, L.F., and Veum, T.L., The dietary protein and/or lysine requirements of barrows and gilts, *J. Anim. Sci.*, 71, 1510, 1993.

43. Henry, Y. and Dourmad, J.Y., Feeding strategies for minimizing nitrogen outputs in pigs, in *Proc. First Int. Symp. on Nitrogen Flow in Pig Production and Environmental Consequences,* EAAP Publication 69, 1993, 138.

44. Van Peet-Schwering, C.M.C., Verdoes, N., Voersmans, M.P., and Beelen, G.M., Effect of feeding and housing on ammonia emission of growing and finishing pig facilities, Research Institute of Pig Husbandry Report P.5.3, 27, 1996.

45. Spiehs, M.J., Whitney, M.H., and Shurson, G.C., Nutrient database for distiller's dried grains with solubles produced from new ethanol plants in Minnesota and South Dakota, *J. Anim. Sci.*, 80, 2645, 2002.

46. Moeser, A.J., Kim, I.B., van Heugten, E., and Kempen, T.A.T.G., The nutritional value of degermed, dehulled corn for pigs and its impact on the gastrointestinal tract and nutrient excretion, *J. Anim. Sci.*, 80, 2629, 2002.

47. Van Kempen, T.A., van Heugten, E., and Moeser, A.J., Dehulled, degermed corn as a preferred feed ingredient for pigs, North Carolina State University Swine Report, 2003.

48. Lee, D.J., Hancock, J.D., DeRoughery, J.M., Maloney, C.A., and Dean, D.W., Use of feed processing technologies and diet formulation strategies to maximize digestibility and minimize excretion of nutrients in finishing pigs, *J. Anim. Sci.*, 79, 182, 2001.

49. Smiricky, M.R., Saddoris, K.L., Albin, D.M., Gabert, V.M., and Fahey, G.C., Improving ileal and total tract digestion of corn and soybean meal-based diets by growing pigs using feed enzymes, steeping and particle size reduction, *J. Anim. Sci.*, 79, 181, 2001.

50. DeCamp, S.A., Hankins, S.L., Carroll, A., Ivers, D.J., Richert, B.T., Sutton, A.L., and Anderson, D.B., Effect of ractopamine and dietary crude protein on nitrogen and phosphorus excretion from finishing pigs, *J. Anim. Sci.*, 79, 61, 2001.

51. Hankins, S.L., DeCamp, S.A., Richert, B.T., Anderson, D.B., Ivers, D.J., Heber, A.J., and Sutton, A.L., Odor production in stored manure from ractopamine (RAC) fed pigs, *J. Anim. Sci.*, 79, 61, 2001.

52. Spencer, J.D., Allee, G.L., and Sauber, T.E., Phosphorus bioavailability and digestibility of normal and genetically modified low-phytate corn for pigs, *J. Anim. Sci.,* 78, 681, 2000.
53. Spencer, J.D., Allee, G.L., and Sauber, T.E., Grow-finish performance and carcass characteristics of high lean growth barrows fed normal and genetically modified low phytate corn, *J. Anim. Sci.,* 78, 1529, 2000.
54. Kendall, D.C., Richert, B.T., Bowers, K.A., DeCamp, S.A., Herr, C.T., Weber, T.E., Cobb, M., Kelly, D., Sutton, A.L., Bundy, D.W., and Powers, W.J., Effects of dietary manipulation on pig performance, manure composition, aerial ammonia, hydrogen sulfide, and odor levels in swine buildings, *J. Anim. Sci.,* 79(Suppl.), 59, 2001.
55. Hill, B.E., Hankins, S.L., Trapp, S.A., Sutton, A.L., and Richert, B.T., Effects of low phytic acid corn, low phytic acid soybean meal, and phytase on nutrient digestibility and excretion in growing pigs, *J. Anim. Sci.,* 81(Suppl. 1), 37, 2003.
56. Golovan, S.P., Meidinger, R.G., Ajakaiye, A., Cottrill, M., Wiederkehr, M.Z., Barney, D., Plante, C., Pollard, J., Fan, M.Z., Hayes, M.A., Laursen, J., Hjorth, J.P., Hacker, R.R., Phillips, J.P., and Forserg, C.W., Pigs expressing salivary phytase produce low phosphorus manure, *Nat. Biotechnol.,* 19, 741, 2001.
57. Canh, T.T., Aarnink, A.J.A., Verstegen, M.W.A., and Schrama, J.W., Influence of dietary factors on the pH and ammonia emission of slurry from growing pigs, *J. Anim. Sci.,* 76, 1123, 1998.
58. Sutton, A.L., Kephart, K.B., Patterson, J.A., Mumma, R., Kelly, D.T., Bogus, E., Jones, D.D., and Heber, A., Manipulating swine diets to reduce ammonia and odor emissions in *Proc. 1st Int. Conf. Air Pollution from Agric. Operations,* 1996, 445.
59. Mroz, Z., Jongbloed, A.W., Beers, S., Kemme, P.A., DeJong, L., van Berkum, A.K., and van der Lee, R.A., Preliminary studies on excretory patterns nitrogen and anaerobic deterioration of fecal protein from pigs fed various carbohydrates in *Proc. First Int. Symp. Nitrogen Flow in Pig Production and Environ. Consequences,* Wageningen, the Netherlands, 1993, 247.
60. Canh, T.T., Verstegen, M.W.A., Aarnink, A.J.A., and Schrama, J.W., Influence of dietary factors on nitrogen partitioning and composition of urine and feces in fattening pigs, *J. Anim. Sci.,* 75, 700, 1997.
61. Canh, T.T., Sutton, A.L., Aarnink, A.J.A., Verstegen, M.W.A., and Bakker, G.C.M., Dietary carbohydrates alter the fecal composition and pH and the ammonia emission from slurry of growing pigs, *J. Anim. Sci.,* 76, 1887, 1998.
62. Hawe, S.M., Walker, N., and Moss, B.W., The effects of dietary fibre, lactose and antibiotic on the levels of skatole and indole in feces and subcutaneous fat in growing pigs, *Anim. Prod.* 54, 413, 1992.
63. DeCamp, S.A., Hill, B.E., Hankins, S.L., Kendall, D.C., Richert, B.T., Sutton, A.L., Kelly, D.T., Cobb, M.L., Bundy, D.W., and Powers, W.J., Effects of soybean hulls in a commercial diet on pig performance, manure composition, and selected air quality parameters in swine facilities, *J. Anim. Sci.,* 79(Suppl. 1), 252, 2001.
64. Mackie, R.I., Stroot, P.G., and Varel, V.H., Biochemical identification and biological origin of key odor components in livestock waste, *J. Anim. Sci.,* 76, 1331, 1998.
65. Yokoyama, M.T. and Carlson, J.R., Microbial metabolites of tryptophan in the intestinal tract with special reference to skatole, *Am. J. Clin. Nutr.,* 32, 173, 1979.
66. Ward, L.A., Johnson, K.A., Robinson, I.M., and Yokoyama, M.T., Isolation from swine feces of a bacterium which decarboxylates *p*-hydroxyphenylacetic acid to 4-methylphenol (p-cresol), *Appl. Environ. Microbiol.,* 53, 189, 1987.

67. Houdijk, J.G.M., Bosch, M.W., Tamminga, S., Verstegen, M.W.A., Berenpas, E.B., and Knoop, H., Apparent ileal and total tract nutrient digestion by pigs as affected by dietary nondigestible oligosaccharides, *J. Anim. Sci.*, 77, 148, 1999.

68. Hidaka, H., Eida, T., Takizawa, T., Tokanaga, T., and Tashiro, Y., Effect of fructo-oligosacchairdes on intestinal flora and human health, *Bifidobact. Microflora*, 5, 50, 1986.

69. Geisting, D.W. and Easter, R.A., Acidification status in swine diets, *Feed Manage.*, 37, 8, 1986.

70. Risley, C.R., Kornegay, E.T., Lindemann, J.D., Wood, C.M., and Eigel, W.N., Effect of feeding organic acids on selected intestinal content measurements at varying times postweaning pigs, *J. Anim. Sci.*, 70(Suppl. 1), 196, 1992.

71. Canh, T.T., Aarnink, A.J.A., Mroz, Z., and Jongbloed, A.W., Influence of dietary calcium salts and electrolyte balance on the urinary pH, slurry pH and ammonia volatilization from slurry of growing-finishing pigs, ID-DLO Rep. 96-51, 1996.

72. Mroz, Z., Jongbloed, A.W., Vreman, K., Canh, T.T., van Diepen, J.T.M., Kemme, P.A., Kogut, J., and Aarnink, A.J.A., The effect of different cation-anion supplies on excreta composition and nutrient balance in growing pigs, ID-DLO Rep. 96-28, 1996.

73. Mroz, Z., Krasucki, W., and Grela, E., Prevention of bacteriuria and ammonia emission by adding sodium benzoate to diets for pregnant sows, presented at Ann. Mtg. EAAP, Vienna, Austria, 1997.

74. Colina, J.J., Lewis, A.J., Miller, P.S., and Fisher, R.L., Dietary manipulation to reduce aerial ammonia concentrations in nursery facilities, *J. Anim. Sci.*, 79, 3096, 2001.

75. Van Kempen, T.A.T.G., Dietary adipic acid reduces ammonia emission from swine excreta, *J. Anim. Sci.*, 79, 2412, 2001.

76. Van Kempen, T.A.T.G., van Heugten, E., and Trottier, N.L., Adipic acid increases plasma lysine but does not improve the efficiency of lysine utilization in swine, *J. Anim. Sci.*, 79, 2406, 2001.

15 Diet Manipulation to Control Odor and Gas Emissions from Swine Production

O.G. Clark, S. Moehn, J.D. Price, Y. Zhang, W.C. Sauer, B. Morin, J.J. Feddes, J.J. Leonard, J.K.A. Atakora, R.T. Zijlstra, I. Edeogu, and R.O. Ball

CONTENTS

15.1 INTRODUCTION

Gaseous emissions associated with pig production include odorants, toxic or corrosive gases, and greenhouse gases. These emissions originate from the pigs and excreted manure (feces and urine), the latter specifically from dirty surfaces or storage pits in barns, from manure storage facilities, and manure disposal operations such as land spreading. Manipulation of swine diets can reduce undesirable emissions (e.g., hydrogen sulfide) and increase desirable emissions (e.g., methane for energy). Dietary manipulation affects the enteric production of gases by altering the dietary nutrient content and digestibility. Subsequently, the chemical properties of the feces and urine might also change, affecting the emissions that evolve from the manure. Dietary manipulations include the adjustment of the protein, non-starch polysaccharide, or fat fractions of the diet, and the addition of exogenous enzymes (e.g., phytase, xylanase, carbohydrase), hormones (e.g., growth hormone), or metabolic modifiers (e.g., β-agonists, ionophores). Diet manipulations that reduce emissions are usually associated with improved nutrient utilization. Diet manipulations might generate other potential revenue opportunities, such as marketable carbon credits (in the case of reduced greenhouse gases) and energy (e.g., methane or biogas), and might also reduce feed costs. Finally, control of emissions might affect worker health and the environment and thereby improve the sustainability of pig production.

15.2 EMISSIONS FROM PIG PRODUCTION

Emissions from pig production include toxic or corrosive gases such as hydrogen sulfide (H_2S) and ammonia (NH_3), odorants, and greenhouse gases (GHGs). The three sources of these emissions are the pigs, the manure, and combusted fossil fuels associated with energy needs to operate and maintain the pig production facilities. This chapter concentrates on gas, odor, and GHG emissions from pigs and manure, because these can be altered directly using nutritional means.

15.2.1 ODOR

Nuisance odor from swine production is mainly emitted from building ventilation, manure storage, and land spreading of manure. Of more than 160 odorous compounds identified in pig manure slurry,[1] most are compounds that contain phenolic and indolic chemical groups, sulfides, or volatile fatty acids.[2-5] These compounds are largely the products of incomplete anaerobic digestion of protein and carbohydrates, either in the pig's gut or by bacteria in the stored manure, whereas complete metabolism of protein and carbohydrates yields carbon dioxide (CO_2), methane

(CH_4), and NH_3.[2,6-8] Odor within or near a swine barn results from dozens of these compounds acting together.[9] Some evidence exists that a few specific compounds become more important than others in defining the character and intensity of the odor, especially as odors have been attenuated by distance.[10]

The analytical determination of individual malodorous compounds is comparatively easy; however, measurement of odor character and intensity is more difficult. Odorous air can simultaneously contain many odorous compounds, and the human response to these can vary among individuals and the circumstances of odor exposure. Dynamic olfactometry is one method of measuring odor, whereby odorous air is sampled and transported in a container to a remotely located lab or pumped directly to a mobile olfactometer on the barn site. Olfactometry is based on the response of a panel of at least five people selected for their olfactive sensitivity to a reference substance (*n*-butanol). The panelists are isolated from one another and presented with the odor sample at decreasing levels of dilution with neutral air until each panelist is able to correctly differentiate the odorous air stream from neutral air streams. The mean dilution level at which the panelists can distinguish the odor is taken as the detection threshold for that sample.[11,12] The resulting measure is called odor concentration, and is expressed in odor units per cubic meter (OU_E m^{-3}), which describes the average response from a human equivalent to the response elicited by a specified mass of a reference compound evaporated into the same volume of neutral gas.[11]

15.2.2 AMMONIA

NH_3 emission is a major pathway of N loss[13] that is associated with the health of workers and pigs and environmental problems. The reduction of NH_3 emissions has been an area of extensive research, and diet manipulations to alter N excretion patterns are a promising tool for ammonia emission abatement.[14] Although emitted NH_3 can be converted into nitrous oxide (N_2O) and might therefore act as an "indirect" GHG,[15] NH_3 is currently not included in national GHG inventories.[16]

15.2.3 HYDROGEN SULFIDE

Swine barns present the potential problem of toxic gases that are released from stored manure, often inside the barn. A deadly gas that threatens both human and swine health is H_2S. The Alberta Provincial Department of Workplace Health and Safety has set limits on the length of time that workers may be exposed to specific H_2S concentrations. In Alberta, the threshold limit value for H_2S in the air is 5 ppm for an 8-hour exposure and 10 ppm for an exposure up to 15 min.[17] Sulfurous compounds also contribute to odor problems; approximately half of the malodorous compounds from swine manure contain sulfur.[18,19]

The risk from the release of sulfurous compounds from manure is low if the manure is stored aerobically. Trace amounts of a single sulfurous gas (dimethyl sulfide) were detected among gaseous products from manure stored under aerobic conditions.[20] Large quantities of H_2S can, however, be produced by bacterial sulfate reduction and decomposition of sulfur-containing compounds in manure under the anaerobic storage conditions present in most swine barns.[20,21] Although H_2S emissions from undisturbed manure are negligible, the dissolved or suspended H_2S is released rapidly during manure disturbance. The rapid release of H_2S can pose a grave risk to the health and lives of workers and animals when manure storage pits are emptied.[22]

15.2.4 GREENHOUSE GASES

GHGs such as CO_2, CH_4, and N_2O are believed to contribute to climate change.[23] Each gas has a different relative climatic impact, or global warming potential (GWP), which is referenced to the estimated impact of CO_2. The difference in GWP among the gases is due to their differing potential effect on the average net radiation exiting from the atmosphere to space and their atmospheric residency times, which are estimated as 100, 12, and 120 years for CO_2, CH_4, and N_2O, respectively. On a molar basis, the 100-year GWP of CO_2 has been defined as 1 (the reference standard) and estimated as 23 and 296 for CH_4 and N_2O.[23]

GHG emissions are inventoried according to the guidelines of the intergovernmental panel on climate change (IPCC), which includes N_2O and CH_4 emissions and CO_2 emissions from the use of fossil fuels. CO_2 emitted by animals or their manure is not included in such inventories, however, because it is considered to originate from renewable resources and is part of the normal carbon cycle. Following the IPCC estimates, the 1996 Canadian GHG inventory includes 61 Mt CO_2-equivalent from agriculture, or about 9.5% of the national total.[24,25] It is estimated that, in the year 2000, Canadian swine production systems were responsible for about 1.835 Mt CO_2-equivalent total GHG emissions (1.536 Mt CO_2-equivalent excluding CO_2 as per the IPCC inventory guidelines). This is equivalent to about 3% of Canada's agricultural or 0.3% of the country's overall total GHG production. CO_2 production is addressed in this chapter because it comprises a large part of emissions from pigs and can be influenced by dietary manipulations.

The type of GHG emissions and the underlying production processes differ between pigs and manure. GHGs emitted by the pigs are CO_2, originating from the oxidation of carbon-containing compounds, and CH_4, originating from enteric fermentation. Emissions from manure originate from the urinary and fecal excretion of waste products by pigs. Manure emissions are dominated by CH_4, which constitutes about 65% of total GHG emissions from manure.[26] N_2O comprises less than 5% of emissions from stored manure and thus is not emphasized in this chapter.[26] Emissions from barns can be regarded as a combination of emissions from both pigs and manure.

The physiological basis for gas production should be considered in order to effectively manipulate GHG production by pigs. CO_2 is produced *in vivo* during the oxidation of carbon-containing compounds to derive energy for metabolic processes and to create heat for the maintenance of body temperature. Manipulations that increase the efficiency of nutrient utilization of the animal can therefore be expected to decrease CO_2 production. In contrast, CH_4 is produced during the fermentation of nutrients in the gastrointestinal tract, mostly in the large intestine from nutrients that were undigested in the small intestine. Manipulations that limit the influx of nutrients into the large intestine by improving small intestinal digestion or that reduce microbial activity in the gastrointestinal tract will tend to reduce CH_4 production.

Manure nutrient content and composition, the storage environment and management regime influence GHG production from stored manure. Diet manipulation can greatly influence the manure nutrient content and composition. Improved efficiency of nutrient utilization by the pig will decrease the nutrients available in the manure for the generation of emissions.

15.3 DIET MANIPULATION STRATEGIES

Traditionally, researchers have been little concerned with GHG and odor emissions by pigs, instead directing their efforts toward improving production efficiency. GHG emissions, however, appear to be directly related to nutrient efficiency in pigs, suggesting that better production efficiency is accompanied by reduced GHG emissions. Several strategies appear promising to improve nutrient efficiency. First, the intake of excess nutrients can be reduced toward the actual nutrient requirement while maintaining growth performance. Examples of this strategy include the reduction of dietary crude protein (CP) content combined with amino acid supplementation, or split-sex and phase-feeding of pigs.[27,28] A second strategy is to improve small intestinal digestion using exogenous enzymes or other feed additives. Improving nutrient digestion leaves less substrate available for bacterial breakdown and the consequent production of undesirable components such as odorants[9,29] or CH_4. A further strategy is to curb hindgut fermentation by controlling or altering the hindgut microbial population using pro- or antibiotics.

The production of odor, NH_3, H_2S, and GHGs during the anaerobic biodegradation of nutrients excreted in pig manure also makes diet manipulation an appropriate strategy for abatement in the context of manure storage and handling.[8,25] Reduced N excretion, for example, means less waste of dietary N and probably less pollutant N in forms such as NH_3, N_2O, and odorants.

15.3.1 REDUCING DIETARY PROTEIN CONTENT

The primary objective of reducing dietary protein content while supplementing limiting amino acids is to reduce N excretion. Previous work indicates that matching dietary amino acids with the requirements of the pig reduces N excretion[5,8,27,29,30] and odor and GHG emissions from manure,[8,31,32] with no negative effects on pig performance.[33–38] Emission reduction can be achieved without affecting pig performance and might also be cost-effective if the protein reduction is moderate and the existing feeding program does not already optimize the use of synthetic amino acids on an economic basis.[39,40]

Dietary protein reduction is related to the concept of the ideal protein:[41–43] a balance of essential amino acids, without excesses and deficits, that exactly fits the nutritional requirements of the pig. Protein reduction is achieved by replacing protein ingredients like soybean meal with appropriate amounts of synthetic amino acids, such as lysine, methionine, threonine, or tryptophan, the lack of which might otherwise limit performance.[27,44–46] In amino acid–supplemented, low-protein diets, the amount of amino acids in excess of that required is reduced and, therefore, fewer amino acids are available as an energy substrate after deamination.

Dietary protein reduction is achieved by replacing protein ingredients (e.g., soybean or canola meal), with cereal grains containing a large amount of starch. Starch is more efficiently used for fat deposition than are amino acids (0.84 vs. 0.52)[47] and starch contains 40% carbon, while amino acids contain on average 52% carbon.[48] Therefore, reducing dietary protein content reduces carbon content and increases the efficiency of carbon utilization, so that less CO_2 production by the pig and less carbon excretion in the manure can be expected. In addition to decreasing the carbon content of the diet, the exchange of protein ingredients for energy ingredients changes the content and composition of dietary fiber and might also alter the CH_4 production by pigs.

Protein reduction can be accompanied by a reduction in excess dietary sulfur-containing amino acids and thereby also reduce sulfur excretion in manure.[49] Undigested and spilled feed, drinking and cleaning water can all contain sulfate and sulfurous compounds that ultimately contribute to the manure slurry.[21,50] Diet manipulation does, however, effectively reduce the concentration of sulfurous compounds in the manure, thereby directly lowering H_2S emissions.[49]

15.3.1.1 Dietary Protein and Nutrient Excretion

The concept is well established that reducing the level of N ingested by the pig reduces the level of N excretion, provided that digestible amino acids remain correctly balanced. With few exceptions, 1% (absolute) reduction of dietary protein content has been found to reduce N excretion from pigs by approximately 10% (relative) (Table 15.1).[37] This response is similar in both sows and growing pigs.[37] The predominant mechanism of dietary protein reduction is the decrease of urinary N excretion due to reduced amino acid catabolism by the liver.[34] Fecal N excretion might also be reduced due to the replacement of protein sources with highly digestible synthetic amino acid supplements.

Nitrogen excretion from finishing pigs fed low-protein and very low-protein, barley-based diets was reduced by 24 and 48%, respectively (Table 15.1).[37,51] Dietary protein reduction lessened the estimated N excretion from sows by 20%.[52] The nutrient content of manure excreted and NH_3 emitted by finisher pigs fed high- and low-protein diets was subsequently analyzed in a production setting.[53] Manure from pigs fed a low-protein wheat and barley diet had 6% less total nitrogen (nonsignificant) than manure from pigs fed a high-protein control diet, coinciding with a nonsignificant 5% reduction in manure N as NH_3. A 40% reduction in NH_3 emissions was achieved in a previous study.[54]

In finisher pigs fed either restrictively or *ad libitum* in metabolic crates or in a production setting, dietary protein reduction did not reduce carbon excretion.[37,52,54] In sows, however, feeding a low-protein diet was estimated to reduce carbon excretion by 7.1%.[52]

A reduction in dietary protein from 16.8 to 13.9% reduced the sulfur concentration in manure from finisher pigs by 15%.[53] Similarly, selection of low-sulfur feed ingredients reduced total sulfate and sulfur excretion by 30%.[19]

TABLE 15.1
Reductions in Excreted Nitrogen from Reduced Dietary Protein

Dietary Protein Reduction (%)	N Excretion Reduction (%)	Ref.
Each 1%[a]	10 to 12.5	Canh et al.[34]
Each 1%[a]	8.4	Kerr[115]
Each 2%[a]	20 to 25	Pierce et al.[40]
Each 2%[a]	20	Lenis and Jongbloed[27]
17.0	24	Atakora et al.[37]
18.8	32	Simmins et al.[116]
20.0	Up to 35	Möhn and Susenbeth[117]
23.0	28	Sutton et al.[29]
24.0	33	Quiniou et al.[118]
24.3	38.4	Hobbs et al.[7]
29.9	40	Zervas and Zijlstra[119]
30.3	39	Fremaut and Deschrijve[120]
32.0	20	Misselbrook et al.[32]
34.0	48	Möhn et al.[52]

[a] Absolute percentage; values without a letter superscript are relative.

In summary, dietary protein reduction reduced N and S excretion consistently and can be expected to decrease N emissions from manure (e.g., NH_3), but has less effect on carbon excretion.

15.3.1.2 Dietary Protein and Manure Odor

Reduced levels of dietary CP might be associated with lower levels of odor emission from pig manure. Odor emissions from manure are commonly estimated by measuring the concentration of specific odorous compounds or are assessed by olfactometry. Dietary protein reduction decreased the concentrations of most of the odorous compounds in slurry from grower–finisher pigs.[7,55] Results based on olfactometry were less clear. Protein reduction in diets for grower-finisher pigs from 16.8 to 13.9% was not shown to reduce manure odor concentration.[53] Protein reduction from 12.4 to 9.7% in the corn–soy diet of grower pigs reduced odor concentration by 30%[54] and a reduction from 22 to 13% CP reduced odor emissions by 31% for finisher pigs;[55] however, a 3% reduction in dietary protein did not decrease manure odor emissions,[56] nor did manure odor decrease with a reduction from 15 to 0% CP in corn–soy diets fed to barrows.[57]

15.3.1.3 Dietary Protein and Manure pH

Reduced dietary protein is also apparently correlated with lower pH in the manure slurry, due primarily to the relationship between protein level and ammonium concentration in the slurry.[58,59] Slurry pH and urinary nitrogen were lower when pigs were fed protein-reduced diets that were similar to a control diet in dietary electrolyte

TABLE 15.2
Least Square Means of Gas Exchange and Greenhouse Gas Production of Finisher Pigs Fed Conventional or Protein-Reduced Wheat–Barley Based Diets

		Control[a,b] (19.3% protein)	Low Protein[b] (16.0% protein)	Control[a,c] (18% protein)	Very Low Protein[c] (12% protein)
CO_2	(g d⁻¹)	1994	1931	2082	1943
	(Relative %)	100	96.8	100	93.3
CH_4	(g d⁻¹)	25.0	17.6	21.7	17.0
	(Relative %)	100	70.4	100.0	78.3
CO_2-equivalent	(g d⁻¹)	3439	2948	3334	2993
	(Relative %)	100	85.7	100	89.8

[a] Same ingredient composition, but different batches of base ingredients.
[b] Restrictively fed.
[c] *Ad libitum* fed.

balance (dEB) and fibrous content.[59] Reducing dietary protein lowered the pH of manure stored at the pilot scale, in 20,000-L anaerobic storage tanks,[60] and at the bench scale, in 200-L anaerobic vessels.[53] Lowering the pH of the slurry from neutral to 5.5 reduces NH_3 emissions by up to 85%[61-63] and lesser reductions in slurry pH also reduce NH_3 emissions.[64,65]

15.3.1.4 Dietary Protein and Manure H_2S

Until recently, diet manipulation to lower H_2S emissions from stored manure received little attention. Lowering sulfur intake and, consequently, sulfur excretion, reduces H_2S emissions from stored manure. Manipulation of the protein content changed feed sulfur level from 0.34 to 0.24 and 0.15% DM, and reduced manure sulfur concentration from 0.12 to 0.08 and 0.04%, respectively.[49] The manure was stored in closed containers, and the concentration of H_2S in the headspace of the containers was measured daily. Shurson et al.[19] performed studies in which the selection of low-sulfur feed ingredients reduced total sulfate and sulfur excretion by 30% without compromising pig performance. Similarly, a 40% reduction in H_2S emissions was measured in a previous study.[54]

15.3.1.5 Dietary Protein and CO_2 Production

Atakora et al.[37] demonstrated in finishing pigs that moderate reduction of dietary protein from 19.3 to 16.0% in wheat–barley-based diets had no significant effect on CO_2 production, while a drastic protein reduction from 18 to 12%[51] decreased CO_2 production 6.7% (Table 15.2). In pregnant sows, reducing protein from 16.3 to 13.5% reduced CO_2 production by 5.4% (Table 15.3).[52] In lactating sows, low-protein diets (18.2% protein) reduced CO_2 production 2.6% (Table 15.3) as

TABLE 15.3
CO_2 Production per Sow When Fed a Conventional or Protein-Reduced Diet

		Control		Low-Protein Diet	
		Gestation	Lactation	Gestation	Lactation
CO_2^a	(g d^{-1})	3024.2*	6482.1**	2861.2*	6315.5**
	(Relative %)	100.0	100.0	94.6	97.4
CH_4^b	(g d^{-1})	26.7	76.8	16.9	45.8
	(Relative %)	100.0	100.0	63.3	59.6
CO_2-equivalent^c	kg year^{-1}		2015.9		1682.4
	(Relative %)		100.0		83.4

^{a} * Indicates significant diet effect at $P < 0.10$; ** indicates significant diet effect at $P < 0.05$.
^{b} Estimated based on CH_4 production of sows at maintenance: 0.821 g MJ^{-1} and 0.558 g MJ^{-1} metabolizable energy intake for control and low-protein diets, respectively.
^{c} 115 d gestation + 16 d nonpregnant + 23 d lactation = 2.37 reproductive cycles year^{-1}.

compared with conventional diets (21.1% protein). The results were inconclusive regarding to the underlying mechanisms causing the reduction of CO_2 production, because both carbon intake and efficiency of carbon utilization were similar between the conventional and low-protein diets.

15.3.1.6 Dietary Protein and Enteric CH_4 Production

In a series of experiments,[37,51,52] CH_4 was measured for finishing pigs (Table 15.2) and sows at maintenance (Table15.4). For finishing pigs and sows, CH_4 production was reduced 30 and 57%, respectively, by a reduction in protein content of barley-based diets. The CH_4 production was not affected by protein level in corn diets. The CH_4 production was correlated to the dietary content. Neutral detergent fiber (NDF) and acid detergent fiber (ADF) were reduced in the low-protein barley-based diets, but were not different between protein levels in the corn-based diets. The reduction of CH_4 production therefore appears to have been caused not by the reduction in protein intake, but by the associated reduction of fermentable substances, as proposed by Kirchgessner et al.[66] Thus, the reduction in CH_4 emissions was not a direct effect of dietary protein reduction, but was due instead to the reduced fiber content that accompanied the changes in diet composition.

15.3.1.7 Dietary Protein and CO_2-Equivalent GHG Emissions

CO_2-equivalent GHG emissions are the sum of GHG emissions multiplied by their respective GWP. For emissions from pigs, CO_2 and CH_4 production were both included for the purpose of this chapter, but N_2O emissions are negligible. CO_2-equivalent GHG emissions were reduced from finishing pigs fed barley-based diets,

TABLE 15.4
Production of CO_2, CH_4, and CO_2-Equivalent by Nonpregnant Sows Fed Barley- or Corn-Based Diets at Two Levels of Protein

		Barley-Based Diets		Corn-Based Diets		Level of Significance[a]		
Parameter		Control	Low Protein	Control	Low Protein	Diet	Protein	Diet × Protein
CO_2	(g d^{-1})[b]	2912.0	3090.0	2761.0	2906.0	0.26	0.08	
	S.E.	61.0	32.0	58.0	32.0			
	(Relative %)[c]	100.0	106.0	100.0	105.0			
CH_4	(g d^{-1})[d]	34.4	14.8	14.1	18.8	0.03	0.06	0.001
	S.E.	2.2	3.1	2.2	2.2			
	(Relative %)[c]	100.0	43.0	100.0	133.0	0.02	0.16	0.002
CO_2-equivalent	(g d^{-1})[d]	4885.0	3970.0	3592.0	3976.0			
	S.E.	165.0	233.0	165.0	165.0			
	(Relative %)[c]	100.0	81.0	100.0	111.0			

[a] Blank entries indicate nonsignificance.
[b] Means adjusted for feed intake.
[c] Relative to high protein, within types of ingredient.
[d] Least square means.

but not from those fed corn-based diets.[37,51,52] The CO_2-equivalent emitted by sows at maintenance was lower from those fed a low-protein diet than from those fed a conventional barley diet. The overall reduction in CO_2-equivalent emissions was 14.3% for finishing pigs and 16.4% for sows fed wheat–barley–canola meal-based diets. Overall, a 10% reduction in dietary protein reduced GHG emissions from pigs by 10% CO_2-equivalent.

15.3.1.8 Dietary Protein and Manure N_2O Emissions

Reduced concentrations of N in the manure do not seem to affect N_2O emissions during storage, which are usually very low, but might influence N_2O emissions from manure handling, composting, or spreading operations. Clark et al.[53,60] did not detect N_2O emissions from manure stored anaerobically at either laboratory or pilot-scale studies, which corroborates similar research.[67,68] Following manure composting with straw, however, a nonsignificant difference in the N_2O emission rate was measured from manure derived from either low- or high-protein diets: 0.6 g vs. 1.0 g N_2O-N d^{-1} m^{-2} for the low- vs. high-protein diets (13.5 and 16.8% CP).[60] Manure slurry from a low-protein diet resulted in an apparent reduction in soil N_2O emissions after spreading on pasture, as compared to manure from a control diet.[32]

15.3.2 Manipulation of Dietary Non-Starch Polysaccharide

Non-starch polysaccharides (NSP) are a heterogeneous group of plant polysaccharides that are not hydrolyzed by endogenous porcine enzymes.[69] NSP can lower nutrient digestibility by impeding gastric emptying, the passage rate of digesta, and digestion and absorption of nutrients.[70] As a result, transit time might be increased and feed intake reduced.[71] Most NSP are poorly digested in the small intestine so that approximately 80% reach the hindgut of the pigs,[70] there serving as fermentation substrates to increase bacterial populations.[71] The effects of different types of NSP vary; for example, pectin has little effect on nutrient digestibility but increases microbial activity whereas cellulose has the opposite effects.[69]

The effect of NSP on digestion and absorption makes them a target for diet manipulation. Pig diets can be supplemented with NSP-degrading enzymes such as xylanase to break up arabinoxylans, reduce viscosity, and achieve more rapid digestion.[71,72] However, xylanase supplements in swine diets have yielded inconsistent growth rate and nutrient utilization responses.[72-74] An alternative approach, which is not economically viable, is to increase the overall digestibility of the carbohydrate fraction in feed by replacing dietary NSP with starch. The organic solid contents in the manure would thereby be reduced. While reducing NSP has apparent advantages, the *addition* of specific NSP might also have benefits, such as the reduction of manure NH_3 emissions.

15.3.2.1 Dietary NSP and Manure Odor

The production rates of malodorous components, such as VFA, amines, and sulfides, can be changed by adjusting dietary NSP.[5] Growing pigs were fed diets high in either cornstarch or NSP where, in the latter, the cornstarch was replaced by coconut expeller, soybean hulls, or dried sugar beet pulp.[64] For every 100 g of added NSP, up to a maximum of 700 g d^{-1}, the concentration of VFA in the slurry increased by 0.51 g kg^{-1}. Among the three replacement ingredients, soybean hulls caused the highest increase in VFA, while effects of dried sugar beet pulp and coconut expeller were similar.

Manure VFA increased when weanling pigs ate feed containing Jerusalem artichoke,[75] indicating that sources rich in fermentable NSP increased manure VFA content. The manure from the Jerusalem artichoke diets smelled sweeter, was less sharp or pungent, and smelled less of skatole than manure from a control diet. Adding sugar beet pulp to a low-protein diet, however, did not affect hedonic tone or odor dilution threshold.[65]

15.3.2.2 Dietary NSP and Manure NH_3 Emissions

Excess protein entering the large intestine is usually partly converted to NH_3, then absorbed, transformed into urea in the liver, and excreted in urine.[76] Increasing fermentable NSP in pig feed decreases the ratio of urinary to fecal N, because NSP are a fermentable substrate for the intestinal microflora in the hindgut enabling increased microbial protein synthesis. Similarly, infusion of the large intestine with casein and starch decreases urinary N excretion and increases bacterial protein in the feces.[77] Consequently, less NH_3 is absorbed from the hindgut and excreted in urine as urea and more is incorporated into microbial protein and excreted as feces.

The ratio of urinary N (urea) to fecal N (protein) decreases when dietary fermentable NSP are increased.[78–81] The protein excreted in feces is more slowly degraded than urea,[27,82] so manure NH_3 emissions are reduced when pigs are fed fermentable NSP.[83] For example, the ratio of NH_3-N to total N in manure was reduced when wheat and barley diets were supplemented with fermentable NSP (beet pulp), while the ratio increased following xylanase supplementation.[53] The absolute concentration of NH_3-N in manure from the beet pulp diet decreased by 39% and total N decreased by 32%. Supplementation of diets with 2% sucrose thermal oligosaccharide caramel reduced the NH_3-N concentration in stored and fresh manure by 62 and 37%, respectively, without affecting pig performance.[80]

Fermentable NSP and low dietary electrolyte balance (dEB) together reduce manure pH, since the former decrease feces pH and the latter decreases urine pH. Lower manure pH decreases potential NH_3 emissions. Canh et al.[84] confirmed that manure slurry pH could be controlled by adjusting dEB and dietary NSP. Four grower diets, based on barley and wheat, tapioca, barley and tapioca, and sugar beet pulp, were formulated to have equal net energy and CP but different NSP and dEB. The beet pulp diet had the lowest dEB and highest NSP, and resulted in manure with 0.8 units lower pH and 52 to 55% less NH_3 emissions than the other diets. Similar results were reported by Canh et al.[64] and Mroz et al.[85] when feeding diets rich in cellulose, as compared to diets high in cornstarch, hemicellulose, or pectin.

Diet manipulations can be effectively combined to alter N excretion patterns. Urinary N excretion, ammonium, and total N content of fresh manure were reduced when pigs ate a diet with 10% CP, 5% cellulose, and synthetic amino acid supplements.[13] Urinary N excretion from grower pigs dropped by 41 to 66% and total N excretion by 37 to 55% when 5% soybean hulls were included in a low-protein, amino acid–supplemented diet.[86] Finally, NH_3 emissions dropped by 42% when low-protein diets containing sugar beet pulp were fed to grower–finisher pigs in commercial swine barns.[87]

15.3.2.3 Dietary NSP and Enteric CH_4 Production

The amount of NSP in swine diets is clearly linked to enteric CH_4 production.[88] It appears that CH_4 production depends almost solely on the intake of bacterially fermentable substances, and that other factors, such as age and performance, have only minor effects on methanogenesis.[66,89] The CH_4 production increased following increasing levels of sugar beet pulp (which has high NSP content) in swine feed.[90] Finally, grower–finisher pigs produced more enteric CH_4 when fed maize and barley-based diets with 16% dried beet pulp, compared with 8% coarse soft wheat bran.[91]

15.3.2.4 Dietary NSP and Manure CH_4

The amount and type of dietary fiber ingested by pigs might also affect CH_4 emissions from their manure. Increasing dietary fiber from 5 to 30% raised CH_4 emissions from stored pig feces from 71.8 mL g^{-1} to 218.8 mL g^{-1}.[92] CH_4 from manure can be used as an energy source and thereby generate carbon-credits.

15.3.3 Other Dietary Manipulations

Reduction of dietary protein content and manipulation of the carbohydrate fraction of pigs' diets might be promising methods for reducing emissions from pigs and pig manure, but are certainly not the only means of doing so. Most techniques, however, have not been studied in the context of GHGs or odor reduction but, instead, in relation to NH_3 emissions. Target areas for the reduction of enteric emissions by pigs are nutrient digestion and efficiency in the small and large intestine and nutrient utilization in the intermediary metabolism. Diet manipulations that reduce nutrient excretion or otherwise alter manure properties appear promising for reducing emissions from manure.

15.3.3.1 Improving Small Intestinal Digestion

Phytase is sold as a feed additive to reduce phosphorus excretion, and might or might not also increase protein digestibility.[93,94] Phytase addition improved growth and protein deposition in grower–finisher pigs by 15%, while feed efficiency was improved by 10%.[95] As mentioned previously, N_2O and possibly CO_2 emissions might also be reduced in this case. Phytase did not affect nitrogen and energy digestibility, however, and is thus not expected to alter odor or gas emissions,[96] although to our knowledge, it has not been studied in this context.

The addition of organic acids to the diet of pigs is a potential diet manipulation strategy to improve nutrient digestibility. Organic acids added to the diets of weaner and grower pigs reduced stomach pH and increased the retention time of feed in the stomach.[97] Organic acids improved protein digestion[98] and exerted a bacteriostatic effect,[28] indicating that growth performance might consequently be improved; however, experimental results have been variable in this regard.[28] Formic acid, among the acids tested so far, appears to be the most beneficial to pig performance.[98] If organic acids do indeed improve protein digestion, then N excretion can be expected to decrease.[99,97] The primary benefit of organic acids, however, might be improved gut health.

15.3.3.2 Reducing Hindgut Fermentation

Limiting the available nutrient supply to the resident bacterial population can reduce fermentation in the hindgut. The nutrient flow into the large intestine is reduced by feeding regimens that result in an improvement of small intestinal digestion and absorption, such as the inclusion of carbohydrases. A further possibility is to use more digestible feedstuffs in swine diets, e.g., replacing barley with hull-less barley or wheat. Such measures affect dietary fiber content and should reduce CH_4 production by pigs.[66,88]

Limiting microbial activity in the hindgut with antibiotics might reduce CH_4 production by pigs. Tylosin and salinomycin have been tested as feed additives for pigs and improve growth rate and feed conversion efficiency.[100] Salinomycin increased N digestibility, but also increased urinary losses, indicating that improved amino acid absorption did not increase protein deposition.[101] Antibiotics have the opposite effect of fermentable NSP on microbial populations, and might therefore drastically suppress hindgut fermentation and decrease CH_4 production.[90] Indeed, salinomycin decreased CH_4 production in the cecum of pigs under *in vitro* conditions.[102] The addition of dietary fat might also suppress hindgut fermentation and CH_4 production,[103] although this effect is not consistent.[104]

In summary, suppression of microbial activity might decrease CH_4 production by pigs. This might also result in a shift from fecal to urinary N excretion, thereby increasing NH_3 emissions from manure.

15.3.3.3 Metabolic Modification with Exogenous Hormones

Exogenous hormones might be used to manage emissions from swine production. The application of exogenous porcine growth hormone (pST) increased daily gain, reduced feed intake, and improved overall feed conversion efficiency, thereby decreasing emissions per pig.[105,106] Despite increased heat production in pST-treated pigs and a shift toward greater protein and lower fat retention,[107] daily CO_2 production was generally decreased.[105,108] Beta-adrenergic agonists, despite a different mode of action from pST, have similar effects on protein and fat retention.[109] Beta-adrenergic agonists, therefore, will also lead to similar reductions in CO_2 production.

15.3.3.4 Altering Manure Properties

Diet manipulation research has been focused on CP or NSP, but other approaches to lowering pollutant and nonpollutant emissions from slurry have been reported. For instance, replacing $CaCO_3$ with $CaSO_4$ or $CaCl_2$[110] or adding adipic or phosphoric acid to the feed[111] can lower manure pH and reduce NH_3 emissions.[112] Another promising diet modification is the replacement of $CaCO_3$ with Ca-benzoate, which can lower urinary and slurry pH and reduce manure NH_3 emissions by 50%.[64]

Odor emissions from stored slurry can also be changed using dietary manipulations. Weaner pigs were fed a dry diet or diets with feed-to-water ratios of 3:1 and 4:1.[113] Slurry odor was least from the 4:1 diet, likely due to nutrient dilution in the manure. Supplementing weanling diets with copper sulfate (10, 66, and 225 ppm) or copper citrate (33, 66, and 100 ppm) reduced perceived slurry odor.[114]

15.4 CONCLUSIONS

Dietary manipulation can reduce odor and gas emissions from swine production, and can also affect GHGs emitted by pigs (primarily CO_2 and CH_4). Reductions in CO_2 emissions from pigs will not create carbon credits, however, because these emissions are not considered by the IPCC to be anthropogenic. Reduced CO_2 production by pigs seems, however, to accompany improved nutrient utilization. Improved nutrient efficiency can reduce the flux of nutrients to the hindgut, which reduces the generation of CH_4, although NH_3 emissions might increase.

Most diet manipulations affect nutrient excretion and, *ceteris paribus*, changes in manure emissions might thereby be expected on the basis of nutrient flow alone. Diet manipulations might also, however, alter other chemical characteristics of manure, especially pH, electrolyte balance, and the amount and composition of volatile solids. These chemical changes can influence the emission of odorants and gases from manure, including NH_3, H_2S, and GHGs. There could be trade-off effects whereby the emission of some compounds decrease while others increase. When evaluating the utility of diet manipulation for emissions reduction, the effects on the emission rates of a range of key compounds must therefore be considered simultaneously.

REFERENCES

1. O'Neill, D.H. and Phillips, V.R.A., Review of the control of odour nuisance from livestock buildings: Part 3: Properties of the odorous substances which have been identified in livestock wastes or in the air around them, *J. Agric. Eng. Res.*, 53, 23, 1992.
2. Spoelstra, S.F., Volatile fatty acids in anaerobically stored piggery wastes, *Netherl. J. Agric. Sci.*, 27, 60, 1979.
3. Hammond, E.G., Heppner, C., and Smith, R., Odors of swine waste lagoons, *Agric. Ecosyst. Environ.*, 25, 103, 1989.
4. Hobbs, P.J., Misselbrook, T.H., and Pain, B.F., Emission rates of odorous compounds from pig slurries, *J. Sci. Food Agric.*, 77, 341, 1998.
5. Zhang, Q., Feddes, J., Edeogu, I., Nyachoti, M., House, J., Small, D., Liu, C., Mann, D., and Clark, G., Odour production, evaluation and control, Project MLMMI 02-HERS-03, MLMMI Inc., Winnipeg, MB, 2002. Available online at http://www.manure.mb.ca/projects/completed/pdf/02-hers-03.pdf (19 Aug. 2004).
6. Ritter, W.F., Odour control of livestock wastes: State-of-the-art in North America, *J. Agric. Eng. Res.*, 42, 51, 1989.
7. Hobbs, P.J., Pain, B.F., Kay, R.M., and Lee, P.A., Reduction of odorous compounds in fresh pig slurry by dietary control of crude protein, *J. Sci. Food Agric.*, 71, 508, 1996.
8. Sutton, A.L., Kephart, K.B., Verstegen, M.W.A., Canh, T.T., and Hobbs, P.J., Potential for reduction of odorous compounds in swine manure through diet modification, *J. Anim. Sci.*, 77, 430, 1999.
9. Mackie, R.I., Stroot, P.G., and Varel, V.H., Biochemical identification and biological origin of key odor components in livestock waste, *J. Anim. Sci.*, 76, 1331, 1998.
10. Wright, D.W, Eaton, D.K., Nielsen, L.T., Kuhrt, F.W., and Koziel, J.A., Multidimensional gas chromatography-olfactometry for identification and prioritization of malodors from confined animal feeding operations, ASAE Paper 044128, ASAE, St. Joseph, MI, 2004.
11. European Committee for Standardisation (CEN), European standard — Air quality — Determination of odour concentration by dynamic olfactometry, Document no. CEN/TC264/WG2/N222/e, CEN, Brussels, Belgium, 1999.
12. Feddes, J.J.R., Qu, G., Ouellette, C.A., and Leonard, J.J., Development of an eight-panelist single port, forced-choice, dynamic dilution olfactometer, *Can. Biosyst. Eng.*, 43, 6.1, 2001.

13. Sutton, A.L., Applegate, T., Hankins, S., Hill, B., Allee, G., Greene, W., Kohn, R., Meyer, D., Powers, W., and Van Kempen, T., Manipulation of animal diets to affect manure production, composition, and odors: state of the science, National Center for Manure and Animal Waste Management White Paper, Midwest Plan Service, Ames, IA, 2001.

14. Phillips, V.R., Cowell, D.A., Sneath, R.W., Cumby, T.R., Williams, A.G., Demmers, T.G.M., and Sandars, D.L., An assessment of ways to abate ammonia emissions from UK livestock buildings and waste stores. Part 1: Ranking exercise, *Bioresour. Technol.*, 70, 143, 1999.

15. Wulf, S., Vandré, R., and Clemens, J., Mitigation options for CH_4, N_2O and NH_3 emissions from slurry management, in *Non-CO_2 Greenhouse Gases: Scientific Understanding, Control Options and Policy Aspects*, van Ham, J., Baede, A.P.M., Guicherit, R., and Williams-Jacobse, J.G.F.M., Eds., Kluwer, Dordrecht, the Netherlands, 2002, 487.

16. Intergovernmental Panel on Climate Change (IPCC), Revised 1996 IPCC guidelines for national greenhouse gas inventories. Reporting instructions. Vol. 1, IPCC, Geneva, Switzerland, 1997.

17. Atia, A., Hydrogen sulphide emissions and safety, Agdex 086-2, Alberta Agriculture, Food and Rural Development, Edmonton, AB, 2004. Available online at http://www1.agric.gov.ab.ca/$department/deptdocs.nsf/all/agdex8269 (19 Aug. 2004).

18. Spoelstra, S.F., Origin of objectionable odorous components in piggery wastes and the possibility of applying indicator components for studying odor development, *Agric. Environ.*, 5, 241, 1980.

19. Shurson, J., Whitney, M., and Nicolai, R., Nutritional manipulation of swine diets to reduce hydrogen sulphide emissions, University of Minnesota, St. Paul, MN, 1998. Available online at http://manure.coafes.umn.edu/research/sulfide_emmisions.html (19 Aug. 2004).

20. Banwart, W.C. and Brenner, J.M., Identification of sulphur gases evolved from animal manures, *J. Environ. Qual.*, 4, 363, 1975.

21. Arogo, J., Zhang, R.H., Riskowski, G.L., and Day, D.L., Hydrogen sulphide production from stored liquid swine manure: a laboratory study, *Trans. ASAE*, 43, 1241, 2000.

22. Chenard, L., Lemay, S.P., and Laguë, C., Hydrogen sulfide assessment in shallow-pit swine housing and outside manure storage, *J. Agric. Saf. Health*, 9, 285, 2003.

23. IPCC, *Climate Change 2001: The Scientific Basis*, Houghton, J.T., Ding, Y., Griggs, D.J., Noguer, M., van der Linden, P.J., Dai, X., Maskell, X., and Johnson, C.A., Eds., Cambridge University Press, Cambridge, U.K., 2001.

24. Agriculture and Agri-Food Canada (AAFC), Agriculture and agri-food climate change foundation paper, AAFC, Ottawa, ON, 1999. Available online at http://www.agr.gc.ca/policy/environ-ment/eb/public_html/pdfs/climate_change/founda2.pdf (19 Aug. 2004).

25. Lagë, C., Management practices to reduce greenhouse gas emissions from swine production systems, *Adv. Pork Prod.*, 14, 287, 2003.

26. Climate Change Central (CCC), Greenhouse gas emissions and opportunities for reduction from the Alberta swine industry, Discussion paper C3-012, CCC, Calgary, Alberta, 2003. Available online at http://www.climatechangecentral.com/resources/discussion_papers/GHG_Emission_%20Alta_Swine.pdf (19 Aug. 2004).

27. Lenis, N.P. and Jongbloed, A.W., New technologies in low pollution swine diets: diet manipulation and use of synthetic amino acids, phytase, and phase feeding for reduction of nitrogen and phosphorus excretion and ammonia emission — review, *Asian-Aus. J. Anim. Sci.*, 12, 305, 1999.

28. Han, I.K., Lee, J.K., Piao, X.S., and Defa, L., Feeding and management system to reduce environmental pollution in swine production, *Asian-Aust. J Anim. Sci.*, 14, 432, 2001.

29. Sutton, A.L., Kephart, K.B., Patterson, J.A., Mumma, R., Kelly, D.T., Bogus, E., Jones, D.D., and Heber, A., Manipulating swine diets to reduce ammonia and odor emissions, in *Proceedings of the International Conference on Air Pollution from Agricultural Operations,* Publication C-3, Midwest Plan Service, Ames, IA, 1996, 445.

30. Council for Agricultural Science and Technology (CAST), Animal diet modification to decrease the potential for nitrogen and phosphorus pollution, Issue Paper no. 21, CAST, Ames, IA, 2002.

31. Lenis, N.P., Lower nitrogen excretion in pig husbandry by feeding: current and future possibilities, *Netherl. J. Agric. Sci.*, 37, 61, 1989.

32. Misselbrook, T.H., Chadwick, D.R., Pain, B.F., and Headon, D.M., Dietary manipulation as a means of decreasing N losses and methane emissions and improving herbage N uptake following application of pig slurry to grassland, *J. Agric. Sci.*, 130, 183, 1998.

33. Oldenburg, J. and Heinrichs, P., Quantitative Aspekte einer proteinreduzierten Schweinemast, *Lohmann Inf.*, 1, 13, 1996.

34. Canh, T.T., Aarnink, A.J.A., Schutte, J.B., Sutton, A., Langhout, D.J., and Verstegen, M.W.A., Dietary protein affects nitrogen excretion and ammonia emission from slurry of growing-finishing pigs, *Livest. Prod. Sci.*, 56, 181, 1998.

35. Grandhi, R.R., Effect of dietary ideal amino acid ratios, and supplemental carbohydrase in hulless-barley-based diets on pig performance and nitrogen excretion in manure, *Can. J. Anim. Sci.*, 81, 125, 2001.

36. Möhn. S., McMillan, D.J., and Ball, R.O., Low protein diets can be fed to gestating sows without adverse effects, *J. Anim. Sci.*, 80(Suppl. 1), 130, 2002.

37. Atakora, J.K.A., Möhn, S., and Ball, R.O., Low protein diets maintain performance and reduce greenhouse gas production in finisher pigs, *Adv. Pork Production*, 14, A17, 2003. Available online at http://www.banffpork.ca/proc/2003pdf/17U ofA-BallAtakoraMohn.pdf (19 Aug. 2004).

38. Kerr, B.J., Yen, J.T., Nienaber, J.A., and Easter, R.A., Influences of dietary protein level, amino acid supplementation and environmental temperature on performance, body composition, organ weights and total heat production of growing pigs, *J. Anim. Sci.*, 81, 1998, 2003.

39. Möhn, S., McMillan, D.J., and Ball, R.O., Economic assessment of amino acid supplemented low protein diets for sows, *Adv. Pork Prod.*, 13, A13, 2002.

40. Pierce, J.L., Enright, K.L., Cromwell, G.L., Turner, L.W., and Bridges, T.C., Dietary manipulation to reduce the N and P excretion by finishing pigs, *J. Anim. Sci.*, 72(Suppl. 1), 331 (Abstr.), 1994.

41. Agricultural Research Council (ARC), The nutrient requirements of farm livestock. No 3: Pigs, ARC, London, U.K., 1981.

42. Fuller, M.F., McWilliam, R., Wang, T.C., and Giles, L.R., The optimum dietary amino acid pattern for growing pigs. 2. Requirements for maintenance and for tissue protein accretion, *Br. J. Nutr.*, 62, 255, 1989.

43. Wang, T.C. and Fuller, M.F., The optimum dietary amino acid pattern for growing pigs. 1. Experiments by amino acid deletion, *Br. J. Nutr.*, 62, 77, 1989.

44. Gatel, F. and Grosjean, F., Effect of protein content of the diet on nitrogen excretion by pigs, *Livest. Prod. Sci.*, 31, 109, 1992.

45. Lee, P.A., Kay, R.M., Cullin, A.W.R., Fullarton, P.J., and Jagger, S., Dietary manipulation to reduce nitrogen excretion by pigs and its effect on performance, in *Nitrogen Flow in Pig Production and Environmental Consequences,* Verstegen, M.W.A., den Hartog, L.A., van Kempen, G.J.M., and Metz, J.H.M., Eds., Publ. 69, EAAP, Rome, Italy, 1993, 163.

46. Roth, F.X. and Kirchgessner, M., Reducing nitrogen excretion in pigs by optimum dietary protein and amino acid supply, *Züchtungskde*, 65, 420, 1993.

47. van Milgen, J., Noblet, J., and Dubois, S., Energetic efficiency of starch, protein and lipid utilization in growing pigs, *J. Nutr.*, 131, 1309, 2001.

48. Schiemann, R., Nehring, K., Hoffmann, L., Jentsch, W., and Chudy, A., *Energetische Futterbewertung und Energienormen*, Deutscher Landwirtschaftsverlag, Berlin, Germany, 1971.

49. Clark, O.G.., Morin, B., Zhang, Y., Sauer, W., and Feddes, J., Effects of dietary sulphur level on hydrogen sulphide and odour emissions from swine manure, presented at The Science of Changing Climates — Impact on Agriculture, Forestry and Wetlands, Edmonton, Alberta, 19–23 July, 2004.

50. van Kempen T.A.T.G., van Heugten, E., and Powers, W.J., Impact of diet on odor, in *Proceedings of the Maryland Nutrition Conference for Feed Manufacturers*, University of Maryland, College Park, 2002.

51. Atakora, J.K.A., Moehn, S., and Ball, R.O., Very low protein diet for finishing pigs maintains performance and reduces greenhouse gas production, *Can. J. Anim. Sci.*, 84(4), 785, 2004.

52. Möhn, S., Atakora, J.K.A, McMillan, D.J., and Ball, R.O., Low protein diets for sows reduce greenhouse gas production by sows, presented at the 9th International Symposium on Digestive Physiology in Pigs, Banff, Alberta, Vol. 2, University of Alberta, Edmonton, 2003, 329.

53. Clark, O.G., Moehn, S. , Price, J., Edeogu, I., and Leonard, J., Diet manipulation for the reduction of odour and greenhouse gas emissions from pig manure, presented at The Science of Changing Climates — Impact on Agriculture, Forestry and Wetlands, Edmonton, Alberta, 19–23 July, 2004.

54. Kendall, D.C., Lemenager, K.M., Richert, B.T., Sutton, A.L., Frank, J.W., and Belstra, B.A., Effects of intact protein diets versus reduced crude protein diets supplemented with synthetic amino acids on pig performance and ammonia levels in swine buildings, *J. Anim. Sci.*, 76(Suppl. 1), 173, 1998.

55. Hayes, E.T., Leek, A.B.G., Curran, T.P., Dodd, V.A., Carton, O.T., Beattie, V.E., and O'Doherty, J.V., The influence of diet crude protein level on odour and ammonia emissions from finishing pig houses, *Bioresour. Technol.*, 91, 309, 2004.

56. Kendall, D.C., Richert, B.T., Sutton, A.L., Frank, J.W., DeCamp, S.A., Bowers, K.A., Kelly, D., and Cobb, M., Effects of fibre addition. 10% soybean hulls to a reduced crude protein diet supplemented with synthetic amino acids versus a standard commercial diet on pig performance, pit composition, odor and ammonia levels in swine buildings, *J. Anim. Sci.*, 77(Suppl. 1), 176, 1999.

57. Otto, E.R, Yokoyama, M., Hengemuehle, S., von Bermuth, R.D., van Kempen, T., and Trottier, N.L., Ammonia, volatile fatty acids, phenolics, and odor offensiveness in manure from growing pigs fed diets reduced in protein concentration, *J. Anim. Sci.*, 81, 1754, 2003.

58. Sommer, S.G. and Husted, S., The chemical buffer system in raw and digested animal slurry, *J. Agric. Sci.*, 124, 45, 1995.

59. Canh T.T., Aarnink, A.J., Verstegen, M.W.A., and Schrama, J.W., Influence of dietary factors on the pH and ammonia emission of slurry from growing-finishing pigs, *J. Anim. Sci.*, 76, 1123, 1998.

60. Clark, O.G., Leonard, J., Feddes, J., and Morin, B., Measurement of greenhouse gas emissions and odour from swine fed standard and modified diets, CSAE Paper 03-617, CSAE/SCGR, Winnipeg, Manitoba, 2003.

61. Stevens, R.J., Laughlin, R.J., and Frost, J.P., Effect of acidification with sulphuric acid on the volatilization of ammonia from cow and pig slurries, *J. Agric. Sci.*, 113, 389, 1989.

62. Frost, J.P., Stevens, R.J., and Laughlin, R.J., Effect of separation and acidification of cattle slurry on ammonia volatilization and on the efficiency of slurry nitrogen for herbage production, *J. Agric. Sci.*, 115, 49, 1990.

63. Pain, B.F., Misselbrook, T.H., and Rees, Y.J., Effect of nitrification inhibitor and acid addition to cattle slurry on nitrogen losses and herbage yields, *Grass Forage Sci.*, 49, 209, 1994.

64. Canh, T.T., Sutton, A.L., Aarnink, A.J., Verstegen, W.M.A., Schrama, J.W., and Bakker, G.C., Dietary carbohydrates alter the fecal composition and pH and the ammonia emission from slurry of growing pigs, *J. Anim. Sci.*, 76, 1887, 1998.

65. Payeur, M., Lemay, S.P., Godbout, S., Chénard, L., Zijlstra, R.T., Barber, E.M. and Laguë, C., Impact of combining a low protein diet including fermentable carbohydrates and oil sprinkling on odour and dust emissions of swine barns, ASAE Paper 024197, St. Joseph, MI, ASAE, 2002.

66. Kirchgessner, M., Kreuzer, M., Müller, H.L., and Windisch, W., Release of methane and of carbon dioxide by the pig, *Agribiol. Res.*, 44, 103, 1991.

67. Laguë, C., Gaudet, E., Agnew, J., and Fonstad, T.A., Greenhouse gas and odor emissions from liquid swine manure storage facilities in Saskatchewan, ASAE Paper 044157, ASAE, St. Joseph, MI, 2004.

68. Pelletier, F., Godbout, S., Marquis, A., Savard, L.-O., Larouche, J.-P., Lemay, S.P., and Joncas, R., Greenhouse gas and odor emissions from liquid swine manure storage and manure treatment facilities in Quebec, ASAE Paper 044158, ASAE, St. Joseph, MI, 2004.

69. Wenk, C., The role of dietary fibre in the digestive physiology of the pig, *Anim. Feed Sci. Technol.*, 90, 21, 2001.

70. Bach Knudsen, K.E., The nutritional significance of "dietary fibre" analysis, *Anim. Feed Sci. Technol.*, 90, 3, 2001.

71. Bedford, M.R., Mechanism of action and potential environmental benefits from the use of feed enzymes, *Anim. Feed Sci. Technol.*, 53, 145, 1995.

72. Mavromichalis I., Hancock, J.D., Senne, B.W., Gugle, T.L., Kennedy, G.A., Hines, R.H., and Wyatt, C.L., Enzyme supplementation and particle size of wheat in diets for nursery and finishing pigs, *J. Anim. Sci.*, 78, 3086, 2000.

73. Chesson, A., Feed enzymes, *Anim. Feed Sci. Technol.*, 45, 65, 1993.

74. Yin, Y.L., Baidoo, S.K., Schulze, H., and Simmins, P.H., Effects of supplementing diets containing hulless barley varieties having different levels of non-starch polysaccharides with glucanase and xylanase on the physiological status of the gastrointestinal tract and nutrient digestibility of weaned pigs, *Livest. Prod. Sci.*, 71, 97, 2001.

75. Farnworth, E.R., Modler, H.W., and Mackie, D.A., Adding Jerusalem artichoke *Helianthus tuberosus* L. to weanling pig diets and the effect on manure composition and characteristics, *Anim. Feed Sci. Technol.*, 55, 153, 1995.

76. Mosenthin, R., Sauer, W.C., Henkel, H., Ahrens, F., and de Lange, C.F., Tracer studies of urea kinetics in growing pigs. The effect of starch infusion at the distal ileum on urea recycling and bacterial nitrogen excretion, *J. Anim. Sci.*, 70, 3467, 1992.

77. Gargallo, J. and Zimmerman, D., Effect of casein and starch infusion in the large intestine on nitrogen and metabolism of growing swine, *J. Nutr.*, 111, 1390, 1981.

78. Canh, T.T., Verstegen, M.W.A., Mui, N.B., Aarnink, A.J.A., Schrama, J.W., Van't Klooser, C.E., and Duong, N.K., Effect of non-starch polysaccharide-rich by-product diets on nitrogen excretion and nitrogen losses from slurry of growing-finishing pigs, *Asian-Aus. J. Anim. Sci.*, 12, 573, 1999.

79. Zervas, S. and Zijlstra, R.T., Effects of dietary protein and fermentable fibre on nitrogen excretion patterns and plasma urea in grower pigs, *J. Anim. Sci.* 80, 3247, 2002.

80. Nahm, K.H., Influences of fermentable carbohydrates on shifting nitrogen excretion and reducing ammonia emission of pigs, *Crit. Rev. Environ. Sci. Technol.*, 33, 165, 2003.

81. Shriver, J.A., Carter, D.S., Sutton, A.L., Richert, B.T., Senne, B.W., and Pettey, L.A., Effects of adding fibre sources to reduced-crude protein, amino acid-supplemented diets on nitrogen excretion, growth performance, and carcass traits of finishing pigs, *J. Anim. Sci.*, 81, 492, 2003.

82. Mroz, Z., Moeser, J.L., Vreman, K., van Diepen, J.T.M., van Kempen, T., Canh, T.T., and Jongbloed, A.W., Effects of dietary carbohydrates and buffering capacity on nutrient digestibility and manure characteristics in finishing pigs, *J. Anim. Sci.*, 78, 3096, 2000.

83. Godbout, S., Lemay, S.P., Joncas, R., Larouche, J.P., Martin, D.Y., Bernier, J.F., Zijlstra, R.T., Chénard, L., Marquis, A., Barber, E.M., and Massé, D., Oil sprinkling and dietary manipulation to reduce odour and gas emissions from swine buildings: laboratory scale experiment, in *Proc. Livestock Environment VI*, Stowell, R.R., Bucklin, R., and Bottcher, R.W., Eds., ASAE, St. Joseph, MI, 2001, 671.

84. Canh, T.T., Aarnink, A.J.A., Mroz, Z., Jongbloed, A.W., Schrama, J.W., and Verstegen, M.W.A., Influence of electrolyte balance and acidifying calcium salts in the diet of growing-finishing pigs on urinary pH, slurry pH and ammonia volatilisation from slurry, *Livest. Prod. Sci.*, 56, 1, 1998.

85. Mroz, Z., Jongbloed, A.W., Beers, S., Kemme, P.A., DeJong, L., van Berkum, A.K., and van der Lee, R.A., Preliminary studies on excretory patterns nitrogen and anaerobic deterioration of fecal protein from pigs fed various carbohydrates, in *Nitrogen Flow in Pig Production and Environmental Consequences,* Verstegen, M.W.A., den Hartog, L.A., van Kempen, G.J.M., and Metz, J.H.M., Eds., EAAP Publication 69, Pudoc Scientific Publishers, Wageningen, the Netherlands, 1993, 247.

86. Hankins, S.L., Dietary Manipulation Effects on Nutrient Excretion and Performance in Beef Cattle and Pigs, M.Sc. thesis, Purdue University, West Lafayette, IN, 2001.

87. Payeur, M., Lemay, S.P., Zijlstra, R.T., Godbout, S., Chénard, L., Barber, E.M., and Laguë, C., A low protein diet including fermentable carbohydrates combined with canola oil sprinkling for reducing ammonia emissions of pig barns, CSAE Paper 02-503, CSAE/SCGR, Winnipeg, Manitoba, 2002.

88. Jensen, B.B., Methanogenesis in monogastric animals, *Environ. Monitoring Assess.*, 42, 99, 1996.

89. Jensen, B.B. and Jorgensen, H., Effect of dietary fibre on microbial activity and microbial gas production in various regions of the gastrointestinal tract of pigs, *Appl. Environ. Microbiol.*, 60, 1897, 1994.

90. Zhu, J.Q., Fowler, V.R., and Fuller, M.F., Assessment of fermentation in growing pigs given unmolassed sugar-beet pulp: a stoichiometric approach, *Br. J. Nutr.*, 69, 511, 1993.

91. Galassi, G., Crovetto, G.M., Rapetti, L., and Tamburini, A., Energy and nitrogen balance in heavy pigs fed different fibre sources, *Livest. Prod. Sci.*, 85, 253, 2004.

92. Stanogias, G., Tjandraatmadja, M., and Pearce, G.R., Effects of source and level of fibre in pig diets on methane production from pig faeces, *Agric. Wastes*, 12, 37, 1985.

93. National Research Council, Nutrient Requirements for Swine, 10th rev. ed., National Academies Press, Washington, D.C., 1998.

94. Adeola, O. and Sands, J.S., Does supplemental dietary microbial phytase improve amino acid utilization? A perspective that it does not, *J. Anim. Sci.*, 81, E78–85E, 2003.

95. Ketaren, P.P., Batterham, E.S., and Farrell, D.J., Recent advances in the use of phytase enzyme in diets for growing pigs, in *Recent Advances in Animal Nutrition in Australia*, Corbett, J.L., Ed., University of New England, Armidale, Australia, 1991, 166.

96. Oryschak, M.A., Simmins, P.H., and Zijlstra, R.T., Effect of dietary particle size and carbohydrase and/or phytase supplementation on nitrogen and phosphorus excretion of grower pigs, *Can. J. Anim. Sci.*, 82, 533, 2002.

97. Eidelsburger, U., Feeding short-chain organic acids to pigs, in *Recent Developments in Pig Nutrition 3*, Garnsworthy, P.C. and Wiseman, J., Eds., Nottingham University Press, Nottingham, U.K., 2001, 107.

98. Partanen, K.H. and Mroz, Z., Organic acids for performance enhancement in pig diets, *Nutr. Res. Rev.*, 12, 117, 1999.

99. Roth, F.X. and Kirchgessner, M., The role of formic acid in animal nutrition, in *Fifth Forum on Animal Nutrition*, BASF, Ludwigshafen, Germany, 1993.

100. Lindemann, M.D., Kornegay, E.T., Stahly, T.S., Cromwell, G.L., Easter, R.A., Kerr, B.J., and Lucas, D.M., The efficacy of salinomycin as a growth promotant for swine from 9 to 97 kg, *J. Anim. Sci.*, 61, 782, 1985.

101. Moore, R.J., Kornegay, E.T., and Lindemann, M.D., Effect of salinomycin on nutrient absorption and retention by growing pigs fed corn-soybean meal diets with or without oat hulls or wheat bran, *Can. J. Anim. Sci.*, 66, 257, 1986.

102. Marounek, M., Savka, O.G., and Skrivanova, V., Effect of salinomycin on in vitro caecal fermentation in pigs, *J. Anim. Physiol. Anim. Nutr.*, 77, 111, 1997.

103. Christensen, K. and Thorbek, G., Methane excretion in the growing pig, *Br. J. Nutr.*, 57, 355, 1987.

104. Bakker, G.C.M., Interaction between Carbohydrates and Fat in Pigs: Impact on Energy Evaluation of Feeds, Ph.D. thesis, Department of Nutrition for Pigs and Poultry, Institute of Animal Science and Health, Lelystad, the Netherlands, 1996.

105. Etherton, T.D., Wiggins, J.P., Chung, C.S., Evock, C.M., Rebhun, J.F., and Walton, P.E., Stimulation of pig growth performance by porcine growth hormone and growth hormone-releasing factor, *J. Anim. Sci.*, 63, 1389, 1986.

106. Hansen, J.A., Yen, J.T., Nelssen, J.L., Nienaber, J.A., Goodband, R.D., and Wheeler, T.L., Effects of somatotropin and salbutamol in three genotypes of finishing barrows: growth, carcass, and calorimeter criteria, *J. Anim. Sci.*, 75, 1798, 1997.

107. Noblet, J., Herpin, P., and Dubois, S., Effect of recombinant porcine somatotropin on energy and protein utilization by growing pigs: interaction with capacity for lean tissue growth, *J. Anim. Sci.*, 70, 2471, 1992.

108. van der Hel, W., Verstegen, W.M.A., Schrama, J.W., Brandsma, H.A., and Sutton, A.L., Effect of varying ambient temperature and porcine somatotropin treatment in pigs on feed intake and energy balance traits, *Livest. Prod. Sci.*, 51, 21, 1997.

109. Mitchell, A.D., Solomon, M.B., and Steele, N.C., Influence of level of dietary protein or energy on effects of ractopamine in finishing swine, *J. Anim. Sci.*, 69, 4487, 1991.

110. Mroz, Z., Jongbloed, A.W., Vreman, K., Canh, T.T., van Diepen, J.T.M., Kemme, P.A., Kogut, J., and Aarnink, A.J.A., The effect of different cation supplies on excreta composition and nutrient balance in growing pigs, Report 96.028, Institute of Animal Science and Health, Lelystad, the Netherlands, 1996.

111. van Kempen, T.A.T.G., Dietary adipic acid reduces ammonia emission from swine excreta, *J. Anim. Sci.*, 79, 2412, 2001.

112. Aarnink, A.J.A., Sutton, A.L., Canh, T.T., and Verstegen, W.M.A., Dietary factors affecting ammonia and odour release from pig manure, in *Proceedings of the Alltechs 14th Annual Symposium for Biotechnology in the Feed Industry*, Lyons, T.P. and Jacques, K.A., Eds., Nottingham University Press, Nottingham, U.K., 1998, 45.

113. Hobbs, P.J., Misselbrook, T.H., and Pain, B.F., Characterisation of odorous compounds and emissions from slurries produced from weaner pigs fed dry feed and liquid diets, *J. Sci. Food Agric.*, 73, 437, 1997.

114. Armstrong, T.A., Williams, C.M., Spears, J.W., and Schiffman, S.S., High dietary copper improves odor characteristics of swine waste, *J. Anim. Sci.*, 78, 859, 2000.

115. Kerr, B.J., Nutritional strategies for waste reduction management: nitrogen, in *New Horizons in Animal Nutrition and Health*, Longenecker, J.B. and Speers, J.W., Eds., Institute of Nutrition, University of North Carolina, Chapel Hill, NC, 1995.

116. Simmins, P.H., Braund, J., Laurie, A., Pinasseau, J., and Weigel, J., Reduction of nitrogen excretion by pigs between 20 and 50 kg liveweight given low protein diets, *Anim. Sci.*, 62, 661, 1996.

117. Möhn, S. and Susenbeth, A., Influence of dietary protein content on efficiency of energy utilization in growing pigs, *Arch. Anim. Nutr.*, 47, 361, 1995.

118. Quiniou, N., Dourmad, J.Y., Henry, Y., Bourdon, D., and Guillou, D., Influence du potentiel de croissance et du taux proteique du regime sur les performances et les rejets azotes des porcs en croissance-finition, alimentes a volonte, *J. Rech. Porcine Fr.*, 26, 91, 1994.

119. Zervas, S. and Zijlstra, R.T., Effects of dietary protein and oat hull fibre on nitrogen excretion patterns and postprandial plasma urea profiles in grower pigs, *J. Anim. Sci.*, 80, 3238, 2002.

120. Fremaut, D. and Deschrijver, R., Effects of age and dietary level on dry matter and nitrogen contents in manure of growing pigs, *Landbouwtijdschr. Rev. Agric.*, 44, 963, 1991.

Part III

---

*Knowledge Gaps
and Challenges*

16 Identifying and Addressing Knowledge Gaps and Challenges Involving Greenhouse Gases in Agriculture Systems under Climate Change

D. Burton and J. Sauvé

CONTENTS

16.1 INTRODUCTION

Today's food and agriculture system faces ever-widening challenges as it reacts to market trends, new technologies, and growing regulatory pressures. Increasing climate variability adds additional challenges to the management of crops, water, pests, and diseases. It is within this context that the agriculture sector in Canada has been asked to develop a strategy for responding to Canada's commitment to the Kyoto Protocol. This chapter describes some of the activities that have been undertaken over the past 5 years to identify gaps in our understanding of greenhouse gas (GHG) emissions from agriculture and their mitigation as well as current research initiatives to address these gaps.

Canada has committed to reduce GHG emissions 6% below 1990 levels in ratifying the Kyoto Protocol in December 2002. In its Options Paper, the Agriculture and Agri-Food Table of the National Climate Change Process (www.nccp.ca) identified current knowledge gaps as one of the significant impediments to developing an action plan in response to Kyoto. The processes by which research priorities are identified and research initiatives undertaken to address these gaps are often ill defined. Several initiatives have been undertaken at both national and regional scales in an attempt to improve the focus of research efforts in this area. These initiatives include the Climate Change Funding Initiative in Agriculture, the Biological Greenhouse Gas Sources and Sinks Program, the Alberta Greenhouse Gas Science Plan, and the activities of the Expert Committee on Greenhouse Gases and Carbon Sequestration.

16.2 THE CLIMATE CHANGE FUNDING INITIATIVE IN AGRICULTURE

In February 2000 the federal Minister of Agriculture and Agri-Food Canada announced funding of $4 million over 4 years from the Canadian Adaptation and Rural Development II (CARD II) program for a Climate Change Funding Initiative in Agriculture (CCFIA). The Canadian Agri-Food Research Council (CARC) was responsible for delivering the CCFIA for the AAFC Environment Bureau, with the following four goals:

1. Increased Canadian human resource research capacity and expertise in climate change issues in agriculture
2. Research on knowledge gaps in agricultural greenhouse gas emissions
3. Development of industry best practices and technology to reduce agricultural greenhouse gas emissions and increase carbon sequestration potential of agricultural soils
4. Enhanced awareness and improved communication on climate change

Under this initiative, 15 research projects were funded, involving a total of 45 graduate students (in whole or in part) and two research chairs were supported. The research projects funded involved a wide range of institutions and sectors of Canadian agriculture and were specifically asked to address the knowledge gaps identified in the Options Paper. These included studies examining GHG emissions from swine, dairy, and cattle production, nitrogen management practices to reduce GHG emissions, and the influence of landscape on GHG emissions and carbon sequestration. Final reports from these projects can be found on the CARC Web site (www.carc-crac.ca).

16.3 BIOLOGICAL GREENHOUSE GAS SOURCES AND SINKS

Under the Science Implementation Plan of the Climate Change Action Plan 2000, Environment Canada (EC), Agriculture and Agri-Food Canada (AAFC), and other federal government departments are collaborating on a program to enhance understanding of biological GHG sources and sinks (BGSS). AAFC is leading the program to support collaborative field, laboratory, and modeling studies in agriculture by teams of government and university scientists. The program focuses on the support of graduate students to address the critical need for future scientists to be trained in global climate change research. The focus of the research is in four areas:

1. Knowledge-based processes for biological greenhouse gas sources and sinks
2. Measurement and spatial variability of greenhouse gas sources and sinks
3. Modeling of biological greenhouse gas sources and sinks
4. Impact of legume crops on carbon sequestration and N_2O emissions

Initiated in 2002, the program funds 12 research projects across Canada. All are collaborative projects in which graduate students are being jointly supervised by teams of academic and Agriculture and Agri-Food Canada research scientists. Details of individual projects can be found on the CARC Web site (www.carc-crac.ca).

16.4 THE ALBERTA GREENHOUSE GAS SCIENCE PLAN

16.4.1 DEVELOPING A SCIENCE PLAN

The need for a science plan was established in March 2000 during an Alberta-wide GHG forum. Various researchers from the agricultural and scientific community, as well as government, reached consensus that comprehensive on-farm GHG assessments were necessary to definitively determine the source and amount of GHG emissions from individual agricultural operations. Knowledge of where and how much GHGs are emitted from agricultural operations is needed before mitigation strategies can be developed.

16.4.2 WHAT IS AN AGRICULTURAL GREENHOUSE GAS SCIENCE PLAN?

A science plan identifies areas of research that need further scientific study or gaps in current knowledge. A science plan also prioritizes those gaps in order of research importance. The objectives of the agricultural GHG science plan were as follows:

- To gather, evaluate, and synthesize GHG emission estimates for various on-farm GHG sinks and sources to determine the mechanisms behind the uptake and release of agricultural GHGs related to different management practices, soil types, and livestock scenarios
- To develop a science plan that will guide researchers and funding agencies in the establishment of future research priorities

16.4.3 How Was the Agricultural Greenhouse Gas Science Plan Developed?

Alberta Agriculture, Food and Rural Development, along with researchers at the University of Alberta, completed a review of prairie-wide agricultural GHG emissions in five different management areas: soil and crop management, livestock management, manure management, land use and energy, and whole-farm integration.

More than 2600 scientific papers and publications were examined, organized into a bibliographic database, and then summarized into a draft report, titled "Development of a Farm-Level Greenhouse Gas Assessment: Identification of Knowledge Gaps and Development of a Science Plan," completed in spring 2003. Following completion of the "State of Knowledge" report on agricultural GHG research, representatives from the scientific community and government met in June 2003 to identify and prioritize gaps in our knowledge of agricultural GHG emissions. From five different management areas, participants generated a list of gaps and then rated each gap to determine:

- How urgently the research is needed in that area
- How great an impact the research would have

The primary goal was to establish which gaps are critical impediments to the development of an on-farm GHG assessment tool to accurately assess GHG emissions, which reflect actual conditions found on agricultural operations.

16.4.4 What Research Gaps Did the Agricultural Greenhouse Gas Science Plan Reveal?

Graphs were generated from the five management areas to illustrate the relative urgency and potential impact of addressing each of the identified gaps, as perceived by the participants at the workshop. For full report and list of gaps, see "Development of a Farm-Level Greenhouse Gas Assessment: Identification of Knowledge Gaps and Development of a Science Plan."[1] However, funding agencies requested more detail and suggested a strategic roadmap for GHG research be developed. Funding agencies also indicated their desire to focus research funding on current needs and avoid funding research that may simply add to knowledge already gathered through other studies.

In June 2004, university researchers, provincial and federal employees, agricultural producers, and funding agencies met again at a workshop. The approach was

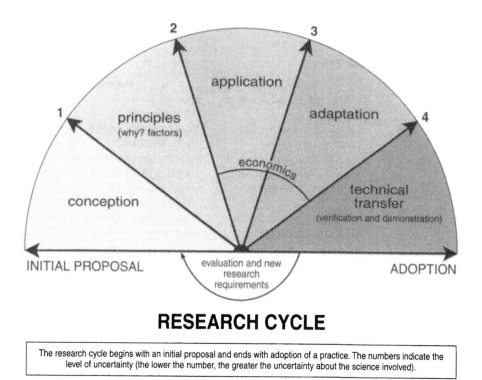

RESEARCH CYCLE

> The research cycle begins with an initial proposal and ends with adoption of a practice. The numbers indicate the level of uncertainty (the lower the number, the greater the uncertainty about the science involved).

FIGURE 16.1 Diagrammatic representation of the research cycle and the five stages of research as described in Table 16.1.

to prioritize management practices for their potential to decrease GHG emissions. The list of management practices analyzed was compiled from lists produced by the CCFIA, Alberta Agriculture GHG Technical Team, and Alberta's Agriculture Policy Framework team. Each management practice was then evaluated for the degree of scientific certainty in predicting the amount of GHG emissions from that practice (from very uncertain to very certain). Participants confirmed that "scientific certainty" could be roughly translated into five stages of research (Figure 16.1, Table 16.1, and Table 16.2).

16.4.5 RESEARCH GAPS THAT ADDRESS HIGH POTENTIAL PRACTICES

Using the management practices with the highest potential to decrease GHG emissions (Table 16.2), the workshop participants identified research gaps for each management practice. The participants were asked to consider how urgent each research gap might be, in comparison with the other research gaps that could be undertaken. This group rating resulted in the "urgency rating."

The "urgency rating" can be interpreted as the approximate time by which the research must be completed (with the assumption that the priority practices should be adopted within approximately 5 years). To ensure that nothing is missed the items

TABLE 16.1
Five Stages of Research

Research Stage	Definition	Example
Conception	Describes and tests the concept or hypothesis	Stage that predicts "what," e.g., what GHGs are emitted
Principles	Describes the principles or factors as a basis for predictability	Stage that answers questions about "why" GHGs are emitted
Application	Applies the theoretical findings to actual field situations (measuring actual results in the field)	Stage that addresses the interaction of factors in an applied setting and tests initial assumptions about economic feasibility
Adaptation	Describes how the findings can be adapted to various settings	Stage that adapts the findings to variances such as scale, landscape, farming practices, and climatic variables; identifies barriers to implementation (including economic)
Tech Transfer	Supports transfer of the technology onto the farm	Stage that includes demonstration projects, education, verification of results

in *italics* were added by the science team after comparison with the list of ideas from the first workshop in 2003.

16.4.6 DEVELOPING A STRATEGIC ROADMAP

The information generated at the second workshop (2004) will be used to develop a strategic roadmap for GHG research. The gaps identified for each of the management practices listed in Table 16.3 will be cross referenced with other research programs such as Institute of Food and Agricultural Sciences Alberta (IFASA), Alberta Agriculture Research Institute's Integrated Crop Management Strategy (AARI-ICM), and the BIOCAP Foundation of Canada to avoid duplication. Multiple occurrences of an identified priority/gap can be seen to confirm its priority. This document should be available from Alberta Agriculture, Food and Rural Development (AAFRD) by spring 2005.

16.5 EXPERT COMMITTEE ON GREENHOUSE GASES AND CARBON SEQUESTRATION

The Canadian Agri-Food Research Council (CARC) maintains a series of expert committees to identify research needs in areas of strategic interest to Canadian agriculture. One of these committees is the Expert Committee on Greenhouse Gases and Carbon Sequestration (ECGHGCS). Through its involvement in the above programs as well as other national and international initiatives, ECGHGCS maintains a list of research gaps. A detailed listing of these research needs is included as part of their annual report to CARC and is available through CARC. A brief overview

TABLE 16.2
Potential Management Practices That Reduce, Remove, or Replace GHG Emissions

Livestock Practices	Research Stage
Include edible oils feedlot diets	Application/Adaptation
Analyze feed and formulate rations to feed livestock a balanced diet	Adaptation
Select for feed efficiency	Adaptation

Manure Practices	Research Stage
Use low-disturbance injection or incorporation of manure within 24 hours	Application
Cover liquid and slurry manure storage systems with straw or synthetic cover	Application
Process liquid or solid manure anaerobically (biodigestors)	Application/Adaptation
Compost manure	Adaptation

Annual Soil and Crop Practices	Research Stage
Soil test periodically before applying fertilizer to ensure applied nitrogen meets crop needs	Application
Include perennial crops in rotations	Application/Tech Transfer
Reduce fallow in rotations	Application/Tech Transfer
Use reduced tillage or no-till seeding of crop	Application/Tech Transfer
Reduce fall nitrogen application and apply nutrients in the spring	Application/Tech Transfer
Use chemfallow instead of summerfallow and/or reduce fallow in rotations	Tech Transfer

Perennial Soil and Crop Practices	Research Stage
Distribute animal manure on pastures uniformly by moving water, shelter, mineral and salt supplements, and temporary fencing periodically and by managing animal density on pastures	Application
Manage forage utilization through timing and frequency of grazing using practices such as controlled rotational grazing on permanent and cropland pasture and controlled grazing on extensively managed native and naturalized pastures and ranges	Principles/Application
Rejuvenate pasture stands using direct seeding, chemical control, seed selection, and fertilization	Application/ Adaptation
Prevent overgrazing by using proper stocking rates as dictated by species, climate, and site-specific soil conditions	Adaptation

Land Use and Energy Practices	Research Stage
Preserve and enhance existing wetlands	Principles
Reduce energy consumption by taking advantage of shelterbelts, solar heating, wind and biogas production	Principles
Convert marginal cropland to pasture, grassland, trees, or wetlands	Adaptation
Energy- and water-efficient retrofits and conservation	Tech Transfer

TABLE 16.3
Research Gaps for Each Potential Management Practice That Reduces, Removes, or Replaces GHG Emissions

Livestock Practices	Time to Complete Research (years)
Evaluation of Western Canadian pasture systems for methane emission	
CO_2/N_2O flux from different hayland, rangeland, pastures; different agro-climatic zones	3.2
Include edible oils in feedlot diets	
Identify drivers for why we get different emission factors/responses	3.4
Quantify level of CH_4 using different sources of fat (canola, sunflower variety, flax, and tallow)	3.7
Economic aspects of using oils	3.9
Analyze feed and formulate rations to feed livestock a balanced diet	
Quantify CH_4 change due to differences in diet	2.5
Select for feed efficiency	
Measure low and high net feed efficiency effects on CH_4 for feeder cattle (University of Alberta, Lacombe, Cattleland)	2.8

Manure Practices	Time to Complete Research (years)
Use low disturbance injection or incorporation of manure within 24 hours	
Evaluate nitrous oxide emissions from injected vs. surface applied manure	3.2
Baseline analysis of raw/incorporated manure vs. composted/surface applied manure	3.4
GHG balance of tanker systems vs. direct injection with dragline systems	4.7
Cover liquid and slurry manure storage systems with straw or synthetic cover	
Baselines for current liquid manure management systems	2.6
Quantities of methane trapped under cover systems — options for utilization	3.3
Process liquid or solid manure anaerobically (biodigestors)	
Barriers to adoption of digestion technology	2.8
Nitrous oxide emissions reductions upon land application of digested liquid manure	2.8
Who owns emissions reductions from anaerobic digestion — policy evaluation	3.0
Compost manure	
Catalog current composting research findings to identify gaps in knowledge	2.0

TABLE 16.3 (continued)

Research Gaps for Each Potential Management Practice That Reduces, Removes, or Replaces GHG Emissions

Compost manure	Time to Complete Research (years)
GHG balance of current manure management baselines vs. implemented compost systems	3.0
Improve overall emissions factors for composting systems	3.0
Manure nitrous oxide reductions from composted manure	3.1
Protocol development for different composting methods used on various farms and farm types	3.5
Effects of additives for compost nutrient stabilization	3.8
Socioeconomic barriers to adoption of composting technologies	4.2

Annual Crop and Soil Practices	Time to Complete Research (years)
More research is required on the fundamental biological processes of N_2O production and consumption	
Biological process identification (de-nitrification and nitrification)	3.3
Quantification of these biological processes	3.7
More fundamental research on C and N cycling in reduced tillage systems (including forages)	4.0
More research aimed at tightening N cycle and reducing residual N in the fall	4.1
Soil test periodically before applying fertilizer to ensure applied nitrogen meets crop needs	
Method development	2.2
Soil sampling protocol	2.4
Define role of soil N test in reducing financial risk and risk of nitrous oxide emission	2.6
Better understanding of soil N test and N_2O emissions under reduced tillage management	2.9
Nutrient use efficiency relating to production, nutrient application and GHG emissions	2.9
Linkage with biological processes (mineralization, de-nitrification)	3.0
Examine N_2O response to fertilizer application for current practices and varieties	3.5
Impact of soil and crop management systems (e.g., minimum tillage)	3.6
Better understanding of residual mineral N following crop production and its role in N_2O production	3.7
Include perennial crops in rotations	
N_2O emissions from legume plow down	3.1
Complete N budget for perennial crops	3.2
More fundamental research on C and N cycling for perennial crops	3.7
Need more information documenting the carbon sequestration of this practice	4.0

TABLE 16.3 (continued)
Research Gaps for Each Potential Management Practice That Reduces, Removes, or Replaces GHG Emissions

	Time to Complete Research (years)
Reduce fallow in rotations	
Need for adaptation of plant species for maintaining permanent green cover in arid regions	4.6
Improved understanding of impact of periods of extended drought on practice of summer fallow	5.0
Use reduce tillage or no-till seeding of crop	
Relationship of reduced tillage with N_2O emissions	2.7
Research into magnitude and duration of carbon gain (saturation point, time course, variance)	3.4
Barriers to adoption	3.7
Optimal seeding technology (e.g., equipment)	4.8
Reduce fall nitrogen application and apply nutrients in the spring	
Nitrogen use efficiency relationships with N_2O emissions	2.8
Research into economic policy and technology transfer to reduce fall N application	3.5
Socioeconomic limitations	3.8
Use chemfallow instead of summerfallow and/or reduce fallow in rotations	
Developing relationships between N flow and N_2O emissions	4.0

	Time to Complete Research (years)
Perennial Crop and Soil Practices	
A lack of information regarding GHG emissions and their management from perennial cropping systems and how this interacts with livestock	
Need more information on N_2O emissions from pastures	3.6
Distribute animal manure on pastures uniformly by moving water, shelter, mineral and salt supplements, and temporary fencing periodically and by managing animal density on pastures	
Document N_2O production associated with urine patches and their distribution	5.3
Manage forage utilization through timing and frequency of grazing using practices such as controlled rotational grazing on permanent and cropland pasture and controlled grazing on extensively managed native and naturalized pastures and ranges	
Need for data in this area (very little available)	3.6
Rejuvenate pasture stands using direct seeding, chemical control, seed selection, and fertilization	
Examine basic understanding of carbon exchange as affected by grazing events	3.7

TABLE 16.3 (continued)
**Research Gaps for Each Potential Management Practice That Reduces,
Removes, or Replaces GHG Emissions**

Rejuvenate pasture stands using direct seeding, chemical control, seed selection, and fertilization	Time to Complete Research (years)
Examine choice of legumes as influenced by climate for pasture rejuvenation	4.8
Prevent overgrazing by using proper stocking rates as dictated by species, climate, and site-specific soil conditions	
Define relationship between stocking rate and sequestration by agro-climate zones	3.8

Land Use and Energy Practices Preserve and enhance existing wetlands	Time to Complete Research (Years)
Quantification of C stocks within different landscape positions (wetland/riparian areas) and soil zones	3.2
Quantify GHG fluxes on wetland–upland transects including natural and cropped	3.3
Basic principle and research on the economic value of wetlands in the ecosystem	3.6
Quantify co-benefits of wetland preservation (groundwater recharge, H_2O quality, nutrient cycling, biodiversity)	3.8
Research into incentives/disincentives on government taxes and assessment	3.9
Reduce energy consumption by taking advantage of shelterbelts, solar heating, wind, and biogas production	
Economics and feasibility of biogas production for different manure systems and farm sizes	2.8
Barriers to adoption on the farm	3.2
Economics	3.6
Convert marginal cropland to pasture, grassland, trees, or wetlands	
Quantify GHG changes (field quantification) with conversion to other land uses (pasture, grasslands, trees, wetlands)	2.8
Verification of agriculture forest land change and implications on GHG and soil carbon dynamics	4.2
Effect of land-use conversion on rural communities (socioeconomic research)	4.8
Long-term economics of agricultural forestry and agroforestry	5.3
Energy- and water-efficient retrofits and conservation	
Economic research into whole system net energy savings (payback period, whole system information packaging)	3.6

of the current research needs identified by the Expert Committee include the following:

There is a need to examine the *cross-cutting issues* relating to economics implications and life-cycle analysis of the adoption of practices to reduce GHG emissions, including the integration of carbon, nitrogen, and phosphorus cycles in nutrient management and GHG mitigation. In this respect there is a need for research to examine whole farm systems and associated watersheds and identify co-benefits of GHG reduction practices at these scales.

There is a need for greater focus on and support for research examining the *adaptation* of agricultural systems to climate change. This should include issues such as the effects of drought, changing pest populations, CO_2 fertilization on crop production, and socioeconomic impacts of climate change. C-CIARN Agriculture has been effective in assessing research needs in this area (www.c-ciarn.uoguelph.ca).

There is a need to develop and disseminate a national system of *measurement, verification, and validation* including improved measurement techniques and standardized protocols. To support this activity there is a need for improved monitoring sources and sinks in Canadian agriculture and a need to better integrate fundamental ecological models in our understanding of these processes.

There also continue to be a number of research issues relating to the mitigation of GHG emissions from Canadian agriculture. In *animal production* the potential for animal genetics, feeding systems, and shelter management to both reduce GHG emissions and increase production efficiencies has not been fully explored and/or this information has not been transferred to managers. For *soils and crops,* including nutrient management and carbon sequestration, there is a need for fundamental research examining key soil processes and components including: the linkage between carbon availability and N_2O production, confirmation of C-sequestration potential of Canadian soils and the management practices that achieve this potential, assessment, and understanding of the methane production and consumption of soils, further research on nitrogen management including improved nitrogen soil tests, assessment of the impact of atmospheric CO_2 fertilization on soil processes, the role of soil carbonate pools in carbon storage, and the potential for genetic manipulation of crops to influence GHG emissions and or C-sequestration. In particular there is a need for better information on grasslands, organic soils, and irrigated land. More research is needed in *manure management* including the potential for manure application technologies, on-farm manure treatment options (composting, solid separation, and anaerobic digestion), and manure storage systems to reduce on-farm GHG emissions.

There is a new and important opportunity for agriculture to contribute to GHG emissions reduction through the production of *bio-fuels and bio-products* — the replacement of petroleum-based processes with agriculturally based stocks. Specifically, research is needed on primary production aimed at products for use in industrial applications including compositional analysis to support high-value component production, development of standards for industrial use, and case studies to understand agronomics and economics. The opportunities for the use of failed or diseased crops and/or animals for energy or other bio-product production need to be examined.

16.6 BIOCAP CANADA FOUNDATION

The BIOCAP Canada Foundation (www.biocap.ca) is developing research networks in the area of agricultural GHG mitigation. These networks include a Landscape Scale Cropping Systems Network, Green Crop Network, Animal Production and Manure Management Network, and Greenhouse Gas Management Canada. The Landscape Scale Cropping Systems network focuses on understanding and quantifying the effects of various crop and landscape management practices on GHG emissions (particularly N_2O) and soil carbon stock changes in agricultural systems. The insights gained from this work will support the National Inventory of GHGs, and offer a critical analysis of those beneficial management practices being put forward to improve the efficiency and sustainability of landscape-scale farm operations. The Green Crop Network will focus on developing or selecting crops and their associated microflora that will improve the sustainability and reduce the environmental footprint associated with crop production. The research objectives of the network are to develop crops that produce fewer N_2O emissions, enhance soil C stocks, flourish in an elevated CO_2 atmosphere, and provide materials for bio-based products. The mandate of the Animal Production and Manure Management Network is to understand and quantify the sources and sinks of GHGs associated with beef, dairy, and pork production. This understanding will be used to identify "best management practices" and new technologies that can mitigate GHGs, while providing additional value to the agricultural producer. Greenhouse Gas Management Canada examines the human dimensions of greenhouse gas management at the local, national, and international levels. For further information, annual reports, and contacts, visit the BIOCAP Web site (www.biocap.ca).

16.7 MOVING FORWARD

This chapter has focused on four examples of efforts to identify and/or address research gaps in our understanding of GHG emissions from agriculture. Clearly, multiple efforts are being, and will continue to be, conducted by various agencies to identify and address knowledge gaps in our understanding of GHG production in agriculture and its mitigation. Communication is key to the success of these efforts, not only in developing and disseminating awareness of the state of science in this area, but also in developing awareness and coordination of the ongoing research initiatives in the area. The ultimate aim of these efforts is to develop practical mitigation practices that will be adopted on the farm. Thus producer/industry input is important in identifying and prioritizing the research gaps as they relate to the development of practical, cost-effective mitigation practices.

ACKNOWLEDGMENTS

We acknowledge input from the Alberta Greenhouse Gas Science Working Group and Canmore I and II workshop participants and the Expert Committee on Greenhouse Gases and Carbon Sequestration in the development and refinement of the research gaps and priorities presented in this chapter.

REFERENCE

1. Alberta Agriculture, Food and Rural Development (AAFRD) and the University of Alberta. Development of a Farm-Level Greenhouse Gas Assessment: Identification of Knowledge Gaps and Development of a Science Plan. AARI Project Number 2001J204, 2003.

17 Knowledge Gaps and Challenges in Forest Ecosystems under Climate Change: A Look at the Temperate and Boreal Forests of North America

P.Y. Bernier and M.J. Apps

CONTENTS

17.1 INTRODUCTION

Vegetation in general and forests in particular form an integral part of the natural carbon cycle. Although most of the long-term (millennia) regulation of atmospheric CO_2 is attributable to exchanges with the oceans,[1] short-term responses to vegetation growth and decay are evident in the multiyear atmospheric CO_2 record.[2] Worldwide, forest ecosystems contain about three times the carbon contained in the atmosphere,[1] and their contribution to an altered carbon cycle has been on both sides of the sink-source equation. Since the start of the industrial revolution in 1850, it is believed that conversion of forested areas to areas with lower carbon densities (fields, pastures, and urbanized areas) has contributed to about 36% of the total anthropogenic emissions to date (155 Pg C released from deforestation vs. 275 Pg C released from fossil fuel (see Table 5 in Reference 3). Recent estimates still show that nearly 25% of total annual anthropogenic emissions can be attributed to deforestation (see Table 7 in Reference 3). However, it is also estimated that terrestrial ecosystems have a net yearly uptake of nearly half of total fossil fuel emissions (2.9 ±1.1 Pg C vs. 6.3 ± 0.4 Pg C; Table 7 in Reference 3), making terrestrial ecosystems a significant component of the global atmospheric CO_2 regulation system.

In a companion paper,[4] we reviewed the mechanisms and uncertainties associated with the terrestrial sink, and have concluded that its future is far from certain. A close watch on how this future unfolds is clearly essential, given the importance of this sink. We also concluded that the quantification of carbon-related costs and benefits from sustainable forest management, and of the impact of climate change on forest carbon sinks and sources requires robust tools for tracking changes in carbon pools, models for predicting their fate, and methods for verifying the estimates over large scales. This conclusion is, of course, not new, and has prompted a significant scientific response over the past decade or so, a period during which there have been dramatic improvements in our ability to quantify the carbon pools of forests and in the scientific understanding of the interaction between climate and forests. In particular, advances have been made in the understanding of landscape- and stand-level processes through the implementation of large-scale manipulative experiments and of biophysical monitoring programs, as well as the development of forest-oriented remote sensing. All of these advances help strengthen sustainable carbon management in our forests. They also improve our capacity to predict the fate of the large carbon pools of the boreal forest under an uncertain climatic future. Future developments in these tools, refinements of models through the inclusion of uncertainties, and better integration between scales in modeling and monitoring efforts will continue to contribute to these ends.

This chapter is intended to briefly take stock of the recent advances in forest carbon science, and to identify knowledge gaps, uncertainties, and underlying challenges related to the interaction between forest carbon and climate. Since gaps should be identified with respect to a desired outcome, we define below a set of questions that will frame the following review of knowledge and the identification of gaps. These questions are especially policy relevant within the context of the Kyoto Protocol and of particular importance to Canada and other forest-rich countries where concerns are emerging with respect to the long-term impact of global changes on the forest resource. These questions are the following: Are current stocks of carbon in forests increasing or decreasing? Can we manage the forests to enhance sinks or reduce sources? Will the mechanisms responsible for the present biotic sink be enhanced, saturate, or reverse sign over time? This chapter focuses particularly on the temperate and boreal forests of North America. Management responses that pertain to mitigation options are treated in Apps et al.[4]

17.2 A SHORT REVIEW OF RECENT ADVANCES

17.2.1 Carbon Budgets and Disturbances

The establishment of the Kyoto Protocol in 1997 has provided a strong impetus for improving our ability to account for past and present changes in forest carbon. Significant advances have been made in this field, leading in Canada to the estimation of carbon stocks in the boreal forest (Figure 17.1) and to development of tools to track or estimate biome-level changes in these carbon stocks.[5,6] These developments have themselves led to the realization that, for the boreal forest, the natural disturbance regime — that is, the frequency, size, and severity of natural disturbances — largely controls the inter-annual to inter-decadal changes in the carbon balance of the boreal forests. Disturbances provide fast pathways for direct release to the atmosphere of carbon previously stored in the various organic components of the ecosystem, but also reset the clock with respect to the carbon uptake capacity of forest ecosystems. The

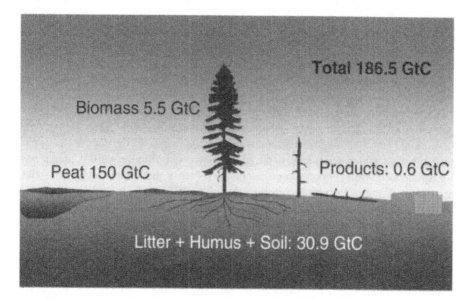

FIGURE 17.1 Estimates of carbon stocks in Canada's boreal forests showing the importance of peatlands in the total carbon content of the forest. (Estimates are from Apps et al.[7] for peat, Apps et al.[8] for forest products, and Kurz and Apps[5] for biomass and dead organic matter carbon.)

large amount of decomposing organic debris left behind by disturbances also gives rise to delayed emissions over a range of timescales. Disturbance dynamics, and fire in particular, have therefore been the focus of much recent research.

Key progress has been made in the analysis of fire statistics,[9] in the construction of historical fire databases from archives and from field observations,[10] and in the quantification of direct fire-related carbon emissions.[11] Of particular importance from a policy perspective is the scale of the inter-annual variability in area burned and in associated greenhouse gas emissions. For example, total area burned in Canada was 0.3×10^6 ha in 1978 and 7.5×10^6 ha in 1989.[11] This large inter-annual variability in burned area is caused by nonlinearities in processes driving the generation of large wildfires, as most fire activity takes place during a few days with extreme fire weather.[12] Small changes in climatic conditions can therefore generate large changes in the fire regime. Direct emissions from fires in Canadian forests were estimated to be 27 ± 6 Tg C yr^{-1} for average 5 years, or equivalent to 18% of Canada's total anthropogenic emissions,[11] but reached 115 Tg C yr^{-1} in high-fire years. Increased fire frequency can therefore provide strong positive feedback to climate change through increased release of CO_2 into the atmosphere.

Changes in the disturbance regime can also generate change in forest composition at the stand and landscape levels, an indirect climate change effect that could be more important to species distribution, migration, and extinction than climate change per se.[13] Estimation of the Fire Weather Index based on the Canadian Global Circulation Model (CGCM) and a $2 \times CO_2$ scenario suggests an increase in fire frequency by 20% or more in most of the central and western boreal forest, but an absence of change or a decrease in fire frequency in eastern Canada, where changes in the precipitation regime and timing of the seasonal warming interact in a different way.[14] Analysis using a $3 \times CO_2$ scenario and the CGCM suggests a 76% average increase in area burned across Canada.[15]

Gains in knowledge have also been made with respect to the relationships between climate and insect-related disturbances and to interactions among disturbance types. For example, knowledge on the historical climate dependence of mountain pine beetle (*Dendroctonus ponderosae* Hopkins[16]) has now been coupled with new climate interpolation methods and climate scenarios to follow the current spread of the insect[17] and predict its possible future expansion. Similar work is under way for spruce budworm (*Choristoneura fumiferana* Clem.), and models suggest possible expansion of outbreaks with a warming climate.[18,19] We are also slowly learning how to tackle the study of interactions among different types of disturbances. For example, Fleming et al.[20] have succeeded in quantifying the probability of fire following epidemic outbreaks of spruce budworm in different parts of the Province of Ontario, Canada. The spectral analysis used to obtain this information required detailed temporal and spatial observations on these disturbances, highlighting the need to maintain and enhance high-quality disturbance-related data sets.

17.2.2 Stand- and Tree-Level Processes

Photosynthesis and respiration are closely coupled to environmental conditions. Long-term changes in the environment must therefore be included in predictions of future forest carbon stocks. In the past, many studies on the interaction between trees and their environment were carried out in totally or partially artificial environments, and often only on seedlings or saplings (e.g., review by Wullschleger et al.[21]). Research emphasis has now moved to observational or manipulative studies at the stand level in natural settings. In the boreal forests of Canada and in other forests around the world, significant knowledge has been gained in the past 10 years about processes at the ecosystem level. Particular emphasis in this chapter is placed on eddy flux covariance measurements, stand-level manipulative experiments, and key transect and laboratory studies.

Flux towers were first installed in forest ecosystems in the early 1990s. In Canada, this deployment was done as part of the BOREAS study.[22] A decade later, many of the original BOREAS sites are still being monitored as part of the BERMS project (http://berms.ccrp.ec.gc.ca/), and the network has recently expanded as Fluxnet-Canada to cover a variety of forest ecosystems and disturbances (http://www.fluxnet-canada.ca/). Two of the key questions being addressed by Fluxnet-Canada are (1) how are disturbances contributing to the carbon budget of Canada's forest, and (2) how sensitive is this budget to the variability in climate. Results to date show that spring temperature and summer drought are key elements of the inter-annual variation in net ecosystem productivity.[23–25] Observational bounds are also being put on post-fire changes in ecosystem carbon stocks (the delayed emissions and regrowth).[26] These experiments and others are shedding light on the mechanisms by which climate change will influence forest ecosystems, and are providing crucial guidance as well as data for model development.

An important body of knowledge is also currently being developed on the ecosystem-level impact of elevated CO_2 through a handful of *in situ* stand-level CO_2 fumigation experiments located in North America and Europe. The most important finding to date is that increasing atmospheric CO_2 concentrations by 200 ppm enhances growth and net carbon accumulation through increased photosynthesis in nearly all systems tested[27–32] without apparent saturation of effect after 3 to 6 years of fumigation. Figure 17.2 shows an example of results obtained with loblolly pine after 3 years of treatment. The effect of increased CO_2 on growth has been reduced only where important drought limitations exist.[33]

The experiments are also helping to determine which other processes within the forest carbon cycle will be affected directly or indirectly by elevated CO_2. Examples of states or processes that appear unaffected by elevated CO_2 in the systems tested are bud phenology,[34] fine root turnover,[35] and leaf area index.[28,30] Examples of processes that are affected by elevated CO_2 are stem respiration,[36] soil respiration,[37] and shoot elongation.[31] Further manipulations in one particular experiment have also shown that additional fumigation with ozone (O_3) eliminates the CO_2-induced gains in growth.[32] The ozone fumigation also interacts with CO_2 to alter the performance of tent caterpillars.[38]

Finally, a number of other experiments are providing information on key uncertainties in the carbon dynamics of the boreal forest. For example, soil warming experiments in mature coniferous and mixed forests have produced increased storage of carbon in biomass and limited loss of soil organic matter.[39,40] Similar conclusions with respect to the general impact of warming on soil carbon storage have been

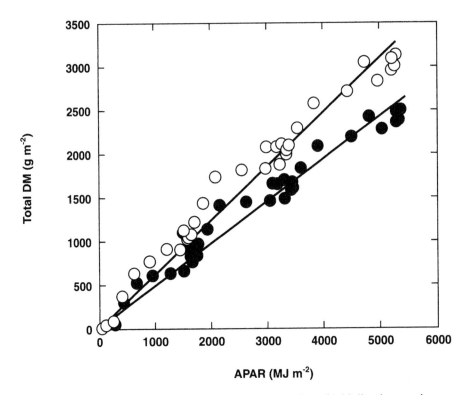

FIGURE 17.2 Cumulative total dry mass (DM) accumulation of loblolly pine stands as a function of absorbed photosynthetically active radiation over a 3-year fumigation period in a FACE study. The open circles represent the mean of three plots maintained at 200 ppm above ambient CO_2 concentration, while the closed circles are the mean values of the three control plots. (From DeLucia, E.H. et al., *Tree Physiol.*, 22, 1003–1010, 2002. With permission.)

reached in studies of climatic transects in Canada.[41] Current ongoing research on soil respiration processes associated with the Fluxnet-Canada network as well as a second soil warming experiment just starting in northern Manitoba should provide further insight into these critical components of climate change response.

17.2.3 LANDSCAPE-LEVEL RESPONSES

Over the past 10 years, our ability to interpret and use multispectral signals from satellite-borne sensors has increased dramatically. Although many problems linked to the application of remote sensing to operational forestry still need to be addressed,[42] remote sensing technologies are making significant contributions to climate change-related applications. Currently, satellite-based observations are used to provide regional or global coverage of forest properties such as land cover,[43] leaf area index,[44–46] and absorbed photosynthetically active radiation.[44,47] They are also increasingly used to monitor the progress of dynamic processes such as disturbances,[48,49] phenology,[50] stress response,[51] and, indirectly, photosynthesis and net primary productivity.[52,53] One of the most notable drivers for advancement in this area has been the recent deployment of the MODIS sensors on board the Terra and Aqua satellites. These sensors now provide near-daily global coverage in spectral bands selected to measure biological and physical processes on land and in oceans.

As a complement to direct field measurements, remote sensing offers two attractive advantages: it can cover large areas, and it can do so repeatedly. These two strengths of remote sensing products can be combined to provide retrospective analysis of vegetation dynamics. One important outcome of such research has been to show significant changes in timing and extent of greening at continental scales in recent decades, suggesting an observable response to recent climatic trends.[54,55] Remote sensing outputs from the MODIS sensor now feed into processes for the production of near-real-time estimates of photosynthesis and net primary productivity.[56] The challenge in this field now lies in the validation of such products (see below).

17.3 GAPS IN KNOWLEDGE

We posed at the outset three questions that relate to the current state of our forests, their long-term response, and the management options for adaptation or mitigation actions. Currently, key areas exist in which progress is needed to better guide forest carbon science and policy and address these questions. Much of the list of research topics proposed by Woodwell et al.[57] in a comprehensive review on biotic feedbacks to the carbon cycle remains pertinent today, although much knowledge has been gained since then. Here, in light of current policy needs, we suggest where the most significant gains can be made to improve our ability to assess and monitor climate change impacts on the forest resource, and to evaluate the impact of our actions on these changes.

17.3.1 PROPAGATING ERROR IN MODELS

A fundamental problem that is seldom addressed in large-scale biological pursuits is the absence of estimation and propagation of errors in models. At the experimental level, we are required by peer review to quantify the uncertainty around estimates so that experimental outcomes can be declared statistically significant or not. However, the same requirement is usually absent from higher-level studies. Estimating and propagating errors through integrative procedures is complex but not always impossible.[58] Such an exercise provides a number of significant benefits.

As a first advantage, quantification of uncertainty in the estimates of lower-order processes permits the intercomparison of these processes and the identification of those on which resources should be spent. A second advantage is that the estimation and propagation of uncertainty in models should make it possible to identify processes that can be left out because their contribution is masked by the overall model error. This is particularly important in scaling-up exercises where temptation is great to include detailed processes — at high data and computational costs — because they are known or thought to be important at finer scales. At coarse scales, gains made by propagating fine-scale elements such as canopy structure or the use of short (e.g., hourly) time steps, are likely lost in the noise due to fine-scale variability or errors in model inputs (e.g., maps of leaf area index or interpolated rainfall).

The third benefit of error propagation is that such analyses carry forward to the decision maker a significant quantity of information that is lost if only mean values are reported. Whereas a mean difference may elicit a particular response from decision makers, the same difference with a confidence interval that includes 0 may well generate a totally different one. Figure 17.3 provides an example of a simulation result on tree growth following thinning, in which the 95% confidence interval reaches the "no-effect" level far before its mean.[59] Such an outcome can be used to estimate when the effect ceases to be significant. It can also be used in a risk management sense to determine the probability of being wrong if a decision is made based on the presumed presence of a difference. Decision makers may tolerate risks far above the 1 and the 5% confidence levels generally used by the scientific community, and thus would truly benefit from the presentation of uncertainties.

Methodologies to quantify and propagate uncertainties in models exist but are often overlooked because of the effort (and expense) involved in their implementation. Variance of soil carbon within and among plots, variance of the errors in the comparison of estimated to measured tree volume increments in permanent sample plots, and interval of confidence in modeled allometric equations are all examples of uncertainties that are within the reach of the analyst. Uncertainties such as future climate conditions or disturbance regimes that are more difficult to quantify can be treated through the use of scenario modeling. In the final analysis, trading off the

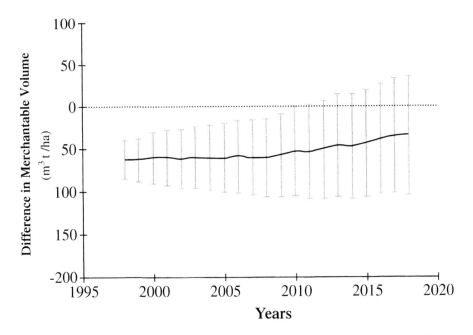

FIGURE 17.3 Predicted evolution of a treatment effect (commercial thinning) on merchantable volume as a difference from a control, with the 95% confidence interval. (From Raulier, F. et al., *Can. J. For. Res.*, 33, 509–520, 2003. With permission.)

possibility to propagate errors for increased model complexity may well be the wrong choice. We therefore recommend pursuing the inclusion of model uncertainties in all analytical efforts related to climate change and forest carbon.

17.3.2 INTERACTION BETWEEN CLIMATE AND DISTURBANCE REGIMES

Global circulation models (GCMs) predict climate anomalies in response to assumed CO_2 emission scenarios. Their predictions therefore incorporate both the uncertainties in the model representation of the climate system and the uncertainties related to the CO_2 emission scenarios. Regional climate scenarios are then down-scaled from GCM outputs with some additional errors injected in these finer estimates. Larger errors are still likely to be incurred, however, when these regional climate scenarios are used to forecast future disturbance regimes. Relationships between environmental variables and drivers of disturbances are often highly nonlinear, such as those captured in the different components of the Canadian Forest Fire Weather Index,[60] or in the proposed climate control on mountain pine beetle populations.[17] Nonlinear relationships with disturbances amplify errors present in climate estimates. Finally, additional biotic interactions such as those between host and parasite play a large role in the downregulation of epidemics and are even more difficult to incorporate in models, even empirically, simply because of the paucity of observations that could be used to fit simple statistical models.

As mentioned above, we now know that forests are in perpetual adjustment to shifts in the disturbance regimes, and that it is these regimes that control to a large extent the age-class distribution of the forest and hence the carbon storage within. In view of the changing environment, predicting with some certainty how stand-replacing disturbances will change under new climatic conditions is a high priority. The benefits of improving the representation of climate–disturbance interactions, or even including their uncertainty, would be substantial for all the reasons cited above for the benefits related to the propagation of uncertainties in models.

Historical studies provide key data and insights needed for the development of predictive models. Fire histories can be mapped and dated to an extent and accuracy that decrease with the remoteness of the site and the time since fire. Nevertheless, considerable progress has been made in the mapping of past fires.[61] Current databases are also capturing recent and current fires using remote sensing[62] and other methods, thus adding to an ever-increasing database of large fires in the boreal forest.[10] For insects, tree ring-based methods are being developed that can be used to reconstruct past insect outbreaks and their effects on forest growth.[61,63,64] Statistical techniques are also helping to unravel the relationship among different types of disturbances.[20] All these methods are slowly helping to build the types of databases that can be used to tackle the climate–disturbance relationships and estimate the first-order uncertainties in our models. In light of the importance of this topic, we recommend pursuing the application of these methodologies to a wider variety of landscapes, forest types, and environments in order to provide a broader data domain for models.

17.3.3 IMPACT OF CLIMATE CHANGE ON NET FOREST GROWTH AND CARBON STOCKS

Across vast landscapes, disturbances are dynamically counterbalanced by the net growth of forests. This net growth is made up of tree-level and stand-level processes that control gross carbon uptake, respiration losses, shedding of tree parts, mortality of individual trees, net accumulation of organic material on the forest floor and within the mineral floor, and loss of dissolved organic and inorganic carbon through leaching. Our ability to predict how the net growth and the attending carbon sequestration in forest stands will behave in the future depends on knowledge of how these individual processes interact with environmental variables related to climate or atmospheric properties such as its CO_2 and ozone concentrations. Significant progress has been made recently with flux tower data, FACE studies, and other such experiments, but significant gaps and uncertainties exist to this day in this area. Three examples of processes in which large uncertainties exist are (1) long-term responses to environmental changes (climate, CO_2, and other atmospheric constituents such as ozone), (2) autotrophic respiration losses, and (3) within-tree carbon allocation.

We have just begun to study stand-level responses to increased CO_2 and ozone.[30,32] Long-term effects in interaction with other limiting factors will remain uncertain and hypothetical for some years to come, but bounds of uncertainty must be quantified. Climate change will also cause significant decoupling between species distributions and optimum growth ranges with unknown consequences to forest growth, in addition to changing the vulnerability of forests to specific natural disturbances.[13] Clearly, long-term assessment of growth impact and of vulnerability to disturbances still requires considerable work.

Significant research is carried out on photosynthesis and its acclimation to temperature and increased CO_2. About 50% of the carbon captured by photosynthesis is respired back to the atmosphere by the plant when the photosynthates are used for growth and maintenance of vegetation structures. Yet, in spite of this large loss, relatively little is known about the mechanisms that control this autotrophic respiration, or the relationship between it and the gross productivity of trees. As an example, in the boreal forests of Canada, there has been only a handful of studies dedicated to the quantification of total respiratory losses from trees.[65-68] As a result, present estimates of percent losses to autotrophic respiration for black spruce, aspen, and jack pine (*Picea mariana, Populus tremuloides,* and *Pinus banksiana*) vary between 35 and 54% of gross photosynthetic uptake. In many models, a fixed value of 45% is assumed.[69]

Once captured by a tree, the residence time of carbon within the biomass will depend on its allocation to the different tree components, and on the turnover rates of these components. This critical computational step in models is closely linked to the estimation of net primary productivity from field measurements as similar assumptions about allocation and litterfall have to be made. Yet, large uncertainties still exist in our ability to quantify allocation and change in allocation with change in the growing environment of the trees. The largest gap in this area is certainly the allocation to belowground processes that involve, in particular, fine roots, root exudates, and mycorrhizal associations. Introduction of field-deployable minirhizotron systems a few years ago is driving the development of fine root productivity databases,[70,71] but much remains to be done in interpreting these field observations, and relating the allocation figures to environmental constraints or conditions.

Gains in these fields are not being made very rapidly. The deployment over the past few years of eddy flux covariance systems in forests across the world will likely rekindle progress in the area of allocation as component fluxes are usually measured along with total ecosystem fluxes and compared with measured changes in ecosystem carbon stocks. The hope is that whole-tree autotrophic respiration estimates will be obtained on key forest tree species and used to improve models of forest productivity. Recent progress on interpreting minirhizotron results[72] may help improve our within-tree allocation estimates. In addition, isotopic-based techniques offer great promise for discriminating and quantifying different inputs and pathways of the belowground portion of the carbon cycle.[73] Finally, transect studies and retrospective analysis of provenance trials and associated ecosystem properties may provide additional insight on long-term responses to climate change. We recommend continued investment of time and resources in basic studies on the response of growth, allocation, and respiration processes to variability or changes in environmental conditions. Co-location of such studies with eddy-covariance measurement systems should be encouraged. We also recommend that these results be provided in ways that are conducive to scaling up from individual field studies to larger land units.

17.3.4 CARBON DYNAMICS OF PEATLANDS

The excellent review by Gorham[74] on northern peatlands and climate change highlights the complex carbon dynamics of peatlands and the importance of their large carbon stocks. Worldwide, peatlands hold an estimated 455 Pg C. In Canada, with only about 10% of the land area, northern peatlands hold approximately four times as much carbon as do upland forests (Figure 17.1). Interactions between peatland carbon and climate are complex and involve carbon fixation and storage, as well as release of CO_2 and CH_4 and dissolved organic carbon as mediated by temperature and water table regimes. Northernmost peatlands are also subject to the thawing of permafrost that causes additional changes to their vegetation and carbon dynamics.[75,76]

Because of the size of their carbon stocks, and because of the close coupling of their carbon exchange mechanisms with climate, peatlands form a very dynamic part of the annual carbon budget in boreal regions. Projected changes in climate may have significant impacts on these exchanges, impacts that must be assessed in order to properly quantify biotic feedback to climate change.[77–79] Peatland response to climate change is still quite uncertain.[80,81]

Improvements to estimates of the carbon balance of peatlands require a mix of measurement and modeling studies. Monitoring of CO_2 and CH_4 exchanges and of environmental variables and water table levels provides essential information for model development and validation.[82] Multiyear records are particularly important to broaden the climatic domain of the observations and pick out particular events with a disproportionate impact on the overall carbon balances. Monitoring of carbon exchanges is being carried out at a small number of peatlands across the world, with three sites in Canada currently providing continuous records of carbon exchange over peatlands.[83] We recommend that peatland measurement programs be expanded to cover a larger variety of climate environments.

17.3.5 VERIFICATION OF SATELLITE-BASED ESTIMATES

Satellite-based remote sensing of vegetation properties has greatly progressed over the past 10 years through both the development of better algorithms for interpreting and using multispectral data from past and existing satellites,[48,84] and the launch of new satellite-borne sensors specifically tailored for the monitoring of vegetation (Terra-Modis). As a result, there has been a significant increase in the type and number of "products" derived from satellite observations. Some are simple interpretations of multispectral data. Others are complex models that use satellite observations as one of their many inputs. For example, one of the more ambitious projects that is currently in operation is the mapping of global terrestrial net primary production on a 1-km^2 and 8-day spatial and temporal resolutions.[56] However, one point that all such products have in common is a relatively low, but improving, level of ground truthing.

It is important both to develop better methods for using satellite-derived inputs and to carry out verification of satellite-derived vegetation-related products. Satellite-derived products are a cost-efficient way of obtaining, over regions, countries, continents, or the globe, the state and trends of simple vegetation-related indicators such as area or change in area by land-cover type, fragmentation, and extent of natural or anthropogenic disturbances to forests. Satellite-driven methodologies are also rapidly making headway into the realm of forest productivity estimates, whether this be net primary productivity, biomass, or biomass increment.[56,84] This area of development offers possible improvements over current methodologies used by policy makers and forest management authorities. However, estimates that previously could be only made by local authorities based on their plot data will, in the not-too-distant future, be supported or challenged by third parties using estimates driven by readily available observations from satellite-based sensors. All the reasons cited above point to the need for the development of a coherent framework for the verification/validation of satellite-based estimates of forest properties. Currently, the Bigfoot project[85] is one of the few projects dedicated to such a validation.

Satellite-driven estimates have specific spatial properties: they usually cover large areas, and are generally made at resolutions that are coarse with respect to field observations. Maps of forest properties from satellite observations usually cover countries, continents, or the globe. Finer-scale observational or modelling studies make use of Landsat-TM at a 30 × 30 m resolution, while broader coverage usually requires a spatial resolution of 1 km^2 or coarser. Field-based verification is challenging even for Landsat-based products because of spatial registration problems or differences in timing between image acquisition and the field visit. Validation of coarser products poses a definite challenge given the large heterogeneity of forests within kilometer-sized pixels.

Flux-tower sites offer some verification capabilities since the tower measurements represent an integration of fluxes across a footprint area of about 1 km^2 under forested conditions. However, the intricacies of eddy covariance measurements and data processing combined with the wind-driven fluctuations in tower footprint decrease the power of such comparisons. Better spatial estimates may be obtained from multiscale frameworks in which multiple layers of information permit the scaling up of detailed field observations and associated uncertainties to larger areas while accounting for the spatial heterogeneity of the forest.[85] Within such frameworks, hierarchical modeling systems[86] can be used to propagate the information to a temporal and spatial scale appropriate for comparison with remote sensing estimates. The development of such verification frameworks or regions is a much-needed step toward the appropriate use of remote sensing products.

Key challenges in such frameworks, however, are the scaling up of soil processes and the measurement of net primary productivity of stands. Soils hold much of the forest ecosystem carbon in a dynamic multicomponent system that processes aboveground and belowground litter into increasingly recalcitrant pools. Even the simple quantification of litter inputs is a major problem, however, not only because of their large spatial and temporal heterogeneity, but also because of the importance of belowground litter, a little-known quantity whose determination we are just starting to tackle using minirhizotron data. In addition, data for ancillary spatial variables required to scale up from point measurements to the landscape are largely nonexistent. For example, although maps of air temperature can be produced with relative ease, drainage properties that control local soil processes are much more difficult to map. Given the large spatial variability of belowground processes, significant gains can therefore be made by improving our understanding of their linkages to easily measured variables, collating databases to verify model outputs and quantify their uncertainties, and developing methodologies for scaling up across time and space.

As mentioned earlier, significant gains have been made in the recent past and work is progressing rapidly on several of these fronts in boreal and temperate forests. Forest soil carbon databases are being developed and used to verify model outputs,[87] and comparisons of estimation methodologies are providing a first test of uncertainties.[88] Results from a network study on litter decomposition[89] are also providing insights into the role of litter quality in decomposition rates, and numerous field projects are quantifying the various components of the soil carbon dynamics in boreal forest settings.[41] Attempts are being made to link soil carbon content to soil survey-type variables so that spatial distributions of soil carbon can be achieved.[90,91] The challenges will be, of course, to move from point estimates to spatial estimates, with appropriate scaling up of the related uncertainties.

A second weak link in the model validation chain is the estimation of net primary productivity (NPP). NPP is a conceptual variable that is defined as yearly gross photosynthesis minus yearly autotrophic respiration losses. NPP is of particular interest to modelers because its estimation is often the first step along the scaling chain from the leaf to the landscape for which field values can be obtained and used for model verification. Most models that are used to forecast how forest carbon stocks will react to changes in climate include estimates of NPP.

In practice, measuring NPP in the field usually involves the measurement of living biomass accrual within a plot as well as the measurement of all plant parts that are shed during the period of measurement, a process that can introduce its own sources of error.[92] A number of terms, such as root exudates, volatile compounds, and pollen production, are assumed to be small, and are typically excluded from measurements. For the sake of simplicity, most modeled-to-measured comparisons are carried out on "aboveground" NPP in which bole, leaf, and sometimes branch production are included but in which allocations to coarse and fine roots are excluded.[93] Part of the reason for this focus is the paucity of data for the belowground components, as pointed out by Gower et al.[94] Even when total NPP estimates are presented, allocation to belowground components, which can account for 30% or more of all NPP,[95,96] are usually estimated as fixed ratios to aboveground productivity. Finally, NPP is more often than not reported only as the annual mean over a period of a few years because measurements of litterfall (needles and branches) correspond to production only over periods of a few years. Correspondence with climate variables would ideally require annual estimates of NPP.

Benefits from improved measurements of NPP are substantial in two respects. The first is that adequate measurements of NPP necessarily entail improved field estimates of allocation to seldom-measured components such as coarse roots, branches, and fine roots. Such new knowledge will necessarily flow into models where carbon allocation to these tree components is often deficient. The second benefit comes in opportunities for model verification. There exist surprisingly few robust total NPP estimates of forest stands worldwide,[94] yet this is the primary verification number for a host of regional and global vegetation dynamics models that form the base of future carbon monitoring networks.

The importance of remote sensing products, either as primary verification or monitoring tools or to supply key spatial information for models, is growing. Because of this, we recommend that comprehensive efforts be made in conjunction with established research sites to develop multiscale frameworks for propagating estimates of key states or processes to scales that are conducive to the validation of satellite products.

17.4 SUMMARY AND CONCLUSIONS

Processes that regulate the capture, release, and sequestration of carbon in forests are key components of the carbon cycle and, as such, are constantly adjusting to changes in climate. This realization has led in the recent past to large advances in our understanding of stand-level and landscape-level processes that control carbon stocks in forests and fluxes between forests and the atmosphere. Areas where advances have been particularly marked are the dynamics of natural disturbances, the interaction between environmental conditions and stand-level processes of carbon exchange, and the remote-sensing applications to landscape-level monitoring and assessment of various forest properties. This chapter has identified broad fields and given specific examples in areas where future advances would have a strong influence on our ability to predict and monitor the impact of climate change on our forest resources. These areas include the propagation of uncertainties in our modeling frameworks, the long-term response of forests to changes in climate and related environmental variables, the influence of climate change on disturbance regimes, the carbon dynamics of peatlands, and new approaches for the development and validation of remote sensing tools.

Given that we can identify the research areas in which advances are most needed, the challenge is now to organize resources around the questions at hand, and to frame all of this in a context that can ultimately help policy and practice. It is also apparent that work done in this area must be multiscale, ideally with functional connectivity between the spatial scales so that information can easily flow from the local or specific to regional assessments. It is also apparent that the climate change issue is cross-sectoral, with implications in the biophysical, policy, and socioeconomic areas. This last point raises the question of linkages to policy and practice, and to the development of frameworks and methods by which these linkages can be achieved at local, regional, and national scales.

Figure 17.4 presents a conceptual picture of how information must flow from process understanding to models, and then to the various policy-level programs that are designed to address the specific questions posed at the beginning of this chapter. Also shown are the five broad areas of gaps or uncertainties (and therefore of research

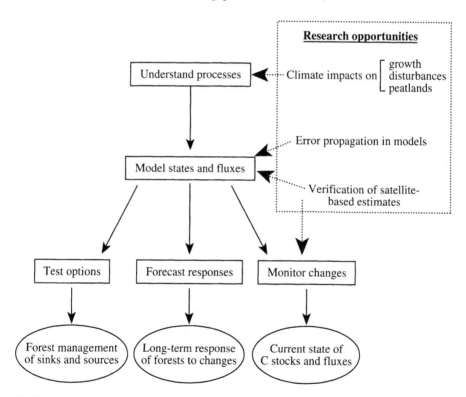

FIGURE 17.4 Flow of information from the basic studies of processes to the provision of answers to policy-related questions (in the ellipses), and linkages to the research opportunities identified in this chapter.

opportunities) identified above. This diagram shows that the development of monitoring tools of regional, national, and global carbon pools or fluxes as well as the capability of forecasting ecosystem responses and the contribution of mitigation or adaptation options ultimately rests on our understanding of processes, and on our inclusion of this understanding in models. Propagation of errors will be of paramount importance in the provision of information to decision makers.

Finally, policy response to forest vulnerability and global changes can be separated into mitigation and adaptation. Mitigation is a relatively recent science focus that has led to consideration of forest carbon as a resource in itself as a direct consequence of the United Nations Convention on Climate Change signed in 1992. However, work on adaptation, which requires in many cases the determination of climatic impact on forest ecosystem processes, encompasses a large part of traditional forest research, and can also be applied to questions that lie outside the purview of climate change research per se. Examples would include the capacity to map forest attributes with remote sensing, or the improvements in forecasting the population dynamics of forest pests. Gains made in such areas therefore serve the additional purpose of improving sustainable forest management. As pointed out by Apps et al.,[4] sustainable forest management may provide a win–win situation with the generation of carbon-based products and the stability of carbon pools at the landscape level. Thus, benefits from climate-change-oriented research should flow into improved sustainable forest management, thereby providing a further incentive for moving forward with research.

ACKNOWLEDGMENTS

We thank the organizers of the meeting, Drs. Jag Bhatti and Mick Price, for providing us the opportunity to reflect upon concepts of uncertainties and gaps in climate change science. We also thank the anonymous reviewer for providing helpful comments on the initial version of this text. Permissions by Tree Physiology and by the Canadian Journal of Forest Research for use of published figures are gratefully acknowledged.

REFERENCES

1. Falkowski, P. et al., The global carbon cycle: a test of our knowledge of Earth as a system, *Science*, 290, 291–296, 2000.
2. Keeling, C.D., Chin, J.F.S., and Whorf, T.P., Increased activity of northern vegetation inferred from atmospheric CO_2 measurements, *Nature*, 382, 146–149, 1996.
3. Houghton, R.A., The contemporary carbon cycle, in *Biogeochemistry*, Schlesinger, W.H., Ed., vol. 8, *Treatise on Geochemistry*, Holland, H.D. and Turekian, K.K., Eds., Elsevier-Pergamon, Oxford, 2003, 473–513.
4. Apps, M.J., Bernier, P.Y., and Bhatti, J.S., Climate change and forest ecosystems, in *Climate Change and Managed Ecosystems*, Bhatti, J.S., Lal, R., Apps, M.J., and Price, M., Eds., CRC Press, Boca Raton, FL, in press.
5. Kurz, W.A. and Apps, M.J., A 70-year retrospective analysis of carbon fluxes in the Canadian forest sector, *Ecol. Appl.*, 9, 526–547, 1999.

6. Chen, J.M. et al., Annual carbon balance of Canada's forests during 1895–1996, *Global Biogeochem. Cycles,* 14, 839–849, 2000.

7. Apps, M.J. et al., Boreal forests and tundra, *Water Air Soil Pollut.,* 70, 39–53, 1993.

8. Apps, M.J. et al., Carbon budget of the Canadian forest product sector, *Environ. Sci. Policy,* 2, 25–41, 1999.

9. Weber, M.G. and Stocks, B.J., Forest fires and sustainability in the boreal forests of Canada, *Ambio,* 27, 313–326, 1998.

10. Stocks, B.J. et al., Large forest fires in Canada, 1959–1997, *J. Geophys. Res. (Atmos.),* 107, doi: 10.1029/2001JD000484, 2002.

11. Amiro, B.D. et al., Direct carbon emissions from Canadian forest fires, 1959–1999, *Can. J. For. Res.,* 31, 512–525, 2001.

12. Flannigan, M.D. and Harrington, J.B., A study on the relation of meteorological variables to monthly provincial area burned by wildfire in Canada 1953–80, *J. Appl. Meteorol.,* 27, 441–452, 1988.

13. Weber, M.G. and Flannigan, M.D., Canadian boreal forest ecosystem structure and function in a changing climate: impact on fire regimes, *Environ. Rev.,* 5, 154–166, 1997.

14. Flannigan, M.D. et al., Future fire in Canada's boreal forest: paleoecology results and general circulation model-regional climate model simulations, *Can. J. For. Res.,* 31, 854–864, 2001.

15. Flannigan, M.D., Logan, K.A., Amiro, B.D., Skinner, W.R., and Stocks, B.J., Future area burned in Canada, *Clim. Change,* 72, 1–16, 2005.

16. Safranyik, L., Effects of climate and weather on mountain pine beetle populations, in *Theory and Practice of Mountain Pine Beetle Management in Lodgepole Pine Forests,* Kibbee, D.L., Berryman, A.A., Amman, G.D., and Stark, R.W., Eds., Symp. Proc., University of Idaho, Moscow, ID, 1978, 77–84.

17. Carroll, A.L., Taylor, S.W., and Régnière, J., Climate change and range expansion by the mountain pine beetle, in *Proceedings: Climate Change in the Columbia Basin,* Columbia Mountains Institute of Applied Ecology, Revelstoke, BC, Canada, 2003, 42–47.

18. Fleming, R.A., Candau, J.N., and Munn, R.E., Influences of climatic change on some ecological processes of an insect outbreak system in Canada's boreal forests and the implications for biodiversity, *Environ. Monit. Assess.,* 49, 235–249, 1998.

19. Williams, D.W. and Liebhold, A.M., Latitudinal shifts in spruce budworm (Lepidoptera: Tortricidae) outbreaks and spruce-fir forest distributions with climate change, *Acta Phytopathol. Entomol. Hung.,* 32, 205–215, 1997.

20. Fleming, R.A., Candau, J.-N., and McAlpine, R.S., Landscape-scale analysis of interactions between insect defoliation and forest fire in central Canada, *Clim. Change,* 55, 251–272, 2002.

21. Wullschleger, S.D., Post, W.M., and King, A.W., On the potential for a CO_2 fertilisation effect in forests: estimates of the biotic growth factor based on 58 controlled-exposure studies, in *Biotic Feedbacks in the Global Climatic System,* Woodwell, G.M. and Mackenzie, F.T., Eds., Oxford University Press, New York, 1995, 85–107.

22. Sellers, P.J. et al., BOREAS in 1997: experiment overview, scientific results, and future directions, *J. Geophys. Res. (Atmos.),* 102, 28731–28769, 1997.

23. Arain, M.A. et al., Effects of seasonal and interannual climate variability on net ecosystem productivity of boreal deciduous and conifer forests, *Can. J. For. Res.,* 32, 878–891, 2002.

24. Barr, A.G. et al., Comparing the carbon budgets of boreal and temperate deciduous forest stands, *Can. J. For. Res.,* 32, 813–822, 2002.

25. Black, T.A. et al., Increased carbon sequestration by a boreal deciduous forest in years with a warm spring, *Geophys. Res. Lett.*, 27, 1271–1274, 2000.

26. Amiro, B.D. et al., Post-fire carbon dioxide fluxes in the western Canadian boreal forest: evidence from towers, aircraft and remote sensing, *Agric. For. Meteorol.*, 115, 91–107, 2003.

27. Calfapietra, C. et al., Free-air CO_2 enrichment (FACE) enhances biomass production in a short-rotation poplar plantation, *Tree Physiol.*, 23, 805–814, 2003.

28. Norby, R.J. et al., Leaf dynamics of a deciduous forest canopy: no response to elevated CO_2, *Oecologia*, 136, 574–584, 2003.

29. Karnosky, D.F. et al., Tropospheric O_3 moderates responses of temperate hardwood forests to elevated CO_2: a synthesis of molecular to ecosystem results from the Aspen FACE project, *Funct. Ecol.*, 17, 289–304, 2003.

30. DeLucia, E.H., George, K., and Hamilton, J.G., Radiation-use efficiency of a forest exposed to elevated concentrations of atmospheric carbon dioxide, *Tree Physiol.*, 22, 1003–1010, 2002.

31. Hättenschwiler, S. et al., Atmospheric CO_2 enrichment of alpine treeline conifers, *New Phytol.*, 156, 363–375, 2002.

32. Isebrands, J.G. et al., Growth responses of *Populus tremuloides* clones to interacting elevated carbon dioxide and tropospheric ozone, *Environ. Pollut.*, 115, 359–371, 2001.

33. Hättenschwiler, S. et al., Thirty years of *in situ* tree growth under elevated CO_2: a model for future forest responses? *Global Change Biol.*, 3, 463–471, 1997.

34. Calfapietra, C. et al., Do above-ground growth dynamics of poplar change with time under CO_2 enrichment? *New Phytol.*, 160, 305–318, 2003.

35. Matamala, R. et al., Impacts of fine root turnover on forest NPP and soil C sequestration potential, *Science,* 302, 1385–1387, 2003.

36. Edwards, N.T., Tschaplinski, T.J., and Norby, R.J., Stem respiration increases in CO_2-enriched sweetgum trees, *New Phytol.*, 155, 239–248, 2002.

37. Andrews, J.A. and Schlesinger, W.H., Soil CO_2 dynamics, acidification, and chemical weathering in a temperate forest with experimental CO_2 enrichment, *Global Biogeochem. Cycles,* 15, 149–162, 2001.

38. Holton, M.K., Lindroth, R.L., and Nordheim, E.V., Foliar quality influences tree-herbivore-parasitoid interactions: effects of elevated CO_2, O_3, and plant genotype, *Oecologia*, 137, 233–244, 2003.

39. Stromgren, M. and Linder, S., Effects of nutrition and soil warming on stemwood production in a boreal Norway spruce stand, *Global Change Biol.*, 8, 1195–1204, 2002.

40. Melillo, J.M. et al., Soil warming and carbon-cycle feedbacks to the climate system, *Science*, 298, 2173–2176, 2002.

41. Lavigne, M.B. et al., Soil respiration responses to temperature are controlled more by roots than by decomposition in balsam fir ecosystems, *Can. J. For. Res.*, 33, 1744–1753, 2003.

42. Holmgren, P. and Thuresson, T., Satellite remote sensing for forestry planning, *Scand. J. For. Res.*, 13, 90–110, 1998.

43. Beaubien, J. et al., Land cover from multiple thematic mapper scenes using enhancement-classification methodology, *J. Geophys. Res. (Atmos.)*, 104, 27909–27920, 1999.

44. Myneni, R.B. et al., Global products of vegetation leaf area and fraction absorbed PAR from year one of MODIS data, *Remote Sens. Environ.*, 83, 214–231, 2003.

45. Chen, J.M., et al., Derivation and validation of Canada-wide coarse-resolution leaf area index maps using high-resolution satellite imagery and ground measurements, *Remote Sens. Environ.*, 80, 165–184, 2002.

46. Fassnacht, K.S. et al., Estimating the leaf area index of north central Wisconsin forests using the Landsat Thematic Mapper, *Remote Sens. Environ.*, 61, 229–245, 1997.

47. Huemmrich, K.F. and Goward, S.N., Vegetation canopy PAR absorptance and NDVI: An assessment for ten tree species with the SAIL model, *Remote Sens. Environ.*, 61, 254–269, 1997.

48. Potter, C. et al., Major disturbance events in terrestrial ecosystems detected using global satellite data sets, *Global Change Biol.*, 9, 1005–1021, 2003.

49. Li, Z., Nadon, S., and Cihlar, J., Satellite detection of Canadian boreal forest fires: development and application of the algorithm, *Int. J. Remote Sens.*, 21, 3057–3069, 2000.

50. White, M.A. et al., Satellite evidence of phenological differences between urbanized and rural areas of the Eastern United States deciduous broadleaf forest, *Ecosystems*, 5, 260–273, 2002.

51. Nichol, C.J. et al., Remote sensing of photosynthetic-light-use efficiency of boreal forest, *Agric. For. Meteorol.*, 101, 131–142, 2000.

52. Turner, D.P. et al., Scaling gross primary production (GPP) over boreal and deciduous forest landscapes in support of MODIS GPP product validation, *Remote Sens. Environ.*, 88, 256–270, 2003.

53. Liu, J. et al., Net primary productivity mapped for Canada at 1-km resolution, *Global Ecol. Biogeogr.*, 11, 115–129, 2002.

54. Bogaert, J. et al., Evidence for a persistent and extensive greening trend in Eurasia inferred from satellite vegetation index data, *J. Geophys. Res. (Atmos.)*, 107, doi: 10.1029/2001JD001075, 2002.

55. Myneni, R.B. et al., Increased plant growth in the northern high latitudes from 1981 to 1991, *Nature*, 386, 698–702, 1997.

56. Running, S.W. et al., A continuous satellite-derived measure of global terrestrial primary production, *BioScience*, 54, 547–560, 2004.

57. Woodwell, G. M. et al., Biotic feedbacks in the warming of the earth, *Clim. Change*, 40, 495–518, 1998.

58. Phillips, D.L. et al., Toward error analysis for large-scale forest carbon budgets, *Global Ecol. Biogeogr.*, 9, 305–313, 2000.

59. Raulier, F., Pothier, D., and Bernier, P.Y., Predicting the effect of thinning on growth of dense balsam fir stands using a process-based tree growth model, *Can. J. For. Res.*, 33, 509–520, 2003.

60. Van Wagner, C.E., Structure of the Canadian Forest Fire Weather Index, Canadian Forest Service, Publ. 1333, 1976.

61. Bergeron, Y. et al., Using dendrochronology to reconstruct disturbance and forest dynamics around Lake Duparquet, northwestern Quebec, *Dendrochronologia*, 20, 175–189, 2002.

62. Amiro, B.D. and Chen, J.M., Forest-fire-scar aging using SPOT-VEGETATION for Canadian ecoregions, *Can. J. For. Res.*, 33, 1116–1125, 2003.

63. Simard, M. and Payette, S., Accurate dating of spruce budworm infestation using tree growth anomalies, *Ecoscience*, 10, 204–216, 2003.

64. Hogg, E.H., Hart, M., and Lieffers, V.J., White tree rings formed in trembling aspen saplings following experimental defoliation, *Can. J. For. Res.*, 32, 1929–1934, 2002.

65. Vose, J.M. and Ryan, M.G., Seasonal respiration of foliage, fine roots, and woody tissues in relation to growth, tissue N, and photosynthesis, *Global Change Biol.*, 8, 182–193, 2002.

66. Malhi, Y. et al., The carbon balance of tropical, temperate and boreal forests, *Plant Cell Environ.*, 22, 715–740, 1999.

67. Ryan, M.G., Lavigne, M.B., and Gower, S.T., Annual carbon cost of autotrophic respiration in boreal forest ecosystems in relation to species and climate, *J. Geophys. Res. (Atmos.)*, 102, 28871–28883, 1997.

68. Lavigne, M.B. and Ryan, M.G., Growth and maintenance respiration rates of aspen, black spruce and jack pine stems at northern and southern BOREAS sites, *Tree Physiol.*, 17, 543–551, 1997.

69. Landsberg, J.L. and Waring, R.H., A generalised model of forest productivity using simplified concepts of radiation-use efficiency, carbon balance and partitioning, *For. Ecol. Manage.*, 95, 209–228, 1997.

70. Box, J.E., Jr., Modern methods for root investigations, in *Plant Roots: The Hidden Half*, 2nd ed., Waisel, Y., Eshel, A., and Kafkafi, U., Eds, Marcel Dekker, New York, 1996, 193–237.

71. Johnson, M.G. et al., Advancing fine root research with minirhizotrons, *Environ. Exp. Bot.*, 45, 263–289, 2001.

72. Bernier, P.Y. and Robitaille, G., The plane intersect method for estimating fine root productivity of trees from minirhizotron images, *Plant Soil*, 265, 165–173, 2004.

73. Hobbie, E.A. et al., Isotopic estimates of new carbon inputs into litter and soils in a four-year climate change experiment with Douglas-fir, *Plant Soil*, 259, 331–343, 2004.

74. Gorham, E., The biochemistry of northern peatlands and its possible responses to global warming, in *Biotic Feedbacks in the Global Climatic System: Will the Warming Feed the Warming?* Woodwell, G.M. and Mackenzie, F.T., Eds., Oxford University Press, New York, 1995, 69–187.

75. Turetsky, M.R., Wieder, R.K., and Vitt, D.H., Boreal peatland C fluxes under varying permafrost regimes, *Soil Biol. Biochem.*, 34, 907–912, 2002.

76. Camill, P. et al., Changes in biomass, aboveground net primary production, and peat accumulation following permafrost thaw in the boreal peatlands of Manitoba, Canada, *Ecosystems*, 4, 461–478, 2001.

77. Weider, R.K., Past, present, and future peatland carbon balance: An empirical model based on 210Pb-dated cores, *Ecol. Appl.*, 11, 327–342, 2001.

78. Campbell, I.D. et al., Millennial-scale rhythms in peatlands in the western interior of Canada and in the global carbon cycle, *Quat. Res.*, 54, 155–158, 2000.

79. Gorham, E., Northern peatlands: role in the carbon cycle and probable responses to global warming, *Ecol. Appl.*, 1, 182–195, 1991.

80. Moore, T.R., Roulet, N.T., and Waddington, J.M., Uncertainty in predicting the effect of climatic change on the carbon cycling of Canadian peatlands, *Clim. Change*, 40, 229–245, 1998.

81. Yu, S.C., Bhatti, J.S., and Apps, M.J., Eds., Long-Term Dynamics and Contemporary Carbon Budget of Northern Peatlands, Carbon Dynamics of Forested Peatlands: Knowledge Gaps, Uncertainties and Modeling Approaches, Northern Forestry Centre Information Report NOR-X-383, Canadian Forest Service, Edmonton, Alberta, 2002.

82. Frolking, S. et al., Modeling northern peatland decomposition and peat accumulation, *Ecosystems*, 4, 479–498, 2001.

83. Bubier, J.L. et al., Spatial and temporal variability in growing-season net ecosystem carbon dioxide exchange at a large peatland in Ontario, Canada, *Ecosystems,* 6, 353–367, 2003.

84. Dong, J.R. et al., Remote sensing estimates of boreal and temperate forest woody biomass: carbon pools, sources, and sinks, *Remote Sens. Environ.,* 84, 393–410, 2003.

85. Reich, P.B., Turner, D.P., and Bolstad, P., An approach to spatially distributed modeling of net primary production (NPP) at the landscape scale and its application in validation of EOS NPP products, *Remote Sens. Environ.,* 70, 69–81, 1999.

86. Reynolds, J.F., Hilbert, D.W., and Kemp, P.R., Scaling ecophysiology from the plant to the ecosystem: a conceptual framework, in *Scaling Physiological Processes: Leaf to Globe,* Ehlringer, J. and Field, C., Eds., Academic Press, New York, 1993, 127–140.

87. Bhatti, J.S., Apps, M.J., and Tarnocai, C., Estimates of soil organic carbon stocks in central Canada using three different approaches, *Can. J. For. Res.,* 32, 805–812, 2002.

88. Banfield, G.E. et al., Variability in regional scale estimates of carbon stocks in boreal forest ecosystems: results from West-Central Alberta, *For. Ecol. Manage.,* 169, 15–27, 2002.

89. Preston, C.M. and Trofymow, J.A., Variability in litter quality and its relationship to litter decay in Canadian forests, *Can. J. Bot.,* 78, 1269–1287, 2000.

90. Davis, A.A., Stolt, M.H., and Compton, J.E., Spatial distribution of soil carbon in Southern New England hardwood forest landscapes, *Soil Sci. Soc. Am. J.,* 68, 895–903, 2004.

91. Tremblay, S., Ouimet, R., and Houle, D., Prediction of organic carbon content in upland forest soils of Quebec, Canada, *Can. J. For. Res.,* 32, 903–914, 2002.

92. Clark, D.A. et al., Measuring net primary production in forests: concepts and field methods, *Ecol. Appl.,* 11, 356–370, 2001.

93. Zheng, D.L., Prince, S., and Hame, T., Estimating net primary production of boreal forests in Finland and Sweden from field data and remote sensing, *J. Veg. Sci.,* 15, 161–170, 2004.

94. Gower, S.T., Kucharik, C.J., and Norman, J.M., Direct and indirect estimation of leaf area index, f_{APAR}, and net primary production of terrestrial ecosystems, *Remote Sens. Environ.,* 70, 29–51, 1999.

95. Li, Z. et al., Temporal changes of forest NPP and NEP in West Central Canada associated with natural and anthropogenic disturbances, *Can. J. For. Res.,* 33, 2340–2351, 2003.

96. Li, Z. et al., Belowground biomass dynamics in the carbon budget model of the Canadian forest sector: recent improvements and implications for the estimation of NPP and NEP, *Can. J. For. Res.,* 33, 126–136, 2003.

18 Knowledge Gaps and Challenges in Wetlands under Climate Change in Canada

B.G. Warner and T. Asada

CONTENTS

18.1 INTRODUCTION

Wetlands are a characteristic element of the Canadian landscape. They occur anywhere supplies of water on the land surface sustain waterlogged conditions and anywhere specialized biotic communities are adapted to extreme variations in soil oxygen between periods of wetness and dryness. The geomorphologic setting of the land and the buildup of sediments in the wetland itself, especially in the case of peat landforms, are important factors controlling water conditions in wetlands. Climate determines the amount of water that enters via precipitation and leaves via evapotranspiration and, more indirectly, influences the supply of water that may enter via surface runoff and groundwater seepage. Human activities can affect the supplies of water entering wetlands by physically modifying the natural shape and size of the wetland and its surrounding watershed, by drawing water directly from the wetland and the sources supplying it, and by altering the vegetation cover, which, in turn, affects the inputs from and losses to the atmosphere. To understand wetland and climate relationships is to understand relationships between climate and water, but these relationships upset the natural balance when human activities interfere and threaten wetlands and their supplies of water.

Several reviews demonstrate Canada's considerable progress toward understanding and managing its wetland resources.[1-5] There is a national classification system, a fundamental first step for defining and recognizing wetlands. Regional inventories have been performed based on concepts and terms in the classification system. Much has been learned about differences in the character and dynamics of wetlands through scientific research and information gathering, which, in turn, wetland managers have used to develop policies governing wetland protection and wise use. These are not small accomplishments considering the extent and great diversity of wetlands in Canada, estimated to be about 18% of the world's freshwater wetlands.[6] Yet there is a rudimentary appreciation for the delicate linkages among climate, water, and wetlands. Wetlands still have not attained the attention they require in assessments of human impacts on the natural environment. This is exemplified by the number of threats to drinking water supplies, water pollution problems, major floods, droughts, fire, and other environmental disasters in recent years in Canada that revolve around climate, water, wetlands, human health, and economy. Wetlands are, or should be, considered central to these issues.

This chapter highlights some topical issues, gaps in our understanding of the role of wetlands in natural and human-dominated landscapes, vulnerability of wetlands and the environment when climate compounds the problems, and what we need to know to better deal with the wetland and water issues in the future. It is by no means complete but is intended to give a sense of the nature of some problems and the gaps in knowledge. Wetland ecotechnology presents the opportunity to direct natural self-organization processes in wetlands to solve problems for the benefit of both the natural environment and society.

18.2 COMMON MISCONCEPTIONS

Wetland specialists have met at intervals to review progress and prioritize future directions and information needs. Several workshops have addressed knowledge gaps and challenges in the science of wetlands for measuring, predicting, mitigating, and adapting to demands of society including consideration of future changes to climate.[7-10] The needs and issues with respect to wetlands identified at these national forums, some more than 10 years ago, still hold to the present. This may, in part, be due to the needs being so diverse and complex that the wetland community is challenged to meet all of them and, in part, from a lack of will owing to a number of common misconceptions about wetlands in Canada (Table 18.1).

18.3 WETLAND CLASSIFICATION AND INVENTORY

Classification is a fundamental prerequisite for all work on wetlands, because it provides a system for defining and recognizing wetlands in the landscape. Two

TABLE 18.1
Some Common Misconceptions about Wetlands in Canada

Perception	Reality
Wetlands are abundant in Canada and are not threatened.	Wetlands are most abundant in the Boreal Ecozone. They are threatened in southern Canada, especially in the most heavily populated and agricultural areas and elsewhere due to dam construction and hydroelectric reservoir development.
Wetlands quality is high and they are not degraded.	True for the most pristine and remote regions, but wetlands closest to inhabited regions are in various degrees of deterioration because of incompatible land uses in and around wetlands, long-term degradation from the historical past, urbanization, agriculture, discharge of chemical and biological contaminants, influx of invasive species.
Wetlands are wastelands and obstacles to development; human society and wetlands cannot live together.	Great progress has been made toward effective management and conservation of wetlands; however, there remains poor appreciation for the value of wetlands, some perceive wetlands as frightening and dangerous places, legislation works against wetlands in some jurisdictions, and there has been a relaxation by groups and organizations to respect the benefits and virtues of wetlands.
Restored and created wetlands are undesirable and unachievable.	Wrong! Many successful restored and created wetlands exist in Canada, some of which are recognized as internationally important.
Wetlands are not linked to climate and are unwanted, because they contribute greenhouse gases that warm the climate.	There is a complex linkage between carbon cycling processes in wetlands and climate. Wetlands not only contribute but also remove greenhouse gases from the atmosphere in ways that vary greatly among different wetland types, among similar wetland types, and within the same wetland basin.
The knowledge base about wetlands is complete and there is no need for further work and research on wetlands.	Wrong! The global scientific community is moving fast to discover new and important attributes about wetlands that need to be explored in Canada. For example, we know almost nothing about aspects of wetland biogeochemistry, hydrology, microbial ecology, and biodiversity in Canada, much less the services and values that wetland might contribute to human society.
The peat moss industry is the only economic value of wetlands.	Wrong! Canada's wetland industry as a whole, of which the peat moss industry is a part, probably contributes as much to the economic well-being of Canadians as most of the other natural resource sectors, yet is largely ignored.

editions of the Canadian wetland classification have been published, the last of which is the culmination of nearly 20 years of work.[11] The classification system captures the range of both peatland and mineral wetland types, including those that are on land and along freshwater and marine coasts. It also includes those influenced by permafrost processes.

Classification systems provide the framework for naming and mapping wetlands in national and regional inventories. It provides the standard "taxonomy" used to name representative wetland types and organize them into groups and subgroups. Unfortunately, the classification system has not been universally adopted across all sectors of the federal government even though the Federal Cabinet accepted the national classification system as the standard to follow when it adopted the Federal Wetlands Policy in 1992.[12]

The 1997 classification system is incomplete and needs refinement. Our concepts of swamp and bog remain confusing as indicated by some discordance between definitions of the classification system and the maps produced by Tarnocai et al.,[6,13] largely due to incomplete understanding of the hydrology of wetlands throughout Boreal Canada. Consideration should be given to explicitly recognizing the condition or state of the wetland at probably the form or type level in the classification (Table 18.2). This would recognize restored and created wetlands, which are not included. Degraded wetlands are problematic because they have characteristics that do not readily fall into the current classification, which is based largely on pristine wetlands. Further work is required in areas such as the following: the classification needs to be reevaluated and field-tested for redundancies, generalities, and incompleteness, and should be brought up to date with current scientific concepts of bog, swamp, fen, marsh, and shallow open water; restored and created wetlands were largely overlooked in the 1997 classification and need to be included in the classification system; in view of the occurrence of corals in Canada's offshore areas, consideration of whether the Canadian definition of wetlands needs to be expanded to be brought in line with the international definition of the Ramsar Convention, which recognizes coral reefs as components of wetlands; the classification should be used as the basis for more regional classification systems as recognized by the federal government; some consideration needs to be given to developing classifications that are based on both scientific and regulatory needs, such as hydrogeomorphic functional classifications widely used in the U.S.;[14] and the Canadian classification should be broadened and modified into a classification system for North American wetlands (i.e., with the U.S., Mexico), and perhaps toward a global classification for the Northern Hemisphere.[15]

Inventories are only as good as the classification system used as the standard reference for naming and differentiating the wetland units. The most comprehensive national inventories are those of Tarnocai et al.[6,13] Most wetland inventories in Canada exclude large areas of marine, brackish, and freshwater coastal wetlands (both marsh and shallow open water), and overlook many restored and created wetlands. The National Wetland Inventory that has been recently launched holds much promise to improve our knowledge of the extent and distribution of wetlands, primarily because it utilizes satellite imagery with modern digital imagery processing techniques.[16] This new initiative must build on the Canadian Wetland Classification system, and perhaps can help to resolve outstanding problems, such as accurate differentiation of *Picea* swamp vs. *Picea* forest in the Boreal Ecozone, and open water wetland vs. lake and ocean in coastal zones. Such challenges can only be confirmed by accurate and supporting field surveys.

Inventory work such as that of Snell[17,18] for southern Ontario needs to be updated and expanded to other regions of the country, where a significant extent of wetland has been converted to alternative land uses since European settlement. Such inventories

TABLE 18.2
Classification Considering State of Wetland Condition

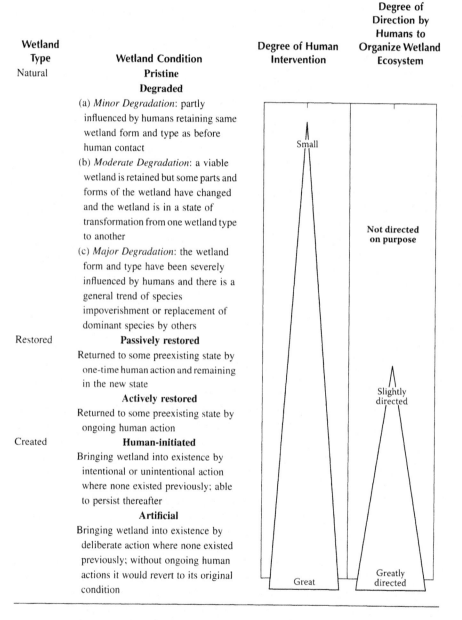

Wetland Type	Wetland Condition	Degree of Human Intervention	Degree of Direction by Humans to Organize Wetland Ecosystem
Natural	**Pristine**		
	Degraded		
	(a) *Minor Degradation*: partly influenced by humans retaining same wetland form and type as before human contact	Small	
	(b) *Moderate Degradation*: a viable wetland is retained but some parts and forms of the wetland have changed and the wetland is in a state of transformation from one wetland type to another		**Not directed on purpose**
	(c) *Major Degradation*: the wetland form and type have been severely influenced by humans and there is a general trend of species impoverishment or replacement of dominant species by others		
Restored	**Passively restored** Returned to some preexisting state by one-time human action and remaining in the new state		
	Actively restored Returned to some preexisting state by ongoing human action		Slightly directed
Created	**Human-initiated** Bringing wetland into existence by intentional or unintentional action where none existed previously; able to persist thereafter		
	Artificial Bringing wetland into existence by deliberate action where none existed previously; without ongoing human actions it would revert to its original condition	Great	Greatly directed

serve to identify priority regions and sites where wetlands are especially vulnerable and where conservation efforts and protection should take precedence (Figure 18.1). Regions where wetlands may not be especially widespread anymore but existed in the

FIGURE 18.1 Map of Canada showing regions where wetlands are most threatened. (Modified after Rubec.[76])

historical past reflect areas of high probability of success and should be targeted for restoration. The problem is compounded, because regions where wetlands are most vulnerable due to historical reasons are also regions where climate-warming scenarios predict extreme warming and drought.[19]

18.4 HYDROLOGICAL LANDSCAPE MODIFICATIONS AND WATER BUDGET FLUCTUATIONS

Our unwise use and poor management of water resources has created problems for wetlands that, if they continue, will only pose more serious threats to wetlands under climate change scenarios. Humans have disrupted the magnitude and timing of natural river flows through land clearance, dam and reservoir construction, diversion and transfer of water across watersheds, exploitation of groundwater aquifers, channelization of streams, and reconfiguration of coastlines in Canada. These activities allow control of water resources to generate electricity, supply water for agriculture and industry, mitigate flooding, navigate rivers and lakes, and use the land and near shore for purposes incompatible with natural conditions.[20,21] Canada ranks third next to the

U.S. and Russia in having the greatest number of major dams.[21] Innumerable wetlands and aquatic habitats have been lost and altered as a result of these landscape changes.

Land clearance by early settlers changed the land from being forest, wetland, and water-dominated to one that is open, drier, and largely devoid of wetlands. Smaller water bodies and streams were obliterated and larger ones became larger, so large that rivers overflowed and changed their channels and water levels rose in the receiving water bodies. Land-use history necessitated the creation of the Conservation Authorities system in the 1940s in Ontario to monitor floods.[22] The loss of wetlands and water bodies on the Prairies prompted the formation of Ducks Unlimited Canada in 1938 to reverse the historical trends and restore wetlands.[22]

Large to small dam construction has increased fivefold since the 1950s. This has resulted in major changes in the volume of water flowing in rivers by water being impounded in upper reaches and significant decreases in volumes reaching lower reaches and the river mouth. Extent of dam construction on a single watershed can be substantial. For example, the Columbia River Basin in Canada and the U.S. has 19 dams on its 2000-km length; only 70 km of the river remains free flowing and natural.[21] What has this kind of control of natural water flow done to wetlands in the watershed? Another serious problem is the change to natural seasonal patterns of river discharge. This has led to an increase in the age of waters reaching the river mouth because it may take water more than a year to reach the mouth compared to days and weeks under natural conditions. The hydroperiods of wetlands in riparian zones have changed, the influx and/or efflux of nutrients have increased, and water and/or nutrients in downstream areas are diminished. The reservoirs themselves have altered natural habitats. Source and sink strength for greenhouse gases have been affected by the reservoirs that can emit large quantities of greenhouse gases.[24] Hydroelectric reservoirs are particularly a problem because not only are they sources of greenhouse gases, but also of methylmercury, which accumulates and biomagnifies in the aquatic food chain.[25,26] Hydroelectric development is the primary threat to wetlands in central parts of Manitoba, Ontario, and Quebec (Figure 18.1).

Wetlands must be a central concern in understanding the impact of hydrological alterations. The obvious impacts are complete displacement and destruction of wetlands, but less known are the less obvious subtle changes to wetland water budgets and the responses by wetland biota. Detailed information needs include the following: undertake detailed evaluation of the state of existing knowledge base and prioritize information needs on impact to wetlands of hydrological alterations; assess impacts of seasonal changes to river runoff cycles on water budgets and hydroperiods of associated wetlands, especially wetlands in riparian and floodplain zones; identify linkages between large-scale hydrological alterations, regional water cycles, wetlands, and climate; determine changes to hydrological conditions, nutrient cycles, and biota in wetlands in upstream vs. downstream reaches; undertake historical comparisons between pre-disturbance and present-day landscapes, wetland ecosystems, and climate, which can be readily done using paleoecological techniques, and long-term monitoring sites need to be established to assess future impacts; and consider conversion to wetland for reservoirs and impoundments when they are decommissioned and taken off-line. Gorham et al.[27] have reviewed the merits of the longer time frames and spatial scales offered by paleoecological approaches in dealing with environmental problems that directly and indirectly affect peatlands. There needs to be much greater use and recognition of the archival record in peatlands as an important management tool for climate change.

18.5 SEDIMENTATION AND WATER QUALITY CHANGES

Overloading of sediments and nutrients due to poor land-use practices is a major threat to wetlands. The gaps and needs in this area with respect to wetlands are complex and have been addressed by the workshops referred to earlier. They are also closely related to issues surrounding hydrological alterations discussed previously. This area has received little attention in the context of wetlands, but likely is a large problem in the vicinity of the mouths of many streams and rivers and along the associated lake shores and ocean coasts of Canada just as it is in other parts of the world. It has likely escaped the attention of Canadians because the problems have not yet reached catastrophic proportions although there may well be accidents waiting to happen. There have been intermittent reports in the Canadian popular press in recent years that report fish kills in our water bodies that are usually attributed to natural cyclic phenomena. There needs to be discussions of nutrients and overall water quality when such fish kills do occur to ensure that poor land-use practices are not contributing to "natural phenomena." Lessons learned from the Mississippi basin issue can help in Canada.

The Gulf of Mexico hypoxia problem in the U.S. serves as a good example for Canadians to illustrate the magnitude of problems of poor land-use management and the catastrophic environmental and economic consequences exacerbated by warm climate. Nutrient loading and sedimentation throughout the Mississippi River basin has been shown to increase significantly in recent decades. A zone of low oxygen levels less than 2 mg l^{-1} has been noted on the continental shelf of the northern part of the Gulf of Mexico.[28–30] Transport of excess nutrients by the Mississippi River has contributed to increased primary production along with warm summers and regional water circulation processes has been found to be responsible. Nitrate concentrations from the Mississippi River show major increases in the 1950s after nitrogen fertilizer came into widespread use and after expansion in other activities such as artificial drainage and hydrological alterations in the basin, such as increased runoff, wastewater discharge from urban settlements, and intensification of agriculture contributed to the problem.[29] Hypoxia is exacerbated by warm climate. Restoration of wetlands is viewed as ways by which nutrients and sediments can be controlled from entering the river and transported downstream to the river mouth.[28] Sedimentation and a decline in water quality in the Great Lakes have been attributed to agriculture and urbanization in adjacent watersheds and efforts are under way to restore wetlands to correct the problem.[31]

18.6 CARBON CYCLING AND CLIMATE

Wetlands contain the largest quantities of carbon in both their living biomass and in their deep peat and sediments in the terrestrial biosphere. Canada contains around one third of the global peatland carbon although uncertainties exist on extent of peatland and carbon in the world.[32] Wetlands in Ontario are estimated to contain about 19% of Canada's wetland carbon and about 6.6% of the wetland carbon in the world.[33] The recent review of McLaughlin[32] presents a good summary of the current state of knowledge of wetland carbon cycling that, despite a focus on Ontario, applies to much of Boreal Canada. Large-scale assessments of general climate change effects on carbon cycling processes are reasonably clear.[33–39] Specific topics that have received considerable attention include the following: net primary production and litter decomposition,[40–42] peat carbon accumulation,[39,43–46] and methane and carbon dioxide production and emission.[47,48] Different models are being developed to scale up peatland carbon to landscape, regional, and global scales.[19,49,50]

There are several important areas that require further research to refine our understanding of carbon pools, fluxes, and cycling processes at different temporal and spatial scales.[32,51] Considerable effort has been directed at quantifying the various carbon pools but gaps still exist. For example, belowground carbon in litter and roots is thought to be large in peatlands covered by vascular plants, yet real estimates are poor at best. There remains incomplete understanding of variability in the nature of temporal and spatial variability of carbon sink and source strength at microtopographic scales. Plant community composition differs markedly and climate is expected to exhibit enough influence to cause vegetation composition to shift both peatland structure and community type. For example, fens are expected to shift into bogs and swamps. Such vegetation changes will vary the nature of carbon being added to the litter and long-term storage. Little is known about how climate will interact with the new litter quality, oxidation–reduction processes, and microbial communities. Recent studies show that smaller-scale intrinsic factors such as water tables, vegetation composition, and litter governed by peatland microtopography and peat formation may be stronger forces in carbon cycling processes than larger scales and longer-term responses to changes in climate.[52,53] Climate is expected to enhance effects of ultraviolet-B (UV-B) radiation on production and decay rates.[54] How will radiation affect carbon cycling and sequestration processes in peatlands?

Much of the attention on wetland carbon cycling and climate has focused on peatlands. Mineral wetlands contain significant quantities of carbon, albeit much less than in peatlands, but in quantities that should not be ignored.[33,55]

18.7 INVASIVE SPECIES

Wetlands seem to be particularly favorable habitats for invasive species and warmer climate may be compounding the problem by allowing more species to survive the winter period than would otherwise under more severe conditions. Invasive species are both non-native species and native species that become harmful to the ecosystem and, in turn, likely cause harm to the economy and humans. The list includes animals (insects, mollusks, fishes, birds, herpitiles, mammals), plants (algae, bryophytes,

TABLE 18.3
Invasive Species Reported in Wetlands in Canada from the Global Invasive Species Database (http://www.issg.org/database)

Scientific Name	Common Name	Alien or Native Status
	Plants	
Cirsium arvense	Canada thistle (herb)	Alien and/or uncertain
Elaeagnus angustifolia	Russian olive (tall shrub)	Alien
Fallopia japonica	Donkey rhubarb (herb)	Alien
Hedera helix	English ivy (woody vine)	Alien
Lespedeza cuneata	Chinese bush-clover (herb)	Alien
Lythrum salicaria	Purple loosestrife (herb)	Alien
Myriophyllum spicatum	Eurasian water- milfoil (herb)	Alien
Phalaris arundinacea	Reed canary grass (herb)	Uncertain
Phragmites australis	Common reed (herb)	Alien and/or native
Potamogeton crispus	Curly pondweed (herb)	Alien
	Animals	
Cyprinus carpio	Carp (fish)	Alien and/or uncertain
Gambusia affinis	Mosquitofish (fish)	Alien
Micropterus salmoides	American black bass (fish)	Alien, uncertain, and/or native
Mustela erminea	Short-tailed ermine (mammal)	Uncertain and/or native
Orconectes virilus	Northern crayfish (crustacean)	Native
Passer domesticus	English sparrow (bird)	Alien
Rattus norvegicus	Norway rat (mammal)	Alien
	Microbes	
Flavivirus spp.	West Nile virus	Uncertain

vascular plants), fungi, and microbes. Unfortunately, there is no complete list of what these species are in Canadian wetlands and aquatic habitats. To compare, a total of 334 non-native aquatic species have been introduced into wetlands in the U.S.[56] The Global Invasive Species Database (http:// www.issg.org) lists 18 taxa (Table 18.3) in Canadian wetlands, but this is only a small fraction of invasive species actually known to affect fresh, brackish, and marine wetlands. This list does not include well-known examples such as *Dreissena polymorpha* (zebra mussel), *Tinca tinca* (tench fish), or *Cabomba caroliniana* (fanwort macrophyte), which are having major impacts on freshwater wetlands.[57] There are at least 150 non-native invasive species in the Great Lakes basin that have been introduced in the 1900s.[57] The vast majority of these live in or close to wetlands. A comparable number of invasive species are associated with marine and estuarine wetlands of the Strait of Georgia in southwest British Columbia. Included are such species as *Sargassum muticum* (brown seaweed), *Zostera japonica* (dwarf eelgrass), *Venerupis philippinarum* (Manila clam), and *Branta canadensis* (Canada goose).[57]

The status of many alien invasives is poorly understood. A good example is *Phragmites australis*, a common wetland plant widely distributed across Canada. It has become a nuisance in the eastern U.S., and is expanding its dominance throughout the Great Lakes Region and parts of eastern Canada. It has been suggested that a more aggressive genotype of European origin has been introduced to North America and is displacing native genotypes, giving rise to development of monospecific wetland communities.[58,59] There is a need to survey *P. australis* in Canada to differentiate native and European genotypes to fully understand the impact of invasion by this plant. Is there a biogeoclimatic difference between the growing conditions of these two races, and is Canada any more or less vulnerable to widespread invasion than other regions at more southern latitudes.

Chen caerulescens caerulescens (Lesser snow geese) is another interesting example where agricultural practices have contributed to changing a native species into an invasive wetland species threatening many Arctic wetlands.[60] The birds have changed their overwintering food source from less nutritional sources in natural marshes on the south coast of the U.S. to agricultural upland sites where food is more nutritional and accessible. The switch in foraging behavior has contributed to overpopulations of geese returning to their summer nesting grounds in coastal marshes in Arctic Canada in numbers the food source in the natural marshes cannot support.

West Nile virus (WNV; *Flavivirus* spp.) has been considered a wetland invasive species because the larvae of mosquitoes inhabit wetland waters and adult mosquitoes are the primary vectors for transmission of WVN from their vertebrate hosts (mostly birds of the Corvidae) to humans. One could debate whether it should be considered a wetland species. There are several strains of *Flavivirus* spp., but the one that has recently appeared in North America and entered Canada in 1999 appears to have a broad range of hosts and adaptability. A number of blood-feeding arthropods (e.g., sand flies, ticks, and no-see-ums) other than mosquitoes are also thought to transmit WNV. A total of ten mosquito species have been found in positive WNV pools in Canada. It is important to determine which mosquito species are more apt to feed on birds to pick up the virus. Two species, *Culex pipiens* and *C. tarsalis*, are thought to be most active in circulating WNV within the bird populations.[61–63] *Culex pipiens* is most common in extreme southern Ontario, Quebec, New Brunswick, and Nova Scotia, and the southwest part of British Columbia. *Culex tarsalis* does not occur in Atlantic Canada, and is confined to extreme southern Ontario, and westwards in the southern part of Canada.[64] If true, then should more widespread *C. perturbans* and *C. territans* be of less concern? Recently, it has been suggested that North American *C. pipiens* is a hybrid form that readily bites both humans and birds, compared to its *C. pipiens* European cousin that tends to bite only birds and not humans. The recent arrival of this hybrid would account for the recent occurrence and greater incidence of WNV in humans in North America than in Europe.[65] What is the *C. pipiens* in North America and what is the habitat preferences and ecology of its larvae compared to those in Europe?

There are a number of wetland species that should be monitored closely because they could become invasive, especially in light of warming winter climates. An example is *Myocastor coypus* (nutria), known to exist in Quebec and British Columbia, but it has not reached the nuisance status in Canada it has in the U.S. *Eichhornia crassipes* (water hyacinth) has been observed in a coastal marsh on Lake Ontario, probably introduced from water gardens or aquaria (Warner, personal observations).

Alien invasives pose a major threat to the biodiversity of Canadian wetlands, some of which are the most diverse ecosystems in the country. Complete biodiversity assessments for wetlands do not exist. Certain groups of organisms are better known for certain kinds of wetlands than others.

Important areas for further work on invasive wetland species are as follows: compile a list of all invasive species in wetlands in Canada, differentiate non-native and native species, and rank those with the greatest threat to wetlands; set up monitoring networks to gauge current and future distributions, ecological interactions, and environmental impacts, and undertake more research to under-stand the biology and ecology of the invasive organisms; be proactive by setting up monitoring networks to assess the potential threat of alien species or native species that may become invasive in the future, especially with the expectation that there may be major range shifts under future climates; conduct economic impact evaluations because invasive species cause major economic harm and pose threats to Canada's wetland industry; and establish a national network of key wetlands for biodiversity assessments.

18.8 WETLAND ARCHIVAL RECORDS

Wetlands, especially peatlands, are unusual in that they contain a record of them-selves preserved in the organic sediments and peat below the surface.[66,67] The peat archives record continuous environmental changes, including climate, that may go back thousands of years, in Canada usually the past 15,000 years or so. These long archival records reconstructed using paleoecological techniques provide a means of extending back farther in time contemporary weather records and other databases and records of environmental conditions. The value of comparing and combining peatland ecosystems on short-time and fine spatial scales of ecologists with the much longer-time and broader spatial scales of paleoecologists have been discussed by Gorham[51] and Gorham et al.[27] Some of these are (1) more observations can be made over greater time and spatial scales, (2) ecosystem processes and responses to environmental changes such as climate, can be clarified that might be difficult to see over shorter timescales, (3) species behavior and community structure can be compared before and after historical period of human disturbances, and (4) human impacts can be assessed and predicted because of the greater time and spatial scales.

The archival record in peatlands has contributed significantly to improving our understanding of belowground carbon storage and cycling processes in peatlands as discussed above. Ecosystem dynamics and succession can be clarified using paleo-ecological techniques. In heavily populated southern Ontario, several peatland types are known to have changed in direct response to human interferences within the historical period and do not represent ecosystems today that have followed "natural" predictable developmental pathways.[68,69] In other cases, early Holocene climate warming and the "Little Ice Age" seem to have affected developmental histories.[69,70] Therefore, if past climate influenced past development trends, how might the present or future climate be affecting these wetlands? Canada contains abundant peatlands and as such offers the potential for reconstructing widespread longer-term climatic and environmental changes for most parts of the country. Compared to lakes, wetlands have not received the attention they deserve.

18.9 WETLAND ECOTECHNOLOGY: THE WAY OF THE FUTURE

Wetland ecotechnology, also referred to as wetland ecological engineering, is defined as working with nature for the benefit of both the natural environment and human society.[71] Wetland ecotechnology is not new to Canada.[72,73] Wetlands have been restored and created to replace wildlife habitat that has been lost ever since Ducks Unlimited Canada began its work in 1938. The first wetlands to treat wastewater and improve water quality were constructed in the late 1970s. Wetlands stabilize soil erosion and shorelines and they have been used to rehabilitate abandoned gravel pits and quarries, mine tailings, contaminated soils, and dumpsites. Naturalistic fish ways that simulate natural channels provide migration passages for fish around dams and in areas where natural channels are unnavigable by fish. Rainwater can be recycled through wetlands in urban area and on rooftops, or in wetland-like holding ponds as storm water. There are thousands of examples in Canada.

Wetland ecotechnology has much potential in Canada so that some of the problems presented here can be avoided or reduced in the future. Further work is needed in the following areas: there is little detailed information to verify how and why various wetland designs work, especially in Canadian climatic conditions; designs are customized to suite specific needs, which need testing and parameterizing into standardized designs; and data are needed to assist regulators and managers with decisions on these nontraditional wetland technologies.

18.10 SUMMARY AND CONCLUSIONS

Wetlands are sensitive to climate changes, especially since they lie between terrestrial ecosystems and open water bodies. A small change in the quantity of water in the system can cause the wetland to become more terrestrial or more aquatic. Humans are responsible for directly altering wetlands by physically changing their shape and size, readjusting the quantity of water reaching them, and altering the biogeochemical transformations in wetlands. Invasive species are a major threat to the biological integrity of wetlands. The stress on wetlands by all of these factors is compounded by climate. Existing wetlands must be protected from further loss and degradation, and wetland ecotechnology can be used to replace lost and enhance existing wetlands. Wetlands affect climate and climate affects wetlands. Do we need to be reminded of the "Dirty 30s" in the southern Prairies when drought conditions dried up all surface water bodies and wetlands? Would this period have been as "dirty" and as dry if decades before extensive surface water bodies, wetlands, and natural vegetation had not been converted into cultivated land?

We should be reminded that wetlands support a large industry in Canada.[74] It has been estimated that over CA\$10 billion is derived from wetlands each year.[75] This value is probably a gross underestimate, because the wetland economy includes four sectors — products and manufacturing, supplies and distribution, services, and a knowledge sector — that all depend on wetlands and contribute to the Canadian economy. Claudi et al.[57] have suggested the cost of control for nine alien species was about CA\$5.5 billion and about CA\$456 million annually. What impact do wetland alien invasives, for example, have on the wetlands industry? What is the economic loss when wetlands are destroyed or altered by poor water management and pollution by contaminants? What role will climate play in exacerbating economic costs? Canada's wetland industry is especially vulnerable because it is diffuse and includes many small, family-run operations in rural and economically disadvantaged parts of the country.

Wetlands are commonly confused with terrestrial and aquatic ecosystems in the landscape. Human activities can dramatically alter the supplies of water entering wetlands. So too can climate affect water exchanges between wetlands and the atmosphere. Wetlands are especially vulnerable when humans alter water supplies on the land and atmospheric inputs by interfering with climate. Too little water may push wetlands into becoming more terrestrial in character and too much water may push wetlands into becoming more aquatic. The association and linkage of wetlands with a number of important environmental issues has probably not been realized due to several misconceptions. Much remains to be done to refine wetland classification schemes. Inventories of Canada's wetland resource are incomplete and need to be accurately evaluated to ensure protection of all wetland resources for future generations. Hydrological modifications of the landscape continue to threaten wetlands and regional water budgets. There is a history of converting wetlands into alternative non-wetland uses, which reduces quantities of water retained on the land and in the vegetation. Wetlands and downstream receiving water bodies continue to be severely affected by sedimentation and nutrient loading. Canadians can learn about the severity and magnitude of the problem from the Mississippi River basin in the U.S. Invasive species continue to pose new surprises and ongoing threats to the biological integrity and biodiversity of wetlands. Human activities are responsible for all of these issues threatening wetlands, which are further exacerbated by climate changes. Unfortunately, Canadians are ill prepared to recognize and deal with the threats to wetlands. Canada is well positioned to utilize more widely wetland eco-technology and nature-sensitive practices to correct historical wrongs and to work in harmony with nature for the benefit of both wetlands and society so that climate will not pose as big a threat as might be possible in the future.

REFERENCES

1. Zoltai, S.C. and Pollett, F.C., Wetlands in Canada: their classification, distribution and use, in *Ecosystem of the World 4B, Mires: Swamp, Bog, Fen and Moor, Regional Studies*, Gore, A.J.P., Ed., Elsevier, Amsterdam, 1983, 245–268.
2. National Wetland Working Group, *Wetlands of Canada*, Polyscience Publishers, Montreal, 1988.
3. Glooschenko, W.A., Tarnocai, C., Zoltai, S., and Glooschenko, V., Wetlands of Canada and Greenland, in *Wetlands of the World 1*, Whigham, D.F., Dykyjová, D., and Hejny, S., Eds., Kluwer Academic, Dordrecht, 1993, 415–514.

4. Zoltai, S.C. and Vitt, D.H., Canadian wetlands: environmental gradients and classification, *Vegetatio*, 118, 131, 1995.
5. Rubec, C.D.A., The status of peatland resources in Canada, in *Global Peat Resources*, Lappalainen, E., Ed., International Peat Society, Helsinki, 1996, 243–252.
6. Tarnocai, C., Kettles, I.M., and Lacelle, B., *Wetlands of Canada*, Agriculture and Agri-Food Canada, Research Branch, Ottawa, 2001.
7. Wedeles, C.H.R., Meisner, J.D., and Rose, M.J., Wetland Science Research Needs in Canada, Wetlands Research Centre, University of Waterloo, 1992, 30 pp.
8. Cox, K., Wetlands: A celebration of life, Sustaining Wetlands Issue Paper 1993–1, North American Wetlands Conservation Council (Canada), Ottawa, 1993.
9. Clair, T.A., Warner, B.G., Robarts, R., Murkin, H., Lilley, J., Mortsch, L., and Rubec, C., Canadian inland wetlands and climate change, in *The Canada Country Study: Climate Impacts and Adaptation*, Vol. 7, Environment Canada, Ottawa, 1998, 190–218.
10. Rubec, C.D.A., Ed., *Wetland Stewardship in Canada*, North American Wetlands Conservation Council (Canada), Report 03-3, 2003, 145 pp.
11. Warner, B.G. and Rubec, C.D.A., Eds., *The Canadian Wetland Classification System*, 2nd ed., University of Waterloo, Waterloo, 1997.
12. Government of Canada, The federal policy on wetland conservation, Ottawa, 1991, 14 pp.
13. Tarnocai, C., Kettles, I.M., and Lacelle, B., *Peatlands of Canada*, Geological Survey of Canada, Open File 3834, Ottawa, 2000.
14. Smith, R.D., Ammann, A., Bartoldus, C., and Brinson, M., Wetlands Research Program Technical Report WRP-DE-9, U.S. Army Corps of Engineers, Washington, D.C., 1995.
15. Scott, D.A. and Jones, T.A., Classification and inventory of wetlands: a global overview, *Vegetatio*, 118, 3, 1995.
16. Leahy, S., Wetlands from space: the national wetland inventory, *Conservator*, 24, 13, 2003.
17. Snell, E.A., Wetland distribution and conversion in southern Ontario, Working Paper 48, Inland Waters and Land Directorate, Environment Canada, Ottawa, 1987.
18. Snell, E.A., Recent wetland loss trends in southern Ontario, in *Wetlands: Inertia or Momentum?* Bardecki, M.J. and Patterson, N., Eds., Federation of Ontario Naturalists, Don Mills, 1989, 183–197.
19. Kettles, I.M. and Tarnocai, C., Development of a model for estimating the sensitivity of Canadian peatlands to climate warming, *Geog. Phys. Quat.*, 53, 323, 1999.
20. Schindler, D.W., Sustaining aquatic ecosystems in Boreal regions, *Conserv. Ecol.*, 2, 1, 1998.
21. Rosenberg, D.M., McCully, P., and Pringle, C.M., Global-scale environmental effects of hydrological alterations: Introduction, *BioScience*, 50, 746, 2000.
22. Richardson, A.H., *Conservation by the People: The History of the Conservation Movement in Ontario to 1970*, Conservation Authorities Branch, Toronto, 1974.
23. Leitsch, W.G., *Ducks and Men*, Ducks Unlimited Canada, Winnipeg, 1978.
24. Vorosmarty, C.J. and Sahagian, D., Anthropogenic disturbance of the terrestrial water cycle, *BioScience*, 50, 753, 2000.
25. Rosenberg, D.M., Berkes, F., Bodaly, R.A., Hecky, R.E., Kelly, C.A., and Rudd, J.W.M., Large-scale impacts of hydroelectric development, *Environ. Rev.*, 5, 27, 1997.
26. St. Louis, V.L., Kelly, C.A., Ducheman, E., Rudd, J.W.M., and Rosenberg, D.M., Reservoir surface as sources of greenhouse gases to the atmosphere: a global estimate, *BioScience*, 50, 766, 2000.

27. Gorham, E., Brush, G.S., Graumlich, L.J., Rosenzweig, M.L., and Johnson, A.H., The value of paleoecology as an aid to monitoring ecosystems and landscapes, chiefly with reference to North America, *Environ. Rev.*, 9, 99, 2001.

28. Mitsch, W.J., Day, J.W., Gilliam, J.W., Groffman, P.M., Hey, D.L., Randall, G.W., and Wang, N., Reducing nitrogen loading to the Gulf of Mexico from the Mississippi River basin: strategies to counter a persistent ecological problem, *BioScience*, 51, 373, 2001.

29. Rabalais, N.N., Turner, R.E., and Wiseman, W.J., Gulf of Mexico hypoxia, a.k.a. "The Dead Zone," *Annu. Rev. Ecol. Syst.*, 33, 235, 2002.

30. Turner, R.E. and Rabalais, N.N., Linking landscape and water quality in the Mississippi River basin for 200 years, *BioScience*, 53, 563, 2003.

31. Environment Canada, *Where Land Meets Water: Understanding Wetlands of the Great Lakes*, Environment Canada, Toronto, 2002, 72 pp.

32. MacLaughlin, J., Carbon assessment in boreal wetlands of Ontario, Forest Research Information Paper 158, Ontario Ministry of Natural Resources, Sault Ste. Marie, 2004, 79 pp.

33. Warner, B.G., Davies, J.C., Jano, A., Aravena, R., and Dowsett, E., Carbon storage in Ontario's wetlands, in Climate Change Research Information Note 1, Ministry of Natural Resources, Sault Ste. Marie, Ontario, 2003, 4 pp.

34. Gorham, E., Northern peatlands: role in the carbon cycle and probable responses to global warming, *Ecol. Appl.*, 1, 182, 1991.

35. Gorham, E., The biogeochemistry of northern peatlands and its possible responses to global warming, in *Biotic Feedbacks in the Global Climatic System: Will the Warming Speed the Warming?* Woodwell, G.M. and MacKenzie, F.T., Eds., Oxford University Press, Oxford, 1995, 169–186.

36. Dean, W.E. and Gorham, E., Magnitude and significance of carbon burial in lakes, reservoirs, and peatlands, *Geology*, 26, 535, 1998.

37. Tarnocai, C., The amount of organic carbon in various soil orders and ecological provinces in Canada, in *Soil Processes and the Carbon Cycle*, Lal, R., Kimble, J.M., Follett, R.F., and Stewart, B.A., Eds., CRC Press, Boca Raton, FL, 1998, 81–92.

38. Tarnocai, C., The effect of climate warming on the carbon balance of cryosols in Canada, *Permafrost Periglacial Proc.*, 10, 251, 1999.

39. Vitt, D.H., Halsey, L.A., Bauer, I.E., and Campbell, C., Spatial and temporal trends in carbon storage of peatlands of continental western Canada, *Can. J. Earth Sci.*, 37, 683, 2000.

40. Vitt, D.H., Growth and production dynamics of boreal mosses over climatic, chemical and topographic gradients, *Bot. J. Linnean Soc.*, 104, 35, 1990.

41. Camill, P., Lynch, J.A., Clark, J.S., Adams, J.B., and Jordan, B., Changes in the biomass, aboveground net primary production, and peat accumulation following permafrost thaw in boreal peatlands of Manitoba, Canada, *Ecosystems*, 4, 461, 2001.

42. Asada, T. and Warner, B.G., Surface peat mass and carbon balance in a hypermaritime peatland, *Soil Sci. Soc. Am. J.*, 69, 549, 2005.

43. Warner, B.G., Clymo, R.S., and Tolonen, K., Implications of peat accumulation at Point Escuminac, New Brunswick, *Quat. Res.*, 39, 245, 1993.

44. Charman, D.J., Aravena, R., and Warner, B.G., Carbon dynamics in a forested peatland in northeastern Ontario, Canada, *J. Ecol.*, 82, 55, 1994.

45. Vardy, S.R, Warner, B.G., Turunen, J., and Aravena, R., Carbon accumulation in permafrost peatlands in the Northwest Territories, Canada, *Holocene*, 10, 273, 2000.

46. Yu, Z., Vitt, D.H., Campbell, I.D., and Apps, M.J., Understanding Holocene peat accumulation pattern of continental fens in western Canada, *Can. J. Bot.*, 81, 267, 2003.

47. Bubier, J.L, Moore, T.R., and Roulet, N.T., Methane emissions from wetlands in the midboreal region of northern Ontario, Canada, *Ecology*, 74, 2240, 1993.

48. Bellisario, J.L., Bubier, J., Chanton, J., and Moore, T.R., Controls on the emission of methane from a northern peatlands, *Global Biogeochem. Cycles*, 13, 81, 1999.

49. Hilbert, D.W., Roulet, N., and Moore, T., Modelling and analysis of peatlands as dynamical systems, *J. Ecol.*, 88, 230, 2000.

50. Frolking, S., Roulet, N.T., Moore, T.R., Richard, P.J.H., Lavoie, M., and Muller, S.D., Modeling northern peatland decomposition and peat accumulation, *Ecosystems*, 4, 479, 2001.

51. Gorham, E., The future of research in Canadian peatlands: a brief survey with particular reference to global change, *Wetlands*, 14, 206, 1994.

52. Belyea, L.R. and Clymo, R.S., Feedback control on the rate of peat formation, *Proc. R. Soc. London*, 268B, 1315, 2001.

53. Belyea, L.R. and Malmer, N., Carbon sequestration in peatland: patterns and mechanisms of response to climate change, *Global Change Biol.*, 10, 1043, 2004.

54. Gehrke, C., Effects of enhanced UV-B radiation on production-related properties of a *Sphagnum fuscum* dominated subarctic bog, *Funct. Ecol.*, 12, 940, 1998.

55. Eulis, N.H., Gleason, R.A., Olness, A., McDougal, R.L., Murkin, H.R., Robarts, R.D., Bourbonnierre, R.A., and Warner, B.G., Prairie wetlands of North America important for carbon storage, *Global Biogeochem. Cycles*, in press, 2005.

56. Benson, A.J., Documenting over a century of aquatic introductions in the United States, in *Nonindigenous Freshwater Organisms*, Claudi, R. and Leach, J.H., Eds., CRC Press, Boca Raton, FL, 1999, 1–31.

57. Claudi, R., Nantel, P., and Muckle-Jeffs, E., *Alien Invaders in Canada's Waters, Wetlands, and Forests*, Natural Resources Canada, Ottawa, 2002.

58. Chambers, R.M., Meyerson, L.A., and Saltonstall, K., Expansion of *Phragmites australis* into tidal wetlands of North America, *Aquat. Bot.*, 64, 261, 1999.

59. Saltonstall, K., Cryptic invasion by a non-native genotype of the common reed, *Phragmites australis*, into North America, *Proc. Natl. Acad. Sci. U.S.A.*, 99, 2445, 2002.

60. Jeffries, R.L., Rockwell, R.F., and Abraham, K.F., The embarrassment of riches: agricultural food subsidies, high goose numbers, and loss of Arctic wetlands — a continuing saga, *Environ. Rev.*, 11, 193, 2003.

61. Sibbald, B., Saskatchewan's West Nile virus surge baffles experts, *Can. Med. Assoc. J.* Available online at http://www.cmaj.ca/news/10_09_03.shtml, September 10, 2003.

62. Artsob, H., West Nile: biology of an emerging *Flavivirus*, in *Proc. West Nile Virus Regulatory Consultative Workshop*, Health Canada, Ottawa, 2004, 1.

63. Centers for Disease Control, *West Nile Virus*. Division of Vector Borne Diseases, Centers for Disease Control and Prevention, Atlanta, GA, 2004.

64. Darsie, R.F. and Ward, R.A., *Identification and Geographical Distribution of the Mosquitoes of North America, North of Mexico*, American Mosquito Control Association, Fresno, CA, 1981.

65. Fonseca, D.M., Keyghobadi, N., Malcolm, C.A., Mehmet, C., Schaffner, F., Mogi, M., Fleischer, R.C., and Wilkerson, R.C., Emerging vectors in the *Culex pipiens* complex, *Science*, 303, 1535, 2004.

66. Warner, B.G. and Bunting, M.J., Indicators of rapid environmental change in northern peatlands, in *Geoindicators: Assessing Rapid Environmental Changes in Earth Systems*, Berger, A.R. and Iams, W.J., Eds., A.A. Balkema, Rotterdam, 1996, 235–246.

67. Chambers, F.M. and Charman, D.J., Holocene environmental change: Contributions from the peatland archive, *Holocene*, 14, 1, 2004.

68. Warner, B.G., Kubiw, H.J., and Hanf, K.I., An anthropogenic cause for quaking mire formation in southwestern Ontario, *Nature*, 340, 380, 1989.

69. Bunting, M.J. and Warner, B.G., Hydroseral development in southern Ontario: Patterns and controls, *J. Biogeogr.*, 25, 3, 1998.

70. Bunting, M.J., Morgan, C.R.M., Warner, B.G., and Van Bakel, M., Pre-European settlement conditions and disturbance of a coniferous swamp in southern Ontario, *Can. J. Bot.*, 76, 1770, 1998.

71. Mitsch, W.J. and Jorgensen, S.E., *Ecological Engineering and Ecosystem Restoration*, John Wiley & Sons, Toronto, 2003.

72. Warner, B.G. and Li, J., Ecological engineering and wetlands. *Can. Civil Eng.*, 15, 16, 1998.

73. Kells, J.A. and Warner, B.G., Emergence of ecological engineering and green technology, *Can. Civil Eng.*, 17, 11, 2000.

74. Warner, B.G., Canada's wetland industry, in Wetlands Stewardship in Canada: Setting a Course Together, Rubec, C.D.A., Ed., North American Wetlands Conservation Council, Report 03-02, 2003, 81–92.

75. Rubec, C.D.A., Lynch-Stewart, P., Wickware, G.M., and Kessel-Taylor, I., Wetland utilization in Canada, in *Wetlands of Canada*, National Wetland Working Group, Polyscience Publishers, Montreal, 1988, 381–412.

76. Rubec, C.D.A., Policy for conservation of the functions and values of forested wetlands, in *Northern Forested Wetlands: Ecology and Management*, Trettin, C.C., Jurgensen, M.F., Grigal, D.F., Gale, M.R., and Jeglum, J.K., Eds., CRC Press, Boca Raton, FL, 1997, 45–59.

Part IV

Economics and Policy Issues

19 Economics of Forest and Agricultural Carbon Sinks

G.C. van Kooten

CONTENTS

19.1 INTRODUCTION

As a result of the Kyoto Protocol (KP) and its so-called "flexibility mechanisms," climate change and mechanisms to mitigate its potential effects have attracted considerable economic and policy attention. A major reason for this attention is that the KP has a complex set of instruments that enable countries to achieve emissions reduction targets in a wide variety of ways, some of which are unlikely to lead to real, long-term reductions in greenhouse gas emissions. One purpose of this chapter, therefore, is to provide an overview of economic reasoning applied to climate change and to illustrate how terrestrial carbon uptake credits (offset credits) operate within the KP framework. Attention is focused on the feasibility of terrestrial carbon sinks to slow the rate of CO_2 buildup in the atmosphere.[1]

I also examine the results of several empirical studies into the costs of carbon uptake in agricultural ecosystems and by forestry activities. For example, Manley et al.[2] examined the costs of creating soil carbon sinks by switching from conventional to zero tillage. The viability of agricultural carbon sinks was found to vary by region and crop, with no-till representing a low-cost option in some regions (costs of less than $15 tC^{-1}), but a high-cost option in others (costs of $100 to $400 tC^{-1}). A particularly relevant finding is that no-till cultivation may store no carbon at all if measurements are taken at sufficient depth. In some circumstances no-till cultivation may yield a "triple dividend" of carbon storage, increased returns, and reduced soil erosion, but in many others creating carbon offset credits in agricultural soils is not cost-effective because reduced tillage practices store little or no carbon. This is particularly the case in the Great Plains. In another study, van Kooten[3] reviewed estimates from 55 studies of the costs of creating carbon offsets using forestry. Lowest costs of sequestering carbon are through forest conservation, while tree planting and agroforestry activities increase costs by more than 200%. The use of marginal cost estimates instead of average cost results in much higher costs for carbon sequestration, in the range of thousands of dollars tC^{-1}, although few studies used this more appropriate method of cost assessment.

I conclude by making the case that, while there remains a great potential for carbon sinks, more attention needs to be paid to post-harvest. In the above research, post-harvest storage of carbon in wood products yielded much lower cost estimates. Nonetheless, the study of post-harvest uses of biomass remains an area that requires greater attention by economists.

19.2 ECONOMIC INSTRUMENTS TO ADDRESS CLIMATE CHANGE AND THE KYOTO PROTOCOL MECHANISM

Economists generally prefer economic incentives over command-and-control regulation, because market incentives are usually better suited to achieving environmental objectives at lower cost than government regulations. In the context of climate change, economic incentives induce firms to adopt technical changes that lower the costs of reducing CO_2 emissions, because they can then sell permits or avoid buying them, or avoid paying a tax. Further, market instruments provide incentives to change products, processes, and so on, as marginal costs and benefits change over time. Because firms are always trying to avoid the tax or paying for emission rights, they tend to respond quickly to technological change.

Whether a quantity or price instrument is chosen should not matter. This can be illustrated with the aid of Figure 19.1. Restricting the amount of CO_2 emissions (focusing on quantity) should lead to the same outcome as an emissions tax (focusing on price). The carbon tax (P in Figure 19.1) determines the level of emissions; if emissions are restricted to C^* and permits are issued in that amount, the permit price should be P, or the same as the tax. The state can choose the tax level (price) or the number of emission permits (quantity), but if all is known the outcome will be the same — emissions will be reduced to C^*.

When abatement costs and/or benefits are uncertain, however, picking a carbon tax can lead to the "wrong" level of emissions reduction, while choosing a quantity can result in a mistake about the forecasted price that firms will have to pay for auctioned permits.[4] Such errors have social costs. If the marginal cost of abatement

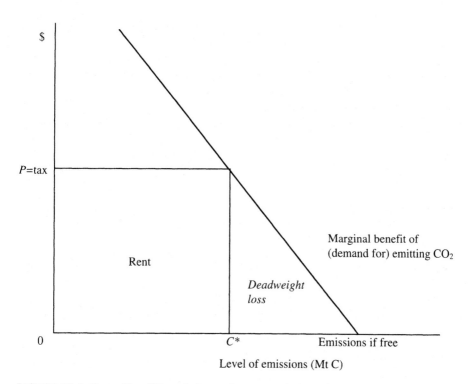

FIGURE 19.1 Controlling CO_2 emissions using economic incentives.

curve is relatively steep but the marginal benefit of abatement rather flat (i.e., damages accumulate slowly), as is likely to be the case with climate change, the costs of relying on permit trading are much higher than those associated with carbon taxes.[4–6] However, as discussed below, the KP relies neither on taxes nor pure emissions trading.

Regardless of how emissions are curtailed, doing so creates a wedge between the marginal costs of providing emission permits (which are effectively zero) and the price at which they sell in the market. This wedge is a form of scarcity rent,[7] with the total unearned rent equal to the restricted level of emissions multiplied by their price (Figure 19.1). The rent represents the capitalized value of the right to emit CO_2, which had previously been free. With a tax, the government captures the rent. With a tradable emissions scheme, the government captures the rent only if emission rights are auctioned off; if emission rights are grandfathered (given to emitters on the basis of current emissions, say), the rent is captured by extant emitters. Those lucky enough to receive tradable emission permits experience a windfall. As a result, governments will be subject to tremendous lobbying pressure in their decision regarding the allocation of permits. Countries that have done the most to reduce emissions in the past may lose relative to ones that made no similar efforts; firms that are high-energy users may benefit relative to those firms that invested in energy-savings technology.

Notice that the rent constitutes an income transfer and not a cost to society of reducing emissions. The authority can distribute the rent any way it sees fit by the method it chooses to allocate emission rights. It can even distribute the rent in ways that provide certain emitters with windfalls not provided to other emitters, if this is what is needed to make the scheme more palatable. However, it can do little about the costs of reducing CO_2-equivalent emissions. Costs are given in Figure 19.1 by the triangle labeled "deadweight loss," which might be considered the minimum cost to society of achieving the emissions target C^*. Costs may well be higher if the wrong policies are implemented. In any event, it is this cost that needs to be compared to the benefits of achieving C^*.

Contrary to the acid rain case (SO_2 emissions from power plants) where emission trading enjoyed great success, the marginal costs of achieving a specified emissions reduction target are not well known. Thus, some economists favor a carbon tax to ensure that costs do not spin wildly out of control. Yet, the international community, fascinated perhaps by the success in reducing SO_2 emissions, opted for a quantity instrument. Two types of quantity instrument are available: permit (allowance) trading and credit trading. They are not the same thing, and I review the merits of each and discuss their implications with respect to carbon sinks.

Under *permit trading* (also known as allowance trading), the authority establishes an aggregate emissions cap (say, C^* in Figure 19.1) and issues emission allowances (permits) of that amount for use and/or trading. This is euphemistically known as "cap and trade." Under *credit trading*, each large industrial emitter (each major source of emissions) is required to meet an emissions target that is usually but not necessarily set below current emissions. The current level of emissions is often referred to as the "baseline." Emission reductions in excess of the prespecified target (reductions in excess of baseline minus target emissions) can be certified as tradable credits. However, other types of credits can also be certified at the discretion of the authority. Importantly, there is no overall cap on emissions and, hence, no guarantee that emissions will not exceed the target.

The Kyoto process began with emission reduction targets and only afterwards considered instruments for implementation. Taxes were rejected as politically infeasible and difficult to coordinate, although individual countries could employ taxes as they saw fit. However, most countries opted not to rely on taxes; for example, Canada's implementation plan makes no mention of taxes whatsoever. Rather than make the effort to "sell" citizens on the notion of carbon taxes, perhaps by reducing income taxes and demonstrating the benefits of the so-called "double-dividend,"[8,9] countries opted for a hodgepodge of means for meeting targets that included possibilities for credit trading. Credit trading of emissions and carbon offsets (e.g., carbon sequestration in sinks as permitted under KP Articles 3.3, 3.4, and 3.7) is seen as a method of achieving KP targets cheaply and efficiently, and individual countries are encouraging the establishment of emission trading schemes that include offsets.

19.3 TERRESTRIAL CARBON SINKS: ISSUES

Land use, land-use change, and forestry (LULUCF) activities can lead to carbon offset credits or debits. Such offsets have taken on great importance under the KP despite the EU-15's initial opposition to their inclusion. As a result, carbon offsets need to be taken into account in any credit trading scheme. The Marrakech Accords to the KP lay out the basic framework for including offset credits.[10] Tree planting and activities that enhance tree growth clearly remove carbon from the atmosphere and store it in biomass, and thus should be eligible activities for creating carbon offset credits. However, since most countries have not embarked on large-scale afforestation and/or reforestation projects in the past decade, harvesting trees during the 5-year KP commitment period (2008–2012) will cause them to have a debit on the afforestation-reforestation-deforestation (ARD) account. Therefore, the Marrakech Accords permit countries, in the first commitment period only, to offset up to 9.0 megatons of carbon (Mt C) each year from 2008–2012 through (verified) forest management activities that enhance carbon uptake (although the amount of carbon sequestered is not verified). If there is no ARD debit, then a country cannot claim the credit. In addition, some countries are able to claim carbon credits from business-as-usual forest management that need not be offset against ARD debits. Canada can claim 12 Mt C year^{-1}, the Russian Federation 33 Mt C, Japan 13 Mt C, and other countries much lesser amounts. These are simply "paper" claims as there is no new net removal of CO_2 from the atmosphere.

In addition to forest ecosystem sinks, agricultural activities that lead to enhanced soil organic carbon and/or more carbon stored in biomass can be used to claim offset credits. Included are revegetation (establishment of vegetation that does not meet the definitions of afforestation and reforestation), cropland management (greater use of conservation tillage, more set-asides) and grazing management (manipulation of the amount and type of vegetation and livestock produced).

One problem with agricultural and to a lesser extent forestry carbon sequestration activities is their ephemeral nature. One study found, for example, that all of the soil organic carbon stored as a result of 20 years of conservation tillage was released in a single year of conventional tillage.[11] Likewise, there is concern that tree plantations will release a substantial amount of their stored carbon once harvested, which could happen as soon as 5 years after first planting due to the use of fast-growing hybrid species. Payments that promote direct changes in land uses for the purpose of carbon sequestration often result in indirect changes in land use that release CO_2, something known as a "leakage." Further, carbon flux from LULUCF activities is extremely difficult to measure and monitor over time, increasing the transaction costs of providing carbon offset credits. Despite these obstacles, many scientists remain optimistic about the importance of terrestrial carbon sinks.[12]

In this section, I examine some issues related to the inclusion of carbon offset credits in a larger emissions trading scheme. Some of these issues are related to the trading scheme itself, but others relate to the costs and benefits of creating offsets — the economic efficiency of relying on carbon sink offsets rather than CO_2-emissions reduction.

19.3.1 ADDITIONALITY, MONITORING, AND LEAKAGES

In principle, a country should get credit only for carbon uptake over and above what occurs in the absence of carbon-uptake incentives, a condition known as "additionality."[13] Thus, for example, if it can be demonstrated that a forest would be harvested and converted to another use in the absence of specific policy to prevent this from happening, the additionality condition is met. Carbon sequestered as a result of incremental forest management activities (e.g., juvenile spacing, commercial thinning, fire control, fertilization) would be eligible for carbon credits, but only if the activities would not otherwise have been undertaken (say, to provide higher returns or maintain market share). Similarly, afforestation projects are additional if they provide environmental benefits (e.g., regulation of water flow and quality, wildlife habitat) not captured by the landowner and would not be undertaken in the absence of economic carbon incentives.

It is often difficult to determine whether an activity is truly additional. For example, farmers have increasingly adopted conservation tillage practices because costs of controlling weeds (chemical costs) have fallen, fuel and certain machinery costs have risen, and new cultivars reduce the impact of yield reductions often associated with conservation tillage. If farmers adopt conservation tillage practices in the absence of specific payments for carbon uptake, they should not be provided with carbon offset credits. If zero tillage is adopted simply because it is profitable to do so, the additionality condition is not satisfied and no carbon credits can be claimed. Likewise, farmers who have planted shelterbelts should not be provided carbon subsidies unless it can be demonstrated that such shelterbelts are planted for the purpose of sequestering carbon and would not otherwise have been planted.

In addition to determining whether a LULUCF project is indeed additional, it is necessary to determine how much carbon is actually sequestered and for how long. Measuring carbon uptake is a difficult task and can be even more difficult if the carbon sink is short-lived. Monitoring and enforcement are costly and measurement is an inexact science in the case of carbon uptake in terrestrial ecosystems. Research studies reporting differences in soil organic carbon (SOC) between conventional and conservation tillage practices find that these depend on soil type, depth to which soil carbon is measured, and other factors.[2] But if SOC needs to be constantly measured and monitored, as appears likely for ephemeral sinks (see below), transaction costs could greatly exceed the value of the carbon sequestered.*

The onus of establishing whether or not certain agricultural practices or tree planting (forest management) programs should receive carbon offset credits extends beyond simply examining the direct LULUCF impact. The direct impact relates to the carbon flux at the site in question. The indirect impact refers to the changes in CO_2 emissions elsewhere that are brought about by the LULUCF activity. In particular, there may be leakages caused by changes/shifts in land use elsewhere and/or changes in emissions, and these need to be set against the direct impacts. Large-scale tree planting programs in Canada, for example, might reduce future lumber prices, thereby causing U.S. forest landowners to harvest trees sooner, or convert land from forestry to agriculture, in anticipation of falling stumpage prices (see, for example, Reference 15). This causes an increase in CO_2 emissions that needs to be offset against the gain in carbon uptake from the original afforestation project.

---

* Little research has been done on estimating transaction costs, although a study by van Kooten, Shaikh, and Suchánek[14] demonstrates that they can be a serious obstacle to adoption of tree planting programs.

Likewise, subsidies to stimulate ethanol production will increase grain prices, thereby providing an impetus to convert land from forest to agriculture at the extensive margin and to increase use of chemical and fuel inputs that emit CO_2-equivalent gases at the intensive margin. Further, as Lewandrowski et al.[11] note, payments to get a landowner to adopt no tillage on one field may be accompanied by the conversion of another field from zero to conventional tillage by the same landowner. Such leakages could substantially offset a project's direct gains in carbon uptake. They also increase the costs of creating carbon offset credits, making them less attractive relative to emission reduction credits.

19.3.2 DISCOUNTING PHYSICAL CARBON

By discounting carbon, we acknowledge that it matters when CO_2 emissions or carbon uptake occurs — carbon sequestered today is more important and has greater potential benefits than that sequestered at some future time. Yet, the idea of discounting physical carbon is anathema to many who would discount only monetary values. But the idea of weighting physical units accruing at different times is entrenched in the natural resource economics literature, going back to economists' definitions of conservation and depletion.[16] One cannot obtain consistent estimates of the costs of carbon uptake unless both project costs and physical carbon are discounted, even if different rates of discount are employed for costs and carbon. To illustrate why, consider the following example.

Suppose a tree-planting project results in the reduction of CO_2-equivalent emissions of 1 tC yr^{-1} in perpetuity (e.g., biomass burning to produce energy previously produced using fossil fuels). In addition, the project has a permanent sink component that results in the storage of 6 tC yr^{-1} for 10 years, after which time the sink component of the project reaches an equilibrium. How much carbon is stored? If all costs and uptake are put on an annual basis, we need to determine how much carbon is actually sequestered per year? Is it 1 or 7 tC yr^{-1}? Clearly, 7 tC are sequestered for the first 10 years, but only 1 tC is sequestered annually after that time. Carbon sequestration, as stated on an annual basis, would either be that experienced in the first 10 years (7 tC yr^{-1}) or in the infinite number of years to follow (1 tC yr^{-1}). Suppose the discounted project costs amount to $1000; these include the initial site preparation and planting costs plus any annual costs (maintenance, monitoring, etc), appropriately discounted to the current period. If a 4% rate of discount is used, costs are $40 yr^{-1} — the amount that, if occurring each year in perpetuity, equals $1000 in the current period. The costs of carbon uptake are then estimated to be $5.71 tC^{-1} if it is assumed that 7 tC is sequestered annually and $40 tC^{-1} if 1 tC is assumed to be sequestered each year. The former figure might be cited simply to make the project appear more desirable than it really is.

Suppose instead we intend to divide the $1000 cost by the total undiscounted sum of carbon that the project sequesters. Since the amount of carbon sequestered is 7 tC yr^{-1} for 10 years, followed by 1 tC yr^{-1} in perpetuity, the total carbon absorbed is infinite, and the cost of carbon uptake would essentially be zero. To avoid an infinite sum of carbon uptake, an arbitrary planning horizon needs to be chosen. If the planning horizon is 30 years, 90 tC are sequestered and the average cost is calculated to be $11.11 tC^{-1}; if a 40-year planning horizon is chosen, 100 tC are removed from the atmosphere and the cost is $10.00 tC^{-1}. Thus, cost estimates are sensitive to the length of the planning horizon, which is not always made explicit in studies.

Consistent cost estimates that take into account all carbon sequestered plus the timing of uptake can only be achieved by discounting both costs and physical carbon. Suppose physical carbon is discounted at a lower rate (say, 2%) than that used to discount costs (4%). Then, over an infinite time horizon, the total discounted carbon saved via our hypothetical project amounts to 112.88 tC and the correct estimate of costs is $8.86 tC^{-1}. Reliance on annualized values is misleading in this case because costs and carbon are discounted at different rates. If carbon is annualized using a 2% rate, costs amount to $17.70 tC^{-1} (=$40 ÷ 2.26 tC). If the same discount rate of 4% is employed for costs and carbon, the cost is $30.20 tC^{-1} (or $8.24 tCO$_2$$^{-1}$) and it is the same regardless of whether costs and carbon are annualized.

The rate at which physical carbon should be discounted depends on what we assume about the rate at which the damages caused by CO_2 emissions increase over time.[17,18] If the damage function is linear so that marginal damages are constant — damages per unit of emissions remain the same as the concentration of atmospheric CO_2 increases — then the present value of reductions in the stock of atmospheric CO_2 declines at the social rate of discount. Hence, it is appropriate to discount future carbon uptake at the social rate of discount. "The more rapidly marginal damages increase, the less future carbon emissions reductions should be discounted" (Reference 18, p. 291). Thus, use of a zero discount rate for physical carbon is tantamount to assuming that, as the concentration of atmospheric CO_2 increases, the damage per unit of CO_2 emissions increases at the same rate as the social rate of discount — an exponential damage function with damages growing at the same rate as the social rate of discount. A zero discount rate on physical carbon implies that there is no difference between removing a unit of carbon from the atmosphere today, tomorrow, or at some future time; logically, then, it does not matter if the carbon is ever removed from the atmosphere. The point is that use of any rate of discount depends on what is assumed about the marginal damages from further CO_2 emissions or carbon removals.

The effect of discounting physical carbon is to increase the costs of creating carbon offset credits because discounting effectively results in "less carbon" attributable to a project. Discounting financial outlays, on the other hand, reduces the cost of creating carbon offsets. Because most outlays occur early on in the life of a forest project, costs of creating carbon offsets are not as sensitive to the discount rate used for costs as to the discount rate used for carbon.

19.3.3 CREDIT TRADING

Perhaps the most important market-based initiative with respect to terrestrial carbon sinks is the establishment of the exchange-traded markets for carbon uptake credits. Through exchange landowners could potentially profit from practices that enhance SOC or carbon in vegetation. But studies indicate that this will require a well-functioning design mechanism for implementing carbon trading. Indeed, emission trading schemes fail not because of a lack of interest, but from a break-down in necessary economic and market conditions, such as imperfect information and high transactions costs. The Chicago Climate Exchange (CCX) was launched early in 2003 as the first North American central market exchange to allow trading of CO_2 emissions between industry and agriculture. Its purpose is to provide price discovery, which will clarify the debate about the costs of emissions reduction and the role of carbon sinks. Carbon sequestration through no-till farming, grass and tree plantings, and other methods will enable farmers to sell carbon credits on the CCX. However, the prices that are "discovered" may not reflect the true costs to society because the CCX is a credit trading scheme as opposed to an allowance trading scheme.[19]

Trading is also possible through CO2e.com, a U.K. exchange for carbon emis-sion offsets that began in April 2002 and subsequently went global.[*] Initially, it provided a market for emissions trading for British firms that held agreements to cut emissions under the U.K.'s climate change levy scheme, for which they receive tax rebates on energy use. Companies failing to meet targets are able to buy credits to offset their above-target emissions. Companies participating in the exchange are hedging their exposure to losing a tax rebate on energy use. As a result, by mid-July 2003, carbon was trading for as much as U.S.\$10.50 tCO_2^{-1}, with transaction sizes in the range of 5,000 to 15,000 tonnes.

CO2e.com now functions as an exchange for trading CERs from Joint Imple-mentation and Clean Development Mechanism projects, and carbon offset activ-ities. Countries and firms can purchase (sell) CERs and removal units (carbon offsets) for delivery in 2010. Trades for delivery in 2010 have been occurring at around U.S.\$4.50 to \$5.50 tCO_2^{-1}, with trades involving 2 to 10 Mt CO_2. Not surprisingly, Canada has thus far been the largest buyer as a result of its commit-ment to domestic large industrial emitters that they would not have to pay more than \$C15.00 tCO_2^{-1} for reducing emissions. CO2e.com also anticipates that it will be able to arrange trades in carbon offsets through the emissions exchange newly established by the European Union.[**] It is not clear, however, how the exchange rate between sink offsets and emission reductions will be established (see below).

A number of other traders in carbon credits can be found on the Internet, including eCarbontrade (www.ecarbontrade.com/ECIAbout.htm), the Kefi-exchange (http://www.kefi-exchange.com/), and CleanAir Canada (http://www.cleanaircan-ada.org), which is government backed. The Kefi-Exchange is a private exchange begun in Alberta by traders with experience in the trading of various commodities on-line, including electricity. However, a CO_2 emissions-trading market appears to present a greater challenge. As pointed out on the Kefi-Exchange Web site:

> The on-going uncertainty of the global endorsement of the Kyoto Protocol has left the future of the KEFI Exchange in limbo.... [T]he actual operation of the exchange cannot

---

[*] Discussion of CO2e.com is based on http://www.co2e.com/trading/MarketHistory.asp (viewed 7 July 2004).

[**] See http://europa.eu.int/comm/environment/climat/emission.htm (viewed 7 July 2004).

proceed without some clarity in the regulation of emissions. As a result of the current stalemate, the KEFI Exchange has opted to move to a "stand down" mode pending a clearer determination of the directions to be taken in Alberta and the rest of Canada in respect to emission reductions.*

Commodity markets, such as the Winnipeg Commodity Exchange, are also looking into trading carbon emissions and carbon sink credits. With all the problems, it is not surprising that trades are few and far between, especially those that involve carbon offsets. Indeed, Australian solicitors McKean & Park, who were asked to make a judgment on the proposed Australian trading system, indicate that any trading in carbon credits is unlikely to occur before 2005. Tietenberg et al.[20] also indicate that there are a significant number of obstacles to overcome before trading can occur, including most importantly a means of verifying emission-reduction and carbon sequestration claims.

Clearly, a market-based approach to carbon sinks will be effective only in the presence of certain market conditions. For example, in order to buy and sell carbon offset credits, it is necessary to have legislation that delineates the rights of land-owners, owners of trees, and owners of carbon, because what any one of these parties does affects the amount of carbon that is sequestered and stored. Without clear legislation, buyers of carbon offsets are not assured that they will get proper credit — their claims to have met their emission reduction targets with carbon credits is open to dispute. Carbon offsets need to be certified, and an overseeing (international) agency with well-defined rules and regulations is needed. It would appear that, currently, those participating in the few exchanges that have been established are doing so despite the risk that carbon offset credits may not deliver because of their ephemeral nature.

19.3.4 The Ephemeral Nature of Sinks

Compared to not emitting CO_2 from a fossil fuel source, terrestrial sequestration of carbon is unlikely to be permanent.** Kyoto is in the process of developing policy for addressing the nonpermanence of terrestrial carbon uptake. Some nations want emissions and removals to be treated identically, so that the removal of a unit of carbon results in a credit just as does a reduction in emissions. Does it matter whether the "removal" from the atmosphere is the result of biological sequestration or a consequence of leaving a CO_2-equivalent unit of fossil fuel in the ground? Some argue that leaving fossil fuels in the ground only delays their eventual use and, as with carbon sequestered in a terrestrial sink, results in the same obligation for the future.[17] Others argue that there is an asymmetry between carbon uptake in a sink and emissions reduction (leaving fossil fuel in the

---

* This quote was originally viewed on 8 May 2003, but had been removed as of 7 July 2004. This is a telling observation about the difficulty of establishing exchanges that take carbon offset trading seriously.
** This is not to suggest that carbon sinks are not worthwhile. Temporary removal of carbon helps postpone climate change, buys time for technological progress, buys time to replace fuel-inefficient capital equipment, allows time for learning, and may lead to some permanent sequestration as the new land use continues indefinitely.[21]

ground).* Whatever the case, carbon sequestered in a sink creates a liability for the future that is not the case with an emissions reduction. As a result, a country will under the KP need to ensure that carbon entering a sink in the 2008–2012 commitment period is somehow covered (or still in place) in a second, third, and later commitment period. Currently, this is not a serious problem for a country because the liability can be factored into a country's self-selected future commitment to emission reductions.

The ephemeral nature of terrestrial carbon uptake can be addressed by providing partial instead of full credits for stored carbon according to the perceived risk that carbon will be released from the sink at some future date. The buyer or the seller may be required to take out an insurance policy, where the insurer will substitute credits from another carbon sink at the time of default. Alternatively, the buyer or seller can provide some assurance that the temporary activity will be followed by one that results in a permanent emissions reduction. For example, arrangements can be put in place prior to the exchange that, upon default or after some period of time, the carbon offsets are replaced by purchased emission reductions. Again, insurance contracts can be used. Insurance can also be used if there is a chance that the carbon contained in a sink is released prematurely. It is also possible to discount the number of offset credits by the risk of loss (so that a provider may need to convert more land into forest, say, than needed to sequester the agreed upon amount of carbon).

Three "practical" approaches to nonpermanence of sinks have been discussed in the literature. One is to specify a conversion factor that translates years of temporary carbon storage into a permanent equivalent. The concept of ton-years has been proposed to make the conversion from temporary to permanent storage.[12,17,22]

Suppose that 1 ton of carbon-equivalent GHG emissions is to be compensated for by a ton of permanent carbon uptake. If the conversion rate between ton-years of (temporary) carbon sequestration and permanent tons of carbon emissions reductions is k, a LULUCF project that yields 1 ton of carbon uptake in the current year generates only 1 k^{-1} tons of emission reduction — to cover the 1 ton reduction in emissions requires k tons of carbon to be sequestered for 1 year.** The exchange rate ranges from 42 to 150 ton-years of temporary storage to cover one permanent ton.

Many observers have condemned the ton-year concept on various grounds. Herzog et al.[17] argue that the value of storage is based on the arbitrary choice of an exchange rate, while Marland et al.[21] point out that the ton-year accounting system is flawed: Ton-year credits (convertible to permanent tons) can be accumulated while trees grow, for example, with an additional credit earned if the biomass is subsequently burned in place of an energy-equivalent amount of fossil fuel (Reference

---

* Even Herzog et al.[17] admit that fossil fuels left in the ground may not be used at some future date if society commits to de-carbonize energy, while carbon in a terrestrial sink always has the potential to be released in the future. The bigger problem of not emitting CO_2 by burning fossil fuels pertains to leakages: Reduced fossil fuel use by some causes others to use more since prices are lower, while lower prices discourage new sources of energy.

** This interpretation is slightly different from the original intent. The original idea is to count a temporary ton as equivalent to a permanent one only if the carbon is sequestered for the full period of time given by the exchange rate. The advantage of the interpretation here is that it enables one to count carbon stored in a sink for periods as short as 1 year (as might be the case in agriculture).

21, p. 266). That is, the ton-year concept could lead to double counting. Yet, the concept of ton-years has a certain appeal, primarily because it provides a simple, albeit somewhat naïve, accounting solution to the problem of permanence. The choice of an exchange rate, or, rather, time frame, is political (which is another reason for its condemnation). Once an exchange rate is chosen, carbon uptake credits can be traded in a CO_2-emissions market in straightforward fashion. Yet, the ton-years approach has been rejected by most countries, primarily because it disadvantages carbon sinks relative to emissions avoidance.[22]

A second approach discussed extensively at Conferences of the Parties has been the potential creation of a "temporary" certified emission reduction unit, denoted TCER. The idea is that a temporary carbon offset credit is purchased for a set period of time (e.g., 1 year or 5 years) expiring thereafter. Upon expiration, TCERs would have to be covered by substitute credits or reissued credits if the original project were continued. Compared to ton-years, monitoring and verification are more onerous because a more complex system of bookkeeping will be required at the international level to keep track of credits. Countries favor this approach over other approaches because they can obtain carbon credits early, while delaying their "payment" to a future date. Since politicians will discount future obligations very highly (essentially ignoring them), carbon offsets are treated as the equivalent of emission reductions.

A third approach to the problem of temporary vs. permanent removal of CO_2 from the atmosphere is to employ a market device that would obviate the need for an arbitrary conversion factor or other forms of political maneuvering. Marland et al.[21] and Sedjo and Marland[23] propose a rental system for sequestered carbon. A 1-ton emission offset credit is earned when the sequestered carbon is rented from a landowner, but, upon release, a debit occurs. "Credit is leased for a finite term, during which someone else accepts responsibility for emissions, and at the end of that term the renter will incur a debit unless the carbon remains sequestered *and* the lease is renewed."[21] In addition to avoiding the potential for double counting, the landowner (or host country) would not be responsible for the liability after the (short-term) lease expires. The buyer-renter employs the limited-term benefits of the asset, but the seller-host retains long-term discretion over the asset.[23]

Rather than the authority establishing a conversion factor, the interaction between the market for emission reduction credits and that for carbon sink credits determines the conversion rate between permanent and temporary removals of CO_2 from the atmosphere. The rental rate for temporary storage is based on the price of a permanent energy emissions credit, which is determined in the domestic or international market. The annual rental rate (q) is simply the price of permanent emission credit (P) multiplied by the discount rate (r), which equals the established financial rate of interest (if carbon credits are to compete with other financial assets) adjusted for the risks inherent to carbon uptake (e.g., fire risk, slower than expected tree growth, etc.). Thus, $q = P / r$, which is a well-known annuity formula. If emissions are trading for \$15 tCO_2^{-1}, say, and the risk-adjusted discount rate is 10%, then the annual rental for a t CO_2 in a terrestrial sink would be \$1.50 tCO_2^{-1}. This would be the selling price for biological carbon uptake, and, like the ton-year concept, it may make terrestrial sink projects less attractive than they might be under some other political solution.

A rental system works best if we are dealing with credit trading as opposed to allowance trading. Under a cap-and-trade scheme, it would be necessary to set not only a cap on emissions from fossil fuel consumption, but also a cap on sinks. In that case, one might expect separate markets to evolve for emissions and carbon sink allowances.

19.4 PROGNOSIS FOR FOREST ECOSYSTEM SINKS

Conservation of forest ecosystems that are threatened by deforestation, enhanced management of existing forests, reforestation of sites that have been denuded earlier, and afforestation are some ways in which carbon offset credits might be earned. The question is: Are carbon offsets created in these different ways competitive with emission reductions? If not, there is little sense in pursuing them, even though they might indeed increase the amount of carbon in forest ecosystems. As noted above, the KP deals with forest (and agricultural) sinks in interesting ways in order to make them attractive as means for enabling countries to attain their KP targets. In theory, carbon flux in terrestrial ecosystems needs to take into account the carbon debit from harvesting trees, or otherwise changing land use (e.g., draining sloughs/swamps), but it also needs to take into account carbon stored in wood product sinks (and exported carbon), and additional carbon sequestered as a result of forest management activities (e.g., juvenile spacing, commercial thinning, and fire control). Even when all of the carbon fluxes are appropriately taken into account (and product sinks are not yet permitted under the KP), it is unlikely that "additional" forest management will be a cost-effective and competitive means for sequestering carbon.[24]

Evidence from Canada, for example, indicates that, for the most part, reforestation does not pay even when carbon uptake benefits are taken into account, mainly because northern forests tend to be marginal.[25] While many of Canada's forests regenerate naturally, only artificial regeneration that is not required by law as a normal part of forestry operations can truly result in carbon offset credits (although the KP currently permits some credits to count that are not additional). Artificial regeneration is costly and returns accrue in the distant future, making such investments unprofitable (Reference 26, p. 395), even when the potential value of carbon offsets is taken into account. However, if short-rotation, hybrid poplar plantations replace natural forests, might forest management result in the creation of carbon offset credits that are competitive with emission reduction credits? Hybrid poplar plantations may also be the only cost-effective, competitive alternative when marginal agricultural land is afforested.[27,28]

To determine the cost-effectiveness of various forest activities in creating carbon offset credits, van Kooten et al.[29] investigated information from 55 studies. A meta-regression analysis of 981 estimates of the costs of creating carbon offsets using forestry yielded some interesting conclusions. Studies were classified into four different types of forestry projects — forest conservation programs that prevent harvesting of trees (and subsequent release of carbon), forest management programs that enhance tree growth, tree planting (afforestation) programs, and agroforestry projects where trees are planted in fields that continue to be used for crop production or grazing. Forest conservation was chosen as the baseline program.

Studies were also classified by three locations: tropics, North American Great Plains, and all other regions, which included mainly studies in the U.S. South, the U.S. cornbelt, the U.S. New England states, Europe, and studies that covered more than one region (including global efforts at estimating costs of carbon uptake). The "other" region was chosen as the baseline.

What factors appear to have an important effect on estimates of the cost of carbon uptake in forest ecosystems? (1) When the opportunity cost of land was taken into account (which was not done in all studies), carbon uptake costs were significantly higher. (2) If a study was peer-reviewed, estimated costs were 10 to 30 times higher. (3) As expected, discounting of operating, monitoring, and other annual costs lowered the overall estimate of sequestration costs. However, discounting of physical carbon did not appear to have a large effect. (4) Studies that included carbon product sinks had lower overall carbon sequestration costs, although inclusion of soil carbon pools did not have a statistically significant effect on costs. (5) Most studies computed only the average cost of carbon uptake; if marginal cost was calculated, it was much larger. (6) Tree planting and agroforestry activities increase costs by more than 200%. (7) Finally, costs in the Great Plains region were significantly lower than those in other regions of the world.

A summary of the costs of carbon uptake in forest ecosystems is provided in Table 19.1. Baseline estimates of costs of sequestering carbon through forest conservation are U.S.\$46.62 to \$260.29 tC^{-1} (\$12.71 to \$70.99 tCO_2^{-1}).[*] When post-harvest storage of carbon in wood products, or substitution of biomass for fossil fuels in energy production, are taken into account, costs are lowest — some \$3.42 to \$18.67 tCO_2^{-1}. Average costs are greater, \$31.84 to \$383.62 tCO_2^{-1}, when appropriate account is taken of the opportunity costs of land. Since the vast majority of studies ignored the ephemeral nature of carbon offsets and all ignored the potential transaction (measuring, monitoring) costs, the costs reported in Table 19.1 are probably an underestimate of the true costs of creating carbon offset credits.

19.5 PROGNOSIS FOR AGRICULTURAL SINKS

Much the same story can be told about agricultural soil-carbon sinks. In order to increase soil organic carbon, farmers need to change their agronomic practices. In drier regions where tillage summer fallow is used to conserve soil moisture, this requires the use of chemical fallow or continuous cropping, or cessation of cropping altogether (i.e., return to grassland). In other agricultural regions, a movement from conventional tillage (CT) to reduced tillage (RT) or no tillage (NT) might increase soil organic carbon. Soil carbon increases by increasing plant biomass entering the soil and/or reducing rates of decay of organic matter. This might be done by switching

---

[*] In Table 19.1, costs are provided on a tC^{-1} basis. They can be converted to a tCO_2^{-1} basis by multiplying by 12/44. Conversely, if emissions trade at \$15 tCO_2^{-1}, then carbon offset credits must trade for \$55 tC^{-1} or less to be competitive.

TABLE 19.1
Predicted Average and Marginal Costs of Creating Carbon Offsets through Forestry Activities (U.S.$2003 tC⁻¹): Various Scenarios[a]

Scenario	Average Costs (if studies not reviewed)	Average Costs (based on peer-reviewed studies)	Marginal Costs (based on peer-reviewed studies)
Baseline (Other regions with Forest Conservation)	8.45	217.01	15,700.48
Other Regions			
Planting	24.80	637.10	46,094.38
Agroforestry	26.65	684.67	49,535.57
Forest management	8.09	207.87	15,039.67
Other Regions with Conservation			
Soil sink	5.35	137.54	9,951.18
Fuel substitution	4.45	114.31	8,270.47
Product sink	2.25	57.74	4,177.41
Opportunity cost of land	46.20	1186.70	85,857.68
Tropics			
Conservation	10.01	257.22	18,609.85
Planting	29.40	755.16	54,635.89
Agroforestry	31.59	811.54	58,714.75
Forest management	9.59	246.39	17,826.59
Tropics with Conservation			
Soil sink	6.35	163.03	11,795.18
Fuel substitution	5.27	135.49	9,803.03
Product sink	2.66	68.44	4,951.50
Opportunity cost of land	54.76	1406.60	101,767.52
Great Plains			
Conservation	5.36	137.68	9,961.14
Planting	15.74	404.21	29,244.49
Agroforestry	16.91	434.38	31,427.74
Forest management	5.13	131.89	9,541.88
Great Plains with Conservation			
Soil sink	3.40	87.26	6,313.51
Fuel substitution	2.82	72.52	5,247.18
Product sink	1.43	36.63	2,650.35
Opportunity cost of land	29.31	752.90	54,472.23

Average costs and marginal costs are determined from the respective regressions provided in van Kooten et al.[29] If the study was peer-reviewed, the dummy variable in the regression is set to 1; otherwise it is 0.

Source: Calculated from information provided in van Kooten et al.[29]

to RT or NT, or replacing tillage summer fallow by continuous cropping or chemical summer fallow. Are such practices worth pursuing, and can they result in significant changes in carbon flux?

Undoubtedly, there are soil erosion benefits from practicing reduced (conservation) tillage and zero tillage. In many cases, lower costs because of fewer field operations offset higher chemical costs since prices of herbicides have fallen in recent years (although there may be higher social costs associated with the environmental spillovers from higher chemical use). As a result of the private benefits, the extent of RT and NT has increased significantly in the U.S. in the past several decades. In 1997, in the U.S., farmers employed conventional tillage on 36.5% of 294.7 million acres (119.3 million ha) planted to cropland; 26.2% was planted using reduced tillage and 15.6% using zero tillage, with other crop residue methods employed on the remaining land (Reference 30, p. 67). Not included were some 20 million acres of land left in tillage summer fallow in drier regions: 22% of all wheat planted in the U.S. in 1997 was part of a wheat-fallow rotation and, in some states, three quarters of all wheat was part of a wheat-fallow rotation.

West and Marland[31] used U.S. data on carbon uptake in soils, production of biomass, chemical and fuel use, machinery requirements, and so on to compare CT, RT, and NT in terms of their carbon flux. They provide a detailed carbon accounting for each practice, concluding that, due primarily to extra chemical use, RT does not differ significantly from CT in terms of carbon uptake benefits, but that NT results in an average relative net carbon flux of -368 kg of C ha^{-1} yr^{-1}, with -337 kg of C ha^{-1} yr^{-1} due to carbon sequestration in soil, -46 kg C ha^{-1} yr^{-1} due to a reduction in machinery operations and $+15$ kg C ha^{-1} yr^{-1} due to higher carbon emissions from an increase in the use of agricultural inputs. While annual savings in carbon emissions of 31 kg C ha^{-1} yr^{-1} last indefinitely, accumulation of carbon in soil reaches equilibrium after 40 years. West and Marland[31] assume that the *rate* of uptake in soil is constant at 337 kg C ha^{-1} yr^{-1} for the first 20 years and then declines linearly over the next 20 years. However, as noted earlier, stored carbon can be released back into the atmosphere in as little as a year when CT is resumed.

Their estimates of carbon uptake by soils in the prairie region of Canada as a result of going from CT to NT vary from 100 to 500 kg C ha^{-1} yr^{-1}.[31] Using these results and discount rates of 2 and 4%, van Kooten[3] estimated that the net discounted carbon prevented from entering the atmosphere as a result of a shift to NT from CT varies from about 4 tC ha^{-1} to at most 12.5 tC ha^{-1}. Compared to forest plantations, the amount of carbon that can potentially be prevented from entering the atmosphere by changing to zero tillage is small.

Research by Manley et al.[2] came to a more pessimistic conclusion even than West and Marland. They found that the costs per tonne of carbon in going from CT to NT are enormous, and may even be infinite in some cases because there may be very little or no addition to SOC, particularly in North America's grain belt. Manley et al. conducted two meta-regression analyses to investigate the potential for the switch from conventional to zero tillage to create carbon offset credits that would be competitive with emission reductions. The first meta-analysis consisted of 51 studies and 374 separate observations comparing carbon accumulation under CT and NT. A particularly important finding was that no-till cultivation may store no

TABLE 19.2
Net Costs of Carbon Sequestered in Going from CT to NT (U.S.$2003 tC⁻¹)

Region	Crop	At Measured Depth of Soil	
		Shallow	Deep
South	Wheat	$10.45	$13.10
	Other crop	$2.02	$2.04
Prairies	Wheat	$390.75	
	Other crop	$153.09	$215.82
Corn Belt	Wheat	$147.55	$193.48
	Other crop	$87.31	$89.73

Note: Converted from U.S.$2001 to U.S.$2003 using the U.S. CPI.

Source: Manley et al.[2]

carbon at all if measurements are taken at sufficient depth. That is, the depth to which researchers measured SOC was important in determining whether there were carbon-sink gains from no-till agriculture. In some regions, including the Great Plains of North America, the carbon-uptake benefits of NT are non-existent. A possible explanation is that, under conventional tillage, crop residue is plowed under and carbon gets stored at the bottom of the plow layer; with no-till, some carbon enters the upper layer of the soil pool, but as much CO_2 is lost from decaying residue as is lost from plowing under conventional tillage.

In a second meta-regression analysis, Manley et al.[2] examined 52 studies and 536 separate observations of the costs of switching from conventional tillage to no-till. Costs per ton of carbon uptake were determined by combining the two results (see Table 19.2). The viability of agricultural carbon sinks was found to vary by region and crop, with no-till representing a low-cost option in some regions (costs of just over $10 tC⁻¹ or about $3 tCO₂⁻¹), but a high-cost option in others (costs of $100 to $400 tC⁻¹). Nonetheless, in some limited circumstances no-till cultivation may yield a "triple dividend" of carbon storage, increased returns, and reduced soil erosion, but in most cases creating carbon offset credits in agricultural soils is not cost-effective because reduced tillage practices store little or no carbon.

Where continuous wheat, reduced (conservation) tillage, and/or zero tillage are already in use, it is difficult to make the case that carbon offset credits are being created — the "additionality" condition is violated. However, if landowners practicing conventional tillage can claim carbon offset credits by making a switch to RT or NT (or to continuous cropping or use of chemical fallow), it will be necessary to extend the claim to extant practitioners of RT, NT, and reduced tillage summer fallow to prevent them from switching back to conventional practices to become eligible claimants in the future (see Reference 11, p.11).

There is a further problem. The advantages of conservation and zero tillage are financial in the sense that there are fewer machinery operations. This cost offsets the cost of increased chemical use and the value of reduced crop yields (which might be small). As more land is put into RT or NT or converted to forestry, and demand for "energy" crops (to produce ethanol, say) increases, crop prices will rise. This will result in a greater loss in revenue from reduced crop yields, making RT and NT less attractive.

19.6 CONCLUSIONS

Although the Kyoto process enables countries to rely on carbon sinks in a major way for meeting their agreed-upon greenhouse gas emission reduction targets, the introduction of carbon uptake in lieu of emissions reduction constitutes a distraction from the real business of addressing anthropogenic causes of climate change. While many argue that terrestrial carbon sinks can serve an important role in the transition to a de-carbonized energy regime, the politics surrounding the creation, verification, and counting of carbon offsets credits under the KP have made this policy instrument much too unreliable to be taken seriously in combating climate change. Parties attempt to gain credits for activities that cannot be considered additional, but are part of business-as-usual practices, such as the spreading adoption of conservation tillage, planting of shelterbelts, and silviculture practices that are required by law or by participation in a forest certification scheme. The measurement, monitoring, and enforcement related to the creation of carbon offset credits is problematic and could result in large transaction costs.

Leakages are often ignored in the calculation of carbon credits, even though leakages lead to a reduction in the total carbon uptake attributed to a project by 50% or more. Leakages are ignored because they are difficult to measure. In practice, this issue is resolved by limiting the parameters of a project, say, the geographic extent of what is to be included, or assuming the project is too small to have an impact on other regions (even when the claimed amount of carbon is large).

Nonetheless, evidence indicates that, even when leakages and transaction costs are ignored, the costs of carbon uptake in forest and, particularly, agricultural sinks are large compared to the costs of emissions reduction. Based on meta-regression analyses, if one considers only the average (let alone marginal) costs of carbon uptake in forest sinks and uses a cutoff of $55 tC^{-1} ($15 tCO$_2^{-1}$) for projects to be competitive with emission reductions, there are no forest activities in any region that meet this threshold if one considers only peer-reviewed studies (see Table 19.1). Likewise, even abstracting from the issue of the depth to which soil carbon is measured, results from meta analyses suggest that only changes in agronomic practices in the U.S. South can sequester enough carbon to make a switch from conventional till to no-till a "project" that is competitive with emission reductions (Table 19.2). Further, the estimates in Tables 19.1 and 19.2 are an underestimate of the true costs of carbon uptake because the studies generally fail to address the temporary nature of carbon sinks.

While the KP permits countries to claim carbon credits associated with questionable sink activities, countries have been less than helpful in attempting to alleviate concerns that the inclusion of sinks in the KP is nothing more than smoke and mirrors. They have opposed any efforts that address the ephemeral nature of sinks in ways that lead to carbon offsets having lower value than emission reductions. Yet, the KP has also failed to treat carbon sinks in a fair and equitable manner. Post-harvest sequestration of carbon in products does not result in carbon credits, even though studies indicate that product sinks play an important role in keeping CO_2 out of the atmosphere. If credit for product sinks is allowed, the value of wood construction will be enhanced thereby reducing reliance on cement, whose production releases large quantities of greenhouse gases.

Finally, recent technological developments in the efficiency of using biomass to produce energy have emerged. These include field-level processes for producing bio-oils from wood fiber and more efficient burners for generating electricity from biomass. This is particularly important in regions where removal of fuel loads is needed to control wildfire, removal of trees damaged by pests such as the mountain pine beetle is warranted, and gathering of crop residues to be burned for electricity is possible. The economics of many of these options as well as other promising means for using biomass to reduce the atmospheric concentration of CO_2 need to be investigated. It will be the inclusion of these activities in the carbon accounting framework that can make biological sinks an attractive option for mitigating climate change.

REFERENCES

1. Beattie, K.G., W.K. Bond, and E.W. Manning, The Agricultural Use of Marginal Lands: A Review and Bibliography. Working Paper 13. Lands Directorate, Environment Canada, Ottawa, Ontario, 1981.
2. Manley, J., G.C. van Kooten, K. Moeltner, and D.W. Johnson, Creating carbon offsets in agriculture through zero tillage: a meta-analysis of costs and carbon benefits. *Climatic Change* 68(January) 41–65, 2005.
3. van Kooten, G.C., *Climate Change Economics*. Edward Elgar, Cheltenham, U.K., 2004.
4. Weitzman, M.L., Prices vs quantities. *Rev. Econ. Stud.* 41(October), 477–491, 1974.
5. Pizer, W.A., Prices vs. Quantities Revisited: The Case of Climate Change. RFF Discussion Paper 98-02. Resources for the Future, Washington, D.C., October 1997, 52 pp.
6. Weitzman, M.L., Landing fees vs. harvest quotas with uncertain fish stocks. *J. Environ. Econ. Manage.* 43, 325–338, 2002.
7. van Kooten, G.C. and E.H. Bulte, *The Economics of Nature: Managing Biological Assets*. Blackwell, Oxford, U.K., 2000.
8. Bovenberg, A.L., and L.H. Goulder, Optimal environmental taxation in the presence of other taxes: general-equilibrium analysis. Am. Econ. Rev. 86(4), 985–1000, 1996.
9. Parry, I., R.C. Williams III, and L.H. Goulder, When Can carbon abatement policies increase welfare? The fundamental role of distorted factor markets. *J. Environ. Econ. Manage.* 37, 52–84, 1999.
10. IPCC, The Marrakesh Accords and the Marrakesh Declaration. Marrakesh, Morocco: COP7, IPCC. Available online at unfccc.int/cop7/documents/accords_draft.pdf, 2001.

11. Lewandrowski, J., M. Peters, C. Jones, R. House, M. Sperow, M. Eve, and K. Paustian, Economics of Sequestering Carbon in the U.S. Agricultural Sector. Technical Bulletin. TB-1909. Economic Research Service, U.S. Department of Agriculture, Washington, D.C., April 2004, 61 pp.

12. IPCC, *Land Use, Land-Use Change, and Forestry.* Cambridge University Press, New York, 2000.

13. Chomitz, K.M., Evaluating Carbon Offsets from Forestry and Energy Projects: How Do They Compare? World Bank, Development Research Group, 2000. Available online at econ.worldbank.org. Accessed June 2003.

14. van Kooten, G.C., S.L. Shaikh, and P. Suchánek, Mitigating climate change by planting trees: the transaction costs trap. *Land Econ.* 78(4), 559–572, 2002.

15. Adams, R.M., D.M. Adams, J.M. Callaway, C.-C. Chang, and B.A. McCarl, Sequestering carbon on agricultural land: social cost and impacts on timber markets. *Contemp. Policy Issues* 11(1), 76–87, 1993.

16. Ciriacy-Wantrup, S.V., *Resource Conservation. Economics and Policies,* 3rd ed. University of California Agricultural Experiment Station, Berkeley, CA, 1968.

17. Herzog, H., K. Caldeira, and J. Reilly, An issue of permanence: assessing the effectiveness of temporary carbon storage. *Climatic Change* 59(3), 293–310, 2003.

18. Richards, K.R., The time value of carbon in bottom-up studies. *Crit. Rev. Environ. Sci. Technol.* 27(Special Issue): S279–S292, 1997.

19. Woerdman, E., Implementing the Kyoto Mechanisms: Political Barriers and Path Dependency. Ph.D. dissertation, Department of Economics, University of Gröningen, Gröningen, the Netherlands, 2002, 620 pp.

20. Tietenberg, T., M. Grubb, A. Michaelowa, B. Swift, and Z. Zhang, International Rules for Greenhouse Gas Emissions Trading: Defining the Principles, Modalities, Rules and Guidelines for Verification, Reporting and Accountability United Nations Publication UNCTAD/GDS/GFSB/Misc.6, 1999. Available online at http://r0.unctad.org/ghg/publications/intl_rules.pdf. Accessed 10 July 2004.

21. Marland, G., K. Fruit, and R. Sedjo, Accounting for sequestered carbon: the question of permanence. *Environ. Sci. Pol.* 4(6), 259–268, 2001.

22. Dutschke, M., Fractions of permanence — squaring the cycle of sink carbon accounting. *Mitigation Adaptation Strat. Global Change* 7(4), 381–402, 2002.

23. Sedjo, R.A., and G. Marland, Inter-trading permanent emissions credits and rented temporary carbon emissions offsets: some issues and alternatives. *Climate Pol.* 3(4), 435–444, 2003.

24. Caspersen, J.P., S.W. Pacala, J.C. Jenkins, G.C. Hurtt, P.R. Moorcroft, and R.A. Birdsey, Contribution of land-use history to carbon accumulation in U.S. forests. *Science* 290(10 November), 1148–1151, 2000.

25. van Kooten, G.C., W.A. Thompson, and I. Vertinsky, Economics of reforestation in British Columbia when benefits of CO_2 reduction are taken into account, in *Forestry and the Environment: Economic Perspectives,* W.L. Adamowicz, W. White, and W.E. Phillips, Eds. CAB International, Wallingford, U.K., 1993.

26. van Kooten, G.C. and H. Folmer, *Land and Forest Economics.* Edward Elgar, Cheltenham, U.K., 2004.

27. van Kooten, G.C., E. Krcmar-Nozic, B. Stennes, and R. van Gorkom, Economics of fossil fuel substitution and wood product sinks when trees are planted to sequester carbon on agricultural lands in western Canada. *Can. J. For. Res.* 29(11), 1669–1678, 1999.

28. van Kooten, G.C., B. Stennes, E. Krcmar-Nozic, and R. van Gorkom, Economics of afforestation for carbon sequestration in western Canada. *For. Chron.* 76(1), 165–172, 2000.
29. van Kooten, G.C., A.J. Eagle, J. Manley, and T.M. Smolak, How costly are carbon offsets? A meta-analysis of carbon forest sinks. *Environ. Sci. Pol.* 7(August), 239–251, 2004.
30. Padgitt, M., D. Newton, R. Penn, and C. Sandretto, Production Practices for Major Crops in U.S. Agriculture, 1990–97. Statistical Bulletin 969. Resource Economics Division, Economic Research Service, U.S. Department of Agriculture, Washington, D.C., September 2000, 114 pp.
31. West, T.O. and G. Marland, A Synthesis of Carbon Sequestration, Carbon Emissions, and Net Carbon Flux in Agriculture: Comparing Tillage Practices in the United States. Environmental Sciences Division Working Paper, Oak Ridge National Laboratory, Oak Ridge, TN, 2001.

Part V

Summary and Recommendations

20 Impacts of Climate Change on Agriculture, Forest, and Wetland Ecosystems: Synthesis and Summary

J.M.R. Stone, J.S. Bhatti, and R. Lal

CONTENTS

20.1 INTRODUCTION

Two important issues are facing humanity at the dawn of the 21st century: (1) human-induced increases in atmospheric concentration of greenhouse gases (GHGs) with anticipated impacts on the climate (increase in the severity and frequency of extreme events, sea level rise, and changes in terrestrial and aquatic biodiversity), and (2) associated impacts on global food security due, among other reasons, to socioeconomic development goals, new technologies, and rapid changes in population and living standards in the developing countries where populations are already under great stress. Both of these issues are interlinked and governed by our management of the Earth's natural resources.

20.2 CLIMATE CHANGE IS REAL

In discussing the anticipated impacts of climate change on agriculture, forestry, and wetlands management it is first useful to describe the global context in which the climate is changing (Chapter 2). Scientists have drawn attention to the significant increase in the atmospheric concentrations of carbon dioxide and other GHGs as a result of the burning of fossil fuels and land-use changes. The threat of anthropogenic climate change has become an issue for governments and the general public at large. As a measure of Canada's fossil fuel emissions, these are equivalent to burning all of Canada's forests every 2 years (Chapter 9). The increases in atmospheric concentrations of GHGs are outside levels seen over the past 400,000 years. They can be expected to alter the climate to a degree that the past will no longer be a useful guide to the future. Indeed, the Intergovernmental Panel on Climate Change (IPCC) in its 3rd Assessment Report (TAR) concluded, "an increasing body of observations gives a collective picture of a warming world."[1]

There is already strong evidence of changes in biological and physical systems. The IPCC concluded in the TAR that "there have been discernible impacts of regional climate change, particularly in temperature, on biological systems in the 20th century."[1] Changes include many that are important for agriculture such as higher minimum temperatures at night, the earlier onset of spring and plant flowering, the lengthening of the growing season, shifts in bird, insect, and other populations, as well as the decline of mountain glaciers with the concomitant decline in runoff.[2] Similar changes have been described in the recent report of the Canadian Council of Ministers of the Environment (CCME) — Climate, Nature, and People: Indicators of Canada's Changing Climate (Chapter 3). As examples, frost-free periods have declined significantly and the blossoming time for trees such as aspen now occurs some 28 days earlier.[3]

Although the IPCC in the TAR concluded that: "There is new and stronger evidence that most of the warming observed over the last 50 years is attributable to human activities," it is important to be somewhat cautious in attributing all observed changes entirely to anthropogenic climate change since there are other stresses involved. Changes in some weather variables such as extreme events — heat-waves, droughts, and floods, for example — are very difficult to detect at the present time, and yet have an important impact on the climate.[4]

Humans have already made a commitment to the future climate because the GHGs emitted into the atmosphere are expected to remain there for many decades. In addition, it will take many years to turn around the development pathways being followed at present. Consequently, humans may be powerless to alter the impacts that are anticipated to occur over the next 30 years. Some adaptation will be essential and this will require significant investments and changes in behavior.[5] Most scientists are convinced that action to reduce emissions is necessary and that there is an urgency to act in order to avoid even worse impacts.

Looking to the future, continental summer droughts, disease, and insect infestations as well as forest fires are all projected to increase; animal and plant productivity are also expected to decline.[4] The actual impact of these changes on agriculture will vary depending on factors such as precipitation changes and other stresses associated with different land-use practices.

However, not all of the impacts of projected climate change will necessarily be negative. Indeed, there have been some recent studies that suggest that mid-latitude agriculture, such as in North America, may benefit from temperature increases of up to 1 to 2°C, although it is still unclear what the attendant moisture stresses may be.[6] On the other hand, forestry is expected to be impacted negatively and may not be able to adapt quite so readily. As an example of what has already occurred, prior to about 1980 the Canadian boreal forest was a net sink for carbon but estimates indicate that since then it has become a small source, due to increases in natural disturbances, which may have been caused by climate change[7] (Chapter 9). At the global scale, research suggests that since the industrial revolution, after which the use of fossil fuels has increased dramatically, the terrestrial biosphere has been a net source. This is something of a surprise and may represent an important finding.

Wetlands have been the forgotten ecosystems. They are very sensitive to climate change and can respond rapidly. Wetlands are estimated to contain about a third of the world's carbon; if this were released, the atmospheric concentration of CO_2 would be at least 50% higher (see Chapter 10). Wetlands are already threatened by human disturbances, for example, through drainage for agriculture, and may become more so due to expected increases in evaporation, although the effects of melting permafrost may confound the final outcome of this complex process.

Changes in agriculture, forestry, and wetland ecosystems will also be affected by the interactions between the climate and the carbon cycle (Table 20.1). There is much that is not understood about the biological processes that govern carbon-stock changes. This was discussed in a report on the scientific basis for separating out the natural and human contributions to carbon-stock changes that the IPCC undertook for the U.N. climate change convention process.[8] Increasing carbon dioxide in the atmosphere is expected to enhance plant growth in some crops, but studies suggest

TABLE 20.1
Atmospheric Concentration of GHGs and the Factors Affecting Them

	CO_2	CH_4	N_2O
Pre-industrial concentration	280 ppmv	0.80 ppmv	288 ppbv
Present-day level (2004)	378 ppmv	1.78 ppmv	310 ppbv
Current annual increase (%) (between 1990 and 1999)	50	90	25
Factors causing increase in emissions			
Agriculture	Deforestration Biomass burning Soil degradation Soil erosion Soil tillage	Rice cultivation Ruminants	Synthetic N fertilizers Animal soil excreta Biological N fixation
Forestry	Fire Biomass burning Land-use change Site preparations	Fire	Synthetic N fertilizers Biological N fixation Fire
Wetlands	Fire Drainage Peat mining Land-use change	Permafrost melting Reservoir creation	Biological N fixation

that this effect declines with increasing concentrations. In addition, while higher levels of carbon dioxide may increase plant mass, there may be a displacement of nitrogen take up which is essential for protein synthesis. The CO_2 fertilization effect is limited by the lack of water and essential elements (e.g., N, P, S, and some micronutrients).[9] Finally, increased temperatures associated with climate change are expected to enhance soil respiration, particularly in boreal soils.

In addition to climate change, other human activities are also exerting an influence on the environment such that they are now overwhelming many of the natural forcing on the ecosystems upon which humans rely for products and services. Such activities include land-use changes and the addition of nitrogen as fertilizer. Humans now control more than 50% of the primary productivity of the planet. These factors are working together and influencing the planet in ways that are not fully understood. Small local perturbations can sometimes have global consequences. As long ago as 1985, a conference in Villach, Austria, concluded, "Many important economic and social decisions are being made today on long-term projects ... based on the assumption that past climate data ... are a reliable guide to the future. This is no longer a good assumption."[10] It is not just the magnitude of the changes that is of concern, but also the rate of change, which is likely to be too fast for many ecosystems and socioeconomic systems to adapt. Some of the changes may be irreversible within several generations and others, such as the loss of species, irretrievable.

The scientific knowledge of the climate system is still not complete and this will remain a challenge for scientists for many generations to come. Earth's climate system is complex, involving many nonlinear components, both natural and human, and we are forcing the system at rates that are likely to produce surprises. Further monitoring and research is needed particularly to strengthen knowledge at the local and regional level knowledge of climatic processes on such indicators as precipitation and extreme events. Despite these continuing challenges, considerable progress has been made in the last few years through concentrated efforts such as the biological carbon program (BIOCAP) initiative.

20.3 IMPACTS OF CLIMATE CHANGE ON AGRICULTURE, FOREST, AND WETLAND ECOSYSTEMS

There are optimum temperatures for plant growth and there is already some evidence that rice crops in the tropics are suffering (see Chapter 5). Crop yields in mid- and high-latitude regions may be less adversely affected by higher temperatures. Farm-level adaptation, including new crop strains, can generally offset the detrimental effects of climate change but poor soil conditions will be a major factor limiting the northward expansion of agricultural crops. The positive impacts of warmer temperature and enhanced CO_2 on the rates of crop maturation and production are expected to mitigate the impact of moisture limitation, so that increased growth rates in grasslands and pastures are generally expected[11] (Chapters 7 and 8). With a doubling of CO_2 concentrations an average increase of about 17% in grassland productivity is anticipated with greater increases in the northern regions. However, some studies

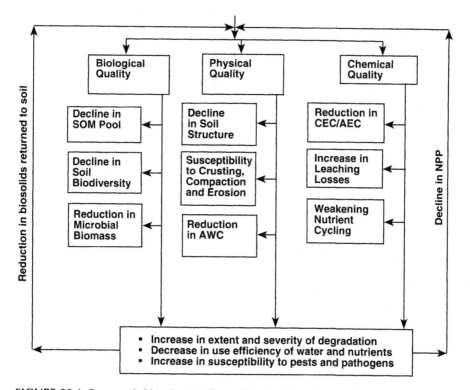

FIGURE 20.1 Some probable adverse effects of projected climate change on world soils.

suggest that under climate change, particularly with extreme weather events, the invasion of alien species into grasslands could reduce the nutritional quality of the grass.[12]

The projected changes in climate are expected to have some adverse impacts on the world's soils (Figure 20.1). Increases in soil temperatures will enhance the rate of decomposition of soil organic matter. Consequently, the soil organic matter pool will decline with adverse impacts on soil structure, plant-available water retention capacity, and cycling of nutrients. All other factors remaining the same, soils with less organic matter content are more susceptible to crusting, compaction, and erosion. However, the impact of projected climate change on soil quality will differ among regions, with more adverse impacts on soils of higher latitudes than in the tropical ecosystems. Agronomic/biomass productivity is also likely to decline because of increased intensity and frequency of drought, reduced nutrient use efficiency, and increased incidence of pests and diseases.

Climate change is expected to present both benefits and challenges to livestock operations. Increases in temperature during winter will reduce the feed requirements, increase survival rate for the young, and reduce energy cost for the farmers (Chapter 12). However, the heat waves during summers could kill the animals and adversely affect the milk production, meat quality, and dairy cow fecundity (Chapters 12, 13, 14, and 15).

The dominant factor to decision making in agricultural ecosystems is the economic net return to the farmer (see Chapter 5). Farmers, as guardians of the land, have experience in adapting to the variability of weather and the climate. Agricultural planners will need to identify options that have benefits not only in moderating the impacts of changes in the climate but that will also benefit farmers' bottom line and, indeed, that for any natural resource manager. Land management practices based on sound scientific recommendations are essential to minimize the adverse effect of climate change and for soil conservation. These arguments have been used in persuading farmers to adopt low or no-till farming since this not only can lead to increases in take-up of soil carbon, and so reducing the atmospheric build-up of CO_2, but can also improve the water-holding capacity of soils, as well as render the soil better able to cope with future climatic change.

Methane emissions from agriculture account for 38% of total greenhouse gas emissions in New Zealand (Chapter 12). Resources diverted by ruminants in producing methane are not being used for growth. Therefore, diets and feed quality are very important factors affecting methane emissions. Furthermore, it has been found that, when looking at total greenhouse gas emissions, cattle raised on pasture produce less than a third as much emissions as those raised in feedlots (see Chapters 13 and 15). Similarly, taking cattle off waterlogged pastures can lead to less nitrogen emissions and reduces damage to the land.

There are already notable climate-related changes in forest growth as a result of different drivers such as increased CO_2, higher temperatures, more water stress, changing nutrient loading and permafrost melting (Chapter 9). The scientific knowledge of the effects of these drivers is limited. With climate warming, it has been suggested that trees will migrate northward and to higher altitudes. However, temperature is not the sole control on species distribution as other factors including moisture conditions, soil characteristics, and nutrient availability may be more important in forest dynamics.[13] Moisture conditions are the most important factor governing the growth of trembling aspen in western Canada.[14] There are also changes in community structure and ecosystem functioning as a result of climate change and other stresses. In addition, changes in disturbance regimes from pests and fires resulting directly and indirectly from human activities are expected to be significant. As an example of the challenges facing the forestry sector, the population of the mountain pine beetle now devastating western Canadian forests currently occupies only a fraction of its potential range.

Yet, there are many uncertainties related to the influence of these drivers on forest ecosystems. For example, although elevated CO_2 concentrations can benefit tree growth, other anthropogenic emissions may complicate this effect. Human-induced increases in ambient concentrations of ground-level ozone (O_3) may lower tree productivity while N_2O may enhance growth in nitrogen-limited boreal forest ecosystems. Furthermore, the positive effects of CO_2 fertilization and nitrogen deposition may be minimal relative to other factors, particularly land-use change.[15] These contrasting results of many studies may be complicated by other factors such as the species studied, age of the tree stand, the length of the study period, and the methodology used.

There are numerous challenges facing the forestry sector (Chapter 16). Among the key needs is the development of tools for the verifiable measurement of carbon stock changes. For a country with as vast a forest area as Canada, this is obviously no simple matter. There are some encouraging techniques now being developed using remote sensing. It is not known whether carbon stocks in Canadian forests are at present increasing or decreasing. Making useful forecasts of future carbon stocks requires an understanding of the cause and effects of fast and slow processes and understanding disturbance regimes (fire and insects). This is a significant modeling challenge. It will also be important to consider the belowground carbon stocks — for example, the change of root respiration with temperature — as well as other processes in the soil and scaling them up. The FLUXNET initiative will provide much of the data required and will examine how climate variability, management practices, and natural disturbance influence carbon cycling in forest and peatland ecosystems. We will also examine how changes in the carbon pools in living biomass and soils might help in the management of GHGs through the short-term sequestration of atmospheric CO_2.

Canada has the second largest area of wetlands and peatlands in the world (Chapters 10 and 17). These wetlands/peatlands were created over the last 8000 years.[16] In terms of GHGs, wetlands could be either a source or sink of CO_2, CH_4, and N_2O. Some of the time, these wetlands may be a sink for one gas and a source for others (Table 20.2). The ability of wetland ecosystems to act as sinks (or sources) of carbon is a delicate and complex balance of ecosystem processes – most largely controlled by climate. On a global scale, wetlands are today a minor source of CO_2 and N_2O while a major source of CH_4 (Table 20.2). Wetlands are subject to change from sink to source due to non-climatic factors such as mining, reservoir creation, agriculture, fire and permafrost melting[17] (Chapter 10). Carbon is sequestered when plant production is greater than decomposition and the export of dissolved organic carbon through stream flow is low. At present, the accumulation of carbon in wetlands is about 13% of their full potential. The worry is that future climate change could convert these stores of carbon into net sources. It is important to recognize that wetlands provide many ecosystem services. Wetlands need to be managed for the benefit of society as a whole — they are a valuable natural resource (Chapter 17).

These examples illustrate the need to take a holistic and integrated approach to addressing climate change. Decisions regarding GHG abatement must be broad in scope and not focus on individual factors in isolation. Reduction in one GHG should not be undertaken at the expense of another. Scientific knowledge must be expanded

---

TABLE 20.2
Wetland Contribution to Global Annual GHGs Emissions (Tg yr^{-1})

GHGs	Wetland Emission	Global Emissions	% Contribution
Carbon dioxide	8.5[19]	7000[1]	0.12
Nitrous oxide	0.133[20]	7.1–12.7[21]	0.8–1.4
Methane[22]	113	540	21

and coupled to include the climate system, the carbon and nitrogen cycles, the hydrological cycle, and ecosystem functioning. Each of these components, which do not operate independently, is important and each is crucial to the proper functioning of agriculture and forestry.

It is important to involve not only the natural science community — biologists, physicists, and chemists — but also experts from the social and economic science disciplines. Furthermore, it is also important to bring in the users and producers in order that the scientific knowledge is transferred and behavior is changed (this is a steep learning curve). There is a strong need to build on current best practices, for example, those used in adapting to today's climate variability as was demonstrated by the experience in Alberta with its GHG science planning process. It is essential to look for win–win options where farmers and forestry managers see an economic advantage for themselves as well as contributing to addressing the global issue of climate change. It is equally important to take advantage of synergies between actions to reduce emissions and those to adapt to the impacts of climate change. Finally, significant investment monitoring and research is required.

One area that illustrates the challenge and could have significant potential in agricultural and forestry management is that of biofuels and bioproducts (Chapter 11). Biological systems have been in the business of managing GHGs and solar energy for more than 400 million years and offer the opportunity to be part of the solution to climate change. The carbon cycle is a natural process that has operated for millennia to maintain the atmosphere at levels that have kept the climate within a habitable range. At present, agriculture and forestry produce an annual harvest of some 143 Mt C/yr which is equivalent to 50% of the biomass needed to meet the nation's current fossil fuel energy demand[18] (Chapter 11). It is important to assess how much of this can be diverted to bioenergy. The biotic carbon is now a tradable commodity, and any forest or agricultural residue should not be considered a waste. Currently, in Canada biomass converted into energy products amounts to only 10 Mt C/year,[18] and this could be increased by utilizing residues from harvest and natural disturbances, which are not currently being used, and by increasing carbon stocks, especially in forests. Such an approach would significantly replace fossil fuel emissions. Canada may need to choose between forest carbon credits and large-scale bioenergy.

20.4 WHAT IS NEXT UNDER CHANGING CLIMATE?

Future climate change and environmental issues need a careful assessment. There are several important issues that need to be addressed. It is necessary to recognize the real threat of climate change and prepare for it. The challenge for scientists is to do good, solid science. But this is not enough; scientists must also be deeply involved in communicating this science and stimulating an informed scientific debate, not one based on narrow vested interests. It is important to communicate the robust findings as well as provide valid estimates of the uncertainties and to be able to communicate the limits of our scientific understanding to decision makers. Determining and communicating uncertainty is essential for policy makers and something that scientists have to learn to do better.

In addition to providing scientific information, it is equally relevant to develop the tools (for example, for measurement), technologies (for mitigation and adaptation), and advice on management practices to farmers and forestry managers. It is time to shift from working mostly to better define the problem of climate change to helping to find solutions to address the problem. There is still much that is to be understood about the functioning of the ecosystems on which our agriculture, forestry, and wetland management relies, and this certainly requires more research.

While good solid science is essential, it must be done in social and economic contexts. There is a need to have much better understanding of the cost and benefit curves. This is not a trivial undertaking. Estimating the costs of addressing climate change requires having a baseline of how the economy would evolve in the absence of action. Estimating the benefits of taking action by avoiding the impacts requires having a good understanding of the regional or local impacts of climate change, other stresses that might be experienced, and the current adaptive capacity. There is a need for a much more complete examination of the economics of using biological carbon sinks — some analyses suggest that conservation tillage has no real gains for the Prairies (Chapter 18).

However, not everything can be expressed in monetary terms. Sound land management practices are essential for soil conservation, an important goal in itself. Similarly, climate change in some parts of the world, such as the Canadian Arctic, is very likely to threaten the very existence of a people and their culture. Ethical considerations, because of the intergenerational and intercultural aspects, will inevitably feature in any consideration of how best to address climate change.

Some argue that enough is known about the threat of climate change to begin to take action to address it. Indeed, the threat may be graver and the need to take action more urgent that previously thought. It is important to recognize that not all impacts will be negative and there will be opportunities. We need to take a balanced approach. Alarmist comments are not an appropriate response but neither is waiting for all the uncertainties to be resolved before acting.

For agriculture, forestry, and wetlands management, the first stage is to consider how to adapt to the anticipated impacts of climate change, while recognizing that these sectors also make contributions to the emissions of GHGs — they are part of the problem (Chapters 11, 15, 16, and 17). Being prepared will mean in part the development of new technologies and the efficient use of existing and new technologies — technologies for adaptation and for mitigation. These would include new cultivars more suited to the changed climate, techniques to enhance carbon sequestration and reduce disturbances, technologies to take greater advantage of the biosphere, and more energy efficient practices. Adopting these new technologies will position Canada better not only to cope with the change in climate but also to remain competitive in international markets for our crops and products. It is essential to get ahead of the curve.

It is important in tackling climate change to recognize the interactions with human activities. This is more than accepting that most of the observed changes in the climate are a result of human actions, such as fossil fuel burning and land-use changes, and exploring deliberate societal and economic options to lessen this perturbation. It is crucial to understand that there is an inescapable interaction between development and climate change. Similar emissions scenarios can be arrived at through different choices of socioeconomic pathways.[1] Choices include population growth, technology development, and addressing equity differences.

Our present development plans may not only be affected by climate change in the future, but there is also the opportunity to modify these development plans in ways that can also address climate change. Our development choices may exert many pressures on environmental, social, and economic systems in agriculture, forestry, and wetland management. Wetlands are being drained, forests being removed, chemicals being added, new species being introduced and, at the same time, being eliminated. Because of the intricate link between climate and development, it may be difficult to tackle climate change as a single silo issue as has been attempted up to now. Rather, it may be necessary to consider the implications of climate change in every development decision and implementation. This may be the lesson of the Kyoto Protocol.

REFERENCES

1. IPCC, *Climate Change 2001: The Scientific Basis, Contribution of Working Group I to the Third Assessment Report of the Intergovernmental Panel on Climate Change,* Houghton, J.T., Ding, Y., Griggs, D.J., Noguer, M., van der Linden, P.J., Dai, X., Maskell, K., and Johnson, C.A., Eds., Cambridge University Press, New York, 2001.
2. Brklacich, M. et al., Implications of global climatic change for Canadian agriculture: a review and appraisal of research from 1984 to 1997; in *Responding to Global Climate Change: National Sectoral Issue,* Koshida, G. and Avis, W., Eds. Environment Canada, Canada Country Study: Climate Impacts and Adaptation, v. VII, 1219, 1998.
3. Barr, A.G., Black, T.A., Hogg, E.H., Kljun, N., Morgenstern, K., and Nesic, Z., Interannual variability in the leaf area index of a boreal aspen — hazelnut forest in relation to net ecosystem production, *Agric. For. Meteorol.,* 126, 237, 2004.
4. Cohen, S. et al., North America, in *Climate Change 2001: Impacts, Adaptation and Vulnerability, Contribution of Working Group II to the Third Assessment Report of the IPCC.* McCarthy, J.J., Canziani, O.F., Leary, N.A., Dokken, D.J., and White, K.S., Eds., Cambridge University Press, New York, 2001, chap. 15.
5. Smithers, J. and Smit, B., Human adaptation to climatic variability and change, *Global Environ. Change,* 73, 129, 1997.
6. Reich, P.B. et al., Ecosystem impacts of elevated CO_2, nitrogen deposition, and plant diversity, *Nature,* 410, 809, 2001.
7. Kurz, W.A. and Apps, M.J., A 70-year retrospective analysis of carbon fluxes in the Canadian forest sector. *Ecol. Appl.,* 9, 526, 1999.
8. IPCC, Report of Experts Meeting on Current Scientific Understanding of the Processes Affecting Terrestrial Carbon Stocks and Human Influences upon Them, July 21–23, Schimel, D. and Manning, M., Eds., IPCC, 2003.
9. Oren, R. et al., Soil fertility limits carbon sequestration by forest ecosystems in a CO_2-enriched atmosphere, *Nature,* 411, 469, 2001.
10. IPCC, *Land Use, Land-Use Change, and Forestry,* Watson, R.T., Novel, I.R., Bolin, N.H., Ravindranath, N.H., Verardo, D.J., and Dokken, D.J., Eds., Cambridge University Press, New York, 2000.

11. Campbell, B.D., Stafford Smith, D.M., and GCTE Pastures and Rangelands Network members, A synthesis of recent global change research on pasture and rangeland production: reduced uncertainties and their management implications, *Agric. Econ. Environ.*, 82, 39, 2000.

12. White, T.A. et al., Impacts of extreme climatic events on competition during grassland invasion, *Global Change Biol.*, 7, 1, 2001.

13. Kimmins, J.P., Importance of soil and role of ecosystem disturbance for sustainable productivity of cool temperate and boreal forest, *Soil Sci. Soc. Am. J.*, 60, 1643, 1996.

14. Hogg, E.H. et al., Factors affecting interannual variation in growth of western Canadian aspen forests during 1951–2000, *Can. J. For Res.*, 35, 610, 2005.

15. Caspersen, J.P. et al., Contributions of land-use history to carbon accumulation in U.S. forests, *Science*, 290, 1148, 2000.

16. Zoltai, S.C., Ecoclimatic provinces of Canada and man-induced climatic change, *Newsletter, Canadian Committee on Ecological Land Classification.* Canadian Forestry Service, Headquarters, Ottawa, 17, 12, 1988.

17. Turetsky, M.R. et al., Current disturbance and the diminishing peatland C sink, *Geophys. Res. Lett.*, 29, 1526, 2002.

18. Wood, S.M. and Layzell, D.B., A Canadian Biomass Inventory: Feedstocks for a Bio-based Economy, Prepared for Industry Canada: Contract 5006125. BIOCAP Canada Foundation, 2003. Available online at http://www.biocap.ca/images/pdfs/BIOCAP _Biomass_Inventory.pdf.

19. Gorham, E., Northern peatlands: role in the carbon cycle and probable responses to climatic warming, *Ecol. Appl.*, 1, 182, 1991.

20. Freeman, C., Lock, M.A., and Reynolds, B., Fluxes of CO_2, CH_4, and N_2O from a Welsh peatland following simulation of water table draw-down: potential feedback to climate change, *Biogeochemistry*, 19, 51, 1993.

21. Davidson, E.A., Fluxes of nitrous oxide and nitric oxide from terrestrial ecosystems, in *Microbial Production and Consumption of Greenhouse Gases: Methane, Nitrogen Oxides, and Halomethanes*, Rogers, J.E. and Whitman, W.B. Eds., American Society for Microbiology, Washington, D.C., 1991, chap. 12.

22. Bartlett, K.B. and Harriss, R.C., Review and assessment of methane emissions from wetlands, *Chemosphere*, 26, 261, 1993.

21 Climate Change and Terrestrial Ecosystem Management: Knowledge Gaps and Research Needs

I.E. Bauer, M.J. Apps, J.S. Bhatti, and R. Lal

CONTENTS

21.1 INTRODUCTION

Atmospheric concentrations of greenhouse gases (GHGs) are increasing as a result of human activities (Chapters 2 and 4). This trend began with settled agriculture, and conversion of natural ecosystems to cropland, rice paddies, and pasture has resulted in measurable increases of CO_2 and CH_4 over the past 8000 and 5000 years, respectively.[1] With the beginning of the industrial revolution, release of CO_2 from the burning of fossil fuel has added a new dimension to this phenomenon, and rates of change have continually increased through the 19th and 20th centuries. Current atmospheric levels of CO_2 are higher than any documented within the past 400,000 years and, without effective mitigation, are projected to reach twice pre-industrial levels by the end of the 21st century (Chapters 2 and 3).

Although it is hard to separate direct effects of climate change from other global change impacts, there is mounting evidence that effects of GHG-related warming can already be detected (Chapters 3 and 4). Increases in the frequency of fire in Canadian forests over recent decades, for example, may be in part attributable to climate warming, and there have been significant regional and continental-scale trends in budding, leaf emergence, and flowering — all phenological events directly controlled by temperature (Chapter 3). The warming expected to accompany future increases in atmospheric GHGs will further accelerate such changes, and agronomic impacts alone have significant implications for global food security, especially given an expected 50% increase in human populations between 2000 and 2050.[2] Other projected effects of climate change include a rapid displacement of major biomes (Chapters 9 and 10) and increased frequency and severity of natural disasters, with serious impacts on human lives and economic systems (Chapters 2 and 3). Given these challenges, understanding of climate-change mechanisms and their effect on biological systems is an important research priority, with land managers and policy makers needing information from scientists to develop effective strategies for adaptation (Chapters 2 and 5) and mitigation (e.g., Chapters 9, 11, 12, 13, 15, and 16).

Much of the human-induced increase in atmospheric GHGs is due to perturbations of the global carbon (C) cycle, with fossil-fuel burning and land-use change (deforestation) as the primary mechanisms (Chapters 4 and 9). So far, therefore, human alterations of the biosphere have been part of the problem, contributing ~25% of all anthropogenic C emissions in the 1990s (Chapter 9). Future cycling of C in the biosphere will be affected by both human land use and climate change, and the complexity of climate-biosphere interactions makes net effects hard to predict (e.g., Chapters 2, 4, 5, and 9). However, human control over global C cycle processes could also make the biosphere part of a climate solution, if changes in land use or management can prevent or offset GHG emissions (Chapters 5, 9, 11, 12, 16).

Drawing on information presented throughout this book, this chapter identifies key knowledge gaps relating to climate and climate-change effects on agriculture, forestry, and wetlands. It further points toward research needed to make management of these ecosystems part of a solution, by identifying gaps in the current understanding of biosphere-based adaptation or mitigation strategies. The list presented here is only concerned with climate change — biosphere interactions, and with questions of land use or management where they intersect with this topic. It cannot tackle the much larger subject of "global change," or strategies for GHG mitigation that are not biosphere based. Further, it focuses on science needed to support economic or policy decision, without making reference to specific market or legislative tools. It also makes no attempt to include knowledge gaps relating to the development of economic or policy mechanisms needed to make biosphere-based GHG mitigation a functional and attractive option. For an introduction to this field, the reader is referred to Chapter 19.

21.2 KNOWLEDGE GAPS AND RESEARCH PRIORITIES

Three overall questions should guide a holistic approach to research into sustainable resource management in a climate change context:

1. Can terrestrial ecosystems, which have helped to moderate the globally coupled C-cycle–climate system within a reasonably narrow domain of CO_2 and climate for at least 420 million years (Chapters 2, 4, and 9), be managed so as to return the system to its previous narrow domain of co-variation?
2. Should such mitigation efforts fail, what will be the new C-cycle–climate domain, and how will terrestrial ecosystems respond? How will agricultural and forest resources be altered, and how will continued increases in fossil fuel emissions affect active C-pools and C-fluxes?
3. If mitigation is feasible (in the sense of question 1), how can agricultural, forest, and peatland ecosystems be managed to best provide a net sink for atmospheric CO_2, provide the needed food and fiber resources, and be a part of the solution for overall improvement of the global environment?

The following sections summarize key research needs to address these overall questions, including different components of the climate system as well as our understanding of current ecological conditions, climate change impacts, and adaptation or mitigation strategies.

21.2.1 THE CLIMATE SYSTEM

1. *What are the processes that determine the role of the biosphere in the climate system, and how can they be represented in global and regional circulation models?* Climate-biosphere interactions are often nonlinear, involving complex interactions between energy, biogeochemical, and H_2O cycles (Chapter 2). The newest generation of GCMs attempt to incorporate biospheric feedbacks by including explicit (although simplified) representations of these processes, but significant challenges remain. These include especially the scaling of exchange processes that occur over short time- (hours and days) and small spatial (plant and stand-level) scales up to the longer (decades and greater) and larger ones (regional and continental) that effect global patterns. In such scaling, responses that cause significant changes in ecosystem physiology (plants and soils) or distribution must be included.
2. *What is the relative importance of different forcing factors in driving observed "recent" changes in temperature?* As discussed in Chapter 2, different climatic forcing factors (solar activity; GHGs; aerosols) vary

independently, and some (e.g., sulfate aerosols from volcanic eruptions) can have a net cooling effect. Disentangling the relative contribution of these forcing mechanisms to past climatic changes — and projections of their likely future variability — are critical to accurate climate projections and associated estimates of uncertainty.

3. *Are trends in means related to the frequency and severity of extreme events, and if so how?* Ecological and societal impacts of climate change depend both on long-term trends in means (i.e., climate), and on changes in daily, intra-annual, and inter-annual patterns of extremes (i.e., extreme weather). Projected increases in the frequency and severity of drought for example (Chapters 2, 3, and 5) may limit agricultural productivity in some regions, while outbreaks of insect pests such as mountain pine beetle (Chapters 3 and 17) may spread beyond their present range with a warming of minimum winter temperatures. Ability to project future trends in extremes as well as trends in means is key to the development of appropriate risk scenarios, especially at regional and local scales (see next point).

4. *How will climatic parameters (especially precipitation and extreme weather conditions) change at regional and subregional scales?* Although most current climate models agree on the expected direction of global trends, projected changes especially in precipitation tend to vary at regional levels (Chapter 2). Moreover, weather patterns (including extreme events) may shift in space and time, even if large-scale regional means remain unchanged. Climate-change impacts such as drought, wild fires, ice storms, and floods are often associated with locally extreme weather, and management is carried out at local scales. To develop adaptation strategies that tailor to biotic and economic realities of specific regions, improved projections (and error estimates) are needed at regional and subregional scales.

5, *Will there be changes in intra-annual climatic patterns?* Intra-annual climatic patterns such as the timing of rainfall and length of growing season affect crop survival and harvesting schedules, with strong implications for the productivity of both agricultural and forest systems (Chapters 3, 5, and 17). Significant changes in the timing of phenological events in recent decades (Chapters 3 and 17) indicate that changes are already occurring, and the ability to forecast intra-annual patterns of climate events and their effects on plant phenology and life cycles is important in developing adaptation strategies (Chapter 3).

6. *How will anthropogenic emissions and atmospheric concentrations of GHGs change over time?* GHG emission scenarios are a key element of uncertainty in climate change projections (Chapter 2). Apart from highlighting the potential range of future climatic trends, emission scenarios are an important component of decision tools, as they allow for a weighing of likely costs and benefits of specific policy decisions. Realistic emissions scenarios have to account for biosphere feedbacks as well as human effects, the latter influenced by both overall population growth and attitudes to land use and fossil-fuel use. Human behavior is complex and

hard to predict, but the ability to do so is critical to emission scenarios and the forecasting of future resource use.

7. *How will current buffering capacities (e.g., oceanic, biotic, pedologic) be altered by changes in climate and further emissions of CO_2 and other GHGs?* As discussed in Chapters 2, 4, and 9, only 40 to 50% of the CO_2 emitted from fossil fuels and land-use change currently remains in the atmosphere. The rest is returned to terrestrial and oceanic sinks, providing an important buffering effect against GHG-related climate effects. The permanence of C taken up by biotic sinks in particular is currently poorly understood, and the potential for further ecosystem C sequestration is likely to decrease with increased temperatures (Chapter 4). Understanding of mechanisms that control the ability of natural systems to buffer anthropogenic emissions is needed to predict future GHG trajectories.

21.2.2 CURRENT STOCKS AND FLUXES

Determining baseline data for terrestrial ecosystem C stocks and fluxes is an essential first step to evaluating climate- or human-induced changes. As discussed in preceding chapters, fundamental gaps in understanding for all sectors still limit our ability to predict the consequences of different management actions on C fluxes in a changing environment. A comprehensive understanding of processes that control terrestrial ecosystem C dynamics and their interactions with the biosphere and hydrosphere is needed to develop recommendations for land managers. Important research domains and questions include the following:

21.2.2.1 C Dynamics of Different Ecosystem Types

1. *What are current ecosystem distributions and their associated C stocks?* Data of this type are needed for most managed and natural ecosystem types, including agricultural land, pastures, woodlots/plantations, forests, and wetlands. Although data exist for some regions and sectors, these are not complete (see, e.g., Chapter 18), and their accuracy is usually not well characterized (Chapters 16 and 17). Without basic data on current ecosystem distributions and extent, C-stock assessments will be inaccurate and projections of change unreliable.

2. *What is the current source/sink status of different ecosystem types, and what are rates of C sequestration under natural conditions (or current management)?* While data on C accumulation and sequestration rates are increasingly available for some ecosystem types and components, sparse data networks and high interannual variability confound the calculation of averages and comparisons between ecosystem types or regions. Comprehensive data on all component C fluxes are known only for a few intensively studied research areas, and are rarely available at the landscape level. In addition to more data collection, there is an urgent need to develop reliable spatial and temporal scaling techniques in order to maximize the usefulness of existing data sets.

3. *What are the factors that control C sequestration in managed and natural ecosystems?* Even in cases where rates of C sequestration are documented, causative factors needed to evaluate the vulnerability of these indicators are often poorly understood, and existing data tend to be inadequate for future change predictions. Dependence of C sequestration on nutrient (N, P, S) and hydrological cycles (Chapter 4) limits our ability to predict effects of climate change on agricultural ecosystems (Chapter 3), and the combined effects of temperature and moisture-related variables on the C-sequestration capacity of peatlands are poorly quantified (Chapter 10).

4. *How sensitive are different ecosystem components to climatic variability, and how can information on total-ecosystem C flux be partitioned into component processes?* Whole-ecosystem measurements of CO_2 exchange (e.g., from eddy-covariance flux towers) examine the net C balance of a site under a given set of climatic and environmental conditions, and are the most direct way to assess the short-term C source/sink status (Chapter 17). However, the response of different ecosystem components to environmental variability is often nonlinear, and understanding of climate-change effects over longer times requires a partitioning of net response into different component processes (e.g., gross and net primary productivity, autotrophic and heterotrophic respiration). Methodologies or models that can partition observed responses and "bridge the gap" between functional levels of ecosystem C cycling are necessary to understand current ecosystem behavior, and to predict future responses.

5. *How important are belowground processes in net ecosystem C exchange?* Belowground processes are hard to observe or measure and have often been neglected in studies of ecosystem C dynamics (e.g., Chapter 5). Few reliable estimates of belowground productivity are available for most ecosystem types, and factors such as the importance of fine-root dynamics or mycorrhizal associations in soil organic matter (SOM) turnover (Chapter 17) or the sensitivity of soil microbial communities to temperature and moisture conditions (Chapter 5) are poorly understood. Belowground processes control rates of ecosystem C and nutrient cycling and are a key component of biosphere-climate interactions.

6. *How do different disturbance events (harvesting, fire and insect defoliation) affect C dynamics?* Ecosystem disturbance leads to biomass C losses that can be minor (e.g., from a low-level insect attack) or severe (e.g., from clear-cut harvesting or stand-replacing wildfire) (Chapters 4 and 9). Beyond immediate C losses, however, disturbance influences many aspects of C and nutrient cycling, and future trajectories can depend on factors such as the fate of dead biomass that remains on-site (Chapters 4 and 9). To fully evaluate the importance of disturbance to ecosystem C dynamics, data are needed on the short- and long-term effects of specific disturbance types on different aspects of C and nutrient cycling (e.g., effects of fire on above- and belowground C allocation, N cycling, or fine root dynamics).

21.2.2.2 Major Non-CO$_2$ Greenhouse Gases

1. *How are N$_2$O emissions from agricultural systems related to hydrology, soil environment, and nutrient cycling?* Nitrous oxide emissions from soils are dependent on hydrology as well as N availability (Chapter 12), and the relative importance of these variables in driving emissions is often poorly understood. More data are needed, for example, on the effect of landscape structure and management practices on N$_2$O emissions, on relationships between fertilization or N-fixation and N$_2$O production, and on the importance of C/N relationships in controlling N$_2$O emissions (Chapter 16). Information of this type is critical to evaluate the full GHG impact of alternative management options, and for the development of mitigation strategies.

2. *What are the factors that control CH$_4$ emissions from wetlands and soils?* Methane is a powerful GHG that is released during anaerobic decomposition in waterlogged soils and wetland systems, with the net flux of CH$_4$ dependent on factors such as temperature, nutrient status, oxygen availability, and water-table depth.[3] Although effects of some of these factors are well documented for some wetland types and systems, their interactions are often poorly understood, and potential trade-offs between lower CH$_4$ and higher CO$_2$ emissions from peatlands under an altered climate (Chapter 3) cannot be adequately quantified.

3. *How are CH$_4$ emissions from ruminant livestock related to dietary composition and genotype?* Although manure can be an important source of CH$_4$ especially in intensive livestock systems, most livestock CH$_4$ emissions are due to microbial digestion of cellulose by either fore- or hindgut fermentation. Many studies have shown clear effects of feed or pasture composition on CH$_4$ production (Chapters 12, 13, and 15), but the influence of different dietary compounds is often hard to separate, and mechanisms that control observed responses are largely unknown (Chapter 12). The same is true for genetic and physiological factors that control differences in CH$_4$ production between individual animals or breeds (Chapters 12 and 13), and all these are basic knowledge gaps that hinder the development of mitigation strategies.

21.2.3 FUTURE IMPORTANCE OF DISTURBANCE

Disturbance and subsequent cultivation or succession are important drivers of landscape patterns of C sources and sinks. In managed terrestrial ecosystems, for example, the current spatial distribution of CO$_2$ sinks may largely reflect historic patterns of land-use change, and areas that act as strong sinks may be recovering from recent anthropogenic or natural disturbance (Chapter 9). Types of disturbance with potentially significant impact on future C emissions include fire (Chapters 3, 9, 10, and 17), pests and diseases (Chapters 3, 5, 9, and 17), extreme climatic events (Chapters 2, 3, and 5), permafrost collapse (Chapters 3 and 10), and human land use/land-use change (Chapters 4 and 9). Key questions relate to the future frequency and severity of different disturbance events, and to their potential interactions and cumulative effects.

1. *What will be the effect of climate change on the frequency and severity of natural disturbance events?* One of the projected effects of climate change is a change in the frequency and severity of natural disturbance events such as pest outbreaks and fire (Chapters 3 and 17). However, the occurrence of such events is highly stochastic and hard to predict accurately in space or time (Chapter 17). To generate appropriate risk scenarios, more data are needed about the role of climate and specific local conditions in influencing the likelihood and severity of different disturbance events (Chapter 5). Resulting probability functions have to be validated wherever possible, and should be incorporated into stochastic models for risk analysis.

2. *How will patterns of anthropogenic disturbance change with increasing population pressure and changes in management strategies?* Humans have already affected many aspects of the C cycle-climate system, and human land use (agriculture/forestry) and land-use change (e.g., deforestation) are strong forcing mechanisms of biosphere GHG dynamics and C stocks (Chapter 4). Effects of anthropogenic disturbance are likely to increase with increasing population pressure, and their accurate forecasting is an important component of future climate projections (Chapter 2) and biosphere C stocks (Chapter 4).

3. *Will there be interactions between disturbance types, and what are the likely cumulative impacts?* Little is known about interactions or cumulative effects of different disturbance types. More data are needed, for example, on effects of management or land-use patterns on the population dynamics of pests, or on interactions between forest susceptibility to disease and fire (Chapter 17). Cumulative effects of multiple disturbances can severely impact C sink potentials of entire ecosystem types (Chapter 10), and strategies to maximize terrestrial C sequestration should be based on a firm understanding of relevant processes and mitigation options.

21.2.4 ECOSYSTEM RESPONSE TO PROJECTED CHANGES

Ability to predict changes in ecosystem behavior resulting from future climate change is crucial to the planning of appropriate adaptation and mitigation strategies. While overall questions are the same for all ecosystem types, there are differences between managed (agriculture and many forests) and unmanaged (most wetlands) systems in both current knowledge and the potential to enhance C-sink capacities through active management. Consequently, key research gaps differ between these sectors.

21.2.4.1 Agriculture and Forestry

Climate change and increasing human populations are combined stressors that challenge policy makers and land managers to ensure food security, especially in developing countries. To support decision processes, improved knowledge is needed of the impacts of climate change in agroecosystems, especially in areas of soil quality

(Chapter 4) and agronomic productivity (Chapter 5). Forests supply human populations with building materials, food, and fuel, and are thought to play an important role in buffering anthropogenic emissions (Chapter 9). At the same time, both the distribution and productivity of these forests will be altered by climate change, a fact that has to be considered in developing adaptation or mitigation strategies. Important knowledge gaps in understanding climate change impacts on agricultural and forest systems are the following:

1. *What are the effects of elevated temperature and precipitation changes on plant growth, life cycles, and productivity?* Variation is a key feature of biological systems, and different species or cultivars differ in overall productive potential, tolerance to temperature and moisture stress, and many life history traits that may be important in a climate change context. Effects of climatic parameters on productivity and life cycles are an important knowledge gap, since they affect the selection of appropriate species/cultivars to maintain productivity under an altered climate (Chapters 3 and 5).

2. *How will these factors impact soil processes such as nutrient dynamics and the structure and functioning of decomposer communities?* Faster rates of nutrient cycling and C mineralization under a warming climate may offset the effect of increased plant production, leading to a net decrease in ecosystem (especially soil) C storage (Chapter 4). Effects of temperature changes have rarely been traced through full biogeochemical cycles, and impacts on decomposer and microbial communities are not well known. All these factors are important in trying to predict effects of future warming on GHG trajectories, or the potential for active management to enhance biosphere C stocks.

3. *How are climate-change effects exacerbated (or mediated) by specific local conditions such as nutrient (N, P, S) limitation, high nighttime temperatures, drought stress, and degraded soils?* At the present time, little is known about interactions between climate change-related variables (temperature, CO_2) and other environmental stressors such as radiation (UV-B), soil degradation, or nutrient limitation (Chapters 3 and 4). Information of this type is needed in order to predict effects of climate change on ecosystem functioning especially at local or regional levels, and to develop risk scenarios and adaptation strategies.

4. *What is the magnitude of the CO_2 fertilization effect, and will it change over time?* As discussed in Chapters 4, 5, and 17, plant responses to enhanced CO_2 differ between species and environmental conditions, and whole-ecosystem studies into CO_2 fertilization have only just begun. To assess whether CO_2 fertilization can partially offset anthropogenic GHG emissions, long-term data are needed to examine the sustainability of increased plant production, possible interactions between CO_2 and temperature effects, and the potential for management to enhance the magnitude and duration of CO_2 fertilization. To obtain such data, current ecosystem-scale studies should be continued wherever possible.

5. *How important are extreme events in controlling the response of agricultural and forest systems to climate change?* Environmental extremes (especially flooding or drought) can destroy entire harvests, and they may limit the potential of some areas to support certain crops (Chapter 5). Data on the sensitivity and resilience of different species to extreme climatic events are needed, for example, to select suitable species for food production or for afforestation (Chapter 9), and to anticipate management costs required to support bioenergy crops in a given region.

6. *How will the geographic distribution of different forest types change under a new climate, and how fast will these changes occur?* As discussed in Chapter 9, the distribution of major forest biomes is expected to shift northwards under a changing climate, but rates of change are hard to predict from current data. To determine (and manage) future forest C-stock or bioenergy potentials, information is needed on the climatic sensitivity of different species and life history stages, on the importance of disturbance in driving range shifts, and on likely response times and natural capacities for dispersal.

21.2.4.2 Wetlands

Northern wetlands (especially peatlands) contain a disproportionate amount of C compared to other ecosystem types and are an active sink for atmospheric CO_2 but a source of CH_4.[4] As discussed in Chapters 10 and 18, C cycling in wetland is intricately linked to hydrological processes, making these ecosystems inherently sensitive to climate change. Unlike agricultural systems and many forests, the large northern wetland areas of Canada and Siberia are mostly unmanaged, and even basic information on their C stocks and dynamics is often lacking. Key knowledge gaps in relation to climate change impacts on the C stocks and GHG source/sink relationships of wetlands are the following:

1. *How will climate change affect wetland hydrology?* Many climate-change projections suggest a drying especially of mid-continental regions (Chapters 2 and 3), i.e., in areas that currently support extensive wetland systems. Direct effects of drying climates on wetland water tables are poorly quantified, and areas where climate scenarios predict extreme future warming and drought tend to be those where human impacts on wetlands have been highest in the past (Chapter 18). The cumulative effects of climate change and human land use on wetland hydrology are important for development of wetland sensitivity ratings and regional assessments.

2. *What will be the effects of increased temperatures and often lowered water tables on productivity and C mineralization in peatlands?* As discussed in Chapter 10, interactions between temperature and water tables and their net effect on plant production and decay in peatlands are poorly quantified. Information of this type is urgently needed to predict C source/sink relationships of peatlands under a changing climate.

3. *How will climate change affect peatland distribution and botanical composition, and what are the consequences of these changes for C cycling?* Changes in peatland distribution and community composition expected under a changing climate have marked implications for future C cycling (Chapters 10 and 18). However, little is known about the potential for plant-driven or hydrological buffering effects, and rates of change (especially C loss from existing deep peat deposits) are impossible to predict from current data. Major knowledge gaps include species response rates, the effect of community change on short- and long-term GHG balances, and the importance of local factors in peatland establishment and disappearance.

4. *How will climate-induced changes in permafrost regimes affect the future GHG balance of peatlands?* Permafrost melt is a widespread phenomenon in peatlands of boreal western Canada (Chapters 10 and 17) and expected to increase in the 21st century (Chapter 3). Localized permafrost melt has been shown to increase C sequestration in peatlands (Chapter 10), but net long-term effects on C storage and GHG dynamics are far from understood. Major knowledge gaps include effects of widespread permafrost melt on peatland hydrology, long-term trajectories of organic matter production and decay at the ecosystem level, and net GHG effects resulting from differential responses of CH_4 and CO_2 to permafrost degradation.

21.2.5 STRATEGIES/TECHNOLOGIES FOR ADAPTATION OR MITIGATION

In spite of uncertainties regarding the magnitude and effects of climate-related changes, scientists and policy makers are faced with the need to develop adaptation and mitigation strategies. Adaptation attempts to limit impacts of climate change through anticipatory or reactive management, for example, by implementing mechanisms that maintain the productivity of forest or agronomic systems (Chapters 3, 5, and 9). Biosphere-based mitigation involves the adoption of management strategies that prevent or offset GHG emissions (Chapters 5, 9, 11, 12, and 16), making the biosphere part of a strategy to prevent further atmospheric GHG increases. The identification of effective technologies for mitigation or adaptation is an important researchable issue.

21.2.5.1 Agricultural and Forest Ecosystems

1. *Development of varieties adapted to specific local conditions.* The development of high-yielding, high-temperature crop varieties is important especially in the tropics, where adaptation to climate change is complicated by added stressors such as soil degradation and increasing population pressure (Chapter 4). In many other regions, adaptation can be achieved by selection from existing cultivars (Chapter 5), but mechanisms are needed to facilitate technology and information transfer to local farmers.

2. *Development of effective technologies for the reduction of CH₄ from livestock.* Altered diets, vaccination/defaunation and development of stock with naturally low CH_4 production are all potential mechanisms to decrease CH_4 emissions from livestock. The relative effectiveness of these different methods needs to be investigated for different livestock systems (e.g., pasture vs. feedlots; Chapters 12, 13, 16), and easy, affordable implementations have to be made available to livestock managers.

3. *Methods of livestock/pasture management that reduce N₂O emissions.* Mitigation options to reduce N_2O emissions from rangeland and pasture systems include dietary manipulation to decrease nitrogen excretion (especially in urine), nitrification inhibitors, soil drainage or removal of stock from wet pasture, and fertilizer management/liming (Chapter 12). The relative efficacy of different methods needs to be investigated for different management systems, and positive side effects (e.g., increased animal efficiency with dietary augmentation; Chapters 12 and 15) need to be fully investigated.

4. *Methods of manure management that minimize GHG impacts.* Animal manure can be a significant source of GHGs, especially in intensive livestock systems. The mitigation benefits of different manure treatment options (application, on-farm treatment systems, composting, or storage) need to be investigated (Chapter 16), including GHG benefits from the use of manure as an on-farm energy source (Chapter 15).

5. *Potential for C sequestration in agricultural and forest soils.* Agricultural and forestry practices that maximize soil C sequestration and retention often have positive effects on both GHG emissions and plant productivity, but their relative effectiveness can differ widely between regions (Chapter 19). Short- and long-term C benefits of management options such as no-till systems (Chapters 4, 5, and 16), various crop rotations (Chapter 16), different harvesting/site preparation methods (Chapters 4 and 9) or conversion of cropland to grassland or forest (Chapters 5, 8, and 9) should be investigated under different climatic and environmental conditions, and any positive side effects (e.g., increased nutrient retention) more fully evaluated.

6. *Strategies to minimize GHG emissions through the entire cycle of production.* Efficient techniques are needed to help trace GHG effects through the entire life cycle of a product or commodity. By identifying "hidden costs," such methods can evaluate the full GHG impact of different building materials (Chapter 9) or land-management practices (Chapter 4), and they can help to avoid unnecessary emissions. In the context of GHG accounting, these methods are needed to do complete cost/benefit analysis and evaluate economic potentials of mitigation options such as biofuels (Chapter 11), on-farm energy production (Chapter 16), or increased use of forest products (Chapters 9 and 19).

7. *Potential for C sequestration in forests and forest products.* Forests are a major biosphere C pool and play a key role in potential strategies to increase terrestrial C sequestration (Chapter 9). Research is needed into the feasibility

and C-sink potential of afforestation projects in different regions, and into mechanisms to reduce deforestation. The potential for increased use of forest products to increase the residence time of sequestered C and avoid fossil fuel emissions during the production of alternative materials (e.g., cement) needs to be more fully evaluated (Chapters 9 and 19).

8. *Potential for bioenergy production.* Use of bioenergy from agricultural or forest systems avoids fossil-fuel emissions and has strong potential as a strategy for GHG mitigation. Research is needed into technologies for the production and processing of bioenergy sources, and into ways to optimally integrate bioenergy with existing fossil-fuel energy (Chapter 11). Analysis of both large-scale (e.g., forest management for biofuels) and small-scale (use of mill/farm waste or livestock CH_4 as an on-site energy source; Chapters 15 and 16) projects is needed to determine the GHG and economic benefits of bioenergy solutions.

9. *What is the permanence of C sequestered in different GHG mitigation strategies?* The permanence of sequestered C is an important consideration in evaluating the effectiveness of mitigation strategies, and is an issue that complicates the handling of offset credits in current trading markets (Chapter 19). Carbon fixed in biomass stocks is quickly returned to the atmosphere when a crop is harvested, unless it is transferred to a longer-lived pool such as forest products. However, forest product C fluxes are hard to trace in space and time, and credit for them is the subject of ongoing negotiation (Chapters 9 and 19). Carbon fixed in soils and wetland sediments has high potential permanence (Chapters 10, 16, and 18), but actual turnover rates are often poorly quantified, and information is needed on the effects of management on long-term retention (e.g., Chapter 19).

21.2.5.2 Wetlands/Peatlands

1. *How can the vulnerability of C in wetlands and peatlands be minimized?* As discussed in Chapters 10 and 18, C cycling in wetlands is intricately linked to hydrological processes, and information is needed on potential mitigation options to limit hydrological impacts of climate change. This is especially true for wetlands already affected by human land use such as agriculture or large-scale industrial activity (e.g., the oil sands development in northern Alberta). Even if such wetlands are not destroyed directly by human activities, cumulative effects may occur from lowered water levels and factors such as increased nutrient input or soil erosion.

2. *How can the C-accumulation potential of drying peatlands be maximized?* Finnish data indicate that increased tree productivity with lowered peatland water tables may help to offset increased soil C losses,[5,6] and afforestation of drying peatlands may be a potential mitigation mechanism under climate change. Data are needed on the economic feasibility and the short- and long-term mitigation potential of such projects in different regions, and on the susceptibility of peatland forest plantations to catastrophic C losses by fire.

3. *How can C accumulation in disturbed peatland sites be restored, and how can new functional wetlands be created?* Active restoration and the re-creation of functional wetlands can reverse effects of human disturbance and restore the GHG and environmental benefits of wetland systems (Chapters 10, 16, and 18). Practical knowledge gaps that limit restoration projects and assessment of their long-term GHG benefits include the long-term sustainability of restored water levels, potential effects of invasive species, and the selection of suitable species complements to maintain wetland function (or C sequestration) under a changing climate.

21.2.6 METHODOLOGICAL AND INTERDISCIPLINARY ISSUES

Numerous compounding factors complicate the prediction of climate-change effects on ecosystem C dynamics. Interactions between climate change and altered distur-bance regimes, for example, or cumulative impacts arising from the joint action of climatic and anthropogenic change, may produce effects that are hard to predict from current data. Given the complexity of climate–human–C-cycle interactions, there is a strong need for integrated data management, and for across-scale and across-ecosystems studies that look beyond narrow disciplinary interests. Modeling plays a key role in this process because it can simulate complex phenomena, integrate different types of data, and bridge conceptual, temporal, and spatial scales. Key methodological challenges that span different fields of climate-change research are the following:

1. *Development of multimodel, multiscenario approaches to generate climate projections and conduct risk assessment.* Variations in the output from different models reflect uncertainties in our current understanding of rel-evant processes and carbon–climate interactions. As discussed in Chapter 5, a multimodel, multiscenario approach is needed to assess likely effects of climate change at local and regional (management-relevant) scales and to constrain scenario predictions and their associated measures of uncer-tainty.
2. *National databases and frameworks.* There is a pressing need to establish national databases for information such as C stocks for predominant land-cover types and classes. National data banks can be an integral component of regional data banks and can supplement more-detailed local data sets. Data-access protocols to permit use by researchers, land managers, and policy makers must be established.
3. *Development of methods for the measurement and monitoring of C stocks and fluxes in a precise, transparent, and credible manner to support accounting frameworks.* There are many practical obstacles to the effective integration of agricultural, forest, and wetland ecosystems into C account-ing frameworks. These include, for example, the economic impossibility of directly measuring local changes in soil C stocks over the short interval of commitment periods (Chapter 19), and a lack of technologies to accu-rately measure livestock feed intake or GHG emissions (Chapter 12). The

development of practical, cost-effective methods to support accounting frameworks is critical to the success of these economic mechanisms.

4. *Integration of measurements with respect to space and time.* Temporal and spatial scaling techniques are critical to our ability to monitor current ecosystem C distributions and fluxes and to the prediction of climate change impacts. As discussed in Chapter 17, an ability to use large-scale measurements from remote sensing may allow for long-term, cost-effective monitoring of terrestrial C stocks and dynamics. Development of reliable methodologies to achieve this is an important research need. Satellite-derived estimates need to be fully verified against ground data, and new ground data will be required in some areas.

5. *Assessment and propagation of uncertainty.* Uncertainties associated with projections of climate change and climate-change impacts are an important component of risk assessment. Methods are needed to more accurately quantify the uncertainty associated with model estimates, and to propagate errors through coupled climate and biosphere models, as well as other assessment procedures (Chapter 17).

21.3 CONCLUSION

Climate change is real. While not all of the changes currently observed are directly attributable to human causes, several challenges of global significance are directly or indirectly related to climate change. Food productivity and security; the availability and safety of water supplies; catastrophic events and their impacts on the quality of human lives and economic systems — all make climate change a key phenomenon and challenge of the 21st century. The issues discussed in this chapter are a summary of major research needs relating to climate change and its interaction with biosphere processes and management. Beyond the broad conceptual framework established by the subheadings, points are presented in no specific order, and no attempt has been made to rank their relative importance. While this may appear as a major omission to many readers, differences in geography, economies, mandates, and priorities render a single ranking practically impossible. The way this list should be used depends on context.

As an obvious example, differences in the importance of methane in GHG budgets of different countries (Chapter 12) translate directly into differences in the potential impact of CH_4-reduction mechanisms relative to those for CO_2 and N_2O. Similarly, individual agencies within each country have very specific mandates, and consequent priorities will differ when it comes to the allocation of funds. Finally, decisions about adaptation or mitigation strategies in particular are made within economic and political contexts, and such contexts may vary in space and time. Countries like Canada that have large forest resources, for example, may have to weigh the relative benefits of managing forests for timber, C stocks, or biofuel production (Chapters 9 and 11), and "correct" answers (and consequent research priorities) may depend on the relative weighting of long- and short-term benefits.

Given likely limits to how much C can be sequestered into Earth's biota — and its often limited permanence — strategies to offset emissions through terrestrial C fixation will only be effective over limited (decadal to century) timescales.[7] During this critical period, however, they can serve as an important bridging mechanism while technologies to replace fossil fuel–based energy are put into effect. In the long term, the only solution to GHG-related warming is to curb fossil-fuel emissions, and some of the technologies that may help achieve this end are also biomass based (Chapter 11). This means that biosphere management could be an important part of future energy strategies, and afforestation projects in particular may yield both short- and long-term GHG benefits.

Finally, responsible ecosystem management is about more than just GHGs or carbon. A large number of C sequestration projects, for example, may be economically unviable in a GHG context alone (Chapter 19), but many yield ancillary benefits that may justify their implementation. Management practices that increase C retention in degraded soils, for example, yield both GHG benefits and are a long-term investment in soil quality (Chapter 4). Management for multiple landscape attributes such as C stocks, biodiversity, and recreational appeal can complicate assessments of additionality in an accounting context (Chapter 19); it is nonetheless responsible ecosystem management. Economic incentives related to C and GHG accounting are likely to play a key role in future land management strategies, and scientists have to provide policy makers with the information they need to devise incentives that promote ecosystem health and sustainability, as well as achieving effective long-term reductions in GHG emissions.

REFERENCES

1. Ruddiman, W. 2003. The anthropogenic greenhouse era began thousands of years ago. *Climate Change* 61: 261–293.
2. Fischer, G. and Heilig, G.K. 1997. Population momentum and the demand on land and water resources. *Philos. Trans. R. Soc.* (London) B 352: 869–889.
3. Roulet, N. 2000. Peatlands, carbon storage, greenhouse gases and the Kyoto Protocol: prospects and significance for Canada. *Wetlands* 20: 605–615.
4. Walter, B.P., Heimann, M., Shannon, R.D., and White, J.R. 1996. A process-based model to derive methane emissions from natural wetlands. *Geophys. Res. Lett.* 23: 3731–3734.
5. Minkkinen, K. and Laine, J. 1998. Long-term effect of forest drainage on the peat carbon stores of pine mires in Finland. *Can. J. For. Res.* 28: 1267–1275.
6. Minkkinen, K. Vasander, H., Jauhiainen, S., Karsisto, M., and Laine J. 1999. Post-drainage changes in vegetation composition and carbon balance in Lakkasuo mire, Central Finland. *Plant Soil* 207: 107–120.
7. Smith, P. 2004. Soil as carbon sinks: the global context. *Soil Use Manage.* 20: 212–218.

Index